PROPERTY OF
PERSHORE COLLEGE OF HORTICULTURE
LIBRARY

Withdrawn

# AN INTRODUCTION TO THE PHYSIOLOGY OF CROP YIELD

PROPERTY OF
PERSHORE COLLEGE OF HORTICULTURE
LIBRARY

# AN INTRODUCTION TO THE PHYSIOLOGY OF CROP YIELD

ROBERT K. M. HAY
AND
ANDREW J. WALKER
Scottish Agricultural Colleges, Auchincruive, Ayr

Withdrawn
PERSHORE COLLEGE OF HORTICULTURE
LIBRARY
10/93
CLASS 581.1
NUMBER 007851

Copublished in the United States with
John Wiley & Sons, Inc., New York

**Longman Scientific & Technical,**
Longman Group UK Limited,
Longman House, Burnt Mill, Harlow,
Essex CM20 2JE, England
*and Associated Companies throughout the world.*

*Copublished in the United States with*
*John Wiley & Sons, Inc., 605 Third Avenue, New York, NY 10158*

© Longman Group UK Limited 1989

All rights reserved; no part of this publication may be reproduced, stored in a retrieval system, or transmitted in any form or by any means, electronic, mechanical, photocopying, recording, or otherwise without either the prior written permission of the Publishers or a licence permitting restricted copying in the United Kingdom issued by the Copyright Licensing Agency Ltd, 33–34 Alfred Place, London WC1E 7DP.

First published 1989

**British Library Cataloguing in Publication Data**

Hay, Robert K M
An introduction to the physiology of crop yield.
1. Crops. Productivity. Physiological aspects
I. Title II. Walker, Andrew J.
631.5′4

ISBN 0-582-40808-3

**Library of Congress Cataloging-in-Publication Data**

Hay, Robert K. M., 1946–
An introduction to the physiology of crop yield/Robert K. M. Hay and Andrew J. Walker.
p. cm.
Bibliography: p.
Includes index.
ISBN 0-470-21192-X (Wiley, USA only).
1. Crops – Physiology. 2. Crop yields. I. Walker, Andrew J., 1951– . II. Title.
SB112.5.H38 1989 88-18731
633 – dc19 CIP

Set in 10/12 pt Linotron 202 Imprint Roman

Produced by Longman Group (FE) Limited
Printed in Hong Kong

LIBRARY
PERSHORE COLLEGE
AVONBANK
PERSHORE
WORCESTERSHIRE WR10 3JP

This book is dedicated to our parents, in gratitude

**Margaret Hay** 1916–1982
**William Hay**
**Mary Walker**
**Kenneth Walker** 1920–1987

# CONTENTS

# PREFACE

As we neared the completion of this book, the population of the earth reached five thousand million (billion), double that of about 1950. World agricultural production kept pace with, or slightly exceeded, this increase, and indeed the current production of food is sufficient to provide an adequate nutrition for about six billion people (the estimated population for the year 2000). In this sense, agriculturalists have met the demands imposed on them by the increasing population. However, it hardly needs to be repeated here that the available food is unequally distributed, with surpluses in some regions and severe shortages in others. Apparently intractable social, political and economic complexities, at local, national and international levels, maintain this imbalance. The long-term solution to the problem of food shortages in many developing countries is, therefore, an increase in the production of food, for local consumption and sale, on countless small farms. The problems of agriculture in the developed world are, of course, very different, although here again agriculture cannot be considered in isolation, simply as the means of producing food. Decisions which a farmer makes about what he grows are influenced, or even dictated, by the policies of governments attempting to control their countries' political and economic machinery.

This is not the place to enlarge upon the problems of agricultural production throughout the world. Instead, we offer our belief that the introduction of improved crop plants and growing systems should play no small part in solving some of these problems. 'Improved' could mean a variety of changes specific to particular species and environments but, in general, we are referring to plants able to photosynthesize more efficiently in terms of available solar radiation, water use and fertilizer input, able to convert a greater proportion of biomass into the desired product, whether food, fibre, fuel or the precursors of medicinal drugs, or better able to withstand environmental stresses. We further believe that such changes will be brought about as a result of understanding how the plants function, both individually and as crops: 'plant physiology is to agronomy what human physiology is to medicine' (van Overbeek 1976). The rapidly developing techniques of genetic engineering, complementing traditional plant breeding, provide the means; physiology must provide the knowledge required to draw up the specifications.

We have written this book as a text for degree-level teaching, and have assumed that our readers have a basic knowledge of plant physiology, biochemistry and anatomy. We have taken pains to explain ideas which we know, from our own teaching experience, students find difficult. Some of these problems arise because students can be reluctant to accept that, because our understanding of certain topics is incomplete, some experimental observations have more than one (or no) 'explanation'. Other difficulties arise because processes at the crop level are controlled by many factors, the relative importance of which can change with time or with variations in the environment or management. The 'correctness' of different explanations may therefore change according to circumstances. And, of course, our perception of how the crop works must be continually reviewed as new observations are made.

These complications pose problems also for the writer, who tries to be objective, but whose inevitable (but unwitting) bias in interpretation may well be shown by subsequent work to have been misjudged. This is unfortunate, but unavoidable if the work is to draw on the most recent literature, and will at least serve to reinforce the reader's awareness of the multiplicity of influences on the crop plant. However, any unintentional bias in interpretation has been superimposed upon a conscious bias in the selection of topics. Crop physiology is a vast field, and we have chosen to deal only with the above-ground processes fundamental to crop yield. Even so, in trying to keep the book to a manageable size, we have perhaps neglected work which should have been included, and it would be helpful if any such omissions, as well as errors or misinterpretations, are brought to our attention (R K M H, Chapters 2, 6, 7, 8; A J W, Chapters 3, 4, 5, 9). We are grateful to Dr Dale Walters who has collaborated in this project from the start, contributing several valuable sections on the interactions between plant disease and crop physiology.

Parts of the book have been read in draft form by Mr E J Allen, Dr L C Ho, Dr E J M Kirby, Dr P E H Minchin, Dr J R Porter, Dr M J Robson, Dr J H M Thornley and Miss J Woledge, and we are grateful for their valuable comments. Professor D A Baker kindly read the entire manuscript, and to him we are especially indebted. The exacting task of producing the diagrams in their final form fell to Jacqueline Gemmell, Felicity Walker and Ewan McCall, and to them we extend our thanks. Finally, we thank our families for their forbearance during a long period of undeserved neglect.

## REFERENCE

van Overbeek, J. 1976 Plant physiology and the human ecosystem, *Ann Rev Plant Physiol* **27**: 1–17

# ACKNOWLEDGEMENTS

We are grateful to the following for permission to reproduce copyright material:

Academic Press Inc. (London and San Diego) and the authors for figs. 1.1 from fig. 1 (Biscoe & Gallagher 1977), 2.26 adapted from fig. 12 (Ripley & Redman 1976), 4.1 from fig. 5b, p. 297 (Loomis *et al.* 1967), 5.10 from fig. 4, p. 334 (Christy & Swanson 1976) and table 4.1 from table 1, p. 329 (Penning de Vries 1972); the American Association for the Advancement of Science and the authors for fig. 5.8 from fig. 1 (Gifford *et al.* 1984) (c) 1984 by the AAAS and table 4.2 from table 1 (Sinclair & de Wit 1975) (c) 1975 by the AAAS; the American Society of Agronomy Inc. and the Crop Science of America Inc. for fig. 2.23 adapted from fig. 3.5 (Loomis & Williams 1969); the American Society of Plant Physiologists for figs. 3.15 & 3.16 from figs. 1 & 2 (Ehleringer & Pearcy 1983), 3.23 from fig. 1 (Camp *et al.* 1982), 3.24 from fig. 2 (Evans 1983), 3.26 from figs. 1–4 (Thorne & Koller 1974), 3.27 from figs. 1 & 2 (Nafziger & Koller 1976), 3.28 & 29 from figs. 3 & 8A (Azcon-Bieto 1983), 3.31 from fig. 2, (Setter *et al.* 1980a), 3.32 from fig. 2 (Setter *et al.* 1980b), 3.34 from fig. 1 (Boyer & Bowen 1970), 3.35 from figs. 2 & 3 (O'Toole *et al.* 1976), 3.37a from fig. 2B (Hutmacher & Krieg 1983), 4.2 from fig. 1 (McCree & Troughton 1966b), 4.4b from figs. 3 & 4 (Azcon-Bieto & Osmond 1983), 5.9 from fig. 2 (Servaites & Geiger 1974), 5.11 from fig. 1 (Outlaw *et al.* 1975), 5.12 & 13 from figs. 1, 2, 5 & 6b (Fisher *et al.* 1978), 5.15 from figs. 1 & 2 (Rufty & Huber 1983), 5.16 from fig. 1 (Chatterton & Silvius 1979), 5.18 from figs. 1 & 3 (Hammond & Burton 1983), 5.32 from figs. 1 & 2 (Thorne 1981), 5.36 from fig. 1B (Thorne 1982), 5.37 & 5.38 from figs. 7, 1 & 6 (Schussler *et al.* 1984) and table 3.7 from table 1 (O'Toole *et al.* 1976); the Annals of Botany Company and the authors for figs. 2.10 from fig. 2b (Robson & Deacon 1978), 2.12 adapted from fig. 4 (Last 1962), 2.13 from fig. 5 (Watson 1947), 3.39 & 3.40 from figs. 7 & 6 (Robson & Parsons 1978), 4.8 & 4.9 from figs. 1, 4 & 5 (Jones *et al.* 1978), 4.13–15 from figs. 1, 2 & 4 (Robson 1982), 5.2 from fig. 4 (Austin *et al.* 1977a), 5.39 from fig. 3 (Walker & Ho 1977) and tables 4.3 from table 3 (Breeze & Elston 1978), 4.5 from table 1 (Robson 1982), 5.1 from table 2 (Austin *et al.* 1980a); Annual Reviews Inc. for fig. 5.28 from fig. 3 (Thorne 1985) (c) 1985 by Annual Reviews Inc.; Arable Unit, National Agricultural Centre for figs. 6.2 adapted from figs. 5.16–5.19 & 5.26–5.27 (Kirby and Appleyard 1984a), 6.5 adapted from fig. 1 (Thorne & Wood 1982) 6.8a adapted from p. 104 (Widdowson 1979) and 6.11 adapted from fig. 6 (Biscoe & Willington 1984b); the Association of Applied Biologists and the authors for figs. 2.7 (adapted Milford *et al.* 1985), 4.12a from fig. 1b (Wilson 1975), fig. 8.1 (Colvill & Marshall 1984) and tables 4.4a & b from tables 8 & 3 (Wilson 1975) and 6.1 (adapted Tottman *et al.* 1979); Blackwell Scientific Publications, Oxford for fig. 3.46 from fig. 6.2, p. 77 (Smedegaard-Petersen 1984) and table 2.1 adapted from table 4 (Kirby, Appleyard & Fellowes 1982); the British Ecological Society and the authors for figs. 3.7 from fig. 8a (Biscoe *et al.* 1975a), 3.19 from fig. 2b (Biscoe *et al.* 1975b), 4.5–7 from figs. 9, 8 & 6 (Biscoe *et al.* 1975c) and 8.4 from fig. 1

(Parsons *et al.* 1983a); the British Grassland Society for figs. 2.18 adapted from fig. 1 (Jones 1981), 8.5 from fig. 1 (Hodgson *et al.* 1981) & table 4.6 from table 1, p. 211 (Wilson & Robson 1981); the British Plant Growth Regulator Group for figs. 5.19–21 from figs. 3b, 2a & 2b, pp. 71–3 (Farrar & Farrar 1985b) and 5.42 from fig. 1 (Morris 1983); the author, Dr. W. G. Burton for table 7.1 from table 50 (Burton 1966); Butterworths & Co. Publishers Ltd. for fig. 9.5 from fig. 7.5, p. 119 (France & Thornley 1984); Cambridge University Press for figs. 2.1 from fig. 4.7.2, p. 132 (Williams 1975), 2.4 from fig. 6.4, (Dale & Milthorpe 1983), 2.6 from fig. 1 (Hay & Tunnicliffe 1982), 2.15 adapted from fig. 5.6 (Jones & Allen 1983), 2.16b from fig. 1 (Kirby 1967), 2.17 adapted from fig. 4b (Allen & Morgan 1972), 2.19 adapted from fig. 8a (Moorby & Milthorpe 1975), 2.21 adapted from fig. 5.4 Evans *et al.* 1975), 3.12a from fig. 7.9 (Jones 1983), 3.41 & 3.42 from fig. 3 & 6 (Gregory *et al.* 1981), 3.44 from fig. 3A (Walters 1985), 3.45 from fig. 8, p. 24 (Buchanan *et al.* 1981), 5.41 from fig. 5 (Walters 1985), 6.4 adapted from fig. 5 (Kirby 1967), 6.12 from fig. 2 (Easson 1984), 7.6 & 7.7 from figs. 1, 2 & 11 (O'Brien *et al.* 1983), 7.8 & 7.11 from figs. 7 & 13 (Allen & Scott 1980), 7.9 from fig. 1 (Wurr 1974), 7.12 adapted from fig. 2 (Birch *et al.* 1967), 8.2 adapted from fig. 3 (Anslow & Green 1967), 8.10 adapted from fig. 2 (Reid 1970), 9.1 from fig. 1 (Weir *et al.* 1984), 9.2, 9.3a, 9.10 & 9.11 from figs. 1, 2, 4 & 6 (Porter 1984), 9.9 from fig. 4 (Weir *et al.* 1984) and tables 5.3 from tables 1, 2 & 5 (Austin *et al.* 1980b), 5.4 from tables 1 & 4 (Brooking & Kirby 1981), 6.6 adapted from table 1 (Ellis & Kirby 1980), 8.3 from table 5 (Jackson & Williams 1979), 9.1 & 9.2 from App. 1 and table 1 (Weir *et al.* 1984); Chapman and Hall Ltd. and the author, Prof. E. G. Cutter for figs. 2.2 from fig. 3.5 (Cutter 1978), 7.2 & 7.16 from figs. 17.6 & 17.5 (Scott & Wilcockson 1978); Commonwealth Agricultural Bureaux for figs. 2.24 from fig. 1 (Brown & Blaser 1968), 7.10 from fig. 2 (Holliday 1960) & table 2.3 adapted from table 1 (Trenbath & Angus 1975); C.R.C. Press Inc. for table 3.2 from table 3, pp. 235–8 (Eagles & Wilson 1982)/Copyright CRC Press Inc.; Crop Science Society of America Inc. for figs. 2.16a adapted from fig. 2A (Williams *et al.* 1968), 3.37b from fig. 4A (Krieg & Hutmacher 1986) and 4.10b from fig. 5 (Moser *et al.* 1982); C.S.I.R.O. Editorial and Publishing Unit for figs. 2.5 adapted from fig. 3 (Williams & Rijven 1965), 3.5 from fig. 4 (Jones & Osmond 1973), 3.20 from figs. 1 & 3 (Constable & Rawson 1980), 3.21 from fig. 4 (Woodward & Rawson 1976), 3.33a from figs. 2 & 8 (Troughton 1969), 3.33b from figs. 3 & 4a (Ludlow & Ng 1976), 4.3 from fig. 7 (Sale 1974), 5.4 from fig. 1 (Singh & Jenner 1984), 5.6 from fig. 1a (Stockman *et al.* 1983), 5.7 from fig. 4 (Fischer & HilleRisLambers 1978), 5.23 & 5.24 from figs. 1c & 5 (Bremner & Rawson 1978), 5.26 from fig. 1 (Jenner 1970), 5.30 from fig. 2 (Jenner & Rathjen 1978), 5.31a from fig. 1 (Offler & Patrick 1984), 5.31b from table 1 (Patrick & McDonald 1980) and tables 3.9 from table 1 (Moorby *et al.* 1975), 5.2 from table 1 (Fischer & HilleRisLambers 1978), 5.5 from table 2 (Singh & Jenner 1982), 8.4 from table 4 (Stern & Donald 1962a); Czechoslovak Academy of Sciences, Prague for fig. 3.36 from fig. 1 (Krampitz *et al.* 1984); Department of Agriculture and Fisheries for Scotland, Agricultural Scientific Services for fig. 6.7 from data; Elsevier Science Publishers for fig. 2.14 adapted from fig. 2 (Keating *et al.* 1982) and table 6.4 from table 1 (Hay 1986); European Association for Potato Research for figs. 2.20 from fig. 3 (Khurana & McLaren 1982), 7.4 adapted from fig. 2 (Headford 1962) and 7.14 adapted from fig. 5 (Schippers 1968); the author, Dr. J. Grace for fig. 3.9a from fig. 3.5a, p. 43 (Grace 1983); Gustav Fischer Verlag, Stuttgart and the author, Dr. W. Rademacher for fig. 5.25 from fig. 2b & c (Rademacher & Graebe 1984); The Controller of Her Majesty's Stationery Office for figs. 6.1 (Large 1954), 6.8b (Lidgate 1984), tables 6.8 & 6.9 (MAFF 1983); International Potash Institute, Bern for figs. 5.1 & 5.3 from figs. 4 & 5 pp. 68–9 (Stoy 1980); International Thomson Publishing Ltd. for fig. 5.5 from fig. 4 (Biscoe & Gallagher 1978); Dr. W. Junk bv., Publishers, The Hague for figs. 3.43 from fig. 1 (Ayres 1979) and 5.40 from fig. 8 (Ho *et al.* 1983); the author, Dr. A. J. Keys for fig. 3.13 from fig. 7.1, p. 142 (Keys & Whittingham 1981); Longman Group (UK) Ltd. for fig. 5.17b from figs. 6.16 & 6.17, p. 243 (Hall *et al.* 1974); the editor, *The New Phytologist* for fig. 3.4 from fig. 2 (Jones 1973); Martinus Nijhoff Publisher, The Netherlands for fig. 3.22 from figs. 1 & 2 (Araus *et al.* 1986); Oxford University Press for figs. 2.3 redrawn from fig. 6 (Sunderland 1960),

2.8 & 2.9 from figs. 3 & 7 (Gallagher & Biscoe 1979), 3.18 from fig. 1 (Bird *et al.* 1977), 5.29 from fig. 3 (Ho & Gifford 1984), 6.6 adapted from fig. 5 (Kirby & Faris 1970), 7.5 from fig. 4 (Goodwin 1967), 9.6 from fig. 1b & c (Marshall & Biscoe 1980) and table 3.8 from table 1 (Lawlor & Fock 1977); Packard Publishing Co. for figs. 3.17 & 3.30 and tables 3.4 & 3.5 (Edwards & Walker 1983); the author, Dr. A. J. Parsons for fig. 8.3; Pergamon Books Ltd for table 5.6 (Holligan *et al.* 1973; Fung 1975); Plenum Press Inc. and the author Dr. W. Day for figs, 9.3b, 9.4 & 9.8 from figs. 1b, c & d (Weir *et al.* 1985); Pudoc, Centre for Agricultural Publishing and Documentation and the authors for figs. 3.8 from fig. 2, p. 182 (Uchijima 1970), 4.10a from fig. 2, p. 84 (Vos 1979) and 8.9 adapted from fig. 5 (Van Burg *et al.* 1980); Royal Botanical Society of the Netherlands, Leiden for fig. 3.11 from fig. 13 (Bange 1953); Royal Society of London and the author, J. L. Monteith for fig. 3.3 from fig. 1 (Monteith 1977a); Royal Society of New Zealand for fig. 5.27 from fig. 3b (Jenner 1974); Society of Agricultural Meteorology of Japan for fig. 3.9b from figs. 3 & 4 (Yabuki & Miyagawa 1970); Society of Chemical Industry, London and the author, Dr. K. Gales for fig. 6.14 from fig. 1 (Gales 1983); the editor, *Span* and the author, Dr. A. J. Parsons for fig. 8.7 adapted from fig. 1 (Parsons 1985); Springer Verlag (Heidelberg and New York) and the authors for figs. 3.12b from 1a (Von Caemmerer & Farquhar 1981), 3.25 from fig. 1 (King *et al.* 1967), 3.38 from fig. 6 (Von Caemmerer & Farquhar 1984), and 5.33–35 from figs. 1, 5 & 2 (Wolswinkel & Ammerlaan 1983); the author, Prof. J. Warren Wilson for fig. 2.25 from fig. 3 (Warren Wilson 1959); the Director of the Welsh Plant Breeding Station for fig. 4.12b from fig. 2 (Wilson 1976); Westview Press Inc for fig. 8.6 from fig. 1, p. 602 (Bircham & Hodgson 1983b); the author, Miss J. Woledge for fig. 8.11.

# CHAPTER 1 INTRODUCTION

*And then a queer thought came to her there in the drooked fields, that nothing endured at all, nothing but the land she passed across, tossed and turned and perpetually changed below the hands of the crofter folk since the oldest of them had set the Standing Stones by the loch of Blawearie and climbed there on their holy days and saw their terraced crops ride brave in the wind and sun. Sea and sky and the folk who wrote and fought and were learned, teaching and saying and praying, they lasted but as a breath, a mist of fog in the hills, but the land was forever, it moved and changed below you, but was forever, you were close to it and it to you, not at a bleak remove it held you and hurted you.*

(From *Sunset Song*, Lewis Grassic Gibbon 1932).

This book is an introduction to the physiological processes determining the yield which can be harvested from a stand of crop plants, processes which are common to all crop species, whether grown for direct consumption by humans (cereals, seed legumes, potatoes, vegetables), for indirect consumption via livestock (grass, forage legumes, oilseeds), or for industrial purposes (production of fibre, alcohol, fuel). Although crop physiology is firmly founded upon plant physiology and biochemistry, it has now emerged as a relatively distinct subject, for a number of reasons. First, the range of species studied is restricted, compared with that studied by plant physiologists and ecologists, with the greatest emphasis, to date, being placed upon cereals (wheat, rice, maize, barley). However, this specialization is offset by the unique depth of study of individual species at all levels from subcellular biochemistry to field-scale agronomy. Furthermore, crop physiology involves complementary studies of plants growing singly and in stands, in the field and under controlled conditions. Secondly, in the field, concepts such as leaf area index have led to the idea of the crop leaf canopy as a single functional unit, rather than an assemblage of individual plants. This, in turn, has permitted the development of new micrometeorological methods for measuring mass and energy exchange between atmosphere and canopy (e.g. canopy net photosynthesis).

Growing a crop is an exercise in energy transformation, in which incident solar radiation is converted to more useful forms of chemical potential energy located in the harvested parts (e.g.

starch in cereal grains and potato tubers; lipids in oilseeds). To achieve this transformation, it is necessary for the crop to carry out the following three processes in sequence:–

(a) interception of incident solar radiation by the leaf canopy;
(b) conversion of the intercepted radiant energy to chemical potential energy (the latter conveniently expressed in terms of plant dry matter);
(c) partitioning of the dry matter produced between the harvested parts and the rest of the plant.

The yield ($Y$) of a crop over a given period of time can, therefore, be expressed by the equation:

$$Y = Q \times I \times \varepsilon \times H \qquad [1.1]$$

where $Q$ is the total quantity of incident solar radiation received over the period

$I$ is the fraction of $Q$ which is intercepted by the canopy

$\varepsilon$ is the overall photosynthetic efficiency of the crop (i.e. the efficiency of conversion of radiant to chemical potential energy), commonly expressed in terms of the total plant dry matter produced per unit of intercepted radiant energy

$H$ is the fraction of the dry matter produced which is allocated to the harvested parts. This is really the harvest index of the crop stand, although it should be emphasized that it is normally expressed in terms of above-ground production, excluding the root system.

The amount of solar radiation incident upon unit area of cropland ($Q$) per day, which depends upon daylength and the diurnal pattern of irradiance, varies regularly with season and latitude, and irregularly with altitude and short-term weather factors such as cloudiness. Variation in $Q$ is an aspect of environmental physics, and since it is admirably covered in a number of other textbooks (e.g. Monteith 1973; Woodward and Sheehy 1983), it is not studied in depth here.

The extent to which the canopy intercepts the available radiation ($I$) depends not only upon the crop leaf area displayed per unit of soil surface area (leaf area index) but also upon characteristics such as leaf angle and the arrangement of the leaves in space (the canopy structure or architecture). The factors controlling leaf growth and canopy development are reviewed in detail in Chapter 2. As shown by the pioneering studies of Watson in the 1940s, variation in $I$ accounts for most of the differences in yield between sites and seasons in temperate regions, because $\varepsilon$ and H are relatively constant in the absence of severe stress (drought, disease). This is clearly illustrated by Fig. 1.1, which shows that the total dry-matter production of three contrasting crop species, on a weekly basis, was linearly related to the quantity of radiant energy intercepted during that week (compare with Figs 3.3 and 7.16 where the same effect is shown over the entire growing season). Thus, since canopy development is limited primarily by temperature, potential yield is lost in spring simply because the temperatures during the preceding weeks have been too low for the development of leaf area to intercept the available radiation (e.g. Figs 2.13 and 2.19).

The scientific literature concerned with $\varepsilon$, the photosynthetic efficiency, which is reviewed in Chapters 3 and 4, is much more extensive than that dealing with leaf and canopy development, in spite of the fact that $\varepsilon$ varies much less than $I$ in the field. There are at least three reasons for this concentration of interest upon photosynthesis and related processes (respiration, photorespiration). First, the reactions of photosynthesis are intriguing in themselves since they are unique to green plants and are the primary source of energy for virtually all food chains. Secondly, it is important to be able to account for the fact that the rates of component processes of photosynthetic efficiency vary between species, and are influenced substantially by variation in environmental factors, whereas $\varepsilon$ varies little between crops and species (note the linear relationships in Figs 1.1, 3.3, and 7.16). Thirdly, it has become clear that in certain crops, notably grasses, there is limited scope for improvement of yield by variation in $I$ and $H$, and that increased production can come only from increases in $\varepsilon$.

$H$, the partitioning of dry matter to the harvested parts, is undoubtedly the least under-

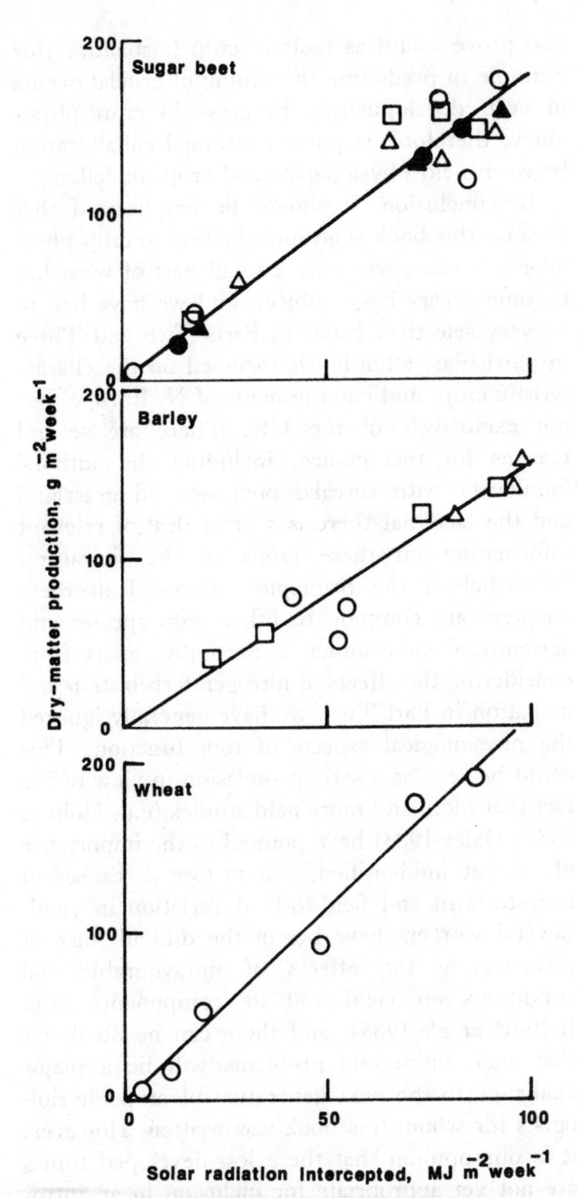


**Fig. 1.1** Relationships between dry-matter production and intercepted solar radiation, based on weekly measurements, for three crop species grown in the Midlands of England. The different symbols indicate different crops (from Biscoe and Gallagher 1977).

stood of the factors in eqn [1.1]. Much of the work on this topic, which is reviewed in Chapter 5, is aimed at understanding the fundamental physiology of assimilate translocation within the plant, because so little is known about the mechanisms controlling the allocation of dry matter among the various competing sinks – leaves, roots, stems, maintenance of mature tissues and defence against stress and disease – even when the harvested parts are morphologically distinct structures such as fruits or tubers. The situation becomes even more complex when the harvested part (the leaves) constitutes both the source and sink for assimilate, as is the case for forage crops. However, the harvest index (the net result of assimilate partitioning) has now been recognized as a crucial characteristic, and the steady increase in the grain yield of wheat and barley during the twentieth century, with the introduction of new cultivars, is now seen in terms of a progressive increase in harvest index, with little or no increase in total crop dry-matter production. This has important implications for breeding programmes since the harvest index of the latest cultivars is approaching the theoretical maximum value.

In Part One, we have drawn examples from a range of crop species to illustrate aspects of radiation interception, photosynthetic efficiency and dry-matter partitioning. In Part Two, we study three contrasting sets of temperate crops (cereals, predominantly wheat and barley; potatoes; grassland) in more detail to find out how variation in crop management (plant population density, planting date, nitrogen fertilization, irrigation, method of harvesting) can affect crop yield. Two important aspects are considered in some detail. First, although the husbandry of a crop is designed to maximize yield at a given level of inputs, it is also important to ensure that the quality of the material harvested is also appropriate. Accordingly, attention is paid to the physiological aspects of cereal grain weight, the size distribution of harvested potato tubers, and the digestibility of the dry matter produced by grassland, as well as other indices of quality. Secondly, the maximum potential yield of each type of crop is examined, bearing in mind that the maximum economic yield will, in the majority of cases, be much lower.

In Part Two, the opportunity is also taken to examine some fruitful fields of controversy in crop physiology. For example, in Chapter 6, the long-held view that the components of cereal grain yield were mutually compensating (an idea which came from studies on the effect of variation in plant population density) is shown to be not generally tenable in the light of more recent work; indeed, it is incompatible with the very high yields which can be harvested from cereal crops. In Chapter 7, the hypothesis that potato tuber yield was controlled by intra-plant competition for assimilate at tuber initiation is also shown to be not generally correct, and in Chapter 8, we emphasize the importance of livestock in studies of the physiology of grass swards. However elegant and rigorously performed, experiments involving frequent cuts of herbage cannot simulate the action of the grazing animal. In considering these controversies, it becomes clear that the rapid progress of crop physiology since 1970 has been the result of improvements in field equipment to monitor the environment and crop responses, and the realization that increased understanding can come only from detailed studies of the growth, development and physiology of crops in the field, often under very difficult working conditions. Experimental work under controlled conditions must always be seen as an adjunct to field work.

However, it is totally impractical to do all the field experiments necessary to investigate the effects of variation in site, soil, climate and management, and it is here that the need for crop simulation models arises. Provided that there exists sufficient field data to simulate the growth and/or function of a stand of crop plants under a given set of conditions, it is then possible to alter the variables in the model to simulate a range of possible experiments. For example, Chapter 9 describes the AFRC wheat model, which was constructed to simulate the development, function and, ultimately, the grain yield of winter wheat crops growing in the UK. Such models have the added value that they identify areas of ignorance in crop development and physiology as well as in agrometeorology and soil science, and they can also prove useful as tools in crop husbandry (for example in predicting the timing of crucial events in crop development). Progress in crop physiology, therefore, requires continued collaboration between field physiologists and crop modellers.

In conclusion, it should be emphasized that because this book is an introduction to crop physiology, it can cover only a small part of what has become a very large subject, and we have had to be very selective. First, in Parts Two and Three in particular, attention is focussed on the characteristic crops and environments of N. Europe, but not exclusively of the UK. There are several reasons for this choice, including the authors' familiarity with cereals, potatoes and grassland and the fact that there is a great deal of relevant information on these crops in the literature. Nevertheless, the principles discussed in these chapters are common to other crop species and agricultural environments. Secondly, apart from considering the effects of nitrogen fertilization and irrigation in Part Two, we have generally ignored the physiological aspects of root function. This could be seen as a serious omission in view of the fact that more and more field studies (e.g. Holmes 1982; Gales 1983) have pointed to the importance of, as yet unidentified, soil factors as causes of farm-to-farm and field-to-field variation in yield. Several workers have begun the difficult task of investigating the effects of unfavourable soil conditions on yield and its components (e.g. Belford *et al.* 1985), and there can be no doubt that such intractible problems will be a major challenge to the next generation of crop physiologists for whom this book was written. However, it is our opinion that these less-developed topics are not yet appropriate for inclusion in an introductory text.

# PART I FUNDAMENTAL PRINCIPLES OF CROP PHYSIOLOGY

# CHAPTER 2 INTERCEPTION OF SOLAR RADIATION BY THE CROP CANOPY

*It is probable . . . that the development of successful systems of crop husbandry largely depends on discovering empirically the optimal conditions for leaf area production in a particular environment.*

(Watson 1952)

Most of the solar radiation absorbed by a crop canopy is intercepted by its leaf blades, although leaf sheaths, petioles, stems and reproductive structures can make considerable contributions under certain conditions. Because of this, it has become customary to express the capacity of a crop to intercept solar radiation by its leaf area index ($L$) – the area of leaf lamina (one surface only) per unit of soil surface area. The fact that $L$ is dimensionless (e.g. $m^2$ of leaf area $m^{-2}$ of land area) has suggested to some that it should be thought of as the number of layers of leaves in a crop. Whichever definition is used, $L$ is a property of the stand of plants rather than of the individual plants of the crop; however, a full understanding of the genetic, environmental and management factors controlling L can come only from consideration of the growth of individual leaves and plants.

## 2.1 THE LIFE-HISTORY OF A LEAF

Leaves begin life as a regular series of primordia, localized outgrowths on the sides of the apical dome of a vegetative shoot (e.g. Figs 2.1 and 2.2). The factors controlling the relative positions of these primordia (which also determine the arrangement of fully-developed leaves on the stem, the phyllotaxis), and the time interval between successive primordia (the plastochron) are not fully understood, but the control is, presumably, mediated by plant growth substances. Leaf initiation begins in the developing seed when it is still attached to the parent plant; for example,

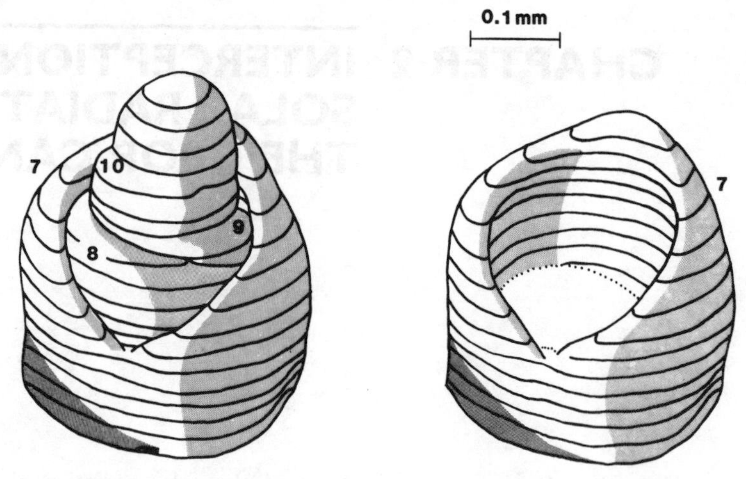


**Fig. 2.1** Three-dimensional reconstruction of the vegetative stem apex of a wheat plant which has just initiated the tenth leaf primordium. The first six fully- and partly-developed leaves have been removed to reveal the expanding blade and sheath of leaf 7. The horizontal lines, which have no structural significance, indicate spacings of 20 $\mu$m (from Williams 1975).

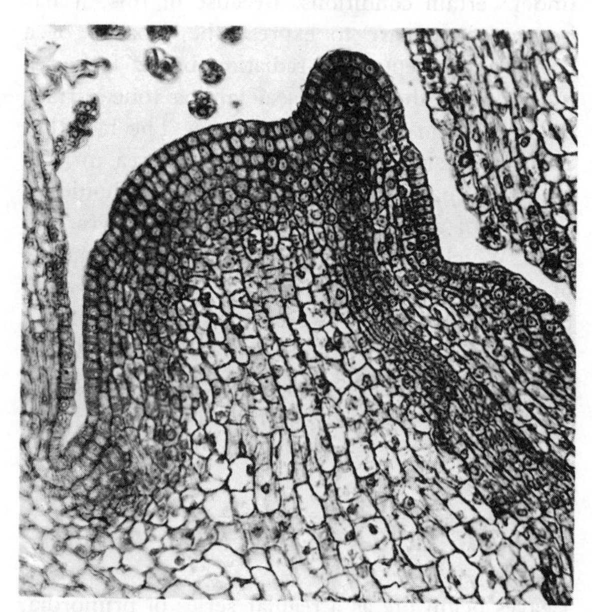

**Fig. 2.2** Longitudinal section of the vegetative apex of an aerial stem of potato cv. King Edward showing one leaf primordium × 230 (from Cutter 1978).

three leaf primordia are already present in the embryo of a wheat grain at sowing (Stern and Kirby 1979). In contrast, because the sprouts of seed potatoes contain many more leaf primordia (e.g. up to 26 primordia and partly-developed leaves in a 2-cm long sprout, Milthorpe 1963), the development of the potato leaf canopy consists largely of the expansion of already-initiated leaves.

After initiation, the developing leaf enters a phase of growth dominated by cell division which, in dicotyledons, continues at least up to the stage of leaf unfolding. During this phase, an increasing proportion of the cells of the leaf does not take part in division because of differentiation, for example, into vascular tissues. However, in most of the species which have been studied in detail, almost exclusively under controlled conditions, the rate of increase of cell number, leaf length, area and fresh weight is approximately exponential (e.g. Fig. 2.3). The pattern of leaf growth in the fluctuating environment of the field is much less regular.

In dicotyledons, leaf unfolding tends to be associated with a decline in cell division which, although subsequently more localized in plate and marginal meristems (e.g. Fig. 2.4), does persist up to a late stage of leaf growth; for example, cell division in spinach leaves can continue until the leaves have achieved 30 to 50 per cent of their final size (Saurer and Possingham 1970). Because of the very large numbers of cells involved, these relatively localized areas of cell division tend to produce a very high proportion of the final number of cells in the mature leaf (Dale 1976). Nevertheless, the great increase in leaf area during

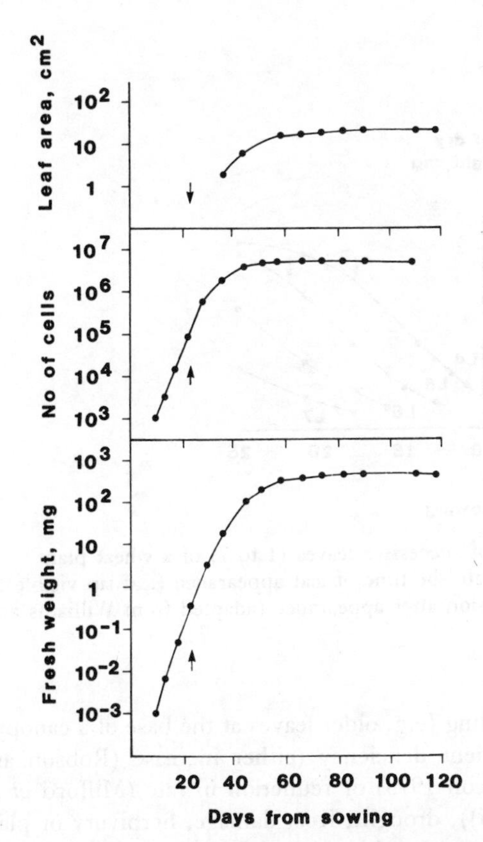


**Fig. 2.3** Time courses of fresh weight, cell number and laminar area of one of the second pair of leaves of a sunflower (*Helianthus annus*) plant growing under constant controlled conditions. The arrows indicate the time of leaf unfolding. Note that cell division is completed before the achievement of maximum fresh weight and leaf area (adapted from Sunderland 1960).

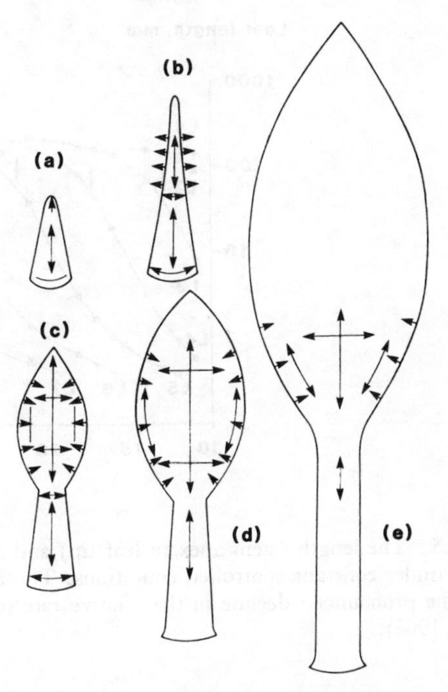


**Fig. 2.4** The sites and directions of growth at different stages (a to e) in the expansion of a (hypothetical) dicotyledon leaf. The final shape of the lamina and petiole are determined by localized cell division and expansion in the directions indicated by arrows (from Dale and Milthorpe 1983).

laminar expansion (Fig. 2.3) is caused primarily by cell expansion (10- to 35-fold increase in mean cell volume, according to species; Dale 1982).

The pattern of leaf growth in cereals and grasses is rather different (Langer 1979). The phase of growth during which most of the leaf cells are involved in division is short, and cell division is thereafter restricted to intercalary meristems above and below the leaf ligule (generating blade cells above and sheath cells below). Cell expansion also begins at an early stage, with the result that when the tip of a growing grass leaf emerges from the enclosing sheath of the previous leaf (leaf appearance), its cells are already fully expanded and mature. The rate of expansion of visible leaf area in grasses and cereals, therefore, depends upon the rates of cell division and expansion in the intercalary meristems (the leaf extension zone), which are enclosed within previous leaf sheaths and situated below, or near to, the soil surface until carried upwards as a consequence of stem extension during the later stages of reproductive development (Hay 1978, 1986; Kemp 1980). In simple terms, the mature regions are pushed up by the growth of the leaf extension zone. In general, the rate of increase of the area of a leaf decreases after its appearance, but this later phase of expansion can account for as much as 80 per cent of the final blade area (Fig. 2.5; note the logarithmic scale). The changes in function associated with leaf ontogeny are considered in subsequent chapters (e.g. sect. 3.4.3; 8.4.1).

Senescence and death are also integral processes in the life history of a leaf, being at least partly under genetic control (Thomas and Stoddart 1980). In a rapidly growing crop, senescence

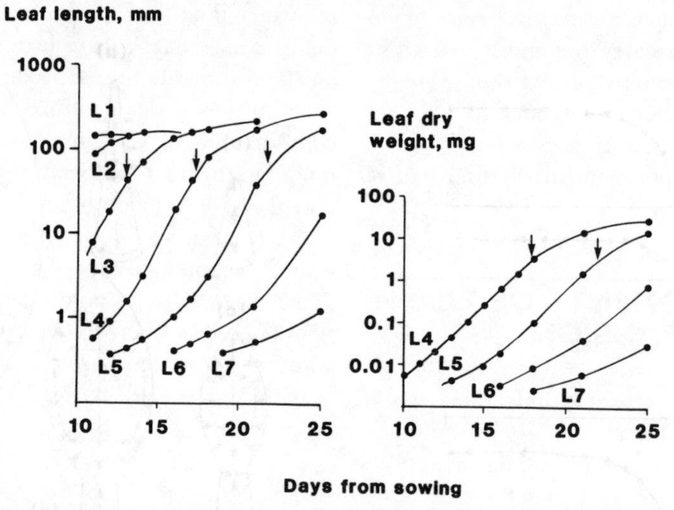


**Fig. 2.5** The length (stem apex to leaf tip) and dry weight of successive leaves (1 to 7) of a wheat plant grown under constant controlled conditions. The arrows indicate the time of leaf appearance (leaf tip visible). Note the pronounced decline in the relative rate of leaf expansion after appearance (adapted from Williams and Rijven 1965).

normally begins shortly after a leaf has attained full size and involves a decrease in photosynthetic activity (sect. 3.4.3) and the mobilization and export of organic nitrogen to developing leaves. The rate of leaf death is basically dependent upon accumulated temperature but can be affected by shading (e.g. older leaves at the base of a canopy), nutrient deficiency (either increase (Robson and Deacon 1978) or reduction in rate (Milford *et al.* 1985)), drought, frost damage, herbivory or plant disease. In the case of grasses and cereals, for example, the regular pattern of leaf appearance,

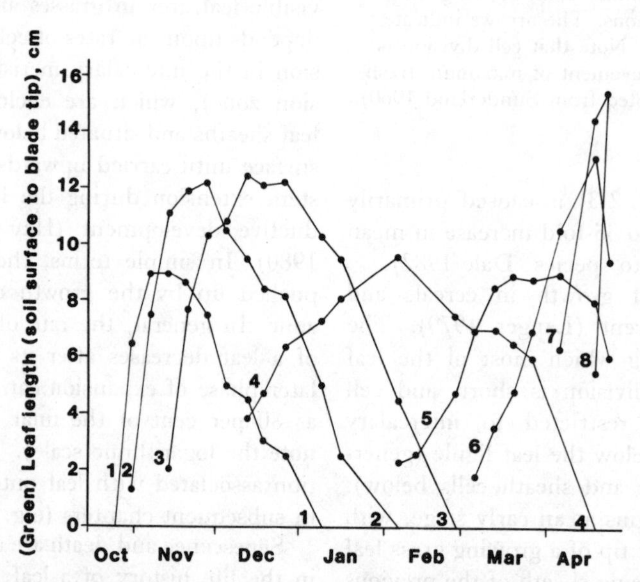


**Fig. 2.6** Appearance, extension and senescence of successive leaves (1 to 8) of a wheat plant cv. Maris Hustler growing in the field in N. England, 1978–9. The falling section of each curve was constructed from visual assessments of green leaf not affected by chlorosis, senescence or physical damage, chiefly freezing damage (from Hay and Tunnicliffe Wilson 1982).

expansion and senescence means that each mainstem or tiller tends to carry not more that three green leaves during vegetative growth (e.g. Davies 1977; Fig. 2.6). In other crops, such as potatoes and sugar beet, there is a progressive increase in the number of leaves per stem in the first half of the growing season.

## 2.2 THE COMPONENTS OF PLANT LEAF AREA EXPANSION

For many crop species growing under field conditions, the rate of production of leaf primordia is greater than the rate of leaf unfolding or appearance, with the result that, during the vegetative phase of growth, there is an accumulation of leaf primordia in the apical bud (e.g. Bunting and Drennan 1966). For example, when a winter wheat mainstem shows unequivocal signs of the change from vegetative to reproductive development in early spring (i.e. at the double ridge stage, Fig. 6.2, Table 6.4), there will be about five more leaves still to appear. Even in cases where there is no accumulation of primordia, the phyllochron (the interval between the appearance or unfolding of succeeding leaves) is commonly longer than the plastochron. Consequently, the rate of initiation of leaf primordia is not normally a factor in determining the leaf area of a plant.

In growing crops, therefore, the leaf area of a plant at a given time is determined by:

(a) the date of crop emergence;
(b) the rate of leaf production; i.e. unfolding (dicotyledons) or appearance (Gramineae and other monocotyledons);
(c) the rate of leaf expansion after unfolding/appearance;
(d) the duration of leaf expansion;
(e) the rate of branching or tillering; and
(f) the rate of leaf senescence, removal or damage (determining leaf duration);

where each of (a) to (f) can be under genetic, environmental or agronomic control.

There exists a wealth of information on the influence of environmental factors on these components of leaf growth, especially (b), (c) and (d) (reviewed by Dale 1982; Terry *et al.* 1983). However, much of this information is of doubtful relevance for the expansion of crop leaf area in the field because of the unrealistic conditions imposed on the test plants; in particular, most of the earlier experiments, under controlled conditions, involved combinations of very low levels of irradiance and moderate to high temperatures, and even in more recent work, radiation levels are generally very modest, as a consequence of the constraints of growth cabinet design (e.g. 90 W $m^{-2}$, Fletcher and Dale 1977). As pointed out by Monteith and Elston (1983), there is an urgent need for more comprehensive measurements of leaf area development in the field, particularly in dicotyledonous crops.

### 2.2.1 CROP EMERGENCE

The rate of germination and pre-emergence growth of crop plants is normally under the control of soil temperature (Hegarty 1973). The date of crop emergence is therefore determined largely by planting date modified by subsequent variations in temperature. This is discussed in more detail in sections 6.3 and 7.4.

### 2.2.2 LEAF PRODUCTION

It has been demonstrated for a range of crop species under controlled conditions and in the field that the rate of unfolding of leaves from the terminal bud is controlled solely by air temperature, provided that crop growth is not limited by severe water or nutrient stress (e.g in field bean, Dennett *et al.* 1979; in sugar beet, Milford *et al.* 1985). There is little evidence to suggest that other environmental factors, such as solar radiation, water supply or nutrient status, within their normal field ranges, have an important influence on leaf unfolding.

Work on graminaceous crops has also shown that leaf appearance is closely related to temperature but here, since the stem apex and leaf extension zone are near to the soil surface, the controlling factor is the temperature of the surface layers of soil (e.g. Hay and Tunnicliffe Wilson 1982). In practice, since air and soil temperatures tend to be closely linked, good linear relationships have been established between cereal mainstem leaf number and accumulated (screen) air temperature above a base temperature, which is

**Table 2.1** Base temperatures (°C) and rates of appearance of barley mainstem leaves (leaves/°C day) in the field in the UK, estimated by maximizing $r^2$ obtained by linear regression analysis

| | Sowing | | | |
|---|---|---|---|---|
| | *Mid-Sept.* | *Mid-Oct.* | *Mid-Nov.* | *Mid-Feb.* |
| *1979–80* | | | | |
| Base temperature | 4.3 | 1.5 | 1.3 | 0 |
| Rate of leaf appearance ($\times 10^{-3}$) | 17 | 13 | 15 | 18 |
| *1980–81* | | | | |
| Base temperature | 5 | 1.8 | 0.6 | −3 |
| Rate of leaf appearance ($\times 10^{-3}$) | 23 | 12 | 12 | 10 |

(Kirby, Appleyard and Fellowes 1982)

estimated by stepwise linear regression analysis (Table 2.1) (Baker *et al.* 1980; Delécolle and Gurnade 1980). The response of the plant to temperature (i.e. number of leaves appearing per unit of accumulated temperature) appears to be fixed at crop emergence, possibly in response to the rate of change of daylength (Baker *et al.* 1980, wheat; Kirby *et al.* 1982, barley). However, studies under controlled conditions, using a range of imposed rates of daylength change, have failed to confirm these field effects, indicating that plant response to temperature is determined by other seasonally varying factors (Kirby *et al.* 1983). In addition, there is some evidence to suggest that base temperatures for cereal leaf appearance can vary with sowing date (Table 2.1) and that, in some cases, the temperature relations of leaf appearance can be modified by ontogeny (e.g. Robson and Deacon 1978).

### 2.2.3 LEAF EXPANSION

Although there is evidence from studies under controlled conditions (e.g. Figs 2.3 and 2.5) that the rate of leaf expansion decreases after unfolding/appearance, this latter phase of growth (which can involve 80 per cent of the total growth in length, 95 per cent of dry weight and more than 99 per cent of blade area (Dale and Milthorpe 1983)) determines the rate of increase of photosynthetic leaf area. For this reason, the majority of field studies of leaf area expansion are restricted to the post-unfolding/appearance period (but note that the rate of growth during the earlier phase largely determines the rate of leaf production – sect. 2.2.2).

From the outset, it should be stressed that there appears to be a genetically-determined upper limit to the size of a leaf at any node; this size is governed by ontogeny, and environmental factors such as temperature, water supply, irradiance and nitrogen fertilizer act to modify the ontogenetic drift in leaf size. Changes in leaf size and shape during the development of a plant are very common both in wild species and crop plants (Fitter and Hay 1987); for example, each successive leaf of a sugar beet plant tends to be larger up to a certain node (around leaf 10, e.g. Fig. 2.7)

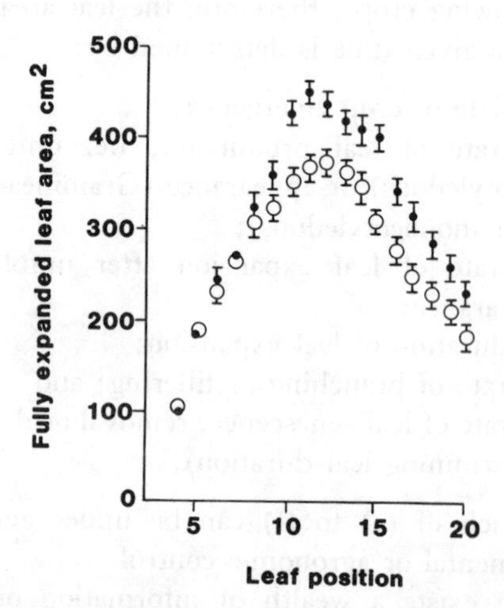


**Fig. 2.7** Ontogenetic changes in the laminar area of fully-expanded leaves of sugar beet cv. Bush Mono G grown in the field in south-east England under standard conditions in 1978 (○) and 1982 (●). See also Fig. 2.11 (adapted from Milford *et al.* 1985).

beyond which the size decreases progressively. The detailed studies of barley by Kirby *et al.* (1982) show how ontogeny can interact with sowing date to give considerable variation in leaf dimensions at a given node.

Like the two previous components, leaf area expansion is also controlled by environmental temperature (commonly expressed as accumulated temperature to deal with the fluctuating conditions in the field) in the absence of severe stress (drought, shading, mineral deficiency). In the case of dicotyledons, the relevant temperature is air temperature (e.g. Auld *et al.* 1978). In grasses and cereals, leaf expansion is controlled by the temperature of the surface layers of soil surrounding the leaf extension zone during vegetative development (e.g. Peacock 1975a; Fig. 2.8), or by air temperature for leaves expanding during reproductive development when stem extension has carried the leaf extension zone above the soil surface (e.g. Gallagher *et al.* 1979). For example, in temperate cereals, the relationship between leaf extension and accumulated temperature has been found to be linear over the normal temperate range (0–25 °C) with base temperatures near 0 °C (e.g. Gallagher 1979; Hay and Tunnicliffe Wilson 1982). The upper temperature limits of these relationships remain to be explored.

Increased rates of leaf expansion at higher temperatures are generally associated with shortening of the duration of leaf expansion, which is conveniently defined as the time elapsing between the achievement of 5 and 95 per cent of final leaf area (e.g. in barley, Gallagher 1979; field bean, Auld *et al.* 1978). However, there is little experimental evidence to support the suggestion that these two responses compensate for one another to give a final leaf area at a given node which is independent of growing temperature (Monteith 1977b; Gallagher 1979). This is not surprising, since experiments on leaf growth tend to be carried out under sub-optimal conditions, whereas compensation would be likely to occur only when the leaf was growing to its full genetic potential (Monteith 1977b). In many experiments, in the field or under controlled conditions, leaf size has been found to vary markedly with temperature; for example, there are several reports of larger leaves: (a) at lower temperatures, due to a longer duration of expansion (e.g. in field bean, Auld *et al.* 1978); or (b) at intermediate temperatures (e.g. in wheat, Milthorpe and Moorby 1979). The most recent studies using a range of species and under varying climatic conditions (reviewed by Monteith and Elston 1983) indicate that the duration of leaf expansion is much less variable, and more closely related to temperature, than is the rate of expansion, which has been shown to be more sensitive to variation in nitrogen nutrition and water supply.

As noted earlier, the overall control of leaf expansion by temperature can be modified by

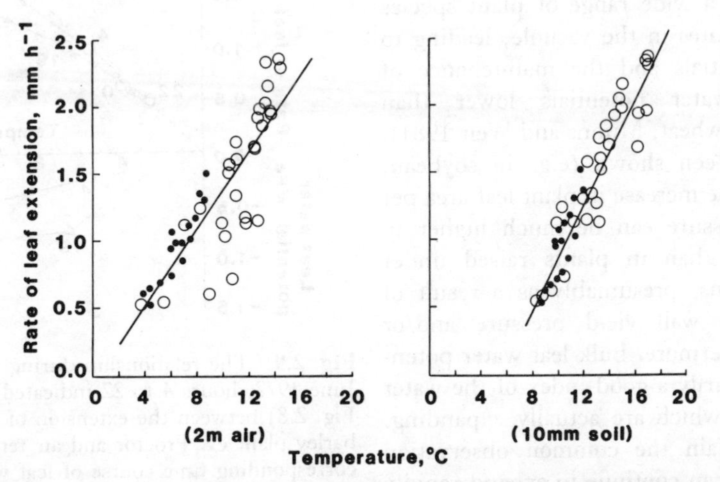


**Fig. 2.8** Relationships between the rate of extension of the sixth leaf of a barley plant cv. Proctor and air or soil temperature in the field in the Midlands of England in 1972. Each point represents a measurement period of 1 hour at night (●) or during the day (○), using an auxanometer (from Gallagher and Biscoe 1979).

other environmental factors. The primary step in leaf cell expansion is the loosening of the cell wall structure, which then yields under cell turgor pressure to give larger cell volumes, initially by the synthesis of new wall material and the influx of water. Subsequent processes include the rapid synthesis of new cytoplasmic materials and the uptake of additional mineral nutrients. The exposure of plants growing under controlled conditions to even mild water stress (corresponding to leaf water potentials of 0 to −0.5 MPa; Hsiao 1973) tends to cause a progressive loss of cell turgor pressure which, in turn, results in a progressive decline in the rates of cell and leaf expansion. Since water stress does not appear to affect the duration of leaf expansion in a consistent manner (e.g. Karamanos *et al.* 1982), the net effect of reduction in the rate is a reduction in the area of the fully-expanded leaf. Many of the earlier studies of the effects of water stress (mainly under controlled conditions, reviewed by Hsiao *et al.* 1976) indicated that cell expansion ceased at leaf water potentials as high as −0.3 to −0.4 MPa, levels which are experienced for prolonged periods even under temperate or irrigated cropping. Although several investigations showed evidence of compensatory growth when water stress was relieved (e.g. at night), these findings suggested that the water relations of a crop were a dominant factor in the development of canopy leaf area, even in humid temperate zones.

More recent work has shown that, under water stress, the cells of a wide range of plant species can accumulate solutes in the vacuole, leading to lower solute potentials and the maintenance of turgor at leaf water potentials lower than −0.5 MPa (e.g. in wheat, Munns and Weir 1981). Secondly, it has been shown (e.g. in soybean, Bunce 1977) that the increase in plant leaf area per unit of turgor pressure can be much higher in field-grown plants than in plants raised under controlled conditions, presumably as a result of differences in cell wall yield pressure and/or extensibility. Furthermore, bulk leaf water potentials are not necessarily a good index of the water status of the cells which are actually expanding. These factors explain the common observation that crop canopies can continue to expand rapidly during periods of water stress. For example, Gallagher and Biscoe (1979) demonstrated that, during a diurnal cycle of water stress (Fig. 2.9), the rate of barley leaf extension did not fall significantly below the expected value, based on leaf response to temperature, until the leaf water potential had fallen below −0.5 MPa; furthermore, the leaves achieved 60 to 70 per cent of the predicted extension rate even at leaf water potentials of −1.5 to −1.8 MPa. Responses to water stress will presumably vary among species and according to the previous drought history of the crop, but these findings do indicate that the influence of water supply on the development of leaf canopies of temperate crops has perhaps been overstressed in the past (Monteith and Elston 1983). (Note that the photosynthesis of field-grown plants is also more tolerant of water stress than those grown under controlled conditions, section 3.4.5.)

Since leaf expansion involves the synthesis of substantial quantities of cell wall and cytoplasm materials, it is possible that supplies of carbohydrate, organic nitrogen or other constituents could become rate-limiting (e.g. under shading stress, low levels of fertilizer application, mineral

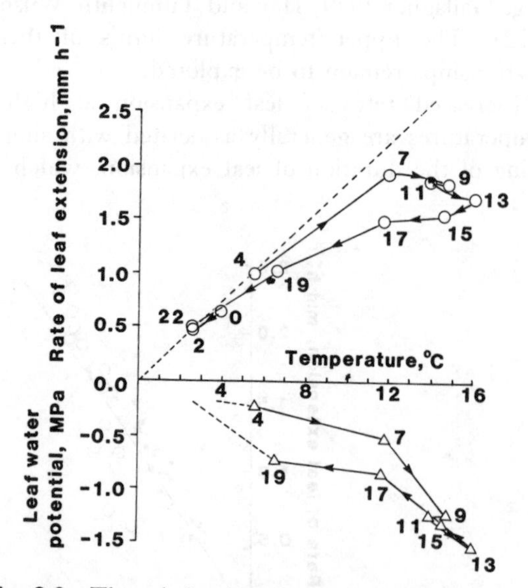


**Fig. 2.9** The relationship during a single day (1 June 1972, hours 4 to 22 indicated) in the field (see Fig. 2.8) between the extension of the eighth leaf of a barley plant cv. Proctor and air temperature, and the corresponding time course of leaf water potential. The upper dashed line indicates the expected response to temperature in the absence of water stress (from Gallagher and Biscoe 1979).

deficiency). At present there is no evidence that carbohydrate supply does influence leaf expansion rates under normal circumstances; for example, Kemp and Blacklow (1980) found that the rates of extension of leaves 3 to 5 of field-grown wheat plants were unaffected by a nine-fold difference in the hexose concentration in the leaf extension zone. The hexose levels below which inhibition would be expected (1 mg $g^{-1}$ fresh weight) were considered to be abnormally low. Sambo (1983) has come to a similar conclusion from studies of the leaf extension zone in pasture grasses.

In contrast, plant nitrogen status has a profound influence on the rate (rather than the duration) of leaf expansion and on final leaf size in all crops (e.g. in ryegrass, Fig. 2.10). Although there are reports of cell size differences, these effects are caused primarily by increased leaf cell number (i.e. enhanced cell division) as illustrated, for example, by the classic measurements of sugar beet leaves by Morton and Watson (1948) (Fig. 2.11). There are also many reports of increases in the number as well as the size of leaves carried by fertilized crop plants; these are almost invariably caused by increased branching or tiller survival (see sect 2.2.4) and not the result of stimulation of leaf unfolding/appearance (e.g. in potatoes, Humphries and French 1963). Under different conditions, each of the other essential mineral nutrients can influence leaf expansion (reviewed by Langer 1979; Terry *et al.* 1983); however, under intensive agriculture, where optimum levels of P, K and trace elements are maintained by regular fertilizer application, nitrogen is by far the most important nutrient controlling canopy development.

Finally, although the literature contains a wealth of information on the responses of growing leaves to different levels of irradiance (e.g. reviewed in Dale 1982), most of these reports have little relevance for crop plants growing in the field because of the very low levels of radiation employed. For example, the majority of experiments involve comparisons of the effects of two or more energy levels, the highest commonly not exceeding 100 W $m^{-2}$, in spite of the fact that this corresponds to a very dull day in the field during the growing season (e.g. Woodward and Sheehy 1983; Hay 1985). Overall, there is little evidence that irradiance has a direct influence upon canopy development in crops growing under normal

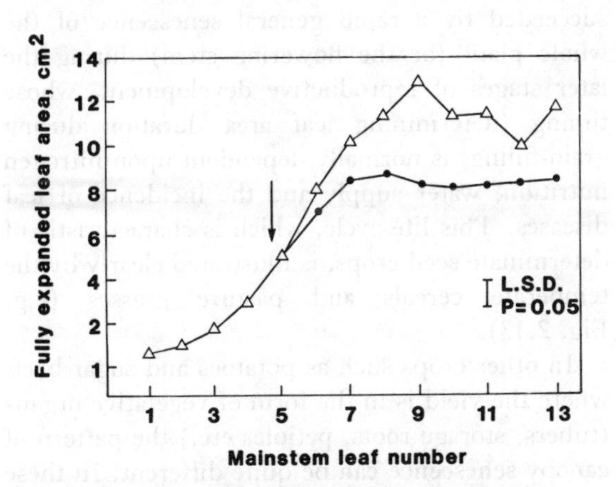


**Fig. 2.10** The influence of nitrogen supply on individual laminar area of leaves of S24 perennial ryegrass grown in simulated swards under controlled conditions. The arrow indicates the time at which the nitrogen supply was changed from uniform irrigation with 300 ppm nitrogen to 3 ppm (●) or 300 ppm (△) (from Robson and Deacon 1978).

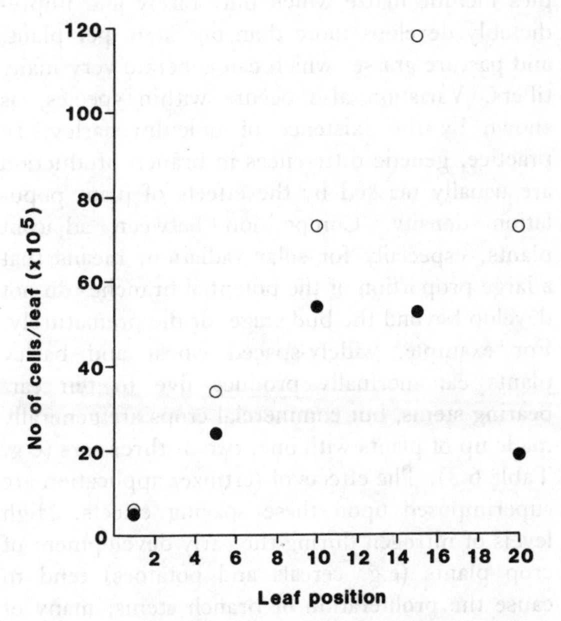


**Fig. 2.11** The influence of nitrogen supply on the number of cells per lamina of leaves of sugar beet grown out of doors in pots containing 10.5 kg of soil. The 'low nitrogen' plants (●) received no fertilizer nitrogen, and depended on the nitrogen mineralized by the soil, whereas the 'high nitrogen' plants (○) received 1 g nitrogen per pot in the form of $(NH_4)_2SO_4$ (from data of Morton and Watson 1948).

conditions; however, high levels can have an important indirect influence upon leaf expansion by increasing leaf temperature and causing a lowering of leaf water potential (e.g. Gallagher and Biscoe 1979). Finally, it should be noted that photoperiod can have a pronounced effect upon cell and leaf dimensions, particularly in temperate grasses (e.g. Hay and Heide 1983).

### 2.2.4 BRANCHING AND TILLERING

The mainstems of most crop species have the potential to produce a large number of secondary branch stems, thereby increasing both the rate of expansion and the ultimate extent of the individual plant leaf canopy. As demonstrated in detail in Part Two, the extent of branching (tillering in the Gramineae) in a given crop is largely determined by management and, in particular, by choice of species or cultivar, by plant population density and by nitrogen fertilizer application.

The capacity to produce branches does vary considerably among crop species; extreme examples include maize which only rarely and unpredictably develops more than one stem per plant, and pasture grasses which can generate very many tillers. Variation also occurs within species, as shown by the existence of uniculm barley. In practice, genetic differences in branch production are usually masked by the effects of plant population density. Competition between adjacent plants, especially for solar radiation, means that a large proportion of the potential branches do not develop beyond the bud stage, or die prematurely. For example, widely-spaced wheat and barley plants can normally produce five to ten ear-bearing stems, but commercial crops are generally made up of plants with one, two or three ears (e.g. Table 6.3). The effects of fertilizer application are superimposed upon these spacing effects. High levels of nitrogen during the early development of crop plants (e.g. cereals and potatoes) tend to cause the proliferation of branch stems, many of which die before the crop achieves maturity, particularly in the case of cereals. However, the timing of nitrogen application to cereal crops can also affect the number of tillers which survive to bear ears (sect. 6.4.2). These complex interrelationships among population density, nitrogen fertilizer application, canopy development and crop yield are explored in more detail in Part Two. Finally, grazing or cutting management has a profound influence upon the tillering of pasture grasses (e.g. Table 8.1).

### 2.2.5 LEAF SENESCENCE, REMOVAL AND DAMAGE – LEAF DURATION

The influence of leaf senescence on the rate of expansion and decline of crop plant leaf area in the field is, with some notable exceptions (e.g. Littleton *et al.* 1979), generally very poorly documented. This is at least partly a consequence of the difficulty of recognizing the time at which a given leaf has begun to senesce or, what is more important, has ceased to contribute to the dry matter production of the plant. Most of the available information on the physiology of senescence has come from studies of non-cultivated plant species, grown under controlled conditions.

In the field, many crop species display sequential leaf senescence (Wareing and Phillips 1981). Newly-expanded leaves compete with older leaves for solar radiation, mineral nutrients and assimilate, with the result that the leaves begin to senesce in sequence according to age. In the case of the Gramineae, this results in a relatively constant number of green leaves per stem, and leaf area per stem can increase only if leaf size increases at higher nodes. This pattern is succeeded by a rapid general senescence of the whole plant (or the flowering stem) during the later stages of reproductive development, whose timing (determining leaf area duration during grain-filling) is normally dependent upon nitrogen nutrition, water supply and the incidence of leaf diseases. This life-cycle, which is characteristic of determinate seed crops, is illustrated clearly by the temperate cereals and pasture grasses (e.g. Fig. 2.13).

In other crops such as potatoes and sugar beet, where the yield is in the form of vegetative organs (tubers, storage roots, petioles etc.) the pattern of canopy senescence can be quite different. In these cases, although sequential leaf senescence does occur, the number of leaves per stem increases, at least during the first half of the season. Renewed production of leaves can occur during the season, for example after frost, drought or lodging, and unless the crop is rapidly defoliated in late summer or early autumn (e.g. by blight or frost) the decline of canopy leaf area is less rapid than in cereals (Fig. 2.13). As we shall see in Chapter 7, the pattern of canopy expansion and senescence in the potato crop is determined largely by management, e.g. by the choice of early or late cultivars, by the use of seed tubers of different physiological ages, by nitrogen fertilization, by fungicide sprays and by harvest date, which commonly occurs before the natural senescence of the crop.

In any season, leaf senescence can be accelerated by a range of environmental factors including mineral deficiency, drought, high temperatures, frost and wind. The pattern of canopy development can also be disrupted, in a complex manner, by herbivory, either by pests (arthropods, birds, mammals) or by livestock. Defoliation reduces plant leaf area directly but, in some cases, it can lead ultimately to increased leaf area; for example, the removal of younger leaves can prolong the life of older leaves, but, more important, damage to mainstems can promote branching. Crop plants are generally most sensitive to defoliation during reproductive development, when the plant or flowering stem has lost the potential to initiate new leaves, but they can tolerate considerable losses during vegetative growth. For example, the common practice in the past of grazing autumn-sown cereals in winter and early spring was thought to have a relatively small influence on final crop yield. Recent work in Australia has indicated that the grain yield penalty is strongly dependent upon the growth stage of the crop at the time of grazing (Dann *et al.* 1983). On the other hand, decisions about the extent and timing of defoliation, by cutting or grazing, are a major part of grassland management (sect. 8.2).

Infection of crop plants by leaf-inhabiting fungal pathogens (e.g. rusts and mildews) tends to reduce leaf area duration, but can also have an important influence upon canopy development. For example, Last (1962) showed that the reduction in the leaf area of mildewed barley plants was caused primarily by depression of tillering (Fig. 2.12) although there was also a reduction in leaf area per tiller. It is not known whether these effects were caused by reduced supplies of assimilate or nitrogen, or to a more direct interference with plant function. In another study, wheat

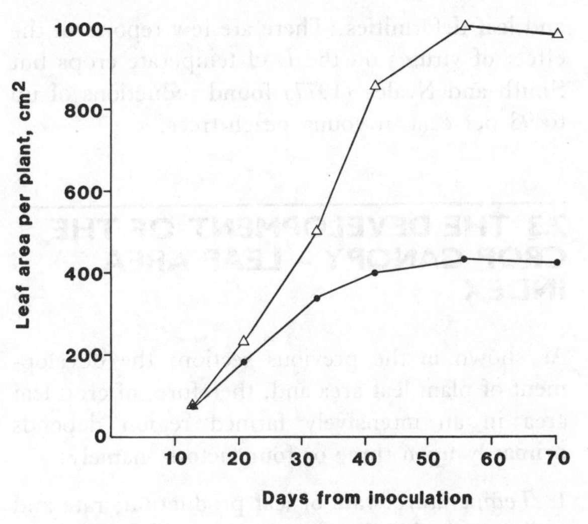


**Fig. 2.12** Time courses of leaf area development of control (△) and mildew-infected (●) barley plants grown in an unheated glasshouse at Rothamsted in July. Differences between the treatments are significant at $p < 0.01$ from day 40 (adapted from Last 1962).

plants infected with yellow rust had smaller leaves, which senesced prematurely (Doodson *et al.* 1964). These two studies were carried out under controlled conditions and there is a general lack of precise information on leaf growth in diseased plants in the field. However, Jackson and Webster (1981) have shown that in barley cultivars which are susceptible to *Rhynchosporium secalis*, the rate of leaf appearance and leaf life span can both be reduced by infection. As discussed in section 6.4.1, there has been increased interest over the last decade in the use of prophylactic fungicide sprays to increase the leaf area duration of cereal crops.

Leaf growth can also be affected by root-infecting fungi. For example, Macfarlane and Last (1959) showed that infection of young cabbage plants by the club root fungus, *Plasmodiophora brassicae*, resulted in smaller leaf area per plant (lower rate of leaf unfolding and smaller individual leaves), and *Verticillium* infection of potatoes has been associated with very substantial reductions in leaf area duration (46–82 per cent; Harrison and Isaac 1968). Other important biotic factors include plant viruses which can cause serious disruption of leaf growth and function, leading to lower plant leaf areas, smaller leaves and leaf deformities. There are few reports of the effect of viruses on the $L$ of temperate crops but Smith and Neales (1977) found reductions of up to 93 per cent in young peach trees.

## 2.3 THE DEVELOPMENT OF THE CROP CANOPY – LEAF AREA INDEX

As shown in the previous section, the development of plant leaf area and, therefore, of crop leaf area in an intensively farmed region depends primarily upon three or four factors, namely:

1 *Temperature*: rate of leaf production; rate and duration of leaf expansion;

2. *Nitrogen status*: leaf size and longevity; plant branching and survival of branches;

3. *Population density*: early season crop leaf area (e.g. low in widely-spaced crops); competition effects (plant branching, leaf shading);

4. *Water supply*: leaf size and longevity (less important in many temperate and irrigated areas);

with secondary controls imposed by environmental stresses and hazards (frost, high temperatures, wind, herbivory, disease etc.). The interplay of these factors can be illustrated by studying the patterns of crop leaf area index development in different crops under a range of management systems.

### 2.3.1 SEASONAL DEVELOPMENT OF LEAF AREA INDEX

The seasonal patterns of leaf area index in annual crops are well illustrated by Watson's (1947) original measurements of winter wheat, spring barley, maincrop potatoes and sugar beet growing at Rothamsted (Fig. 2.13). In the cereal crops, leaf growth was depressed by low temperatures until April/May, when there was a rapid increase in $L$ up to a sharp peak in June/July. After ear emergence, the canopy underwent equally rapid senescence such that $L$ had fallen to zero by harvest in August. The wheat crop showed an early advantage over barley, and an earlier and larger maximum $L$, mainly because the crop, established in autumn, was able to respond more quickly to favourable spring temperatures, irrespective of other soil physical conditions; in contrast, the spring barley could not be sown until the soil had dried sufficiently to permit the necessary cultivations. It should be emphasized that, due to the low levels of nitrogen fertilizer applied at that time (1934–43), the maximum values of $L$ achieved were very modest ($\leqslant 3$) and insufficient to intercept all the incoming solar radiation apart from a few days in May/June (see sect. 2.4.2).

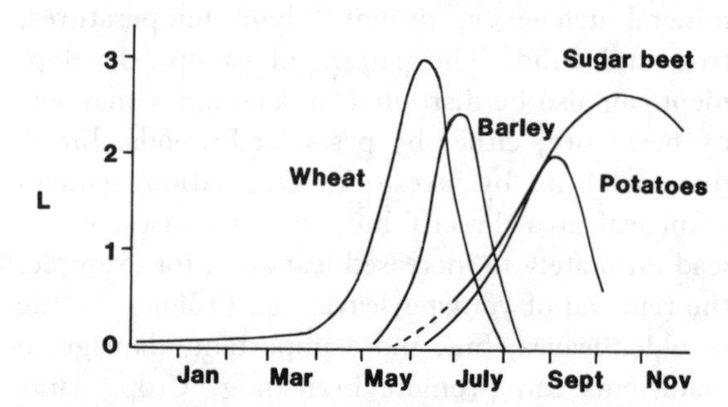


**Fig. 2.13** Seasonal patterns of leaf area index of crops grown at low levels of fertilization at Rothamsted in the 1940s (from Watson 1947).

Compared with the cereals, the growth of the leaf canopy in potatoes and sugar beet was delayed as a consequence of later planting dates (for example, to avoid frost damage in potatoes). For these essentially vegetative crops, there was no abrupt leaf senescence related to reproductive development, and when the crops were harvested in October and November, they still carried a significant amount of green leaf area. Canopy development in biennial/perennial arable crops can be illustrated by cassava/manioc whose starchy storage roots are harvested in the second or subsequent years after planting. Under irrigation in the tropics (e.g. in Queensland, Fig. 2.14), $L$ is closely related to temperature, resulting in almost complete defoliation in winter.

In addition to these seasonal effects, there are large genetic differences between crops in their potential for canopy development. For example, the higher rates of dry matter production in spring by winter rye, as compared with wheat, are the result of higher leaf area indices rather than differences in net assimilation rate (Hay and Abbas Al-

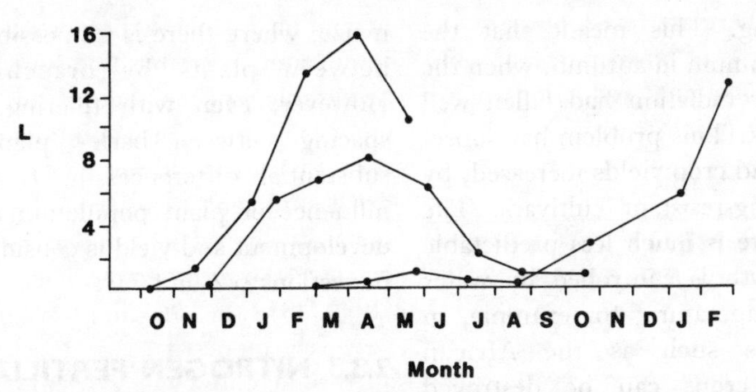


**Fig. 2.14** The seasonal pattern of leaf area index of three plantings of cassava (*Manihot esculenta* Crantz.) in Queensland, Australia (adapted from Keating *et al.* 1982).

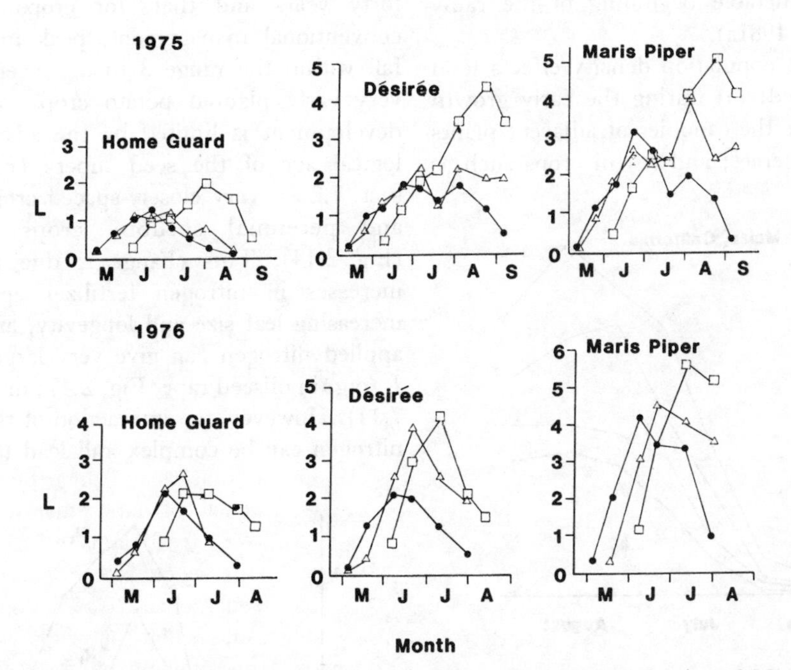


**Fig. 2.15** Seasonal patterns of leaf area index of three plantings (● mid-March, △ end March/beginning April, □ end April) of three varieties of potato in Pembrokeshire, W. Wales, in two contrasting seasons (1975, 1976) (adapted from Jones and Allen 1983).

Ani 1983). Very large differences can also be observed within a single species, for example, among commercial potato cultivars (Fig. 2.15).

## 2.3.2 SOWING DATE AND PLANT POPULATION EFFECTS

Sowing date can have a profound influence upon the course of leaf area index development. Delayed sowing of spring cereals causes an acceleration or telescoping of crop development, resulting in a tendency for lower maximum $L$ values. The opposite tends to hold for late-planted potato crops which develop very much larger and longer-lasting canopies (Fig. 2.15). (The complex interrelationships among date of planting, canopy development and the yields of cereals and potatoes are explored fully in sections 6.3 and 7.4.) The sugar beet crop provides a further valuable example; at the time of Watson's measurements (Fig. 2.13), it was customary to plant late to avoid crop losses due to bolting induced by low temperatures in spring. This meant that the canopy reached its maximum in autumn, when the levels of incident solar radiation had fallen well below the summer peak. This problem has subsequently been solved, and crop yields increased, by the breeding of bolting-resistant cultivars. The influence of sowing date is much less predictable in regions where growth is controlled by water supply rather than temperature; for example, in seasonally-dry regions such as the African savannas, early-sown crops can be destroyed entirely at the seedling stage, or their potential for canopy development reduced significantly by the failure or unpredictable beginning of the rainy season (e.g. Hay 1981a).

Seed rate/plant population density effects tend to be most marked: (a) during the early growth of the crop before the canopies of adjacent plants have begun to interact; and (b) in crops such as maize, where there is no possibility of filling gaps between plants by branching (Fig. 2.16a). However, even with tillering, variation in the spacing between barley plants can result in substantial differences in $L$ (Fig. 2.16b). The influence of plant population density on canopy development and yield is considered in more detail in sections 6.2 and 7.3.

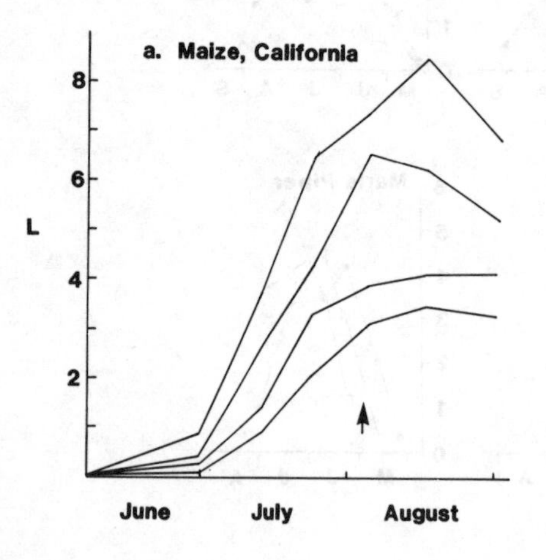


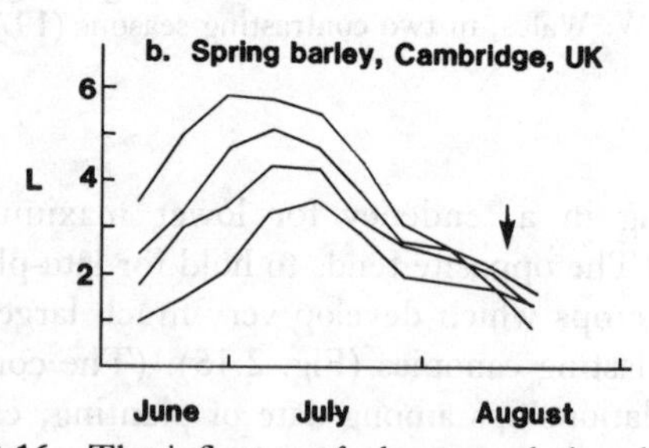


**Fig. 2.16** The influence of plant population density on the seasonal pattern of leaf area index of two species of cereal. The target densities for maize were 1.8, 3.5, 7 and 13 × $10^4$ plants $ha^{-1}$, and for barley 1, 2, 4 and 8 × $10^6$ plants $ha^{-1}$. The arrows indicate tassel (maize) or ear (barley) emergence. See also Fig. 6.4 (adapted from Williams *et al.* 1968; Kirby 1967).

### 2.3.3 NITROGEN FERTILIZER

Examination of Figs 2.13 to 2.16 reveals that crop $L$ values have increased considerably over the last forty years and that, for crops grown under conventional management, peak indices tend to fall within the range 3 to 6. Exceptions include very early planted potato crops, where canopy development is limited by the advanced physiological age of the seed tubers (Figs 2.15, 7.3; sect. 7.2.2), very closely-spaced crops (Fig. 2.16) and perennial shrubby crops like cassava (Fig. 2.14). This change is due principally to increases in nitrogen fertilizer application. By increasing leaf size and longevity, and branching, applied nitrogen can give very large increases in $L$ (e.g. in oilseed rape, Fig. 2.17; in potatoes, Fig. 7.11). However, the interaction of these effects of nitrogen can be complex and lead to unpredicted

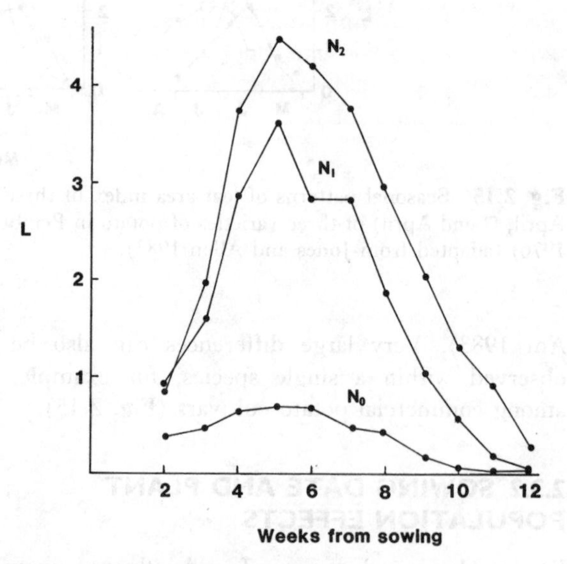


**Fig. 2.17** The seasonal pattern of leaf area index of crops of oilseed rape grown at Cambridge, UK in 1969 at three different levels of nitrogen fertilization ($N_0$ – 0, $N_1$ – 105, $N_2$ – 210 kg N $ha^{-1}$, (adapted from Allen and Morgan 1972).

results. For example, the application of 195 kg N $ha^{-1}$ to a range of wheat cultivars resulted in much higher *L* values at anthesis, when compared with 15 kg N $ha^{-1}$, but there were also differences in canopy structure; the very much larger flag leaves in the higher nitrogen treatment constituted a greater proportion of *L* than in the lower nitrogen treatment, indicating a higher rate of senescence in the lower leaves and sheaths of the more highly fertilized crop (Pearman *et al.* 1978). (See also sect. 6.4, 7.5 and 8.3.)

### 2.3.4 HARVEST DATE, PESTS, DISEASES AND ENVIRONMENTAL STRESSES

Each of these factors can act to reduce the magnitude or duration of *L*. As noted earlier, it is common practice to harvest tuber and root crops before canopy senescence is complete (e.g. Figs 2.13, 2.14, and 2.15). There are also well-documented studies of the partial defoliation of crops by frost or drought (e.g. Radley 1963). However, the influence of herbivory, diseases and pests on *L* in the field has, in general, been studied in much less detail. Exceptions to this generalization include the studies of Simkin and Wheeler (1974) and Jenkyn (1976) which demonstrated that the time courses of *L* in mildewed and protected barley crops differed mainly as a result of the more rapid senescence of the unsprayed crop. Similar results have been reported for *Verticillium* infection of the potato crop (Harrison and Isaac 1968).

### 2.3.5 GRASSLAND SWARDS

Because the sampling and measurement of the total leaf area of grassland swards is an arduous and time-consuming task, there have been few long-term records of *L* under contrasting management systems. Jones (1981) provides an unusually full account of the seasonal changes in *L* for continuously grazed and infrequently cut stands of perennial ryegrass (Fig. 2.18). In the grazed sward, the removal of herbage masked the influence of environmental factors, and *L* fell gradually from 3 to 1 during the growing season, whereas the values for the cut sward were much more variable, fluctuating between 0 and 9, with parallel changes in radiation interception. Elsewhere, *L* values of 1 and 3 are taken to be characteristic of heavily- and lightly-grazed grassland, respectively (Parsons *et al.* 1983a). This topic is considered in greater detail in section 8.2.

## 2.4 CANOPY ARCHITECTURE AND INTERCEPTION OF SOLAR RADIATION

### 2.4.1 SEASONAL PATTERNS OF INTERCEPTION

If the leaf canopy is to fulfill its role of intercepting solar radiation, it is important for the crop to be grown in such a way that the annual cycle of *L* matches the seasonal variation in incident solar radiation (e.g. Fig. 2.19). If they do not match, potential yield will be lost as a conse-

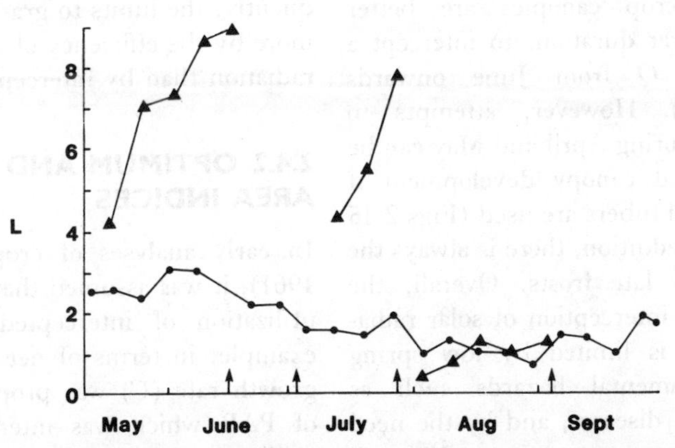


**Fig. 2.18** The influence of the method of defoliation on the seasonal pattern of leaf area index of swards of S23 perennial ryegrass growing in Southern England (▲ cut four times per season at the dates indicated by arrows – note that the data for the first regrowth are omitted; • continuously grazed by sheep) (adapted from Jones 1981).

quence of unintercepted radiation or wasteful investment of dry matter in excessively large leaf canopies during periods of low irradiance. The spectral composition of solar radiation is discussed in section 3.2, but it is important to establish at this point that only 50 per cent of incident solar radiation (i.e the energy received at wavelengths from 400–700 nm) can be used by the photosynthetic apparatus. This fraction is called the photosynthetically-active radiation (PAR). The remaining energy is of no value in photosynthesis and, if absorbed, serves only to increase the temperature of the leaf. The processes governing energy exchange between leaves and the atmosphere are discussed in full in Monteith (1973) and Fitter and Hay (1987).

In north-temperate regions such as the British Isles, the annual fluctuation in incident solar radiation ($Q$; eqn 1.1) follows a broad peak, with maximum values in June and substantial receipts of radiation from April to September (e.g. Fig. 2.19). Assuming that an $L$ of at least 3 is required for the complete interception of all incoming PAR (sect. 2.4.2), it is clear that the major crops are not particularly well-adapted to maximize radiation interception. In cereals, the highest values of $L$ do coincide with the seasonal maximum in $Q$ (Figs 2.13 and 2.16) but the duration of the canopy is very short and substantial quantities of PAR are not utilized. On the other hand measures such as heavy fertilizer application, taken to increase interception by stimulating early growth and prolonging leaf life, can also result in excessively high peak values of $L$. In general, potato crop canopies are better adapted, by their longer duration, to intercept a large proportion of $Q$ from June onwards (Figs 2.15 and 2.19). However, attempts to increase interception during April and May can be hampered by restricted canopy development if physiologically old seed tubers are used (Figs 2.15 and 7.3; sect. 7.2); in addition, there is always the risk of defoliation by late frosts. Overall, the potential for increased interception of solar radiation by arable crops is limited by low spring temperatures, environmental hazards such as frost, drought and leaf diseases, and by the need for the crop to be mature at a time when soil conditions permit the necessary harvest operations (sect. 6.3 and 7.4). Other factors come into play

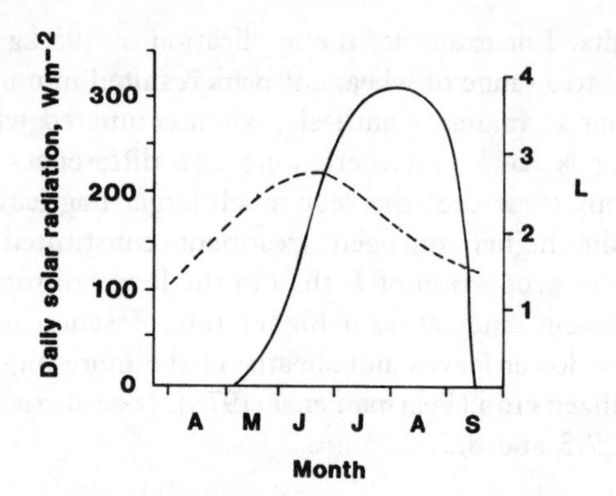


**Fig. 2.19** Seasonal patterns of mean daily irradiance (----) and leaf area index (——) of a potato crop growing in the Midlands of England (adapted from Moorby and Milthorpe 1975).

in different climates; for example, the interception of the very high levels of PAR in sub-tropical regions of Australia by the potato crop is severely restricted by high midsummer temperatures, and even with two crops in each growing season, none of the radiation received during January and February is utilized (Moorby and Milthorpe 1975). Black (1964) gives a particularly clear account of the role of drought in limiting interception and dry-matter yield of stands of subterranean clover in South Australia.

Temperate grassland provides an important exception to this pattern. Under appropriate cutting or grazing management, a continuous canopy can be maintained throughout the year, resulting in the interception of a high proportion of the available PAR (e.g. Fig. 2.18). Consequently, the limits to grassland production are set more by the efficiency of utilization of intercepted radiation than by interception (sect. 8.6).

## 2.4.2 OPTIMUM AND MAXIMUM LEAF AREA INDICES

In early analyses of crop growth (e.g. Donald 1961), it was assumed that at a given efficiency of utilization of intercepted PAR (expressed, for example, in terms of net assimilation rate), crop growth rate ($C$) was proportional to the fraction of PAR which was intercepted, and reached a ceiling when interception was complete. In an imaginary crop with large, thick, undissected and horizontal leaves, complete interception of incoming

PAR would be achieved at an $L$ value close to 1; however, for real crops, in which there is a wide range of leaf size, shape, thickness and angle, it has been found that an $L$ of at least 3 is generally required for the interception of 90–95 per cent of incoming radiation (e.g. for potatoes, Fig. 2.20; for wheat, Hipps *et al.* 1983). It was further assumed that, at values of $L$ greater than 3–4, there would be a tendency for older, shaded leaves in the lower parts of the crop canopy to reach their light compensation point, and become sinks rather than sources of current assimilate. This analysis of interception, which is discussed in greater detail in section 4.1, led to the concept of optimum leaf area index ($L_{opt}$) at which $C$ achieves its maximum value under the prevailing conditions; below $L_{opt}$, growth rate would be dependent upon $L$ and would be depressed owing to incomplete interception of the available solar radiation, whereas above $L_{opt}$, $C$ would be depressed due to increased respiratory losses (lowered net assimilation rate) (Brown and Blaser 1968).

There are a few crops for which this very simplified analysis does predict the pattern of growth rate with some success; for example, the kale crop can show a distinct $L_{opt}$ near 3 (Watson 1958). However, this pattern of response is unusual and it is much more common for $C$ to increase up to a critical value of $L$ at which interception is complete ($L_{crit}$, which is normally in the range 3–5, but can be higher, for example, in highly erectophile canopies, sect. 2.4.3), above which a relatively constant, maximal value of $C$ is maintained (e.g. for wheat, Fig. 2.21; for clover, Fig. 4.45b). The reason for this pattern of response is explained fully in section 4.1; briefly, there is little evidence to support the thesis that shaded leaves constitute important respiratory sinks for current assimilate and it is now generally assumed that overall rates of crop respiration are more closely related to the rate of gross photosynthesis than to biomass or leaf area. For example, McCree and Troughton (1966b) demonstrated that, in stands of white clover, the increase in crop respiration rate with increasing $L$ was small beyond $L$ values of 2–3 (Fig. 4.45b, see also Fig. 2.21), and it is presumed that the fall in $C$ above the proposed $L_{opt}$, found in earlier experiments (e.g. Donald 1961), was a consequence of the death and loss of shaded leaves (Monteith and Elston 1983).

Although this explains the existence of $L_{crit}$ rather than $L_{opt}$ values, it does not explain why $L_{crit}$ values can vary considerably between crop species and between different cultivars of the same species (Evans *et al.* 1975). To do this, it is necessary to study the rate of net photosynthesis of individual leaves at different levels of irradiance, up to and beyond light saturation, as well

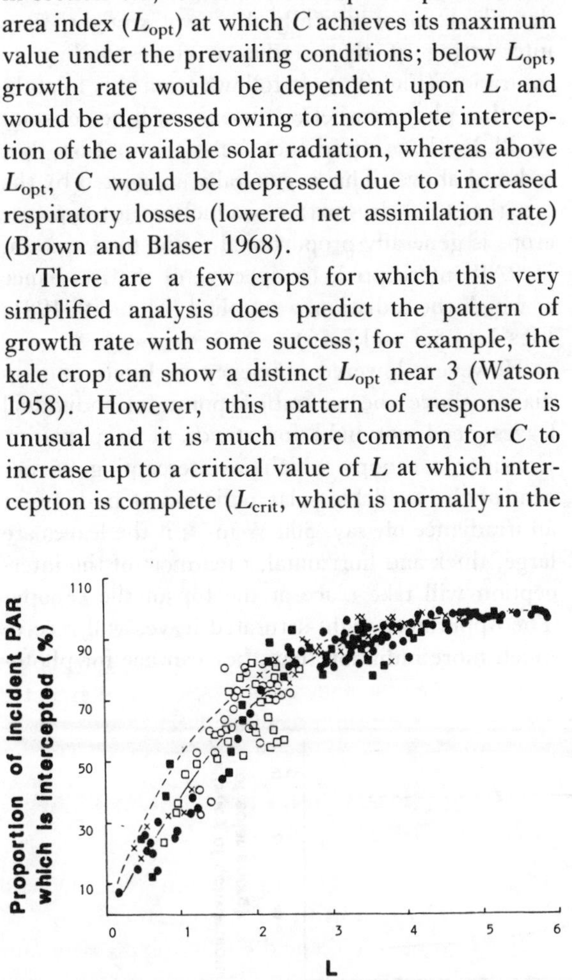

**Fig. 2.20** The interception of incident PAR by potato canopies of differing leaf area index. The different symbols indicate varying treatments in different seasons (1979, 1980) in the Midlands of England, using a single variety, Pentland Crown (from Khurana and McLaren 1982).

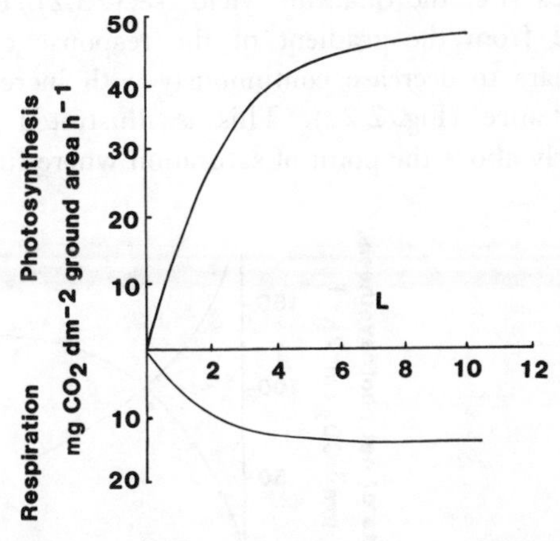


**Fig. 2.21** The relationship between crop growth rate (i.e. photosynthesis–respiration, expressed per unit of ground area) and leaf area index of wheat crops receiving approximately 400 W m$^{-2}$ of total solar radiation. The curves were constructed from measurements in the field and under controlled conditions (adapted from Evans *et al.* 1975).

as the architecture of the crop canopy. In particular, it is important to know how the leaf area index of a crop is arranged vertically, bearing in mind that the leaves of the canopy will not be illuminated uniformly, with the topmost leaves tending to receive much higher levels of irradiance than those lower in the canopy.

### 2.4.3 LEAF PHOTOSYNTHESIS AND CANOPY ARCHITECTURE

In studies of individual leaves of temperate species, the rate of net photosynthesis increases with irradiance, but the slope of the response curve normally fitted to the experimental data decreases steadily, giving an asymptote at the irradiance corresponding to light saturation. A typical fitted response curve is shown in Fig. 2.22 which indicates that, under the prevailing conditions, sugar beet leaves would be saturated at 150 W $m^{-2}$ PAR, or approximately 300 W $m^{-2}$ total solar radiation. Similar values for light saturation have been found for the leaves of other temperate crop species (e.g. 250 W $m^{-2}$ for wheat; Monteith 1981). Thus, if the leaves at the top of a sugar beet or wheat canopy were held at right angles to the direction of the incoming solar radiation, they would be saturated for a large part of each day during the growing season. In contrast, the photosynthetic efficiency of the leaves (i.e. the quantum yield, sect. 3.2), calculated from the gradient of the response curve, appears to decrease continuously with increasing irradiance (Fig. 2.22). This is illustrated most clearly above the point of saturation where further increments of PAR give no increase in the rate of net photosynthesis. The additional input of solar energy to the leaf surface serves only to increase the energy load to be dissipated by transpiration, re-radiation and convection, and to increase the risk of thermal stress. Thus, by this analysis, higher rates of net photosynthesis for individual leaves appear to be associated inevitably with lower levels of efficiency of use of intercepted radiation.

Recently, Monteith (1981) has proposed that the fitting of curves to much of the published experimental data is not justified, and that the response of individual leaf photosynthesis can be described more usefully by two straight lines intersecting at the irradiance corresponding to saturation (Fig. 9.6). It follows from this hypothesis that photosynthetic efficiency will be constant at all levels of irradiance below saturation, but reduced above. This is generally supported by the fact that total dry-matter production in temperate crops is generally proportional to the total amount of PAR intercepted, irrespective of the irradiance at which the radiation is supplied (Monteith 1981; Figs 1.1, 3.3 and 7.16).

However, lowered efficiency at levels of irradiance above the saturation point for individual leaves need not hold for *stands* of crop plants. Consider a canopy which is intercepting 95 per cent of the available solar radiation at midday, at an irradiance of, say, 500 W $m^{-2}$; if the leaves are large, thick and horizontal, then most of the interception will take place at the top of the canopy. The uppermost, light-saturated leaves will receive much more radiation than they can use for photo-

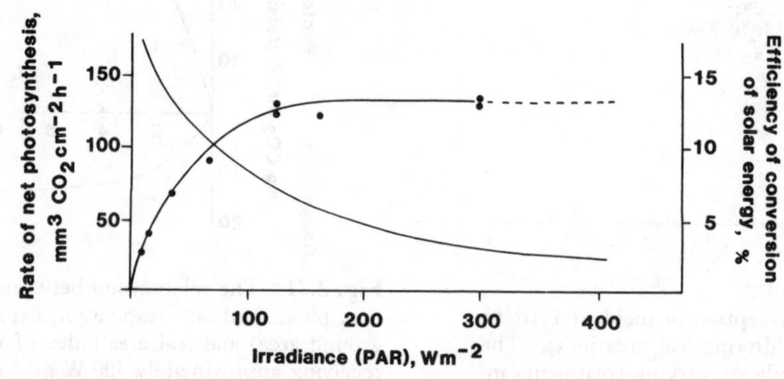


**Fig. 2.22** The relationship between the rate of net photosynthesis of a sugar beet leaf and incident irradiance (curve with data points). The descending curve indicates the efficiency of conversion of solar to chemical potential energy in the form of plant dry matter (from Gaastra 1958).

WORCESTERSHIRE WR10 3JP
LIBRARY
PERSHORE COLLEGE
AVONBANK
PERSHORE

synthesis, whereas the poorly-illuminated lower leaves will be unable to contribute much to dry-matter production. Consequently, the photosynthetic efficiency of the canopy will be low and potential dry-matter production will be lost.

If, on the other hand, the leaves are more inclined to the vertical, then the overall efficiency of the canopy to utilize intercepted radiation, and crop growth rate, can increase for two reasons. First, the angled leaf blades will intercept a smaller proportion of the incoming radiation at high solar elevations (i.e. the projected leaf areas of the angled leaves which can be seen by an observer looking vertically down into the canopy will be smaller than for horizontally-disposed leaves of equal area). An increase in the angle of the leaves in relation to the horizontal will, therefore, lead to an improvement in photosynthetic efficiency at the top of the canopy as the irradiance incident upon the leaf surface is reduced to the saturation point. If the solar radiation incident upon these leaves is further reduced by increasing leaf angle, then their rate of photosynthesis will be reduced without any significant gain or loss of efficiency (Fig. 9.6). Secondly, PAR which is not intercepted at the top of the canopy will become available for photosynthesis by the lower leaves, which can now contribute to the productivity of the crop stand, at the same efficiency as the more highly illuminated leaves at the top of the canopy. Overall, this analysis (which takes no account of the decrease in leaf photosynthetic potential with age, sect. 3.4.3) suggests that for temperate crops (using the $C_3$ pathway of photosynthesis; sect. 3.4.2) the highest growth rates will be achieved when all the incident solar radiation is intercepted by a crop canopy whose leaf blades are disposed in space in such a way that no leaf is more than just saturated.

The relationship between the rate of net photosynthesis and irradiance is quite different for the leaves of $C_4$ plants, which are of predominantly tropical or sub-tropical origin and, therefore, have evolved under high irradiance. Typical response curves (e.g. for maize, Fig. 2.23) are steeper, and virtually linear, up to levels of irradiance (600–700 W $m^{-2}$) which are well above those corresponding to saturation in $C_3$ species; at higher levels, the slope of the response curve does diminish steadily, but saturation is not normally

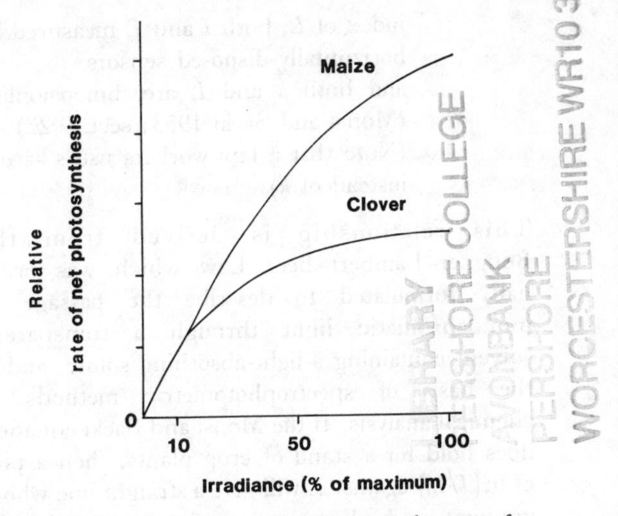


**Fig. 2.23** Typical relationships between the rate of net photosynthesis and incident irradiance for individual leaves of maize ($C_4$ photosynthesis) and white clover ($C_3$) (adapted from Loomis and Williams 1969).

reached within the terrestrial range of irradiance (sect. 3.4.2). Consequently, the efficiency of individual leaf photosynthesis will be reduced only at very high irradiance, and it can be predicted that leaf angle will have a smaller influence upon canopy photosynthesis in $C_4$ species than in $C_3$ (Table 2.3; Trenbath and Angus 1975; Evans and Wardlaw 1976).

Although very important, leaf angle is only one of a number of factors governing the penetration of PAR into the crop canopy; others include leaf surface properties (affecting reflection), leaf thickness (transmission), leaf size, shape and degree of dissection, phyllotaxis and the vertical stratification of leaf area (direct penetration of radiation) as well as the elevation of the sun and the proportions of direct and diffuse solar radiation. In more elementary treatments of canopy interception, it has become customary to combine the plant and canopy characteristics affecting interception into a single composite property, the extinction coefficient, $k$, which is defined by the Monsi and Saeki Equation:

$$I = I_o e^{-kL}$$

where $I_o$ is the irradiance above the crop canopy
$I$ is the irradiance at a point in the canopy above which there is a leaf area

index of $L$, both $I$ and $L$ measured by horizontally-disposed sensors
and both $k$ and $L$ are dimensionless (Monsi and Saeki 1953; sect. 9.22)
(Note that a few workers use $s = e^{-k}$ instead of $k$)

This relationship is derived from the Bouguer–Lambert–Beer Law which was originally formulated to describe the passage of monochromatic light through a transparent solvent containing a light-absorbing solute, and is the basis of spectrophotometric methods of chemical analysis. If the Monsi and Saeki equation does hold for a stand of crop plants, then a plot of $\ln\{I/I_0\}$ against $L$ will give a straight line whose gradient is $k$. Even in its simplest form, this relationship has been found to give satisfactory descriptions of the penetration of radiation into the canopies of a variety of crops (e.g. clover, pasture grasses and barley, Fig. 2.24), yielding values of $k$ which can be used to compare the patterns of interception between cultivars, species, environments and management systems. In more complex forms of the equation, allowance can be made, for example, for the non-random arrangement of leaves in space (Acock *et al.* 1970) or for the elevation of the sun (Niilisk *et al.* 1970).

It is important to note at this point that the Monsi and Saeki equation does not take into account the fraction of incident solar radiation which is lost from the crop stand by reflection, or the fraction which is transmitted through the

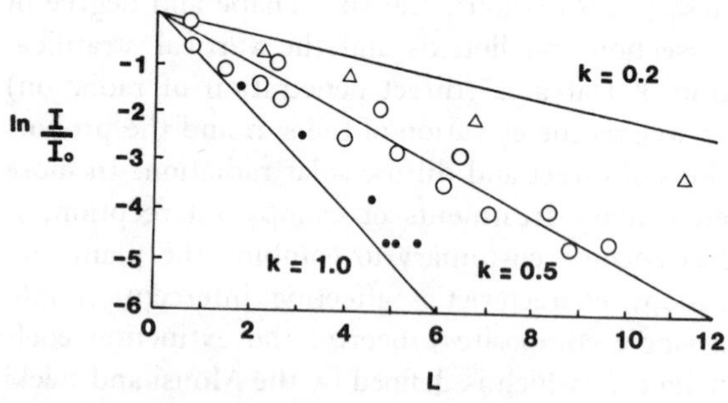


**Fig. 2.24** Relationships between radiation interception and leaf area index for stands of white clover (●), pasture grasses, *L. perenne* and *D. glomerata* (○) and barley (△). The straight lines represent solutions of the Monsi and Saeki equation using different values of $k$ (adapted from Brown and Blaser 1968).

leaves of the canopy. In the case of reflection, the amount of PAR lost at the top of the canopy will be relatively small even with horizontally-disposed leaves (0–10 per cent, depending upon wavelength, Fitter and Hay 1987); with more angled foliage, a considerable proportion of the reflected PAR will subsequently be intercepted by the canopy. The proportion of incident PAR which is transmitted by a leaf is also normally small or negligible (0–5 per cent, Fitter and Hay 1987; sect. 9.2.2) but transmitted radiation has undergone changes in quality (spectral composition), which can have important implications for the development of plant tissues growing under the crop canopy (e.g. branches/tillers and weeds) (Smith 1981).

The experimental verification of the Monsi and Saeki equation, and its application to crops in the field, are laborious processes, involving the measurement of the irradiance at a series of levels above and within the canopy (commonly using tube solarimeters), and stratified measurements of leaf area and angle. In most cases, these series of measurements must be replicated at a large number of stations and continued over a considerable period of time, to allow for the variability of the crop canopy in the field and ontogenetic changes in leaf characteristics; the techniques used for leaf measurement are largely direct and destructive, although some non-destructive, but not necessarily less arduous, methods have been developed (e.g. the point quadrat method of Warren Wilson 1959). In spite of these practical difficulties, the approach has been widely adopted and has led to a considerable amount of interest in the vertical arrangement of canopy leaf areas.

Two contrasting examples of canopy architecture (Fig. 2.25) serve to emphasize the fact that crops can differ widely in overall height, in the vertical distribution of $L$ and in leaf angle, as well as in total $L$. In the case of ryegrasses, the combination of low $L$ in the top strata of the canopy and the vertical disposition of the youngest leaves ($k = 0.5$ for *L. perenne*, Fig. 2.24), means that PAR can penetrate deeply into the crop stand, and canopy photosynthesis can be spread over a large area of leaf (e.g. up to 10 units of $L$ for *L. rigidum*, Stern and Donald 1962a). In contrast, the top layers of the rather shorter clover canopy contain a larger proportion of the total $L$

than in the grass canopy, and the leaves are disposed in a near-horizontal plane (resulting in $k$ values of 0.9 to 1.0 for *T. repens*, Fig. 2.24). Consequently, a much larger fraction of incoming PAR is intercepted at the top of the canopy and photosynthesis tends to be distributed over a smaller area of leaf (e.g. up to 6 units of $L$ for *T. subterraneum*, Stern and Donald 1962a). The architecture and radiation interception of a more uniform ryegrass/clover sward at a higher level of fertility is illustrated in Fig. 8.11. Variations of the basic histogram approach to the graphical representation of canopy architecture (Fig. 2.25) include the use of cumulative leaf area indices (e.g. cotton and sorghum, Niilisk *et al.* 1970), continuous curves of leaf area (e.g. maize, Ross 1970) and the inclusion of some index of leaf age (e.g. wheat, Hodanova 1970), whereas other authors have included photosynthetic stem areas (e.g. red clover, Nösberger and Joggi 1981) or dead leaves which are still involved in interception (e.g. range grasses, Ripley and Redman 1976).

In spite of the fact that canopy extinction coefficient studies have led to a considerable interest in the vertical stratification of crop leaf area, it is important to emphasize that the Monsi and Saeki equation does not contain a height or distance term; i.e. the relationship does not distinguish between canopies in which the same $L$ is distributed over different crop heights, for example due to differences in internode length. Although the distance between layers of leaves can have an important influence upon the distribution of radiation within the crop stand (Loomis and Williams 1969), upon the fate of radiation reflected within the canopy, and upon the interception of solar radiation from low solar elevations, it is usually ignored in elementary treatments. However, in more advanced theoretical treatments of canopy architecture, variation in the distance between layers of leaves can be allowed for by computing the leaf area density (i.e. the leaf area per unit volume of canopy) of each layer (e.g. Ross 1970; Monteith 1976, *passim*).

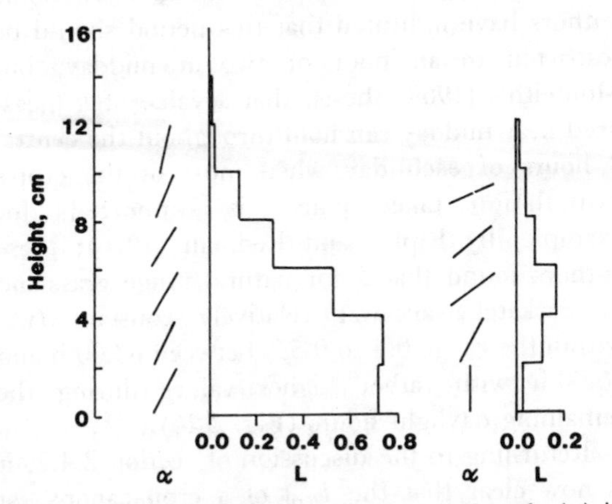


**Fig. 2.25** Leaf area index and mean leaf angle (α) in successive horizontal layers of a mixed stand of perennial ryegrass (left) and white clover (right), measured using the point quadrat technique (from Warren Wilson 1959).

Representative values of $k$, measured at high solar elevation, for stands of a range of crops are presented in Table 2.2. In general, the measured values are closely related to mean leaf angle (α), measured from the horizontal plane; for example, the exceptionally low extinction coefficients recorded for *Gladiolus*, which permit the penetration of PAR through several layers of leaves, are associated with a near-vertical leaf inclination throughout the canopy. Low values are also found in stands composed of rosette plants with highly-angled leaves. However, $k$ is much less sensitive to other canopy characteristics as demonstrated by the fact that very high values (0.9) are shown by clover and sunflower canopies which differ very considerably in overall height, leaf size and leaf shape; the common factor is the rather horizontal disposition of the leaves which leads to substantial interception of solar radiation at the top of the canopy.

**Table 2.2** Representative values of canopy extinction coefficient ($k$) measured in stands of a variety of crop species

| *Species* | $k$ |
|---|---|
| *Gladiolus* | 0.16–0.30 |
| Radish (rosette) | 0.30–0.42 |
| Flax/linseed | 0.37–0.42 |
| Perennial ryegrass, rice, maize | 0.40–0.70 |
| Temperate cereals | (0.20*)0.50–0.75 |
| Sunflower | 0.66–0.92 |
| Clover | 0.90–1.00 |

* Exceptionally low values for seedlings with vertically-disposed leaves.

(From a range of sources reviewed by Brown and Blaser 1968; Trenbath and Angus 1975; Monteith 1976)

The use of a single mean value of $\alpha$ in the interpretation of canopy extinction coefficients is clearly an oversimplification; for example, Fig. 2.25 shows that there are large (and opposing) trends in leaf angle in stands of ryegrass and clover, and that $\alpha$ can vary by more than 45° within a clover canopy. Such variations are common to the canopies of most crop species (Monteith 1976). How, then, can straight-line relationships be obtained, for example in Fig. 2.24, when it would be predicted from the distribution of $\alpha$ that $k$ would vary considerably between the top and bottom of the canopies measured? There are two important reasons for the relative constancy of $k$ values within canopies; first, the interception of solar radiation tends to be dominated by the upper half of the canopy, where leaf angle tends to vary less than towards the bottom of the stand (e.g. Fig. 2.25). Secondly, the uppermost leaves tend to receive solar radiation from a range of elevations and angles, especially under diffuse lighting, whereas, because of the filtering activity of these upper leaves, the lower strata tend to receive radiation from almost directly overhead. Thus, for example, the effect of the rather horizontally-disposed lower leaves of a ryegrass sward in increasing the value of $k$ will be counteracted by the fact that radiation from high elevations penetrates the canopy more easily (Trenbath and Angus 1975).

The pattern of leaf angle displayed by a crop can be depicted quantitatively using a cumulative frequency distribution of $\alpha$. Most of the crops which have been studied in sufficient detail (reviewed by Trenbath and Angus 1975) conform approximately to one or other of the following idealized distributions:

erectophile: ($\alpha$ predominantly greater than 60° to the horizontal) e.g. *Gladiolus*, many grass and cereal species.

plagiophile: ($\alpha$ predominantly between 30 and 60°) e.g. sugar beet, rape.

planophile: ($\alpha$ predominantly less than 30°) e.g. clover, cucumber, bean, sunflower.

The spherical distribution shown by some maize cultivars is a variation of the erectophile pattern in which the distribution of leaf angles is similar to the distribution of angles on the surface of a sphere. Care should be exercised in classifying crop stands according to this scheme because the pattern of leaf angle can be affected by ontogeny (Uchijima 1976); for example, young erectophile grass canopies can change to give a more planophile distribution as a result of the development of longer, more lax, leaves later in the season (e.g. Table 2.2).

The Monsi and Saeki equation has played a central part in the development of theories of canopy architecture but there are several problems associated with its application to real crops in the field. As indicated earlier, its applicability depends upon a random arrangement of leaves, whereas phyllotaxis and planting pattern commonly lead to a distinct clumping of foliage. Other difficulties associated with the arrangement of leaves in space include vertical variation of leaf angle, $k$, and inter-layer distance. However, there is a more difficult problem; although the relationship was formulated to describe the interception by the canopy of solar radiation from high elevations, the original measurements of $k$ were made during periods of bright, but cloudy, weather when the solar radiation received by the crop stand was mainly diffuse. Most of the subsequent measurements in Europe and North America, however, have been made under clear skies, where direct irradiance accounted for a substantial proportion of the incident radiation. It is clearly important to know over what period of each day the $k$ values so obtained hold for the crop canopy. Some authors have indicated that this period should be restricted to an hour or two at midday, but Monteith's (1969) thesis, that a value of $k$ measured near midday can hold throughout the central 8 hours of each day when most of the crop's assimilation takes place, is supported, for example, by Ripley and Redman (1976); these authors found that $k$ for natural range grassland in Saskatchewan was relatively constant (i.e. within the range 0.4 to 0.55) between 08.00 h and 16.00 h with rather higher values during the remaining daylight hours (Fig. 2.26).

Returning to the discussion of section 2.4.2, it is now clear that the $L_{crit}$ of a crop canopy, at which full interception is achieved and $C$ reaches its maximum value, will depend strongly upon $k$, since this determines how effectively PAR can be distributed through the canopy at irradiances

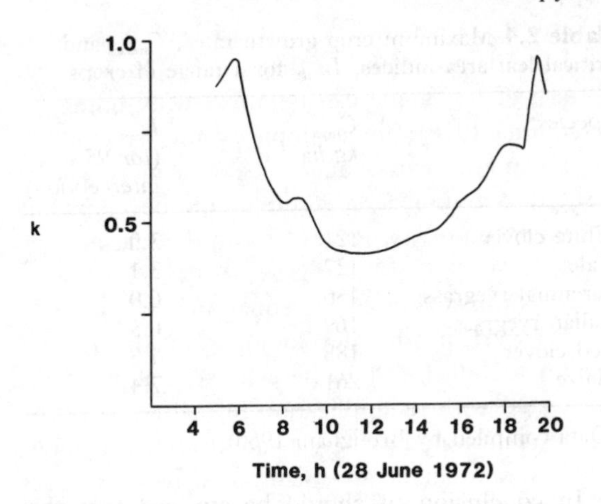


**Fig. 2.26** The diurnal variation of canopy extinction coefficient ($k$) in a stand of native prairie grasses in Saskatchewan, Canada, 28 June 1972 (adapted from Ripley and Redman 1976).

below that required to saturate photosynthesis. Indeed, Monteith and Elston (1983) propose that there is a close inverse relationship between $L_{crit}$ and $k$ such that $kL_{crit}$ is a conservative factor, varying little between cultivars and species. Thus $L_{crit}$ values as low as 2 have been recorded for pure stands of white clover (Fig. 4.45b) compared with a range of values from 4 (Fig. 2.21) to greater than 9 for wheat (reviewed by Evans *et al.* 1975).

## 2.4.4 OPTIMUM CANOPY ARCHITECTURE

Over the past three decades, there has grown up, through the development of theoretical models (e.g. Blackman 1962) and computer simulations (e.g. Monteith 1965; Duncan 1971), the concept of an optimum canopy architecture which, by maximizing $L_{crit}$, will lead to high rates of canopy photosynthesis and crop growth (Hawkins 1982). This ideal structure is commonly thought to resemble that of the ryegrass stand shown in Fig. 2.25, but with horizontal leaves at the bottom of the canopy to intercept deeply-penetrating solar radiation from directly overhead. The 'degree of ideality' of such models is considered below, but an important consequence of these studies has been the realization that different architectures may prove to be optimum: (a) at different stages of the development of a given crop stand; and (b) at different levels of irradiance. For example, it is clear that highly-erectophile canopies will be much less effective than those with more horizontal leaves at intercepting PAR early in the growing season when $L$ is low; thus, during the early stages of growth, when levels of irradiance are low to moderate, and there is less risk of saturation of photosynthesis, a smaller investment of assimilate in horizontally-disposed leaves will result in the same degree of interception as a larger investment in more vertical leaves. The overall conclusion is, therefore, that the highest growth rates would be anticipated for crops which, at appropriate plant densities, showed ontogenetic changes from a seedling planophile to a more mature erectophile architecture; this, of course, is the basic pattern of development of many temperate cereal cultivars. It is also worth noting that canopy architecture can have important implications for assimilate partitioning and harvest index (sect. 5.3 and 6.4.1; Hawkins 1982).

There have been few valid field tests of the ideality of canopy models, and most investigations have taken the form of an evaluation of the influence of mean leaf angle ($\alpha$) on short-term net photosynthesis or on final crop yield. Such experiments are particularly difficult to perform because they should involve the comparison of crop stands which are identical in all respects except for leaf angle; this has led, for example, to the physical manipulation of individual leaves using attached weights. In a detailed review, Trenbath and Angus (1975) have shown the results to be somewhat equivocal, with several experiments, especially on rice, demonstrating the clear superiority of more erectophile canopies, whereas other investigations have given inconclusive results. These differences are probably due, in part, to experimental difficulties but it is not surprising that clear results are obtained for rice since it is a $C_3$ species grown under high irradiance. With a few notable exceptions, investigations of final crop yield have generally shown significant benefits associated with more erectophile canopies (Table 2.3). However, care should be exercised in the interpretation of the published maize results; as it is a $C_4$ crop it would not have been predicted that leaf angle would have a pronounced influence upon yield, but plant density and leaf angle can interact, through shading, to affect the synchrony

**Table 2.3** The effect of mean leaf angle on the growth and yield of crops

| *Species* | *% advantage of erectophile type on* | |
|---|---|---|
| | *Crop growth rate* | *Grain yield* |
| Barley ($C_3$) | +19 | −5 to +10 |
| Wheat ($C_3$) | — | −10 to +19 |
| Rice ($C_3$) | +52 | +34 to +68 |
| Maize ($C_4$) | — | −18 to +41 |
| Sugar beet ($C_3$) | +53 | — |
| Tea ($C_3$) | +108 | — |

(Data compiled by Trenbath and Angus 1975)

**Table 2.4** Maximum crop growth rates, $C_{max}$, and critical leaf area indices, $L_{crit}$, for a range of crops

| *Species* | $C_{max}$ *kg ha$^{-1}$ d$^{-1}$* | $L_{crit}$ *(for 95% interception)* |
|---|---|---|
| White clover | 121 | 3.0 |
| Kale | 127 | 3.1 |
| Perennial ryegrass | 156 | 6.0 |
| Italian ryegrass | 169 | 6.5 |
| Red clover | 188 | 4.8 |
| Maize | 261 | 7.4 |

(Data compiled by Brougham 1960)

of tassel and silk emergence, thereby influencing the success of pollination (Lambert and Johnson, 1978).

More recently, Austin *et al.* (1976) demonstrated that under UK conditions in 1974/5, canopy net photosynthesis in erect-type wheat cultivars was consistently higher than in lax cultivars especially at high $L$, owing to greater penetration of PAR into the canopy, leading to greater contributions to canopy photosynthesis by, and longer longevity of, lower leaf layers. However, due to differences in the quantities of stem reserves of assimilate translocated to the ear, there were no consistent differences in ultimate grain yield between the two canopy types. Furthermore, from this, and subsequent work by Innes and Blackwell (1983), it is clear that some of the effects observed in this type of experiment may be caused by differences in crop–water relations. Taking a different approach to the evaluation of crop canopies, Brougham (1960) showed that, in comparisons between species, crop growth rate was closely related to $L_{crit}$ (Table 2.4).

In conclusion, it should be stressed that this chapter has concentrated almost exclusively upon canopies of crop plants grown as monocultures. Setting aside the important interrelationships between crop plants and weeds, it is important to remember that grassland swards are generally composed of mixtures of species, sometimes of contrasting canopy architecture (compare ryegrass and clover, Fig. 2.25; Rhodes 1973) which compete for PAR, as well as for other resources. Considering grasses alone, combinations of cultivars or species with differing but complementary architectures can be devised which give yields which are higher than the individual components grown as monocultures (for example in ryegrass mixtures under frequent cutting, Rhodes 1969). The physiology and yields of inter- and intra-specific mixtures are considered in more detail in section 8.4. On a global scale, there is a wide range of mixed- and inter-cropping systems involving two or more species with complementary canopies growing together in intimate mixtures (e.g. maize and *Phaseolus* beans grown together in many parts of Africa, where the climbing bean uses the maize stem as a bean pole).

# CHAPTER 3 PHOTOSYNTHETIC EFFICIENCY: PHOTOSYNTHESIS AND PHOTORESPIRATION

*Practically everything that we see about us has involved photosynthesis at some stage or other. The gardener often talks about 'feeding' plants when he applies fertilisers and the notion that plants derive their nourishment from the soil is one that is commonly held. They do not. Plants take up minerals from the soil, they derive their nourishment from the air.*

(G Edwards and D A Walker 1983.)

## 3.1 INTRODUCTION

The rate of net photosynthesis, or dry-matter production, of a plant or crop is determined by the rates of gross photosynthesis, photorespiration and 'dark' respiration. In other words, these processes ultimately determine the efficiency with which intercepted photosynthetically active radiation (PAR, of which perhaps 85 per cent is absorbed) effects the conversation of $CO_2$ into crop dry matter. Net $CO_2$ exchange may be summarized as:

$$P_n = P_g - PR - R \quad [3.1]$$

where $P_n$ is net, or apparent, photosynthesis, $P_g$ is gross, or true, photosynthesis, $PR$ is photorespiration, and $R$ is 'dark' respiration. Net photosynthesis, or dry-matter production, is therefore the balance between one process taking up $CO_2$ (gross photosynthesis) and two processes releasing $CO_2$ (photorespiration and 'dark' respiration). The importance to dry matter production of the relative rates of photorespiration, 'dark' respiration and gross photosynthesis is therefore obvious. It is also clear that the scope for increasing dry matter production can only be recognized if these processes can be accurately measured. Before the importance of these processes to crop dry-matter production is examined in detail, therefore, the problems associated with their measurement, particularly of photorespiration and 'dark' respiration, will be discussed. We shall also consider here the effects of light on 'dark' respiration. The bulk of the chapter then deals with photosynthesis and photorespiration, considered together because they are so intimately related. Respiration, in terms of the requirement of the crop to support growth and maintenance, is dealt with in Chapter 4.

## 3.1.1 CLARIFICATION OF THE MEANING AND RELATIVE MAGNITUDES OF PHOTORESPIRATION AND 'DARK' RESPIRATION

Gross photosynthesis, the potential for dry-matter production, is normally arrived at by addition when net photosynthesis, photorespiration and 'dark' respiration are measured in terms of $CO_2$ exchange, although it can be determined directly by short-term measurements of $^{14}CO_2$ fixation. Unfortunately, there are formidable problems of interpretation in any attempts to measure $P_n$, $PR$ and $R$ simultaneously, largely because the determination of $PR$ and $R$ inevitably requires their isolation from each other and/or from photosynthesis, and this ignores the interrelationships which exist between these processes. Figure 3.1 illustrates these interrelationships in terms of the appropriate biochemical pathways of carbon and nitrogen metabolism.

This problem is exemplified by attempts to measure 'dark' respiration. It has often been assumed that the rate of 'dark' respiration in the light is identical to that in the dark, and can therefore be determined simply as the $CO_2$ efflux in the dark. However, there is considerable controversy about the behaviour of 'dark' respiration in the light, with some authorities believing that it is inoperative in photosynthesizing tissues in the light while others report that it is stimulated. It is now widely held that 'dark' respiration does continue in the light, although its rate is somewhat changed (Graham 1980). This important question

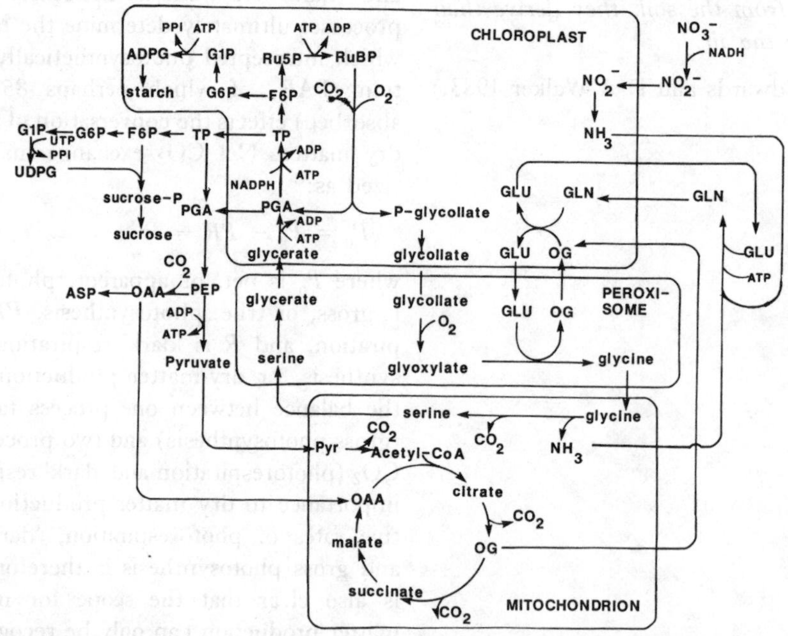


**Fig. 3.1** Photosynthetic, photorespiratory and respiratory carbon metabolism. Reactions inside the chloroplast, peroxisome and mitochondrion are distinguished from those in the cytosol.
Abbreviations: ADPG, adenosine diphosphoglucose; ASP, aspartate; F6P, fructose-6-phosphate; G1P, glucose-1-phosphate; G6P, glucose-6-phosphate; GLN, glutamine; GLU, glutamate; NADH, reduced nicotinamide adenine dinucleotide; NADPH, reduced nicotinamide adenine dinucleotide phosphate; OAA, oxalacetate; OG, 2-oxoglutarate (= α-ketoglutarate); PEP, phosphoenolpyruvate; PGA, 3-phosphoglycerate; PPi, inorganic pyrophosphate; Pyr, pyruvate; Ru5P, ribulose-5-phosphate; RuBP, ribulose-1,5-bisphosphate; TP, triose phosphate (= dihydroxyacetone phosphate or glyceraldehyde 3-phosphate); UDPG, uridine diphosphoglucose; UTP, uridine triphosphate.

will be discussed more fully below, but for the present it should be noted that 'dark' respiration is a misnomer. In fact, the term 'dark' is not only redundant but also misleading, reinforcing the notion that respiration does not occur simultaneously with photosynthesis and photorespiration.

Whatever the misgivings, respiration in terms of $CO_2$ exchange must be measured in the dark and assumptions made about its rate in the light. Photorespiration can be determined with more confidence. There are several methods, the most reliable of which relies on the inhibition of photorespiration when the oxygen concentration is reduced from the ambient atmospheric concentration of 22 per cent to 1 or 2 per cent. The resulting increased net photosynthesis can then be directly attributed to the elimination of photorespiration, since respiration is unaffected by these changes in oxygen concentration.

The problems of measuring respiration in the light, and to a lesser extent photorespiration, limit our knowledge of the contribution of each process to net photosynthesis and, indeed, of the magnitude of gross photosynthesis. This deficiency in our physiological knowledge contrasts with the now well-established understanding of the biochemistry of photorespiration and respiration, in terms of the metabolic pathways and to some extent the control of this metabolism. Indeed, precise definitions of photorespiration and respiration can be given in biochemical terms, and it is perhaps appropriate to include them at this point, if only to stress that we are dealing with two distinct processes which are similar only in the broad sense that each results in $CO_2$ production through the oxidation of substrates. Graham's (1980) definitions will be followed. Photorespiration is the light-dependent oxidation of a 2-carbon metabolite (mainly glycollate) derived from photosynthesis. This is not to say that photochemical reactions are involved in photorespiration itself; the requirement for light is a consequence only of the dependence of glycollate synthesis on the energy and reducing power (ATP, NADPH, reduced ferredoxin) generated by photosynthetic electron transport, which of course is powered by the energy of sunlight. Respiration is defined as the oxidation of substrates (carbohydrates, proteins, fats) by way of glycolysis, the oxidative pentose phosphate pathway, and the tricarboxylic acid cycle linked to oxidative electron transport pathways of the mitochondria. It occurs in the dark and in the light.

Clearly, photorespiration and respiration are biochemically distinct processes. They are also to a large extent spatially separated within the cell. The reactions which comprise photorespiration occur in the chloroplasts, peroxisomes, mitochondria and cytosol, whereas those of respiration are confined to the cytosol and mitochondria (Fig. 3.1). Ultimately, however, it is the functional, or physiological, difference which is most important. Respiration results in energy being made available for cellular work, whereas photorespiration is energetically wasteful. The possible functions of photorespiration will be considered later (sect. 3.4.1).

Because of the assumptions which must be made about the simultaneous operation of photosynthesis, photorespiration and respiration, problems may arise when attempts are made to generalize from a particular set of measurements, especially from laboratory to field. This is because the processes themselves and their interrelationships are strongly affected by environmental conditions and by the age and previous environmental history of the plants. Clearly, it is misleading to state 'typical' values for the rates of photorespiration and respiration. On the other hand, eqn [3.1] has little meaning unless the relative magnitudes of the components are indicated. Zelitch (1980) suggests that photorespiration accounts for about 50 per cent of the net $CO_2$ assimilated by a number of $C_3$ species, although the wide range of figures around this mean value reflects the variety of species, conditions and methods of assay employed. Keys and Whittingham (1981) suggest a figure of 25 per cent for $C_3$ crops growing in the field in the UK. Respiratory losses can be considerable: 45 per cent of the carbon fixed by a sward of perennial ryegrass during a twelve-week period is lost in respiration (Robson 1973), and a similar percentage is lost during the growth of wheat plants (see Morgan and Austin 1983).

Care must be taken in the interpretation of these figures because they are not directly comparable. It may seem illogical to express photorespiration as a percentage of net photosynthesis; photorespiration is, after all, an integral part of net photosynthesis. However, photorespiration is invariably measured in relation to net photosynthesis, and the figure of 25 per cent given above simply means that if photorespiration were eliminated, net photosynthesis would increase by 25 per cent. This is indicated in Fig. 3.2, where the rate of net photosynthesis is given the value of 100 units. Photorespiration is so intimately coupled to photosynthesis that in studies of crop respiration, references to the amount of carbon fixed usually imply the total amount fixed less that lost by photorespiration; this is true for the work cited above. The figure of 45 per cent quoted above for respiration therefore means $0.45\ (P_g - PR)$. Thus, in Fig. 3.2, where net photosynthesis has been set at 100 units, respiration must be given a value of 82 units ($R = 0.45\ (207 - 25) = 82$). Net photosynthesis, photorespiration and respiration then become 48, 12 and 40 per cent of $P_g$, respectively.

It is worth considering these figures in relation to laboratory measurements of single-leaf $CO_2$ exchange, particularly with regard to the respiration estimates. Statements to the effect that the rate of photorespiration is up to five times greater than that of respiration are not uncommon (e.g. Zelitch 1980), but at first sight contradict the picture presented in Fig. 3.2. However, although often unqualified, such statements should not be taken out of context. They invariably refer to attempts to measure photosynthesis, photorespiration and respiration simultaneously in single, well-illuminated leaves, often at comparatively high temperatures. They do not pretend to offer a carbon 'balance-sheet', in which photosynthetic carbon gains are set against photorespiratory and respiratory losses over a protracted period, which is the aim of the crop experiments referred to above. Obviously, long-term experiments include regular periods of darkness during which respiration is the only component of $CO_2$ exchange. Moreover, the $CO_2$ exchange of communities of plants includes contributions from stems and deeply shaded leaves in which, even in the light periods, the photosynthetic and photorespiratory components are comparatively low. If root respiration is included, the balance is tipped even further in favour of respiration. Thus, Fig. 3.2 is not necessarily at odds with laboratory studies which show the rates of photorespiration to be much greater than those of respiration. The need for caution when comparing laboratory and field experiments is clear, however.

## 3.1.2 RESPIRATION IN THE LIGHT

This controversial and confusing aspect of plant metabolism has been reviewed by Graham (1980). It is not yet possible to say unequivocally what happens to respiration in the light, but Graham tentatively concludes that it is 'a significant part of the carbon lost to the plant'. Its existence, role and magnitude are therefore important issues in any consideration of dry-matter production.

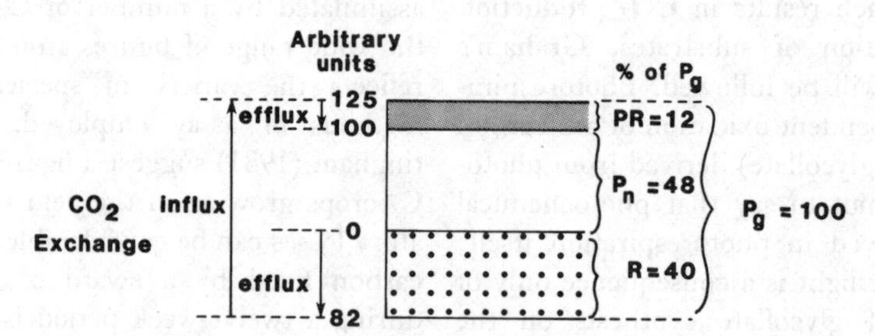


**Fig. 3.2** The components of gross photosynthesis ($P_g$) in a field-grown crop, assuming that photorespiration ($PR$) equals 25 per cent of net photosynthesis ($P_n$) and respiration ($R$) equals 45 per cent of carbon fixed [$= 0.45(P_g - PR)$]. From eqn [3.1],

$$100 = P_g - 25 - [0.45(P_g - 25)],$$

from which $P_g = 207$.
The magnitude of each process is given in arbitrary units of $CO_2$ exchange and as a percentage of gross photosynthesis.

It should be remembered that the complex biochemistry of respiration serves two functions: first, to produce high-energy compounds; and secondly, through various respiratory intermediates, to provide carbon skeletons for new compounds. Respiration in the dark, when there is no fixation of energy by the plant, presumably satisfies both requirements. In the light, however, photosynthetic phosphorylation in the chloroplasts provides a supply of ATP and reduced NADP, the energy and reducing power needed for the subsequent reduction of $CO_2$ to carbohydrate. Some of this assimilatory power can be translocated out of the chloroplast to the cytosol, and renders respiration unnecessary for the supply of energy in the light. In illuminated green tissues, therefore, respiration is required only as a source of carbon skeletons for synthetic reactions. Two inferences can now be made. First, the rate of respiration in the light is likely to be lower than that in dark; and secondly, the difference between the two rates will be related to the metabolic activity, and therefore age, of the tissue, the requirement for respiration being greater in younger, growing tissue supporting high rates of synthesis.

The different synthetic activities, and presumably variations in the extent of photosynthetic phosphorylation, among the tissues which have been studied must be responsible for much of the disagreement over the magnitude of respiration in the light. In addition to these inter-specific differences, there will be considerable spatial and temporal variation in respiratory behaviour among the individual leaves of a crop in the field, reflecting the differences in age, micro-climate and environmental history of the leaves. Furthermore, endogenous rhythms may be responsible for diurnal variation in respiration rate within individual leaves.

A metabolic need for respiration in the light can therefore be identified, although considerable variation in the estimates of its rate are to be expected. This variation makes it difficult to generalize over the magnitude of respiration in the light and, indeed, gives licence to some authorities to question its existence. However, the general conclusion of Graham's comprehensive review has already been noted, and the reader should refer to this work for details of the many experiments upon which it is based. A brief summary will be made here, however, because the assumption that photosynthetic tissues respire at the same rate in the light as they do in the dark is fundamental to many investigations of crop respiration. The physiological experiments, ingenious variations on a theme of $CO_2$ exchange, have produced the most equivocal results: some give evidence of complete inhibition of respiration in the light, while others show it to be unaffected. Clearly the pitfalls of interpretation mean that this type of experiment should not be considered in isolation.

On the other hand, the biochemical approach gives a more coherent picture, at least for the operation of the tricarboxylic acid (TCA) cycle. The basic approach of many experiments has been to study carbon flow through the TCA cycle after feeding $^{14}C$-labelled substrates (e.g. $CO_2$, glucose, pyruvate, acetate) or intermediates (e.g. citrate, succinate, fumarate, malate, oxalacetate). The general conclusion of this work is that the TCA cycle continues in the light in both photosynthetic algae and the leaves of higher plants. Moreover, studies of the intramolecular redistribution of $^{14}C$ provide strong evidence that the rate of respiration in the light is the same as that in the dark. Thus, Marsh *et al.* (1965) showed that in the green alga *Scenedesmus* $^{14}C$ from [1-$^{14}C$] acetate, [2-$^{14}C$] acetate, and [3-$^{14}C$] pyruvate passed through the expected sequence of intermediates of the TCA cycle, in the appropriate positions in the molecules, at equivalent rates in light and dark. Contradictory evidence can be cited, but this usually derives from short-term experiments in which the extent of labelling in TCA cycle intermediates was monitored within up to five minutes of supplying $^{14}C$-substrate. Sequential sampling over three or four hours suggests that there is indeed inhibition of the TCA cycle during the first few minutes of illumination, but that this is a temporary phenomenon. Less attention has been paid to the effects of light on glycolysis and the oxidative pentose phosphate pathway, but the available evidence suggests that each operates in the light, probably only in the cytosol.

The problems in the interpretation of gas exchange experiments caused by the interrelationships between photosynthesis and respiration have already been mentioned, and are discussed further by Brooks and Farquhar (1985). One particular aspect of these interrelationships has also undoubtedly been responsible for confusion in the interpretation of some of the biochemical experiments. It appears that carbon fixed in the light by photosynthesis is largely unavailable for glycolysis and the TCA cycle until it has been converted into storage reserves, emphasizing the danger of short-term $^{14}CO_2$-feeding experiments. The longer-term experiments strongly suggest that the TCA cycle operates in the light on substrates derived from the storage reserves. However, it is interesting to note that in the presence of small amounts of added ammonium ion ($NH_4^+$), the flow of carbon is diverted from the Calvin cycle through PEP to pyruvate and subsequent entry to the TCA cycle via acetyl-CoA (Fig. 3.1). This results in the synthesis of carbon skeletons for amino acid production at the expense of sucrose synthesis. This shift in metabolism is not merely a mass action effect caused by the availability of $NH_4^+$ for amino group formation, but is considered to result from specific activation of the enzyme pyruvate kinase (PEP → pyruvate) and inhibition of sucrose phosphate synthetase (fructose 6-P + UDPG → sucrose + UDP), although other points of control may also operate (Bassham, Larson and Cornwell 1981). Clearly, the ammonium ion plays an important part in controlling the partition of photosynthetically fixed carbon between amino acid synthesis for protein assembly and sucrose synthesis for subsequent storage or translocation.

In summary, therefore, it is concluded that the biochemical pathways which comprise respiration continue to function in the light. The production of energy by this means is less important than the provision of carbon skeletons for synthetic reactions, such as amino acid and lipid synthesis. α-ketoglutarate (2-oxoglutaric acid) plays a key role here because its amination to produce glutamate is an important route by which carbon skeletons leave the TCA cycle and may subsequently be incorporated into other amino acids. It should be noted that, if intermediates of the cycle, such as α-ketoglutarate, are drained off in this way, there must be other means of providing the oxaloacetate required for the continued entry of acetyl-CoA into the cycle. This is achieved by two replenishing, or anaplerotic, reactions: in the cytosol, PEP can be carboxylated to oxaloacetate; and in the mitochondrion, pyruvate can be decarboxylated to malate (Fig. 3.1).

## 3.2 PHOTOSYNTHETIC EFFICIENCY

Photosynthetic efficiency can be expressed in a number of ways. At the molecular level, the maximum efficiency can be determined under carefully controlled laboratory conditions, for algae or leaf cells in suspension, by measuring the amount of absorbed light required for the fixation of each gram molecule (mole) of $CO_2$. At the other end of the scale, it may be expressed as the photosynthesis, or dry-matter production, of a crop in relation to the available PAR averaged over the season. In the latter case, it will be influenced by the ability of the canopy to intercept radiation, by the depredations of the weather, weeds, pests and diseases and by the quality and nutrient status of the soil.

Note that the term 'light' refers to the part of the solar radiation spectrum which can be detected by the human eye, the wavelength range 400–700 nm. This is also the spectrum which activates photosynthesis, and for this reason we use the terms 'light' and 'PAR' interchangeably. However, it is important to recognize that photosynthesis and human vision do not respond identically to different wavelengths within the 400–700 nm spectrum (i.e. they have different spectral responses) (McCree 1981).

Light, or PAR, is somewhat enigmatic, since it has characteristics which suggest that it is propagated through space both in the form of waves and as a stream of particles, or discrete packets of energy, called quanta or photons. The energy content of a quantum is inversely related to its wavelength, such that a quantum of blue light (wavelength 400 nm) contains $4.97 \times 10^{-19}$ joules and a quantum of red light (680 nm) contains $2.92 \times 10^{-19}$ J. When red light is absorbed by chlorophyll, each quantum has sufficient energy to raise an electron to a higher energy state which is stable enough to initiate the photochemical events of photosynthesis. The extra energy of each quantum of blue light, although effecting an initially higher level of excitation, is rapidly dissipated as heat, and thus red and blue light effectively impart the same amount of energy for photosynthesis. We shall therefore use red light as the basis of our discussion of photosynthetic efficiency.

One quantum of PAR is a very small amount of energy, and it is convenient to use instead Avogadro's Number (the number of molecules per gram molecule) of quanta, referred to as a mole of quanta or simply a mole quanta. Thus, a mole of red quanta has an energy content of $(6.02 \times 10^{23}) \times (2.92 \times 10^{-19}) = 1.758 \times 10^5$ J = 176 kJ. The photosynthetic photon flux density (PPFD) is often expressed in units of mole quanta, abbreviated to mol (as in $\mu$mol m$^{-2}$ s$^{-1}$), and although the term Einstein has been widely used to mean the energy content of a mole quanta, it is not an SI unit and its use is going out of fashion. The supply of PAR is also commonly expressed as irradiance (units, W m$^{-2}$ = J m$^{-2}$ s$^{-1}$), which does not take into account the differing energy contents of different wavelengths. For this reason, there is no single relationship between PPFD and irradiance although, as a guide, full sunlight (PAR) is about 2200 $\mu$mol m$^{-2}$ s$^{-1}$, or 500 W m$^{-2}$.

The photochemical events of photosynthesis provide the assimilatory power, in the forms of NADPH and ATP, necessary for the reduction of $CO_2$ to $CH_2O$ (carbohydrate). Three molecules of ATP and two of NADPH are required for each molecule of $CO_2$ reduced: for each mole of $CO_2$ this is equivalent to 525 kJ. This means that at least 3 mole quanta of (red) light (3 × 176 = 528 kJ) are required for the reduction of each mole of $CO_2$ (or three quanta per molecule of $CO_2$). The minimum theoretical quantum requirement is therefore 3. However, this theoretical figure assumes that all of the absorbed quanta are used for $CO_2$ reduction, whereas some energy is lost by chlorophyll fluorescence and some is dissipated as heat after being absorbed by molecules other than the photosynthetic pigments. Such considerations raise the theoretical minimum quantum requirement to 8. In practice, when the quantum requirement is determined by simultaneously measuring $CO_2$ uptake by a leaf and the absorbed photon (quantum) flux density, photorespiratory and respiratory losses of $CO_2$ raise the requirement to 15–20, depending on species and temperature. The reciprocal of this, i.e. the number of moles of $CO_2$ fixed per mol quanta of PAR, is known as the quantum yield. One gram molecule of $CH_2O$ is equivalent to 468 kJ, since one gram molecule of glucose ($C_6H_{12}O_6$) yields 2809 kJ when burned in a calorimeter. A quantum requirement of 15, equal to a quantum yield of 0.067, therefore represents an efficiency of 468/15 × 176 = 17.7 per cent.

Seventeen per cent is therefore the maximum efficiency with which absorbed PAR can be converted into $CH_2O$ by single leaves in the laboratory. Efficiencies for crops in the field, in relation to incident PAR, are much lower for several reasons. First, about 15 per cent of the PAR is unavailable for photosynthesis because it is reflected or transmitted through the leaves without being absorbed. Secondly, for $C_3$ crops at least, the efficiency of leaves exposed to high irradiance will be reduced because of light saturation of photosynthesis (sect. 2.4.3): Monteith (1977a) suggests an approximate reduction of about 30 per cent to account for this. Thirdly, the photosynthetic efficiency of a growing crop must be expressed on a 24-hour basis, taking into account losses of $CO_2$ by respiration in the dark: about 40 per cent of the carbon fixed may be lost in this way. These considerations, which result in an overall efficiency of 6.3 per cent, are summarized in Table 3.1.

**Table 3.1** The conversion of incident PAR into $CH_2O$ in a $C_3$ crop

| | |
|---|---|
| Incident photosynthetically active radiation (PAR) | 100 |
| 15% loss due to reflection and transmission | 85 |
| Maximum efficiency of conversion of PAR into $CH_2O$, 17.7%* | 15 |
| 30% loss due to light saturation of upper leaves | 10.5 |
| 40% loss due to respiration in the dark | 6.3 |

* Takes into account loss of efficiency due to photorespiration and respiration in the light.

This efficiency of about 6 per cent is realistic only when the interception of PAR is complete and when environmental and endogenous conditions are optimum. Thus, the maximum crop growth rates of a range of species, attained when these criteria are satisfied, represent efficiencies of this order (Table 3.2). The variation around this

**Table 3.2** Some high short-term crop growth rates and estimates of efficiency of utilization of photosynthetically active radiation (PAR)

| *Crop* | *Location* | *Length of period over which measurements were made (days)* | *Maximum growth rate* ($g\ m^{-2}\ d^{-1}$) | *Average total radiation* ($J\ cm^{-2}\ d^{-1}$) | *Utilization of PAR (%)* |
|---|---|---|---|---|---|
| Bulrush millet (*Pennisetum typhoides*) | Northern Territory, Australia | 14 | 54.0 | 2132 | 8.9 |
| | | 7 | 44.0 | 2575 | 6.0 |
| Sudan grass (*Sorghum vulgare*) | California, USA | 35 | 51.0 | 2884 | 6.2 |
| Maize (*Zea mays*) | Japan | 21 | 51.6 | 2031 | 8.9 |
| | California, USA | 10 | 38.0 | 3072 | 4.3 |
| Sugar cane (*Saccharum* spp) | Hawaii | 90 | 37.0 | 1676 | 7.7 |
| Rice (*Oryza sativa*) | Japan | 21 | 35.8 | 2032 | 6.2 |
| Tall fescue (*Festuca arundinacea*) | Great Britain | 6 | 43.6 | 2201 | 6.9 |
| Sugar beet (*Beta vulgaris*) | Czechoslovakia | 12 | 31.0* | 1229 | 8.8 |
| | Japan | 21 | 27.8 | 1726 | 7.3 |
| Cocksfoot (*Dactylis glomerata*) | Great Britain | 6 | 40.5 | 2201 | 6.4 |
| Timothy (*Phleum pratense*) | Great Britain | 6 | 36.4 | 2201 | 5.8 |
| Ryegrass (*Lolium perenne*) | Great Britain | 30 | 28.0 | 1983 | 4.9 |
| Soybean (*Glycine max*) | Japan | 21 | 26.7 | 1212 | 7.7 |
| Subterranean clover (*Trifolium subterraneum*) | Australia | 14 | 22.9 | 2801 | 2.9 |
| Kale (*Brassica oleracea*) | Great Britain | 14 | 21.0 | 1597 | 4.6 |

* Includes roots.

The efficiency of utilization of PAR was estimated on the basis that plant dry matter has an energy content (heat of combustion) of 17.5 kJ $g^{-1}$ (Monteith 1977a) and that PAR is 50 per cent of total radiation. (After Eagles and Wilson (1982), where sources of data are cited.)

figure (2.9–8.9) serves to remind one that the loss factors in Table 3.1 are generalizations. For example, the leaves of $C_4$ crops will be much less susceptible to light saturation than $C_3$ crops, while within the $C_3$ group the extent of light saturation will be affected by canopy structure (sect. 2.4.3). It must be emphasized, however, that the efficiencies in Table 3.2 were attained over relatively short periods (the Hawaiian sugar cane crop, maintaining an efficiency of 7.7 per cent over 90 days, is exceptional in this respect).

Photosynthetic efficiency in terms of available

PAR throughout the whole growing season will be less, because of periods of incomplete interception, for whatever reason (sect. 2.4.3) and because the plants themselves will inevitably be subjected to environmental stresses which reduce their capacity to make use of intercepted PAR. In temperate climates, for example, photosynthesis can be limited by low temperatures in the winter and by water deficits in the summer. The overall effect of such environmental stresses, and perhaps of endogenous limitations, is to reduce the photosynthetic efficiency of 'average-to-good' yields to 2–4 per cent, and of course much greater reductions will occur when crops are grown in poor soils, with less efficient husbandry or more extreme weather. When the conditions for growth are optimized as far as possible, as in many field experiments, crops well-adapted to a particular environment have a very similar efficiency. This can be inferred from the linear relationship between final dry matter yield and cumulative intercepted total solar radiation, which has been demonstrated for a number of crops (Fig. 3.3; cf. Figs 1.1 and 7.16): Monteith (1977a) calculated that on average, dry matter is accumulated at a rate of 1.4 g per MJ of total solar radiation intercepted. Assuming that plant dry matter has an energy content of 17.5 kJ $g^{-1}$, this represents an efficiency of conversion of PAR into dry matter of 4.9 per cent.

Relationships such as that depicted in Fig. 3.3 demonstrate the overriding importance of radiation interception for dry-matter production. The slope of the line indicates the efficiency with which intercepted radiation is converted into dry matter: for the diverse crops analysed by Monteith, this varied by no more than 15 per cent. Clearly, when other environmental constraints are minimized, as in these experiments, the capture of light becomes the overall limiting factor. However, the imposition of stress, such as that occasioned in Britain by the hot, dry summer of 1976, causes deviation from the common line (Fig. 7.16). Differences in efficiency might also be expected when plants having differing rates of photorespiration and respiration are compared, or when photosynthesis is limited by sink demand. Furthermore, the linear relationship shown in Fig. 3.3 is unlikely to hold for temperate perennial crops, such as grasses, in which a high $L$ is present throughout the winter and early spring when photosynthesis and growth are limited by low temperature; the photosynthesis of grass canopies is also characterized by pronounced variations in rate with changes in plant development (see sect. 8.1).

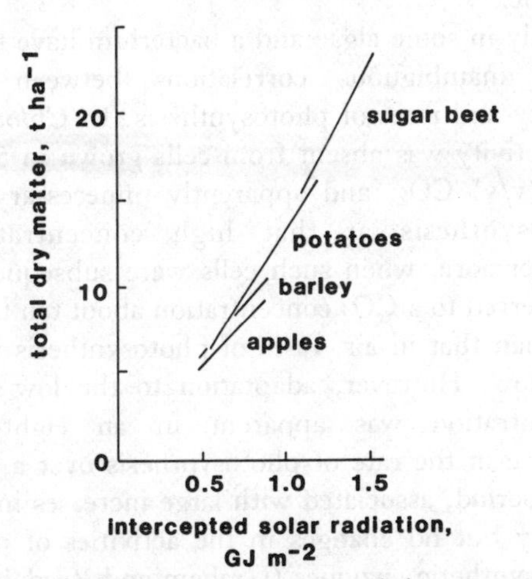


**Fig. 3.3** Relationship between total dry matter at harvest and intercepted radiation throughout the growing season (from Monteith 1977a).

## 3.3 PHOTOSYNTHESIS AS A DIFFUSIVE PROCESS

### 3.3.1 INTRODUCTION

The movement of $CO_2$ into the leaves, and as far as the chloroplasts, is largely by diffusion (albeit facilitated in various ways). It is therefore appropriate to treat photosynthesis as a diffusive process, and this approach provides a useful means of analysing the effects of endogenous and environmental factors on photosynthetic efficiency. The necessary gradient of $CO_2$ concentration is established by the relatively high concentration of $CO_2$ maintained within a few millimetres of the leaves by the turbulent mixing of the atmosphere, and by the consumption of $CO_2$ in the chloroplasts. Photosynthesis can therefore be treated as a diffusive process and considered in terms of Fick's Law, which relates the rate of diffusion of a gas or solute to the concentration gradient. Thus,

$$F = -D\,\frac{(C_1 - C_2)}{\Delta z} \qquad [3.2]$$

where $F$ is the flux density (i.e. the rate of diffusion through unit surface area; cf. flux density of radiation), $D$ is the diffusion coefficient (which can be considered constant for a given gas in air at STP), and $C_1$ and $C_2$ are the concentrations of the substance in question (in this case, $CO_2$) separated by a distance $\Delta z$. By convention, a flux towards a region of lower concentration is indicated by the negative sign.

However, $CO_2$ does not diffuse simply as a gas along the whole length of its pathway from the atmosphere to the chloroplasts; at some point it dissolves in the films of water surrounding the mesophyll cells, and subsequent diffusion is in the liquid phase. Consequently, the diffusion coefficient varies enormously over the pathway of $CO_2$ movement; for example, the diffusion coefficient for $CO_2$ in water is about 10,000 times less than that in air. However, movement of $CO_2$ in the liquid phase is enhanced, or facilitated, in various ways which, in $C_3$ plants, to some extent substitute for the $CO_2$ concentrating function of the $C_4$ cycle (sect. 3.4.2).

There is considerable controversy over the form(s) in which inorganic carbon crosses the plasmalemma and the chloroplast membrane of higher plants, because when $CO_2$ dissolves in water some bicarbonate ($HCO_3^-$) forms, the equilibrium mixture ($CO_2$ + $HCO_3^-$) depending on the pH. However, the diffusion across these membranes of $HCO_3^-$, a relatively large and negatively charged molecule, would be more restricted than that of $CO_2$, unless facilitated by some form of transport mechanism. There is evidence from isolated pea mesophyll protoplasts that $HCO_3^-$ can cross the plasmalemma, by means of an unidentified carrier-mediated transfer mechanism (Volokita *et al.* 1981), and such a system may also be responsible for the transport of $HCO_3^-$ across the chloroplast membrane. Indeed, Poincelot and Day (1976) have suggested that the higher rate of net photosynthesis of sunflower plants compared with spinach may be attributed to their more rapid transfer of $HCO_3^-$, associated with a higher activity of chloroplast membrane ATPase. However, the weight of experimental evidence, reviewed recently by Colman and Espie (1985), indicates that it is mainly $CO_2$ which crosses the plasmalemma and the chloroplast membrane.

### 3.3.2 THE IMPORTANCE OF CARBONIC ANHYDRASE

It has long been assumed that carbonic anhydrase (CA) has a function in photosynthesis, often that it facilitates in some way the diffusion of $CO_2$ to the sites of carboxylation. However, the evidence for its participation in photosynthesis is only circumstantial, a frustrating conclusion about the enzyme which has its highest activity in green leaves, where it is second in abundance only to RuBP carboxylase (Reed and Graham 1981). CA catalyses the reversible hydration of $CO_2$ according to the equation

$$CO_2 + H_2O \rightleftharpoons H^+ + HCO_3^-, \qquad [3.3]$$

the equilibrium mixture of $CO_2$ and $HCO_3^-$ depending on the pH. The consequences of this reaction, namely the regulation of the relative amounts of $HCO_3^-$ and $CO_2$ and the production of protons, mean that many roles can be attributed to it. For example, Graham and Reed (1971) suggested that CA was the focal point in the control of photosynthesis, influencing photophosphorylation, the Hill reaction, carboxylation and the ionic fluxes of the chloroplast. However, evidence for these and other functions is inconclusive, and the reader is referred to the review by Reed and Graham (1981) for a full assessment. Nevertheless, the circumstantial evidence is persuasive, and it is worth considering briefly the claim that CA should be considered a photosynthetic enzyme.

Only in some algae and a bacterium have there been unambiguous correlations between CA activity and rates of photosynthesis. In *Chlorella*, CA activity was absent from cells grown in 5 per cent (v/v) $CO_2$, and apparently unnecessary for photosynthesis at that high concentration. Furthermore, when such cells were subsequently transferred to a $CO_2$ concentration about ten times less than that in air, rates of photosynthesis were very low. However, adaptation to the low $CO_2$ concentration was apparent in an eight-fold increase in the rate of photosynthesis over a two-hour period, associated with large increases in CA activity but no changes in the activities of other photosynthetic enzymes (Graham and Reed 1971; Reed and Graham 1977). This increase in the rate of photosynthesis did not occur in the presence of the inhibitor of CA activity, acetazolamide ('Diamox'), although this piece of evidence is weakened by the finding that acetazolamide also inhibits photosynthetic electron transport (Reed and Graham 1981). Nevertheless, it is clear that the concentration of $CO_2$ in the growth medium regulates the activity of CA.

Until recently, confirmation of this finding was restricted to algae, but the repressive effect of high $CO_2$ concentrations on CA activity in mature leaves of bean (*Phaseolus*) has now been reported (Porter and Grodzinski 1984). Mature, but photosynthetically active, leaves had lost 95 per cent of their CA activity after seven days' growth in a $CO_2$-enriched (1200 vpm) atmosphere. If it is assumed that this level of CA would limit photosynthesis under normal $CO_2$ concentrations, then this observation could explain previous reports that plants grown in $CO_2$-enriched atmospheres have lower rates of photosynthesis in ambient $CO_2$ than plants grown in air.

Of course, only plants growing in glasshouses are likely to experience such large variations in $CO_2$ content. We have already noted that under normal circumstances there is a large excess of CA in leaves, and it is pertinent to question whether plants growing without $CO_2$ enrichment are ever subjected to conditions which might reduce the CA activity to levels which limit photosynthesis. Such evidence is slight but, in water-stressed cotton plants, a progressive reduction in maximum rate of photosynthesis (estimated by the uptake of $^{14}C$ by thin leaf slices in solution) was paralleled by reductions in CA activity and, to a lesser extent, by RuBP carboxylase activity (Jones 1973; Fig. 3.4). Similarly, Downton and Slatyer (1972) showed that for whole leaves of cotton plants grown under a range of temperature regimes, there was a general similarity between the effects of temperature on the rate of photosynthesis and on CA activity; there was little effect of growth temperature on the activity of RuBP carboxylase.

Support for the involvement of CA in the photosynthesis of thin leaf slices of cotton comes from Jones and Osmond (1973), who showed that the addition of CA to the bathing solution reduced the apparent Michaelis constant ($K_m$, the concentration of free $CO_2$ which gives half the maximum rate of photosynthesis) from a mean value of 80 $\mu$M to about 10 $\mu$M (Fig. 3.5). Similarly, Bird *et al.* (1980) and Okabe *et al.* (1980) have reported CA-induced reductions in the *in vitro* $K_m$ ($CO_2$) for RuBP carboxylase extracted from leaves of wheat and spinach.

Of course, a correlation between the overall rate of photosynthesis and the activity of a particular enzyme does not necessarily imply causality. As Jones (1973) points out, it could be that reduced enzyme activity in water-stressed plants merely reflects the extent to which the various photosyn-

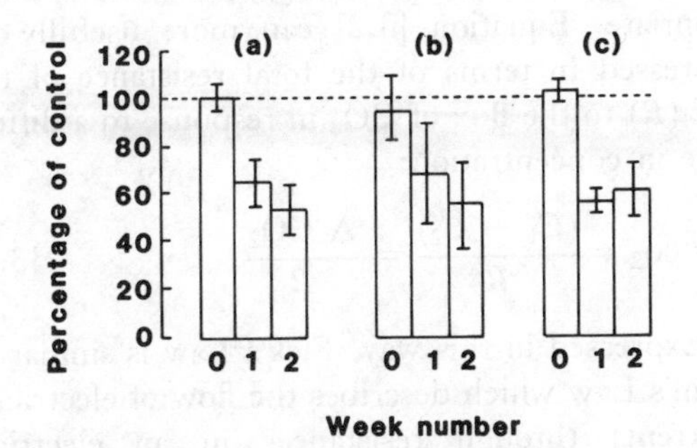


**Fig. 3.4** Effects of 0, 1 or 2 weeks of water stress on (a) $^{14}C$ fixation by leaf slices; (b) carbonic anhydrase activity; (c) RuBP carboxylase activity of cotton leaves. Values are expressed on a leaf area basis, as percentages of controls. The 90 per cent confidence limits are shown (after Jones 1973).

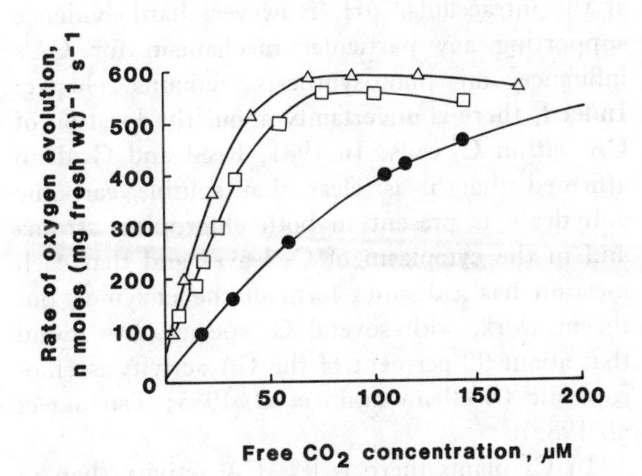


**Fig. 3.5** Effect of carbonic anhydrase at concentrations of 0.0625 mg ml$^{-1}$ (□——□) and 0.2 mg ml$^{-1}$ (△——△) on the response of photosynthetic oxygen evolution of cotton leaf slices to $CO_2$ concentration; (●——●) control (from Jones and Osmond 1973).

thetic controls are coupled: when stomata are closed, high activities of some enzymes are unnecessary, and substrates and energy may be conserved by reduced enzyme turnover. Certainly, in unstressed plants, CA has a very high rate of turnover (Jacobson *et al.* 1975). The same argument can be applied to associations between CA activity and photosynthesis under other circumstances, but the greater the range of such evidence the more likely is a causal relationship to be responsible. For example, CA was found to be absent in etiolated leaves of wheat, *Phaseolus vulgaris*, and soybean, but was induced during greening in the light, in parallel with chlorophyll and RuBP carboxylase activity (see Tsuzuki *et al.* 1985). Zinc deficiency in rice has resulted in increases in the $CO_2$ compensation point and in several photorespiratory enzymes, associated with decreases in the activity of CA, which has zinc at its active site (Seethambaram *et al.* 1985). It is suggested that enhanced photorespiration in zinc-deficient plants results from the low CA activity and a consequent reduction in the availability of $CO_2$ at the sites of carboxylation (sect. 3.4.1).

The facilitation of $CO_2$ transport would therefore appear to be a consequence of the rapid hydration of $CO_2$ within the cell. Edwards and Walker (1983) have outlined such a role for CA, based on the rapid diffusion of $HCO_3^-$, which would be present at relatively high concentrations at the intracellular pH. However, hard evidence supporting any particular mechanism for CA's influence on photosynthesis remains elusive. Indeed, there is uncertainty about the location of CA within $C_3$ cells. In 1981, Reed and Graham affirmed that 'it is clear that soluble carbonic anhydrase is present in both chloroplast stroma and in the cytoplasm of $C_3$ leaves and that each location has a distinct form of the enzyme', but recent work, with several $C_3$ species, has found that about 90 per cent of the CA activity is chloroplastic (Seethambaram *et al.* 1985; Tsuzuki *et al.* 1985).

In $C_4$ plants there is less CA activity than in $C_3$ plants, but most of it is in the cytosol of the mesophyll cells. It therefore appears that CA is not required for the operation of the Calvin cycle in the bundle sheath cells, which are supplied with high concentrations of $CO_2$ by the $C_4$ $CO_2$-concentrating system (Fig. 3.14). This again is circumstantial evidence that in the photosynthetic cells of $C_3$ plants and in the mesophyll cells of $C_4$ plants, CA participates in the supply of $CO_2$ to the sites of carboxylation. The significant difference between these two groups is that the initial carboxylation in the $C_4$ mesophyll cells is catalysed by phosphoenolpyruvate carboxylase in the cytosol and uses $HCO_3^-$ as the substrate, whereas $CO_2$ is the substrate for RuBP carboxylase.

### 3.3.3 THE RESISTANCE ANALOGY

The intracellular diffusion of $CO_2$ (as $HCO_3^-$) therefore appears to be facilitated by the rapid hydration of $CO_2$, catalysed by CA. In addition, it is likely that the purely molecular diffusion may be aided by cytoplasmic streaming, which rapidly transfers molecules from one side of a cell to the other. If the streaming is sufficiently rapid, turbulence will be generated, producing a mechanical mixing which mimics on a lesser scale that which occurs in the bulk atmosphere outside the leaf. Fick's Law (eqn 3.2) can be seen as a form of the generalized relationship:

$$\text{flux density} = \text{proportionality constant} \times \text{driving force} \qquad [3.4]$$

with the force provided by the concentration difference. However, the heterogeneous nature of the pathway of $CO_2$ movement into and within the leaf, and the facilitation of diffusion discussed above, means that a proportionality constant $D/\Delta z$, based on a molecular diffusion coefficient but quantitatively very different from it, is inappropriate. Equation [3.2] can more usefully be expressed in terms of the total resistance of the leaf ($R$) to the flux of $CO_2$ in response to a difference in concentration:

$$F_{CO_2} = \frac{(C_1 - C_2)}{R} = \frac{\Delta\ CO_2}{R} \qquad [3.5]$$

Expressed in this way, Fick's Law is similar to Ohm's Law which describes the flow of electricity (current) through resistances in an electrical circuit:

$$\text{current} = \frac{\text{potential difference}}{\text{resistance}}$$

Gaastra (1959) introduced this analogy to describe the movement of $CO_2$ into a leaf, and subsequently techniques for evaluating the individual resistances to this movement have allowed the identification of those parts of the pathway most limiting to photosynthesis. The units of resistance are easily derived from eqn [3.5]: if the flux density of $CO_2$ is measured in g $cm^{-2}$ $s^{-1}$ and the concentration difference in g $m^{-3}$, then resistance must have units of s $cm^{-1}$. It is clear from eqns [3.2] and [3.5] that for a given rate of $CO_2$ uptake the reciprocal of resistance ($1/R$) is equal to $D/\Delta z$, in which $D$ is analogous, but not equal, to the molecular diffusion coefficient. The units of resistance may also be derived from this relationship: the diffusion coefficient has units of, for example, $cm^2$ $s^{-1}$, giving $D/\Delta z$ units of cm $s^{-1}$. The proportionality constant in eqn [3.4] can therefore be represented as either $D/\Delta z$, in which case it is referred to as a conductance, or, more commonly, as $1/R$.

#### 3.3.3.1 THE BOUNDARY LAYER RESISTANCE AND TURBULENCE

The individual resistances are shown diagrammatically in Fig. 3.6. The first diffusive resistance is that offered by the boundary layer of undisturbed air which envelopes each leaf, and which separates it from the surrounding turbulent air. In fact, the nature of the boundary layer may be best appreciated by contrasting it with this turbulent air. Turbulence is a vital agent in the exchange of matter (e.g. water vapour or $CO_2$) and energy (e.g. heat) between a crop and the atmosphere. It is a consequence of the circular movements of discrete, and transitory, packets of air (eddies), caused by the frictional drag imposed on air movement by, for example, the ground or a crop. It results in the rapid transfer of air from one horizontal layer to another, and is vital for the replenishment of $CO_2$, depleted by photosynthesis, within the canopy. It should be remembered that the ambient $CO_2$ concentration of about 330 vpm (0.033 per cent v/v) is considerably below the optimum for photosynthesis, and the maintenance of high rates of canopy photosynthesis requires the continual mixing of large volumes of air. The air above a photosynthesizing crop is depleted of $CO_2$; and within the canopy, $CO_2$ concentration profiles demonstrate a marked minimum at that horizon corresponding to maximum rates of photosynthesis. As would be expected, the extent of this depletion varies during the day as irradiance changes (Fig. 3.7). Turbulence is more pronounced at higher windspeeds, as indicated in Fig. 3.8 by the difference in $CO_2$ depletion of the air above a maize crop on windy and still days.

Turbulent transfer is therefore responsible for the movement of $CO_2$ from the atmosphere into

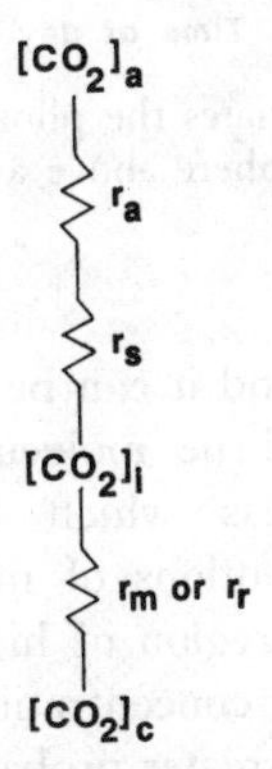


**Fig. 3.6** The resistances to $CO_2$ movement into and within a leaf. $[CO_2]_a$, $[CO_2]_i$ and $[CO_2]_c$ are the concentrations of $CO_2$ in the air, in the intercellular spaces and at the sites of carboxylation in the chloroplasts. $r_a$, $r_s$ and $r_m$ ($r_r$) are the boundary layer, stomatal and mesophyll (residual) resistances.

Equation [3.5] can thus be re-written as:

$$P_n = \frac{[CO_2]_a - [CO_2]_c}{r_a + r_s + r_m} \qquad [3.6]$$

in which $P_n$ is the rate of net photosynthesis (e.g. ng $CO_2$ $cm^{-2}$ $s^{-1}$).

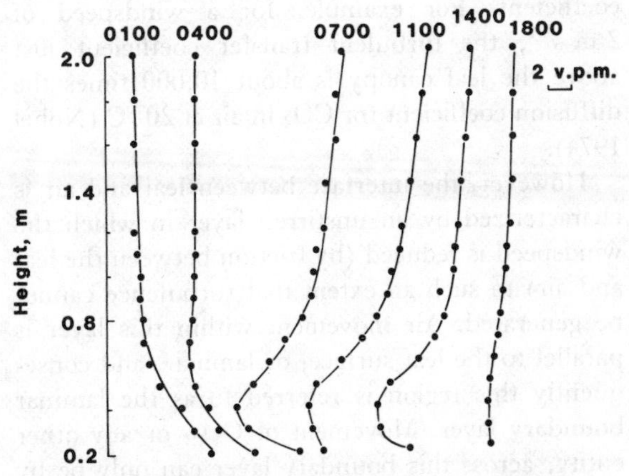


**Fig. 3.7** Profiles of $CO_2$ concentration measured above and within a barley crop at various times during 19 June 1972. Note that the soil as well as the atmosphere is a source of $CO_2$ (from Biscoe *et al.* 1975a).

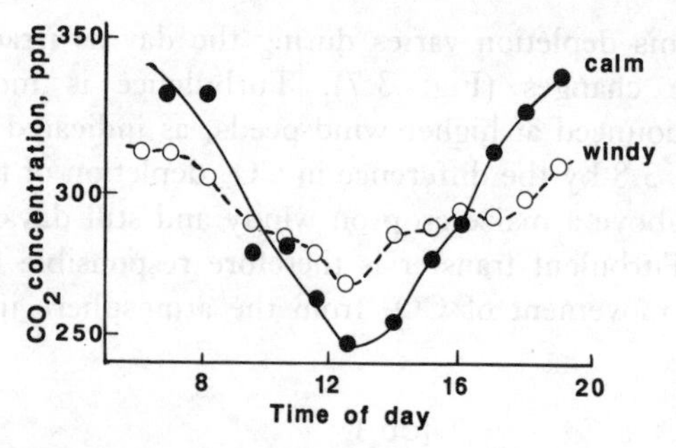


**Fig. 3.8** Wind alleviates the photosynthetic depletion of $CO_2$ in the atmosphere above a maize canopy (after Uchijima 1970).

the crop canopy, and it can be regarded as accelerated diffusion. True molecular diffusion is a spontaneous process which results from the random thermal motions of molecules. The net movement from a region of higher concentration to one of lower concentration occurs simply because there is a greater probability of molecules moving from the concentrated to the dilute region than *vice versa*. The same argument applies to turbulent transfer, but with the random motion of individual molecules supplanted by that of individual eddies which, of course, comprise enormous numbers of molecules. The essential characteristic of diffusion, i.e. movement down a concentration gradient of the entity in question, is retained, but the molecular diffusion coefficient is now replaced by the analogous, but much greater, turbulent transfer (or eddy diffusion) coefficient. For example, for a windspeed of 2 m $s^{-1}$, the turbulent transfer coefficient just above the leaf canopy is about 10,000 times the diffusion coefficient for $CO_2$ in air at 20 °C (Nobel 1974).

However, the interface between leaf and air is characterized by an unstirred layer in which the windspeed is reduced (by friction between the leaf and air) to such an extent that turbulence cannot be generated. Air movement within this layer is parallel to the leaf surface, or laminar, and consequently this region is referred to as the laminar boundary layer. Movement of $CO_2$, or any other entity, across this boundary layer can only be by molecular diffusion, and consequently the thickness of this region, characterized as the boundary layer resistance, determines the extent to which the transfer of $CO_2$ from the bulk atmosphere to the surface of the leaf is impeded.

The thickness of the boundary layer, and hence the boundary layer resistance ($r_a$), depends on windspeed, leaf size and, to a lesser extent, leaf structure and shape. The lower the windspeed and the larger the leaf, the thicker will be the boundary layer (Table 3.3). Superimposed on these main effects are those caused by structural features: a dense covering of hairs helps maintain the unstirred layer, whereas sparse hairs, prominent veins and serrated edges on leaves tend to increase turbulence (Grace 1983). The formula used for the calculation of $\delta$ in Table 3.3 is one of several developed from hydrodynamic theory for laminar flow over flat surfaces; Grace (1983) and Nobel (1974) give alternative related formulae. It is clear from the explanations of eqns [3.2] and [3.5] that for a given difference in concentration of $CO_2$ across a boundary layer, the resistance to $CO_2$ diffusion will be equal to $\delta/D$. $r_a$ may also be determined experimentally from measurements of evaporation from damp blotting paper replicas of leaves (for details, see Woodward and Sheehy 1983; Coombs *et al.* 1986).

**Table 3.3** Thickness of the unstirred air layer (laminar boundary layer) adjacent to a leaf ($\delta$, cm) and the corresponding boundary layer resistance ($r_a$, s $cm^{-1}$), for a range of leaf sizes and windspeeds

| *Leaf size* | *Windspeed (u, cm $s^{-1}$)* | | | | | |
|---|---|---|---|---|---|---|
| | *10* | | *100* | | *1000* | |
| *(d,cm)* | $\delta$ | $r_a$ | $\delta$ | $r_a$ | $\delta$ | $r_a$ |
| 1 | 0.21 | 1.28 | 0.07 | 0.14 | 0.02 | 0.13 |
| 5 | 0.46 | 2.87 | 0.15 | 0.91 | 0.05 | 0.29 |
| 15 | 0.80 | 4.98 | 0.25 | 1.57 | 0.08 | 0.50 |

$\delta$ was calculated as $0.65 \times (u/d)^{-0.5}$ (after Jones 1983), in which d is a characteristic dimension which depends on the leaf's shape: for parallel-sided leaves, such as those of grasses, it is the downwind width. $r_a$ was then calculated as $\delta/D$ (see text) were D is the diffusion coefficient of $CO_2$ in air (0.16 $cm^2$ $s^{-1}$ at 20 °C).

The windspeeds given in Table 3.3 cover the range normally experienced by leaves in the field: 10 cm $s^{-1}$ (= 0.22 mile $h^{-1}$) is described as 'calm'

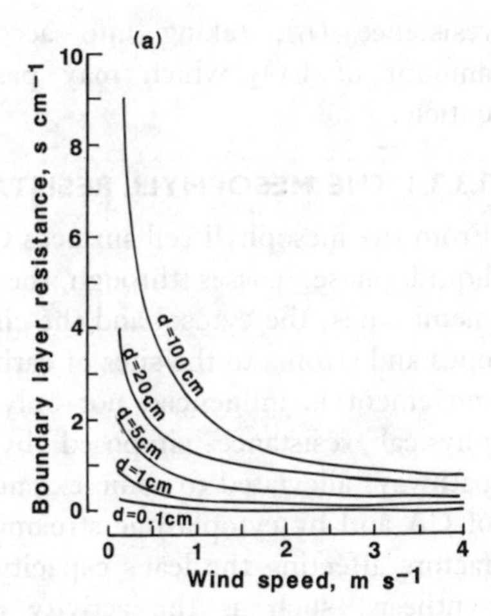


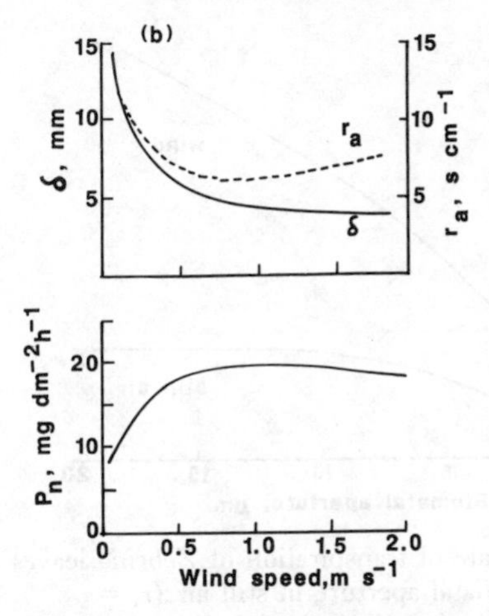


**Fig. 3.9** (a) Calculated effect of windspeed on the boundary layer resistances of leaves with dimensions (*d*) from 0.1 to 100 cm (from Grace 1983). (b) Effect of windspeed on the observed boundary layer thickness (δ) the boundary layer resistance ($r_a$) and the rate of net photosynthesis ($P_n$) of cucumber leaves (after Yabuki and Miyagawa 1970).

in the Beaufort Scale, whereas 1000 cm s$^{-1}$ represents a 'fresh breeze', which would cause small trees to sway. Obviously, higher windspeeds occur, but since closure of stomata may occur at windspeeds of more than 700 cm s$^{-1}$ (Grace 1977), their effect on $r_a$ becomes irrelevant. Also, the effect of increasing windspeed on the thickness of the boundary layer is much more pronounced over a range of low windspeeds, as theoretical calculations (Fig. 3.9a) and actual measurements (Fig. 3.9b) demonstrate.

It should be noted, however, that the effects of partial closure of stomata may be offset by the improved $CO_2$ supply which results from increased windspeeds. Clearly, $r_a$ is considerably reduced by even modest windspeeds, and under natural conditions its importance as a limitation to photosynthesis might be thought correspondingly slight. However, there is a logarithmic relationship between height and windspeed, which means that leaves deep within a crop canopy may experience windspeeds an order of magnitude lower than those at the top of the canopy (Fig. 3.10). Diffusion of gases into, and out of, such leaves would therefore be severely limited by $r_a$ over a wide range of stomatal apertures, as Bange (1953) realized over 30 years ago (Fig. 3.11). A complication which does not apply to Fig. 3.11, which refers only to transpiration, is

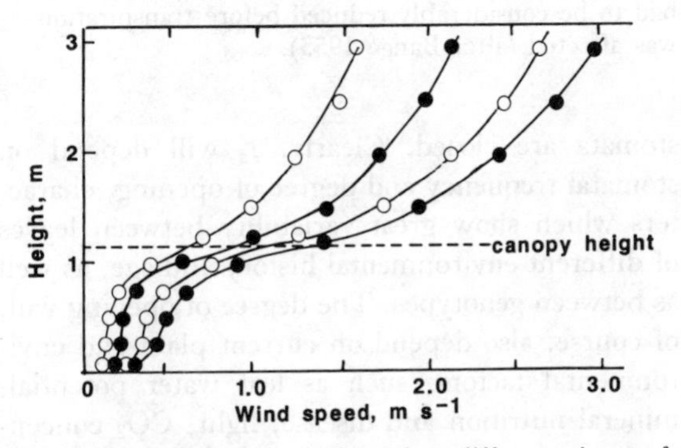


**Fig. 3.10** Profiles of windspeed at different times of the day in and above a bean (*Vicia faba*) crop (after Thom 1971).

that low levels of PAR within the canopy will reduce the significance of $r_a$ if light becomes more limiting than $CO_2$ supply.

### 3.3.3.2 THE STOMATAL RESISTANCE

Theoretical calculations and direct measurements show that $r_a$ can vary from an almost negligible value in turbulent air to about 3–4 s cm$^{-1}$ in still air. Under most circumstances, particularly for crops growing in the field, it is therefore less important than the stomatal resistance, $r_s$, which ranges from about 1 s cm$^{-1}$ when stomata are wide open to values in excess of 50 s cm$^{-1}$ when

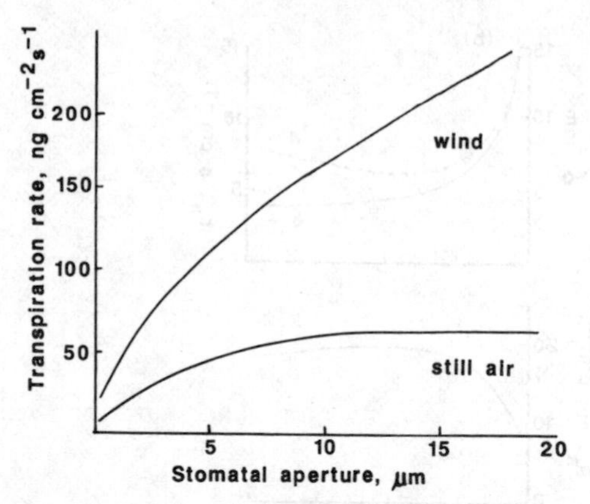


**Fig. 3.11** The rate of transpiration of *Zebrina* leaves in relation to stomatal aperture in still air ($r_a$ = 2.0 s cm$^{-1}$) and moving air ($r_a$ = 0.1 s cm$^{-1}$). The stomatal resistance was much greater than $r_a$ in moving air (even when fully open, $r_s$ was 1.0 s cm$^{-1}$), and transpiration rate was therefore closely related to stomatal aperture. In still air, $r_a$ was twice $r_s$ when the stomata were fully open, and stomatal aperture had to be considerably reduced before transpiration was affected (after Bange 1953).

stomata are closed. Clearly, $r_s$ will depend on stomatal frequency and degree of opening, characters which show great variability between leaves of different environmental history and age, as well as between genotypes. The degree of opening will, of course, also depend on current plant and environmental factors, such as leaf water potential, mineral nutrition and disease, light, $CO_2$ concentration, temperature and humidity. The responses of stomata to water stress and humidity are summarized in section 3.4.5.1, but for further details the reader is referred to recent reviews or books (e.g. Willmer 1983; Mansfield and Davies 1985).

$r_s$ can, in fact, be determined from the application of diffusion theory to anatomical measurements (e.g. Jones 1983), but it is commonly measured experimentally by diffusion porometers (for details, see Woodward and Sheehy 1983; Coombs *et al.* 1986). $CO_2$ which diffuses from the leaf surface through the stomata continues to diffuse as a gas through intercellular spaces to the mesophyll cells where it dissolves in the surface films of water. It is the resistance to this diffusion of $CO_2$ as a gas which is normally referred to as the stomatal resistance, although it is sometimes called the gaseous resistance ($r_g$) or simply leaf resistance ($r_l$), taking into account the small amount of $CO_2$ which may pass through the cuticle.

#### 3.3.3.3 THE MESOPHYLL RESISTANCE

From the mesophyll cell surfaces $CO_2$, now in the liquid phase, passes through the cell walls, cell membranes, the cytosol and the chloroplast envelopes and stroma to the sites of carboxylation. This movement is influenced not only by the direct, physical resistances imposed by the diffusion pathway, alleviated to some extent by the activity of CA and by cytoplasmic streaming, but also by factors affecting the leaf's capacity for net photosynthesis, such as the activity of the primary carboxylating enzymes, the photochemical efficiency and the rates of photorespiration and of respiration in the light. These influences comprise the composite mesophyll resistance ($r_m$), sometimes referred to as the intracellular resistance ($r_i$) or residual resistance ($r_r$); the latter name arises because it is usually determined by difference when the total resistances, $r_a$ and $r_s$ are known.

This complex term can therefore be resolved, at least conceptually, into a transport component ($r_t$) and a 'biochemical' component ($r_x$). However, it is difficult to separate these two components: in principle, $r_t$ can be estimated as the distance between mesophyll cell walls and chloroplasts divided by an effective diffusion coefficient for $CO_2/HCO_3^-$ in water (cf. estimates of $r_a$), although the uncertainties about $CO_2$ movement within cells give little confidence to estimates of either term. For this reason, Edwards and Walker (1983) note, with characteristic candour, that $r_m$ is sometimes called the ignorance resistance. Jones (1983) suggests that $r_t$, with a value of around 0.2 s cm$^{-1}$, contributes little to $r_m$, which typically has values of 2–10 s cm$^{-1}$ ($C_3$ plants) and less than 2 s cm$^{-1}$ ($C_4$ plants) (Wilson 1973; Jones 1983): the reasons for this difference between plant types will be considered later (sect. 3.4.2). Raven and Glidewell (1981) similarly conclude that the major part of the mesophyll resistance can be attributed to the activity of RuBP carboxylase, and that the intracellular transport resistance is greatly reduced by the involvement of CA.

The 'biochemical' component of $r_m$ differs from the transport component, and from $r_s$ and $r_a$, in that it does not represent a physical resistance to diffusion; it is a consequence of, among other things, enzyme kinetics, for which the resistance analogy is inappropriate. How then can $r_x$ (or, indeed $r_m$, if $r_x$ is its dominant component) be interpreted, other than merely as a name and value assigned to the unaccounted for part of the denominator in eqn [3.6]? $r_m$ can, in fact, be interpreted in terms of a carboxylation efficiency (Ku and Edwards 1977) and, since it is clear that reduced efficiency in the use of available $CO_2$ impairs net photosynthesis, its use as a 'resistance' in eqn [3.6] is justified. The relationship of $r_m$ to the efficiency of carboxylation will now be examined in order to establish that $r_m$ does have a recognisable physiological identity.

Since the resistances to $CO_2$ assimilation are linked in series, the relationship of each one to assimilation can be treated separately. Thus, referring to Fig.3.6,

$$P_n = \frac{[CO_2]_i - [CO_2]_c}{r_m} \qquad [3.7]$$

When $r_m$ is estimated from eqn [3.7] a value must be assumed for $[CO_2]_c$: this is now usually assumed to be the $CO_2$ compensation concentration ($\Gamma$), at which $P_g$ exactly balances $R + PR$ so that $P_n$ is zero. Thus,

$$P_n = \frac{[CO_2]_i - \Gamma}{r_m} \qquad [3.8]$$

or

$$r_m = \frac{[CO_2]_i - \Gamma}{P_n} \qquad [3.9]$$

The photochemical component of $r_m$ will be eliminated under high irradiance, and the carboxylation efficiency (CE) will be reflected in the slope of the line relating net photosynthesis ($P_n$) to increasing $[CO_2]_i$ in the linear, $CO_2$-limited part of the curve (Fig. 3.12a). Therefore:

$$CE = \frac{P_n}{[CO_2]_i - \Gamma} \qquad [3.10]$$

Comparison of eqns [3.9] and [3.10] reveals that $r_m = 1/CE$.

It should be noted that this definition is more restricted than that given earlier (p. 46), in that it does not include a photochemical component; $r_m$ as defined by eqn [3.9] depends on the transport resistance in the liquid phase ($r_t$) and the carboxylase activity (the biochemical component, or

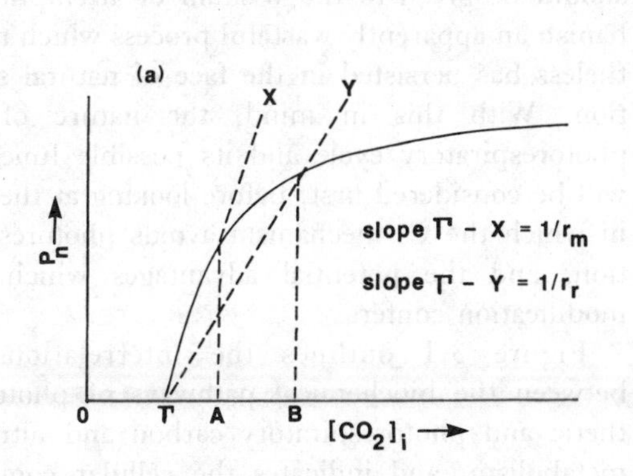


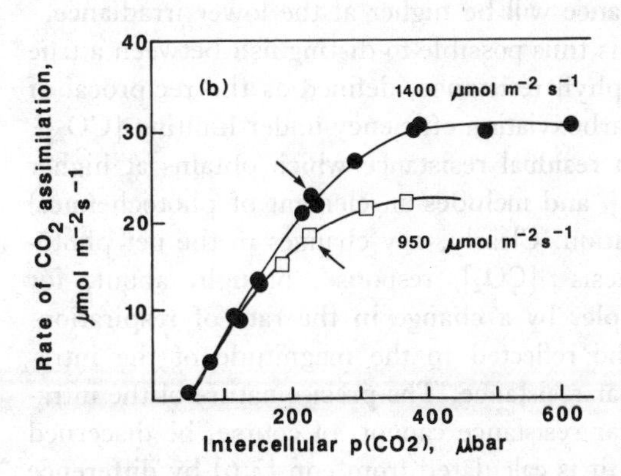


**Fig. 3.12** (a) The response of net photosynthesis to the intercellular $CO_2$ concentration, and the dependence of the intracellular resistance on $[CO_2]_i$. The initial slope of the response, in the range of $[CO_2]_i$ from Γ to A, defines the carboxylation efficiency; $r_m$ is the reciprocal of this slope. At higher $[CO_2]_i$, e.g. B, the mean slope is reduced, and its reciprocal gives the residual resistance rather than the 'true' mesophyll resistance. See text for details (after Jones 1983).

(b) The response of net photosynthesis of *Phaseolus vulgaris* leaves to the intercellular partial pressure of $CO_2$ at different photosynthetic photon flux densities (PPFDs). The reduction in PPFD reduced the $p(CO_2)$ at which the transition from the linear to the curved part of the $P_n$ : $[CO_2]_i$ response occurred from 220 μbar to 150 μbar. The arrows indicate the points obtained at an external $p(CO_2)$ of 330 μbar. The intracellular resistance would therefore be higher at the lower PPFD. The partial pressure of $CO_2$ ($p(CO_2)$, μbar) is approximately equal to mole fraction ($[CO_2]_i$, $\mu l\ l^{-1}$) or volume fraction ($[CO_2]_i$, vpm) (after von Caemmerer and Farquhar 1981).

carboxylation resistance, $r_x$). However, the response of net photosynthesis to $[CO_2]_i$ shows a characteristic transition from the initial, linear, $CO_2$-limited part, whose slope is believed to be proportional to RuBPcarboxylase activity, to a curved region over which increasing $[CO_2]_i$ has progressively less effect. It is suggested that at these higher, non-limiting intercellular $CO_2$ concentrations, photosynthesis is limited by the rate of regeneration of RuBP, which in turn depends on the capacity for electron transport and hence, among other things, on the amount of absorbed light (Farquhar *et al.* 1980; von Caemmerer and Farquhar 1981). If the $[CO_2]_i$ is such that the rate of photosynthesis is on the curved part of the net photosynthesis : $[CO_2]_i$ response, and in plants well adapted to their environments this may well be the case (Farquhar and Sharkey 1982; Jones 1983), the intracellular resistance derived from eqn [3.9] will be higher because the mean slope of the curve is lower (Fig. 3.12a). This effect is illustrated by Fig. 3.12b, which shows that when irradiance is reduced, the transition in the photosynthetic response from $[CO_2]_i$ limitation to limitation by RuBP regeneration occurs at lower $[CO_2]_i$. For a given $[CO_2]_i$ above the transition concentration, the intracellular resistance will be higher at the lower irradiance.

It is thus possible to distinguish between a true mesophyll resistance, defined as the reciprocal of the carboxylation efficiency under limiting $[CO_2]_i$, and a residual resistance which obtains at higher $[CO_2]_i$ and includes an element of photochemical limitation. Clearly, any changes in the net photosynthesis : $[CO_2]_i$ response, brought about, for example, by a change in the rate of respiration, will be reflected in the magnitude of the intracellular resistance. The precise nature of the intracellular resistance cannot, of course, be discerned when it is calculated from eqn [3.6] by difference when all the other terms are known (or assumed); for this, estimates of $[CO_2]_i$ are necessary, as explained above, and details of its estimation and measurement are available (e.g. Sharkey 1985). However, when considering experiments involving intracellular resistances, it is important to have in mind the physiological basis of the terms under discussion, as will be clear, for example, from a consideration of the effects of leaf age or of water stress on photosynthesis.

## 3.4 ENDOGENOUS AND ENVIRONMENTAL FACTORS AFFECTING PHOTOSYNTHESIS

### 3.4.1 PHOTORESPIRATION

A definition of photorespiration and a general assessment of its significance to the carbon economy of a crop have already been given (sect. 3.1.1). The realization that the elimination of photorespiration in $C_3$ plants allows increases in net photosynthesis of up to 50 per cent (demonstrated simply by lowering the oxygen concentration to 2 per cent) has stimulated much research into ways of reducing photorespiration chemically or through breeding. $C_4$ plants, on the other hand, have negligible rates of photorespiration as a result of the evolution of a specialized biochemical and anatomical organization. Therefore, in considering the potential advantages to be gained from the elimination of photorespiration, it is worth comparing the productivities, or rates of canopy photosynthesis, of $C_3$ and $C_4$ crops. Care must be taken in the interpretation of such comparisons, however, because of the differences in the environments to which these two groups of plants are adapted. At the same time, due regard should be given to the wisdom of attempting to banish an apparently wasteful process which nonetheless has persisted in the face of natural selection. With this in mind, the nature of the photorespiratory cycle and its possible functions will be considered first, before looking at the way in which the $C_4$ mechanism avoids photorespiration, and the potential advantages which this modification confers.

Figure 3.1 outlines the interrelationships between the biochemical pathways of photosynthetic and photorespiratory carbon and nitrogen metabolism, and indicates the cellular compartments involved. The photorespiratory cycle (variously referred to as the $C_2$ cycle, the glycollate pathway or the photorespiratory carbon oxidation cycle) is initiated in the chloroplasts by the oxygenation of RuBP to form the two-carbon compound, glycollate. This is transported to the peroxisomes and there converted into the amino acid, glycine, which in turn is transported to the mitochondria and decarboxylated to give serine, carbon dioxide and ammonia. Carbon is diverted back into the Calvin cycle, however, via the conversion of serine to glycerate in the peroxisomes and further metabolism to phosphoglycerate in the chloroplasts. Details of the biochemistry of photorespiration and of the history of its elucidation may be sought elsewhere (e.g. Tolbert 1980; Edwards and Walker 1983); the brief summary given here provides an adequate background to a consideration of the roles of photorespiration.

It is now generally believed that the oxygenation of RuBP is an unavoidable consequence of the active-site chemistry of RuBP carboxylase (which in this context must therefore be called RuBP carboxylase-oxygenase or, sometimes, RuBisCO): carboxylation and oxygenation of RuBP are catalysed at the same site on the enzyme, and are therefore competitively inhibited by oxygen and carbon dioxide, respectively (Andrews and Lorimer 1978). The extent of each reaction therefore depends on the relative concentrations of $CO_2$ and oxygen at the active site, and also on the temperature. If carboxylation is thus inevitably accompanied by some oxygenation, the major function of photorespiration is clear: it is a scavenging process which diverts some of the carbon initially lost as glycollate back into the Calvin cycle. Up to 75 per cent of the carbon which leaves the Calvin cycle as a result of RuBP oxygenation is recovered in this way (Fig. 3.13). The greater part of photorespiratory metabolism must therefore be regarded as beneficial, a means of partially making good the wasteful consequences of RuBP oxygenation. Increases in net photosynthesis are likely to result, therefore, only from suppression of this oxygenase activity, as happens when the ambient oxygen concentration is reduced; suppression of the subsequent glycol-

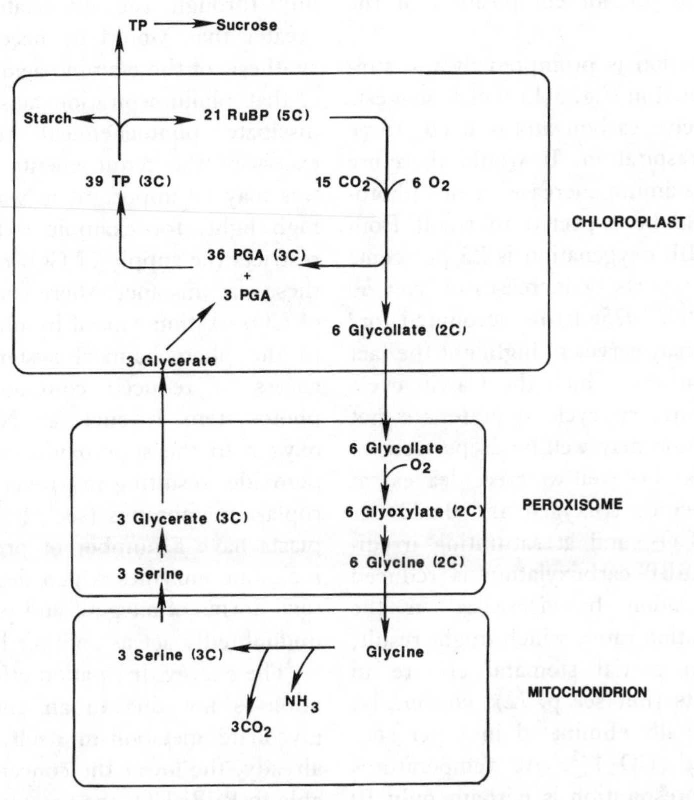


**Fig. 3.13** The stoichiometry of photosynthesis and photorespiration. The number of carbon atoms per molecule of each compound is shown. The return of three PGA molecules to the Calvin cycle represents a 75 per cent recovery of the carbon diverted out of the Calvin cycle as glycollate. However, the relative rates at which the Calvin cycle and the photorespiratory cycle operate are not fixed, and the stoichiometry therefore varies. Abbreviations are as in Fig. 3.1 (after Keys and Whittingham 1981).

late metabolism is equally likely to be counterproductive.

Two questions immediately arise. The first concerns the evolution of an enzyme whose primary function in catalysing the fixation of $CO_2$, a relatively scarce substrate, is partially undone by the presence of oxygen, a much more prevalent alternative substrate. Note that the concentration ratio of $CO_2 : O_2$ dissolved in water at 25 °C, and in equilibrium with atmospheric concentrations, is 1 : 24. The atmospheric concentrations, in fact, show even greater dominance by oxygen, whose solubility in water is less than that of $CO_2$. However, most authorities suggest that RuBisCO's catalytic properties were established through evolution in atmospheres which contained much more $CO_2$ and much less oxygen, when the oxygenase activity would have been insignificant. It is thus ironic that photosynthetic plants themselves have been responsible for the progressive change towards the present composition of the atmosphere.

The second question is prompted by the stoichiometry represented in Fig. 3.13 which suggests that, for every twelve carbon atoms fixed, three are lost by photorespiration. It would therefore appear that the maximum increase in net photosynthesis that could be expected to result from suppression of RuBP oxygenation is 25 per cent; how then can the reports of increases of over 50 per cent (e.g. Zelitch 1975a,b) be accounted for? This apparent anomaly serves to highlight the fact that the relative rates at which the Calvin cycle and the photorespiratory cycle operate are not fixed. Photorespiration may well be 25 per cent of net photosynthesis in well-watered leaves at 20–25 °C, in 21 per cent oxygen and 0.033 per cent (330 $\mu l\ l^{-1}$) $CO_2$, and at saturating irradiance. However, RuBP carboxylation is reduced relative to oxygenation by decreases in the $CO_2 : O_2$ concentration ratio, which might result, for example, from partial stomatal closure in water-stressed plants (but see p. 72); conversely, oxygenation is virtually eliminated in 1 per cent oxygen or 1000 $\mu l\ CO_2\ l^{-1}$. At temperatures below 15 °C, photorespiration is perhaps only 10 per cent of net photosynthesis, but may be over 50 per cent at 35 °C. This pronounced effect is thought to be due to the differential response to temperature of the carboxylase and oxygenase reactions: the affinity of the enzyme for $CO_2$ in the carboxylase reaction is reduced almost threefold as the temperature increases over the range 15–35 °C, whereas the oxygenase reaction is little affected (Laing *et al.* 1974). Environmental conditions thus have a considerable effect on the relative magnitude of photorespiration and, therefore, influence the stimulatory effect on net photosynthesis of the suppression of photorespiration.

Thus, if oxygenation of RuBP is unavoidable in normal atmospheres, glycollate metabolism can be viewed simply as a means of alleviating the draining effect of this on net carbon fixation. However, it has also been suggested that photorespiration has beneficial, or even essential, effects in its own right. The production of metabolic intermediates, such as glycine and serine, may be useful, but this cannot be more than a secondary function of photorespiration because the carbon flux through the glycollate pathway is much greater than would be necessary merely for the synthesis of these amino acids. Another possibility is that photorespiration acts as a safety-valve to dissipate photochemical energy generated in excess of the requirements of the Calvin cycle: this may be important in water-stressed plants in high light, for example, when stomatal closure restricts the supply of $CO_2$ to the chloroplasts. In these circumstances there is a danger of inhibition of $CO_2$ fixation caused by photo-oxidative damage to the photochemical system. For example, an excess of reduced compounds associated with photosystem I, such as NADPH, can reduce oxygen to the superoxide radical or to hydrogen peroxide, resulting in irreversible damage to chloroplast membranes (see Halliwell 1981). Chloroplasts have a number of protective mechanisms, including enzymes which degrade these and other toxic forms of oxygen, and photorespiration could undoubtedly act as another line of defence.

The energy dissipation effected by photorespiration is not due to an energy requirement of glycollate metabolism itself. As has been noted already, the lower the concentration of $CO_2$ available to RuBisCO, the greater is the flow of carbon through the glycollate pathway. In the extreme case, the carbon flux through the glycollate cycle will be increased, relative to that through the Calvin cycle, to such an extent that the amount of photorespired $CO_2$ exactly matches that combining with RuBP. Net $CO_2$ fixation is therefore zero, and $CO_2$ is effectively recycled from the glycollate pathway to the Calvin cycle. Thus, the Calvin cycle continues to function, with its attendant consumption of energy, but with no net fixation of $CO_2$. This internal recycling of $CO_2$ also ensures that RuBP carboxylase does not become inactivated, as happens when it is exposed to RuBP in the absence of $CO_2$. Even when the $CO_2$ concentration is high enough to support net carbon fixation, any increase in photorespiration will increase the energy requirement per molecule of $CO_2$ fixed, and thus reduce the quantum yield.

Powles and Osmond (1978) demonstrated that inhibition of photosynthesis did indeed occur when intact, attached bean leaflets were strongly illuminated in $CO_2$-free nitrogen containing 1 per cent oxygen. Under these conditions, photochemical energy could neither be used for $CO_2$ reduction nor dissipated by photorespiration. This photoinhibition was not observed in $CO_2$-free atmospheres containing between 7 and 21 per cent oxygen, presumably because photorespiration generated $CO_2$ and resulted in the integrated operation of the Calvin cycle and photorespiration, and the consequent utilization of energy. This explanation is supported by the fact that there was no inhibition in leaflets in atmospheres containing 1 per cent oxygen and sufficient $CO_2$ to maintain an intercellular $CO_2$ partial pressure of 62 $\mu$bar (6.2 Pa). Photorespiration would thus have been suppressed, but there was apparently sufficient $CO_2$ to support the turnover of the Calvin cycle and the consequent orderly consumption of energy. It is perhaps no mere coincidence that this level of $CO_2$ is the $CO_2$ compensation point of bean leaflets in air, the level at which, under normal circumstances, the internal $CO_2$ partial pressure would equilibrate when $CO_2$ supply was blocked by closed stomata.

## 3.4.2 AVOIDANCE OF PHOTORESPIRATORY CARBON LOSSES IN $C_4$ PLANTS

### 3.4.2.1 THE $C_4$ MECHANISM

$C_4$ plants minimize photorespiration by concentrating $CO_2$ at the site of RuBisCO to such an extent that the oxygenase function is largely suppressed. RuBisCO and the other enzymes of the Calvin cycle are situated almost exclusively in the so-called Kranz cells, which are large, have thick walls and prominent chloroplasts, and form a layer, or wreath (German: *kranz*), around the vascular bundles. The alternative name, bundle-sheath cells, is less precise in that many $C_3$ plants also have a distinct layer of cells around the vascular bundles; here, however, the features which characterize Kranz cells are lacking. The mesophyll cells which surround the Kranz cells are also rich in chloroplasts, but although this tissue is the site of the initial fixation of $CO_2$ in $C_4$ leaves, this does not occur in the chloroplasts; RuBisCO is not the enzyme responsible and sugars cannot be synthesized there.

In fact, the primary carboxylation in all $C_4$ plants is the conversion of phosphoenolpyruvate (PEP) to oxaloacetate, catalysed by PEP carboxylase. This enzyme is restricted to the cytosol of the mesophyll cells, is not present in significant amounts in the Kranz cells, and although it is also present in $C_3$ leaves its activity there is low. The pool-size of oxaloacetate is very small, however, and its turnover very fast (Edwards and Walker 1983), which means that the major labelled compounds first detected after feeding $^{14}CO_2$ to leaves are the immediate products of the metabolism of oxaloacetate. Thus, $C_4$ plants can be grouped according to whether they are predominantly malate-formers or aspartate-formers. It is the translocation of these 4-carbon acids from the mesophyll cells to the Kranz (bundle-sheath) cells, and their subsequent decarboxylation, which brings about the concentration of $CO_2$, which is the salient feature of the $C_4$ syndrome. A more specific classification groups $C_4$ plants according to the enzyme which catalyses the release of $CO_2$ in the bundle-sheath cells: malate-formers use NADP-malic enzyme (NADP-ME types), while aspartate-formers use PEP-carboxykinase (PEP-CK types) or NAD-malic enzyme (NAD-ME types). The metabolism involved is summarized in Fig. 3.14; it should be noted that a given $C_4$ species is predominantly, but not exclusively, either a malate- or an aspartate-former and, furthermore, that other minor pathways of carboxylation and decarboxylation are also used.

The translocation of malate and aspartate, and indeed other metabolites, between the mesophyll

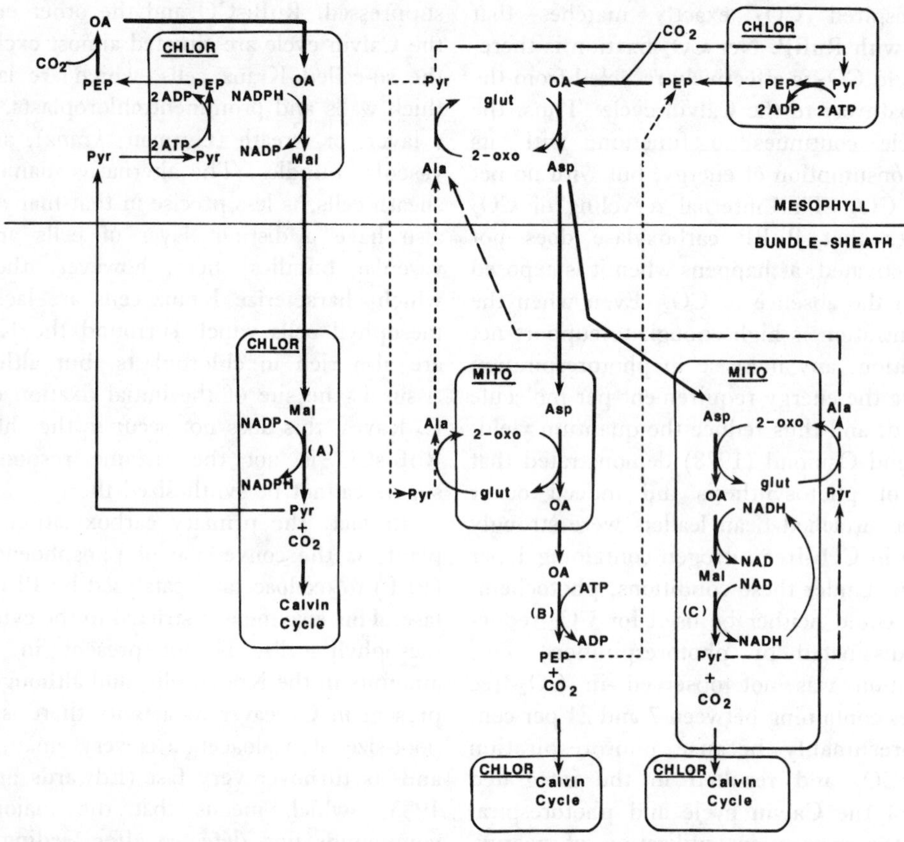


**Fig. 3.14** $CO_2$ transfer from mesophyll cells to bundle-sheath cells in $C_4$ plants.

$C_4$ metabolism involves carboxylation of PEP in the cytosol or chloroplasts of the mesophyll cells and translocation of either malate or aspartate to the bundle-sheath cells, where decarboxylation occurs in the chloroplasts, cytosol or mitochondria. Other metabolites, predominantly pyruvate, alanine and PEP, are also shuttled between the mesophyll and bundle-sheath.

$C_4$ species are classified according to the major pathway of decarboxylation, involving either (A) NADP-malic enzyme (e.g. maize, sugar cane), or (B) PEP-carboxykinase (e.g. *Panicum maximum*) or (C) NAD-malic enzyme (e.g *Amaranthus edulis*). The species also differ in the way in which PEP is regenerated in the mesophyll, and in the associated shuttles of metabolites: pathways restricted to PEP-CK types, ------; pathways restricted to NAD-ME types, ——.

Abbreviations: Ala, alanine; Asp, aspartate; glut, glutamate; mal, malate; 2-oxo, 2-oxoglutarate; OA, oxalacetate; PEP, phosphoenolpyruvate; Pyr, pyruvate; CHLOR, chloroplast; MITO, mitochondrion. Reactions not included within chloroplasts or mitochondria occur in the cytosol.

Note that bicarbonate ($HCO_3^-$) is the actual substrate for PEP carboxylase, and that rapid equilibration between $CO_2$ and $HCO_3^-$ is facilitated by carbonic anhydrase.

and bundle-sheath is assumed to be by diffusion through plasmodesmata, but details are speculative. However, the $C_4$ metabolism, by which $CO_2$ is effectively 'pumped' from mesophyll to bundle-sheaths, nonetheless imposes an energy penalty in terms of the ATP and/or NADPH requirements of particular reactions. These are indicated in Fig. 3.14, and are summarized in Table 3.4 together with the additional energy requirements of the Calvin cycle. The resultant estimates of the total photochemical requirements of the different types of $C_4$ plants will be considered later in relation to measured quantum yields; for the moment, Fig. 3.14 prompts further comparison of the different types of $C_4$ metabolism.

First, the decarboxylation of malate by NADP-malic enzyme generates one NADPH in the bundle-sheath chloroplasts, which partially meets

**Table 3.4** Photochemical requirements in $C_4$ plants per $CO_2$ molecule fixed

| *Group* | *Energy* | *Requirements in $C_4$ cycle* | *Additional requirements for Calvin cycle* | *Total per $CO_2$ molecule fixed* |
|---|---|---|---|---|
| NADP-ME | ATP | 2 | 3 | 5 |
| | NADPH | 1 | 1* | 2 |
| PEP-CK | ATP | 1 | 3 | 4 |
| | NADPH | 0 | 2 | 2 |
| NAD-ME | ATP | 2 | 3 | 5 |
| | NADPH | 0 | 2 | 2 |

* The Calvin cycle requires 3 ATP and 2 NADPH molecules for each $CO_2$ molecule fixed; however, there is a net generation of 1 NADPH during malate decarboxylation by NADP-malic enzyme.
(After Edwards and Walker 1983)

the Calvin cycle's requirement for reducing power. Even so, these bundle-sheath chloroplasts are unable to produce the rest of the necessary reducing power because of poorly developed grana and deficiencies in photosystem II. In fact, the bundle-sheath chloroplasts of the NADP-ME types show a range of granal development from the rudimentary thylakoid stacking of maize to the completely agranal forms in sorghum and sugar-cane. In the latter case, the bundle-sheath chloroplasts produce no reducing power. This deficiency could be overcome if some of the phosphoglycerate produced in the Calvin cycle (see Fig. 3.1) by the carboxylation of RuBP were shuttled back to the mesophyll chloroplasts for the reducing reactions of the cycle, and the resulting triose phosphate (glyceraldehyde 3-phosphate or its isomer dihydroxyacetone phosphate) returned to the bundle-sheath to continue the Calvin cycle. The only enzymes of the Calvin cycle found in the mesophyll cells are, in fact, those responsible for the conversion of PGA to triose phosphates.

Such a shuttle is not necessary in the PEP-CK and NAD-ME types, in which the bundle-sheath chloroplasts have normal granal development and are capable of producing sufficient NADPH for the Calvin cycle. More puzzling in these species is the fate of the NADPH produced in the mesophyll chloroplasts or, indeed, in the PEP-CK types, the function of the chloroplasts at all. One possibility is that most of the reducing power is consumed by nitrate reduction, known to occur mainly in the mesophyll of $C_4$ leaves (Halliwell 1981).

#### 3.4.2.2 QUANTUM YIELDS OF $C_3$ AND $C_4$ SPECIES

$C_4$ metabolism delivers $CO_2$ to the bundle-sheath cells at concentrations thought to be non-limiting for photosynthesis and sufficient to suppress photorespiration. The energy cost of photorespiration is thus avoided, but this advantage is counteracted to some extent by the energy requirements of the $C_4$ pathways. Differences in quantum yield between $C_3$ and $C_4$ plants might therefore be expected to depend on the particular pathway of $C_4$ metabolism and on the rate of photorespiration. Comparisons of quantum yields of fifty-five species, measured under normal atmospheric conditions (330 $\mu l\ l^{-1}$ $CO_2$, 21 per cent oxygen) and at a leaf temperature of 30 °C, showed that there was indeed considerable variation across photosynthetic pathway types, but also, within the $C_4$ species, a marked difference between monocots and dicots (Ehleringer and Björkman 1977; Ehleringer and Pearcy 1983). Under these conditions the $C_3$ dicots, $C_3$ grasses and NAD-ME type $C_4$ dicots had the lowest quantum yields (0.052, 0.053, 0.053, respectively) (Fig. 3.15); in contrast, NAD-ME type grasses had much higher quantum yields (average 0.060), and there was a similar distinction between NADP-ME type dicots and grasses.

Clearly, within the $C_4$ species the quantum yield is influenced by factors other than merely the inherent energy requirements of the particular subgroup pathway. For example, there is little difference in quantum yield between PEP-CK

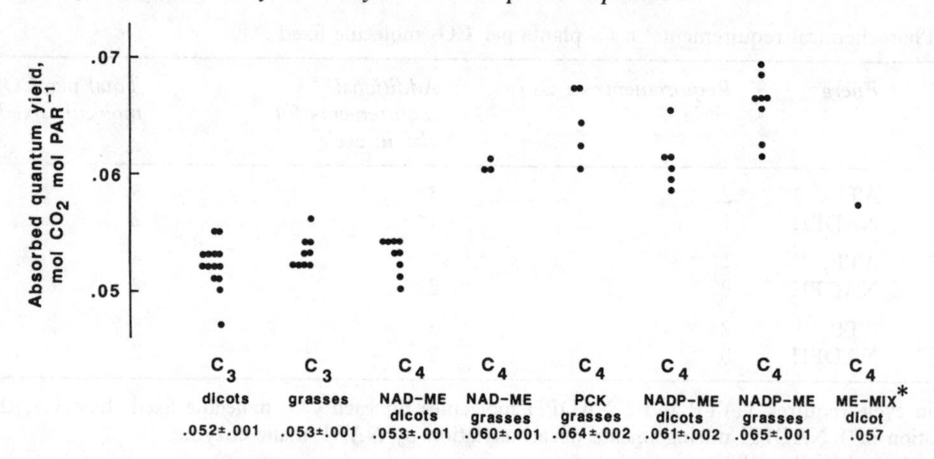


**Fig. 3.15** Absorbed quantum yields for $CO_2$ uptake for $C_3$ and $C_4$ species at a leaf temperature of 30 °C and 330 $\mu l\ l^{-1}$ $CO_2$, 21 per cent $O_2$. Data are classified according to pathway subtype, and each datum is the average for one species. Means and standard errors for each pathway subtype are also presented.
* This quantum yield was obtained from a mixed NAD-ME and NADP-ME dicot: note its intermediate value (from Ehleringer and Pearcy 1983).

type grasses and NADP-ME type grasses, although reference to Table 3.4 suggests that PEP-CK pathway photosynthesis should be more efficient. Ehleringer and Pearcy (1983) suggested that leakage of $CO_2$ from the bundle-sheath cells could affect the quantum yield because refixation of this $CO_2$ into $C_4$ acids would again require photochemical energy. The extent of this leakage is unknown, but may depend on the degree of suberization of the cell walls: suberized lamellae are found only in the NADP-ME and PEP-CK grasses, which might therefore have less $CO_2$ leakage than the NAD-ME grasses or the $C_4$ dicots.

The same argument applies to the limited amount of $CO_2$ which may be generated by photorespiration in bundle-sheath cells and refixed, either in the bundle-sheath chloroplasts or in the mesophyll. The RuBP carboxylase of the bundle-sheath cells has oxygenase activity, and all but one of the enzymes of the glycollate pathway are located in the bundle-sheath (the other being in the mesophyll). There will therefore be some photorespiration if oxygen is present in the bundle-sheath cells, and this is least likely in the NADP-ME species, which have reduced granal development. It is therefore possible to identify sources of efficiency loss which will vary in extent according to particular characteristics of each $C_4$ subgroup pathway. The significance of other causes of reduced efficiency, such as energy consumption for processes not involved in C fixation (e.g. nitrogen assimilation), may also vary between representatives of the different photosynthetic pathways.

The importance of photorespiration in reducing the quantum yield of $C_3$ plants is apparent in experiments in which photorespiration is suppressed by lowering the oxygen concentration from 21 per cent to 2 per cent: at 30°C this treatment increased the quantum yield of a number of $C_3$ species by about 40 per cent, from an average of 0.052 to 0.073 (Ehleringer and Björkman 1977). Since temperature is the factor most likely to affect photorespiration under natural conditions, it will clearly have a large effect on the quantum yields of $C_3$ plants: thus, the quantum yield of oats (*Avena sativa*) was increased from 0.043 to 0.073 when the leaf temperature was reduced from 36 °C to 16 °C; $C_4$ plants, on the other hand, were unaffected over this range of temperature (Ehleringer and Pearcy 1983). Therefore, when the temperature response of quantum yield is compared in $C_3$ and $C_4$ plants, there is a crossover temperature at which the loss of efficiency due to photorespiration in the $C_3$ species is matched by the higher intrinsic energy costs of the $C_4$ species: at this temperature, the quantum yields are the same. Below this crossover temperature, net photosynthesis is more efficient in the $C_3$ species,

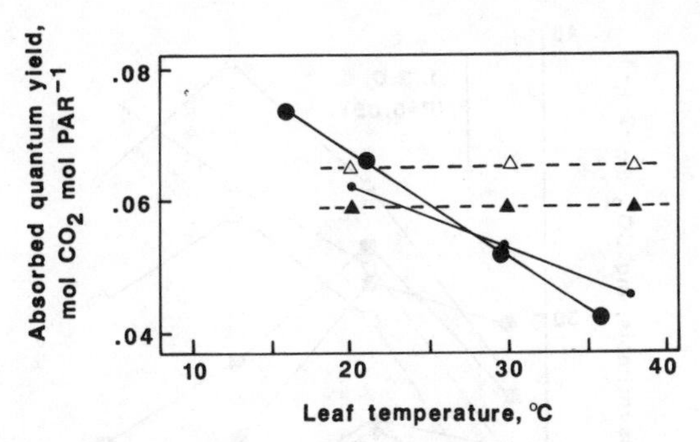


**Fig. 3.16** The leaf temperature dependence of the absorbed quantum yield for $CO_2$ uptake in $C_3$ and $C_4$ grasses. Measurements were made under normal atmospheric conditions (330 $\mu l\ l^{-1}$ $CO_2$, 21 per cent $O_2$). $C_3$ species: ●, *Avena sativa* (oats); •, *Hordeum vulgare* (barley). $C_4$ species: ▲, *Zea mays* (maize); △, *Paspalum vaginatum* (adapted from Ehleringer and Pearcy 1983).

while above this temperature the $C_4$ species has the advantage (Fig. 3.16). Clearly the crossover temperature will depend very much on the species being compared: Ehleringer and Pearcy (1983) found that it ranged between 16 and 28 °C.

### 3.4.2.3 EFFECTS OF IRRADIANCE AND TEMPERATURE ON NET PHOTOSYNTHESIS OF $C_3$ AND $C_4$ SPECIES

The implication of these temperature effects on quantum yield is that at lower temperatures, and when light is limiting, the rate of photosynthesis of $C_3$ plants will be greater than that of $C_4$ plants. Indeed, additional factors, such as inactivation of certain enzymes, reduced enzyme activity and impaired metabolite transport at low temperature and the fact that, in some $C_4$ plants at least, the quantity of RuBisCO is only about a quarter of that in $C_3$ plants, mean that at low temperatures (less than 15 °C) photosynthesis of $C_4$ plants may be less than that of $C_3$ plants even at saturating irradiances (Fig. 3.17; cf. Fig. 2.23). However, the light-saturated rate of photosynthesis in $C_3$ leaves increases only slightly with increased temperature in the range 15–30 °C, because $CO_2$ is limiting and photorespiration is enhanced; at higher temperatures, this maximum rate of photosynthesis is reduced.

Photorespiration increases as temperature increases for two reasons: first, a rise in temperature reduces the solubility of $CO_2$ more than that of $O_2$, reducing the $CO_2/O_2$ ratio at the sites of RuBisCO; and secondly, because of the direct effect of temperature on the carboxylase activity of the enzyme (p. 50). There may, however, be other causes. Lehnherr *et al.* (1985) have shown that at low temperatures (less than 20 °C) the ratio of photorespiration to photosynthesis in white clover is less than would be predicted from the measured *in vitro* oxygenase and carboxylase activities of RuBisCO, and suggest that this could be due to an unidentified $CO_2$-concentrating system, perhaps operating at the plasmalemma or chloroplast envelope. It was not apparent at

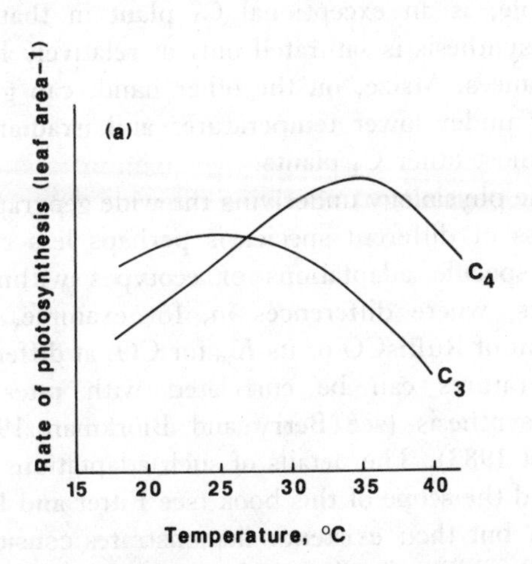


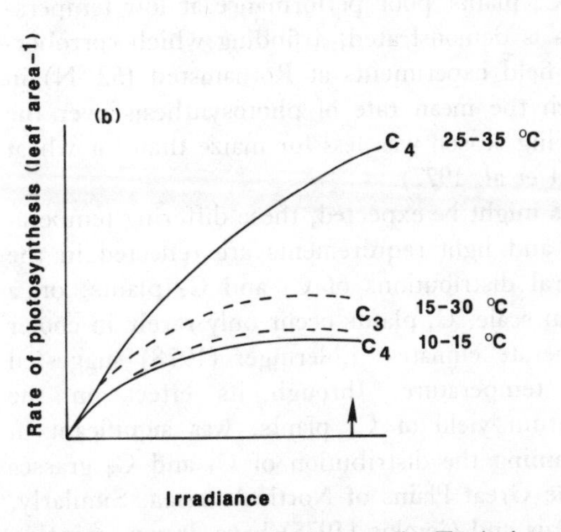


**Fig. 3.17** (a) Typical temperature response curve for photosynthesis in $C_3$ and $C_4$ plants. (b) Typical photosynthetic light response curves for leaves of $C_3$ and $C_4$ plants at different temperatures. The arrow indicates full sunlight (after Edwards and Walker 1983).

30 °C, possibly because increased membrane permeability renders it ineffective. In $C_4$ plants, photorespiration is, of course, largely prevented by the $CO_2$-concentrating mechanism over the range of physiological temperatures: there is, therefore, a pronounced increase in the rate of photosynthesis as temperature increases, with a comparatively high optimum temperature (about 35 °C as compared with 25 °C in $C_3$ plants). At this optimum, light is the major limiting factor, and indeed photosynthesis remains unsaturated up to full sunlight (Fig. 3.17).

Maximum photosynthesis in $C_4$ plants therefore occurs at high temperatures and high irradiances, because the $C_4$ pathway alleviates the effects of low $CO_2$ concentration at the sites of RuBisCO. On the other hand, photosynthesis in $C_3$ plants is maximized at relatively low temperatures and irradiances, and under these conditions may well exceed $C_4$ photosynthesis. This effect of temperature is illustrated by Fig. 3.18. For wheat ($C_3$) the greatest capacity for photosynthesis was developed at day/night temperatures of 13/10°C or 18/14 °C, whereas maize ($C_4$) developed its most effective leaves at 23/18 °C. These growth conditions did not affect the optimum temperature for photosynthesis, which was 13 or 18 °C in wheat and 23 °C in maize. Greater rates of photosynthesis in maize, at higher optimum temperatures, would undoubtedly have occurred if the PPFD had been higher: at 580 $\mu$mol $m^{-2}$ $s^{-1}$ (PAR) it was less than one-third of full sunlight (cf. Fig. 3.17). Nonetheless, the $C_4$ plants' poor performance at low temperatures is demonstrated, a finding which corroborates field experiments at Rothamsted (52 °N) in which the mean rate of photosynthesis over the growing season was less for maize than for wheat (Bird *et al.* 1977).

As might be expected, these differing temperature and light requirements are reflected in the natural distributions of $C_3$ and $C_4$ plants: on a global scale, $C_4$ plants occur only rarely in cooler temperate climates. Ehleringer (1978) suggested that temperature, through its effect on the quantum yield of $C_3$ plants, was significant in explaining the distribution of $C_3$ and $C_4$ grasses in the Great Plains of North America. Similarly, Loomis and Gerakis (1975) have drawn attention to the higher productivity of $C_4$ plants at lower latitudes and the loss of this advantage at latitudes greater than 40–50°. Such geographic distinctions can be made only on a broad scale because $C_3$ and $C_4$ plants may be adapted to increasingly unfavourable environments towards the edges of their natural geographic ranges. The sunflower, for example, is an exceptional $C_3$ plant in that its photosynthesis is saturated only at relatively high irradiances. Maize, on the other hand, can grow better under lower temperatures and irradiances than most other $C_4$ plants.

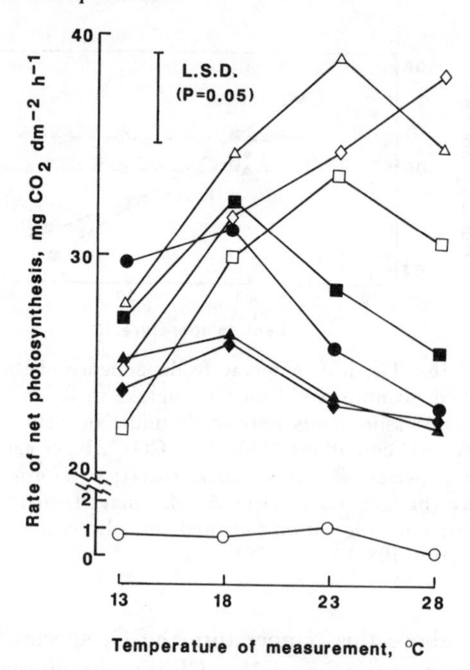


**Fig. 3.18** Effect of temperature on rates of net photosynthesis of leaves of spring wheat (closed symbols) and maize (open symbols), grown at different temperatures: 13/10 °C (○, ●), 18/14 °C (□, ■), 23/18 °C (△, ▲), and 28/22 °C (◊, ♦) (from Bird *et al.* 1977).

The physiology underlying the wide geographic success of different species is perhaps less clear than specific adaptations of ecotypes within a species, where differences in, for example, the amount of RuBisCO or its $K_m$ for $CO_2$ at different temperatures can be correlated with rates of photosynthesis (see Berry and Björkman 1980; Öquist 1983). The details of such adaptations are beyond the scope of this book (see Fitter and Hay 1987), but their existence demonstrates considerable flexibility in the environmental control of photosynthesis. The understanding of such flexibility is important if crops are to be bred for areas in which they do not at present succeed or even for recognizing the existing potential of a crop for a particular environment. As Good and Bell (1980) point out, changes in our requirements of crops, through attempts to extend their geographic distribution or through changing technology or climate, provide most scope for large improvements through breeding.

#### 3.4.2.4 WATER USE EFFICIENCY

The photosynthetic advantage conferred by the $C_4$ pathway under high temperatures and irradiance is reinforced by superior water use efficiency (WUE, grams of water transpired/gram of $CO_2$ fixed). Since water vapour diffuses out of a leaf by the same route that $CO_2$ diffuses into the leaf, loss of water is an inevitable accompaniment of $CO_2$ uptake. For a given stomatal resistance, more water will be lost than $CO_2$ gained because the concentration gradient of water vapour will almost always be greater than that of $CO_2$, water vapour has a higher diffusion coefficient than $CO_2$, and $CO_2$ diffusion is impeded by an additional resistance ($r_m$) not encountered by water vapour (Fig. 3.6). It may be recalled (p. 46) that this resistance reflects the leaf's capacity for net photosynthesis at the cellular level. Thus, a decrease in $r_m$, for a given stomatal resistance, will increase photosynthesis without affecting transpiration, thereby increasing the WUE. In general, $C_4$ plants have lower mesophyll resistances than $C_3$ plants, and a tendency towards higher stomatal resistances (Table 3.5), characteristics which clearly favour a high WUE. The $C_4$ mechanism also results in a lower intercellular $CO_2$ concentration, which consequently gives a higher flux of $CO_2$ into the leaf for a given external $CO_2$ concentration and stomatal resistance. The advantage of $C_4$ plants in this respect is illustrated by measurements from an extensive series of field experiments done between 1911 and 1917 (Table 3.6).

The lower $r_m$ of $C_4$ plants is attributed largely to lack of photorespiration. The reason for the slightly higher $r_s$ of $C_4$ plants is not clear, but may be due to there being fewer stomata per unit area of leaf or smaller stomatal apertures: it is known that $C_4$ plants require more light for maximum stomatal aperture than $C_3$ plants. A high WUE is often particularly important at high temperatures. Increasing temperatures accentuate the $C_4$ plant's advantage in this respect because of the effects of temperature on the carboxylase and oxygenase activities of RuBisCO, and thus on $r_m$.

**Table 3.5** Reported minimal values of mesophyll resistance ($r_m$) and gaseous resistance ($r_a + r_s$) in some $C_3$ and $C_4$ crops

| *Species* | $r_m$ (*s cm$^{-1}$*) | $r_a + r_s$ (*s cm$^{-1}$*) |
|---|---|---|
| $C_3$ | | |
| *Glycine max* (soybean) | 2–3 | 1.3 |
| *Phaseolus* spp. (bean) | 2.6 | 1.1 |
| *Triticum aestivum* (bread wheat) | 2.8 | 1.1 |
| *Solanum tuberosum* (potato) | 5.4 | 2.5 |
| *Medicago sativa* (lucerne, alfalfa) | 2.8 | 1.0 |
| $C_4$ | | |
| *Zea mays* (maize) | 0.7–0.9 | 1.9 |
| *Saccharum officinarum* (sugar cane) | 0.3 | 1.5 |
| *Sorghum bicolor* (sorghum) | 1.8 | 1.8 |
| *Panicum virgatum* (switchgrass) | 2.0 | 2.9 |
| *Pennisetum purpureum* (Napier grass) | 0.6 | 1.5 |

(After Edwards and Walker (1983), where sources of original data are cited)

**Table 3.6** Water use efficiency of some $C_3$ and $C_4$ crops

| *Species* | *g water used/g dry matter produced* |
|---|---|
| $C_3$ | |
| *Hordeum vulgare* (barley) | 518 |
| *Triticum durum* (durum wheat) | 542 |
| *Triticum aestivum* (bread wheat) | 557 |
| *Solanum tuberosum* (potato) | 575 |
| *Avena sativa* (oats) | 583 |
| *Secale cereale* (rye) | 634 |
| *Oryza sativa* (rice) | 682 |
| *Medicago sativa* (lucerne, alfalfa) | 750 |
| $C_4$ | |
| *Panicum miliaceum* (broom-corn millet) | 267 |
| *Sorghum* spp. (sorghum) | 304 |
| *Sorghum vulgare* (sudan grass) | 380 |
| *Zea mays* (maize) | 350 |
| *Setaria italica* (foxtail millet) | 285 |

(After Shantz and Piemeisel 1927)

#### 3.4.2.5 THE POTENTIAL YIELDS OF $C_3$ AND $C_4$ CROPS

We have seen that individual leaves of $C_4$ species, under high irradiances and at their optimal temperatures, have higher maximum rates of net photosynthesis than those of $C_3$ species at their own optimal temperatures. In the closed crop canopy, however, most of the leaves will be subjected to varying degrees of shading, and this will clearly reduce the potential photosynthetic advantage of the $C_4$ crop, even under cloudless skies. The severity of this effect will depend on canopy structure and, in addition to this, other factors come into play to determine canopy photosynthesis, such as the respiratory load and perhaps feedback control of photosynthesis by active sinks (sect. 3.4.4). Considerations such as these prompted Gifford (1974) to compare $C_3$ and $C_4$ species in terms of potential photosynthesis, productivity and yield at the appropriate levels of organization, from the molecular to the whole crop. The published data available to him did indeed suggest that the advantage conferred upon $C_4$ leaves by their lower total resistances to $CO_2$ uptake (Table 3.5) was not apparent in the maximum short-term crop growth rates. However, the maximum recorded annual forage yields of $C_4$ species were two to three times those of $C_3$ species, and Gifford was forced to conclude that this was due to the much longer growing season available to $C_4$ crops in their natural environments.

A number of typographical and interpretative errors have since been identified in the data used by Gifford (Monteith 1978). When these unreliable figures are ignored, a consistent difference between the maximum short-term growth rates of $C_4$ species (range 50–54 g m$^{-2}$ d$^{-1}$) and those of $C_3$ species (range 34–39 g m$^{-2}$ d$^{-1}$) is revealed. Of course, part of this difference may simply be that the $C_4$ plants growing in the tropics benefit from greater insolation than the temperate $C_3$ plants. However, when this was taken into account by comparing the rates of growth of species which had experienced identical irradiances (20 MJ m$^{-2}$ d$^{-1}$ total solar radiation) there remained a clear difference between the photosynthetic efficiency of the $C_4$ crop, maize (4.5 per cent), and that of the $C_3$ crop, rice (3.1 per cent). While the length of the growing season did have the expected effect on the standing dry weights at harvest of both $C_4$ and $C_3$ crops, there was still a difference in the mean crop growth rate over the season: 22.0 ± 3.6 g m$^{-2}$ d$^{-1}$ for the $C_4$ group and 13.0 ± 1.6 g m$^{-2}$ d$^{-1}$ for the $C_3$ group.

This difference is certainly significant in terms of crop yield, but at the same time Gifford's demonstration that the superiority of $C_4$ plants at the biochemical and physiological level is progressively eroded by constraints operating at higher levels of organization remains valid. It is worth noting, however, that the 60–70-fold advantage attributed by Gifford to $C_4$ leaves on the basis of the available values of $K_m$ ($CO_2$ + $HCO_3^-$) for PEP carboxylase and for RuBP carboxylase has now been refuted. The $K_m$ is an inverse measure of an enzyme's affinity for its substrate. PEP carboxylase had long been known to have a low $K_m$ ($HCO_3^-$), while at the time of Gifford's (1974) review, RuBP carboxylase was thought to have a very high $K_m$ ($CO_2$). However, subsequent work has shown that these early assays of RuBP carboxylase were done with the enzyme in a relatively inactive state, and it is now concluded that RuBP carboxylase and PEP carboxylase have similar affinities for $CO_2/HCO_3^-$.

A comparison of the maximum recorded rates of net photosynthesis or dry matter production is the most realistic way of demonstrating consistent differences between $C_3$ and $C_4$ species. This is because mean rates of photosynthesis or dry-matter production result from the combined effect of many factors which may therefore mask any underlying effect of the different photosynthetic pathways. For example, mean rates may well include measurements made for particular species in sub-optimal environments, in which the advantages of each particular photosynthetic pathway cannot be expressed. In addition to this and the influences of canopy structure and respiratory load, the rate of photosynthesis of either individual leaves or whole canopies, of all species, will be affected by leaf age, environmental factors, and perhaps as a result of feedback control by the sinks.

### 3.4.3 LEAF AGE

The relationship between leaf area index and net photosynthesis has been described in section

2.4.2. Once $L_{crit}$ has been attained, and assuming for the moment a favourable environment, a more or less constant rate of net photosynthesis will be maintained as long as the continued production of new leaves compensates for the senescence of older ones. Of course, differences in irradiance will cause day-to-day variations in the rate of photosynthesis, and a progressive decline in net photosynthesis will result from an increasing proportion of non-photosynthetic tissue. Acute or chronic environmental stresses may also take their toll (sect. 3.4.5). Moreover, when leaf production stops, canopy photosynthesis declines as a result of the diminishing photosynthetic area and activity of the ageing canopy.

The net effect of these factors on the photosynthesis of a barley crop is shown in Fig. 4.6. Micrometeorological measurements of $CO_2$ flux showed that the rate of net photosynthesis decreased from a maximum value of about 3.5 g $CO_2$ $m^{-2}$ ground $h^{-1}$ three weeks before anthesis to about 0.8 g $m^{-2}$ $h^{-2}$ shortly before harvest. The effect of senescence was assessed by measuring the photosynthesis of individual leaves using a small leaf chamber connected to a portable gas analysis system. Temperature and irradiance were measured within the chamber, and photosynthetic light response curves were obtained by taking advantage of natural variations in irradiance or by shading the chamber with white paper sleeves. Figure 3.19 illustrates the decline in net photosynthesis per unit leaf area of leaf 8 (immediately below the flag leaf) during part of the grain filling period. This was accompanied by a decrease in the green area of the leaf. The relative contributions of photosynthetic activity and area to this senescent decline in leaf photosynthesis were assessed by comparing changes with time in the rate of net photosynthesis (at an arbitrary irradiance of 100 W $m^{-2}$) and in the green area of the leaf. Initially, the photosynthetic activity decreased while the green area remained constant but, as the leaf aged, green area was lost more rapidly than the decline in activity. Measurements on leaf 7, which of course was older than leaf 8, throughout the period in question suggested that the later phase of senescence was the result of a more rapid decline in activity than in green leaf area.

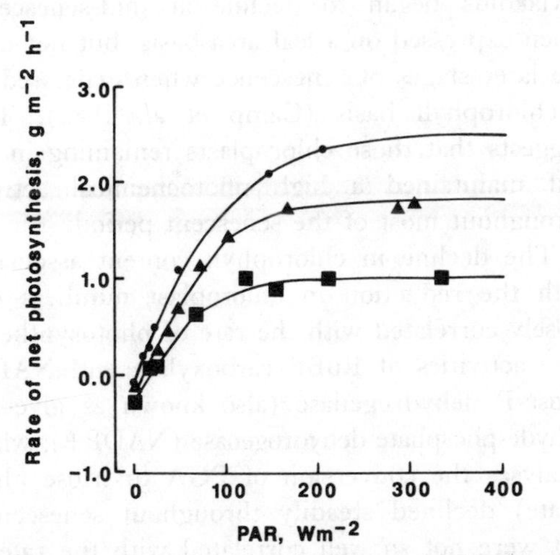


**Fig. 3.19** Net photosynthetic light response curves for leaf 8 of barley plants, measured in the field on 28 June (●), 14 July (▲) and 26 July (■) (from Biscoe *et al.* 1975b).

This senescent decline in the rate of photosynthesis is but part of the overall ontogenetic change in photosynthesis which occurs throughout the life of a leaf. Typically, photosynthetic activity rapidly increases to a maximum at or before full leaf area expansion, followed by a decline which is often linear with time. This pattern and the associated changes in stomatal and residual conductances in cotton are shown in Fig. 3.20. Throughout most of the life of the leaf, the stomatal conductance was about twice the residual conductance, and changes in the conductances were essentially in parallel, with the result that the concentration of $CO_2$ in the intercellular spaces remained constant. These parallel changes are more clearly seen in Fig. 3.21 for soybean plants, but here, where the progress of senescence has been followed more completely (the observations upon which Fig. 3.20 is based ceased while there was still an appreciable rate of net photosynthesis), it is clear that $r_r$ increased much more than $r_s$.

The nature of the interaction(s) between these resistances is not clear. $r_r$ may influence $r_s$ through changes in the substomatal concentration of $CO_2$, or through ABA synthesis (sect. 3.4.5.1, water stress). Alternatively, both resistances may be affected by the availability of an agent produced elsewhere in the plant, such as cytokinins from the roots (Woodward and Rawson 1976; Farquhar and Sharkey 1982; see also sect. 3.4.5). In other experiments there is no obvious parallelism

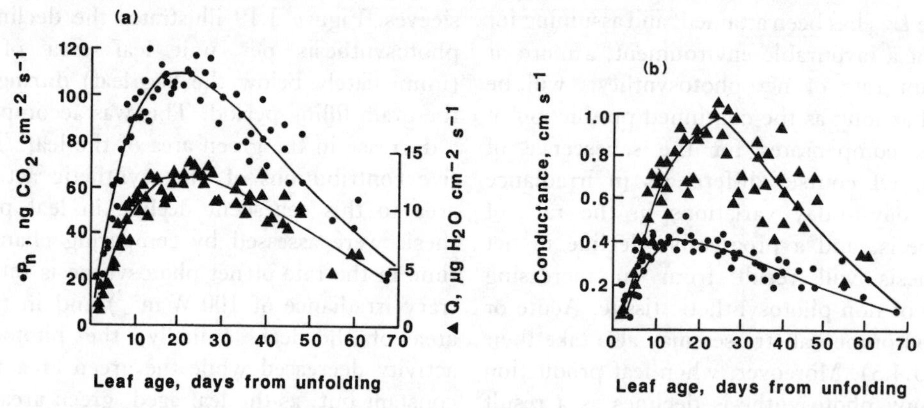


**Fig. 3.20** (a) Changes with age in net photosynthesis, $P_n$ (●), and transpiration, $Q$ (▲), of cotton leaves at 2000 µmol m$^{-2}$ s$^{-1}$ PAR. (b) Changes with age in stomatal (▲) and residual (●) conductances to $CO_2$ flux in cotton leaves at 2000 µmol m$^{-2}$ s$^{-1}$ PAR (after Constable and Rawson 1980).

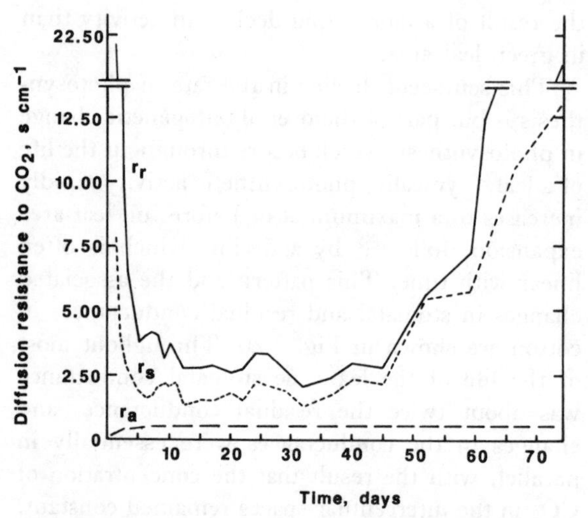


**Fig. 3.21** Boundary layer ($r_a$), stomatal ($r_s$), and residual ($r_r$) resistances to $CO_2$ uptake of leaves on the main stem of soybean plants from leaf emergence until senescence (mean of two plants) (after Woodward and Rawson 1976).

between $r_r$ and $r_s$. Figure 3.22 shows that in field-grown wheat in Barcelona, $r_r$ was always much greater than $r_s$ and increased dramatically during senescence: as a result the intercellular $CO_2$ partial pressure also increased.

Several factors may be responsible for the increased $r_r$ in senescent leaves, including loss of activity of Calvin cycle enzymes and reduced electron transport, associated with the degradation of chloroplasts. Jenkins and Woolhouse (1981) reported a continual decline in electron transport per unit chlorophyll in *Phaseolus vulgaris* leaves, which suggested a gradual degradation of all chloroplasts. In wheat, on the other hand, several lines of evidence suggested that the decline in photosynthesis after full expansion of leaves was related to the loss of whole chloroplasts. For example, although chlorophyll content per unit leaf area declined markedly during the second half of the senescence period, chlorophyll content per chloroplast was constant throughout senescence (Camp *et al.* 1982); similarly, reductions in chloroplast number per cell have also been reported (Peoples *et al.* 1980; Wittenbach *et al.* 1982). Furthermore, electron transport rates in isolated thylakoids began to decline at mid-senescence when expressed on a leaf area basis, but not until the later stages of senescence when expressed on a chlorophyll basis (Camp *et al.* 1982). This suggests that those chloroplasts remaining in the leaf maintained a high photochemical activity throughout most of the senescent period.

The decline in chlorophyll content associated with the reduction in chloroplast numbers was closely correlated with the rate of photosynthesis. The activities of RuBP carboxylase and NADP-triose-P dehydrogenase (also known as glyceraldehyde-phosphate dehydrogenase (NADP+), which catalyses the conversion of PGA to triose phosphate) declined steadily throughout senescence, but were not so well correlated with the rate of photosynthesis which, during the first half of the period, was affected proportionally less than the activities of these enzymes (Fig. 3.23). However,

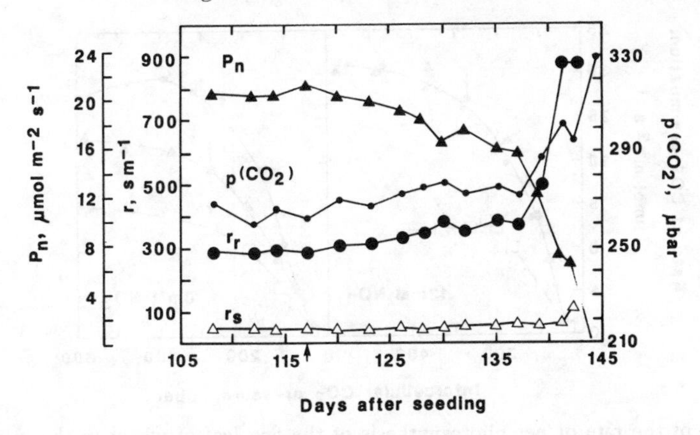


**Fig. 3.22** Changes in the rate of net photosynthesis ($P_n$), the stomatal ($r_s$) and residual ($r_r$) resistances to $CO_2$ uptake, and the intercellular partial pressure of carbon dioxide ($p(CO_2)$) in the flag leaf blade of wheat. The arrow indicates the time of anthesis (after Araus *et al.* 1986).

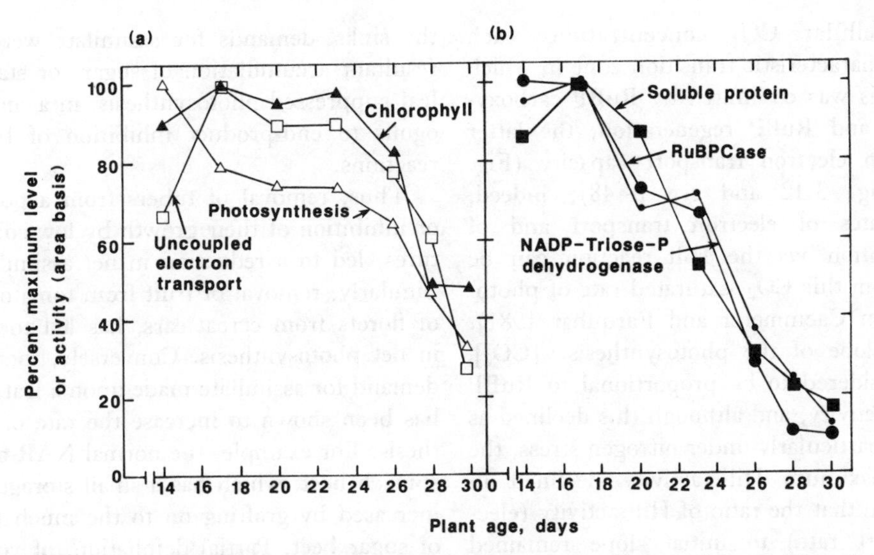


**Fig. 3.23** Changes in total soluble protein, photochemical parameters and activities of selected chloroplast enzymes during senescence of the second leaf of wheat. (a) Changes in: chlorophyll content (▲), uncoupled electron transport rates (□), and the rate of photosynthesis (△). (b) Changes in: total soluble protein (■), RuBP carboxylase activity (●), and NADP-triose-P-dehydrogenase activity (•). Age is measured from the time of planting (from Camp *et al.* 1982).

Makino *et al.* (1983) suggest that the rate of photosynthesis may have been artificially high, since it was measured as $^{14}C$ incorporation by leaf segments supplied with a saturating concentration (10 mM) of $NaH^{14}CO_3$. Certainly, in their experiments with rice, in which the photosynthesis of intact leaves was measured using an open IRGA system, the activity of RuBP carboxylase and other Calvin cycle enzymes was closely correlated with the rate of photosynthesis throughout the lifespan of the leaf.

The approach adopted by Evans (1983) may help to resolve this discrepancy. He measured the rates of net photosynthesis of flag leaves of wheat on three occasions after anthesis at various external $CO_2$ concentrations, from which intercellular $CO_2$ concentrations ($[CO_2]_i$) were calculated. The curves relating net photosynthesis to calcu-

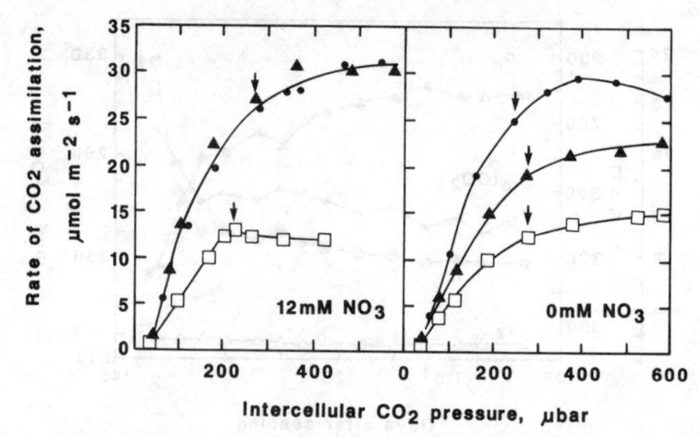


**Fig. 3.24** The response of the rate of net photosynthesis of the flag leaf of wheat to the intercellular partial pressure of $CO_2$. Plants were watered daily with a nutrient solution containing either 12 mM $NO_3$ or no $NO_3$. The curves are shown for three different times: 1 week (●), 5 weeks (▲), and 7 weeks (□) after anthesis. The arrow indicates the operating position at standard ambient conditions (after Evans 1983).

lated intercellular $CO_2$ concentrations each showed the characteristic transition zone in which photosynthesis was co-limited by RuBP carboxylase activity and RuBP regeneration, the latter dependent on electron transport capacity (Fig. 3.24; cf. Fig. 3.12 and see p. 48); indeed, maximum rates of electron transport and of oxygen evolution *via* the Hill reaction can be calculated from this $CO_2$ -saturated rate of photosynthesis (von Caemmerer and Farquhar 1981). The initial slope of the photosynthesis : $[CO_2]_i$ curve is considered to be proportional to RuBP carboxylase activity, and although this declined as leaves aged, particularly under nitrogen stress, the calculated maximum Hill activity declined in proportion, so that the ratio of Hill activity (electron transport rate) to initial slope remained constant. This, and the fact that throughout senescence the rate of photosynthesis at ambient $CO_2$ concentration lay in the transition zone of the photosynthesis : $[CO_2]_i$ curve, suggests that electron transport capacity and carboxylase activity deteriorate in a coordinated way.

### 3.4.4 FEEDBACK CONTROL BY THE DEMAND FOR ASSIMILATE

There is much evidence that the rate of photosynthesis by a leaf is influenced by the demand for assimilate imposed upon it. Early work led to the hypothesis that photosynthesis was controlled by the concentration of assimilate in the leaf; that if the sinks' demands for assimilate were low, the resultant accumulation of sugars or starch in the leaf suppressed photosynthesis, in a manner analogous to end-product inhibition of biochemical reactions.

Thus, removal of tubers from a potato plant, or inhibition of their growth by low soil temperatures, led to a reduction in net assimilation rate. Similarly, removal of fruit from tomato plants, or of florets from cereal ears, has led to reductions in net photosynthesis. Conversely, increasing the demand for assimilate made upon a leaf, or leaves, has been shown to increase the rate of photosynthesis. For example, the normal NAR of leaves of spinach beet, which has a small storage root, was increased by grafting on to the much larger root of sugar beet. Partial defoliation, of young apple trees or of maize, for example, causing the remaining leaves to meet the assimilate requirements of all the sinks, led to increased NAR. Both approaches were used in an experiment with wheat: removing the ear caused a 40 per cent reduction in the rate of net photosynthesis of the flag leaf, whereas subsequent shading of the lower leaves partially reversed this effect, since the flag leaf was then the only source of assimilate (Fig. 3.25). Ontogenetic variations in the rate of photosynthesis have also been interpreted in terms of changing demand for assimilate. Thus, during fruit growth in pea, wheat and apple, high rates of photosynthesis are attained in those leaves primarily supporting fruit growth; and in potato

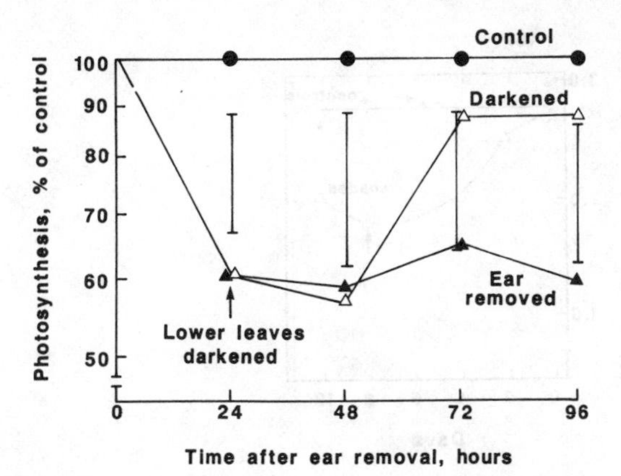


**Fig. 3.25** Daily measurements of the rate of net photosynthesis of flag leaves of wheat after three treatments: (●) control; (▲) ears removed at zero time; (△) ears removed at zero time and lower leaves darkened 24 hours later. The vertical lines indicate 5 per cent LSD values (after King *et al.* 1967).

a two- or three-fold increase in the rate of photosynthesis has been reported to occur at the time of tuber initiation.

### 3.4.4.1 CARBOHYDRATE ACCUMULATION

This and other early work has been reviewed by Neales and Incoll (1968). In many experiments, reduced rates of photosynthesis were associated with increased levels of sucrose and/or starch in the leaves. However, the interpretation of experiments involving defoliation and sink manipulation is confounded by possible unidentified nutritional and hormonal complications, especially in the longer-term experiments measuring NAR. Neales and Incoll therefore concluded that a causal relationship between assimilate accumulation and the rate of photosynthesis had not been established; nor was there a satisfactory mechanism to account for some form of end-product inhibition, although it was suggested that chloroplast function might be impaired by excessive accumulation of starch granules.

Experiments with soybean have lent support to this idea. Thorne and Koller (1974) completely shaded the pods, stems and all but one leaf of soybean plants at the rapid pod-filling stage, in order to increase assimilate demand on the single illuminated leaf (the source leaf). The rate of net photosynthesis of this leaf had increased by about 50 per cent after eight days of this shading treatment, an effect which was reversed when the treatment was discontinued (Fig. 3.26). The gas-phase resistance to $CO_2$ diffusion ($r_a + r_s$) was unaffected by the shading treatment, but the residual resistance decreased to a minimum on the eighth day and increased when the plants were uncovered. Associated with these changes were a reduction in starch content, and increases in sucrose content, inorganic phosphate content and RuBP carboxylase activity in the source leaf.

A number of factors could have affected $r_r$ and hence photosynthesis. Photochemical limitations are discounted since photosynthesis was measured under saturating photosynthetic irradiance (but see the discussion of Azcón-Bieto's work, p. 66). The increased RuBP carboxylase activity could have been partially responsible, stimulated perhaps by a change in the hormonal status of the plant. In addition, there may have been a direct feedback effect, whereby the breakdown of starch in the illuminated leaf, in response to increased export of assimilate, relieves the putative adverse effect of excess starch granules in the chloroplasts. It is suggested that in chloroplasts heavily laden with starch, the noticeable distortion and disorientation of thylakoid membranes could increae the path-length of diffusion, and hence $r_r$.

Nafziger and Koller (1976) confirmed the negative correlation between leaf starch concentration and net photosynthesis rate (Fig. 3.27) in an experiment which attempted to avoid any hormonal influence. Leaf starch concentrations were altered by exposing plants in a controlled environment room to either 50, 300 or 2000 $\mu l\ l^{-1}$ $CO_2$ for 12.5 hours. The rate of net photosynthesis of the youngest trifoliate leaf was then measured at a $CO_2$ concentration of 300 $\mu l\ l^{-1}$, after which it was removed and frozen for subsequent carbohydrate analyses. The concentration of soluble sugars was unaffected by this treatment, and the reduction in net photosynthesis was primarily due to an increased $r_r$ in leaves with a high starch concentration. Again, cause and effect cannot be established, and the supposed effect of starch is through its physical presence. It is perhaps significant, however, that in field-grown soybeans the initial decline in the rate of net photosynthesis, which was one of the first signs of senescence after flowering, was associated with the distortion of chloroplasts and diso-

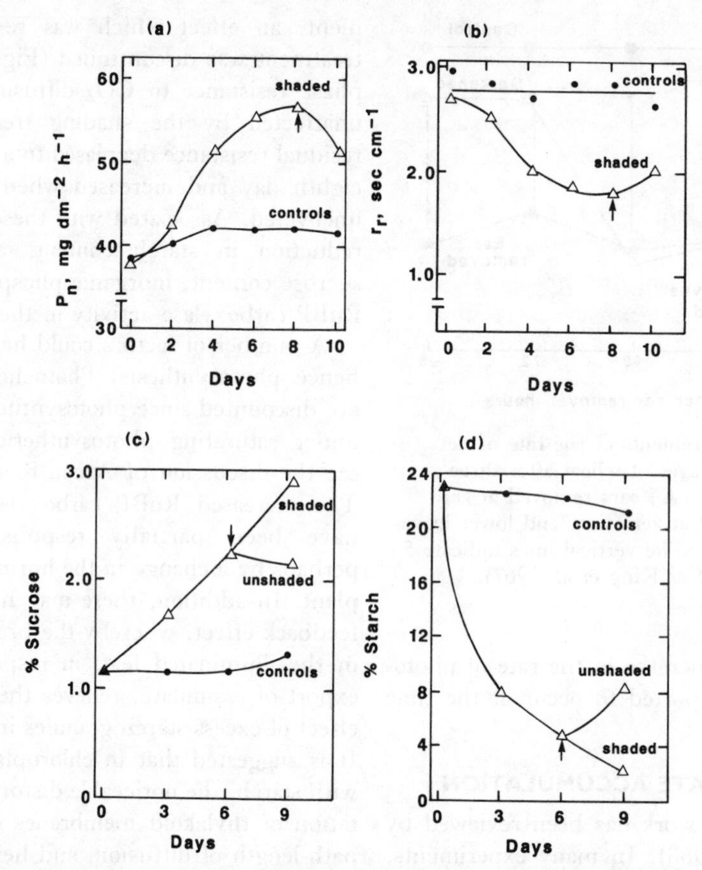


**Fig. 3.26** (a) Light-saturated rates of photosynthesis of source leaves of controls and shaded (see text for details of treatment) soybean plants. Shaded plants were uncovered on the eighth day (arrow). SE = 3.7.
(b) Residual resistances of source leaves of controls and shaded plants. SE = 0.29.
(c) and (d) Sucrose and starch levels in source leaves of controls and shaded plants. Shaded plants were either uncovered on the sixth day (arrow) or continuously shaded. SE = 0.14 (c), 0.81 (d) (after Thorne and Koller 1974).

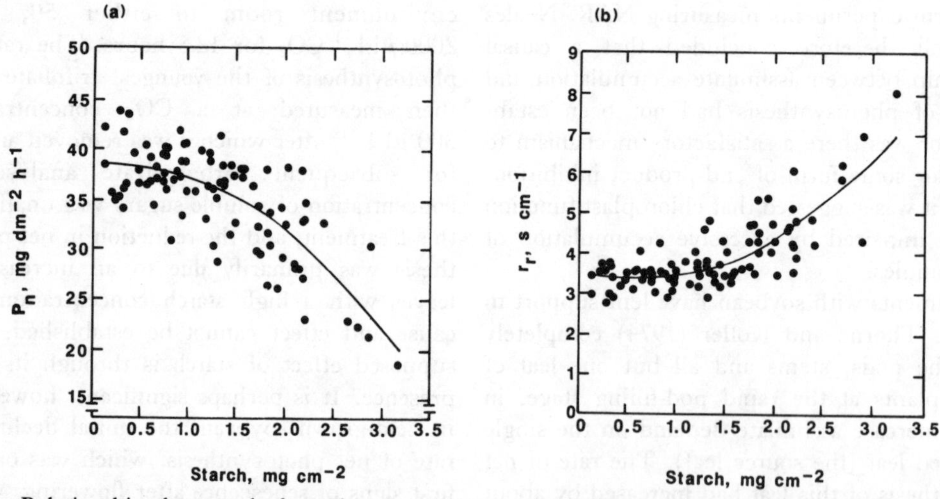


**Fig. 3.27** Relationships between leaf starch concentration and (a) the rate of net photosynthesis, $P_n$, and (b) the residual resistance, $r_r$, of the first trifoliate leaves of 21-day-old soybean plants after 13 hours in the light (after Nafziger and Koller 1976).

rientation of the lamellae resulting from an increase in starch deposition (Wittenbach *et al.* 1980). These changes preceded the large increase in $r_s$ and decrease in RuBP carboxylase activity which occurred later in senescence.

There is thus some evidence, at least circumstantial, which suggests that photosynthesis can be limited by starch accumulation in the leaves. The extent to which this form of control applies to other species, however, is unclear; certainly, in leaves of species which do not accumulate much starch it is unlikely to be important. For example, in *Phaseolus* leaves, net photosynthesis was not inhibited by starch accumulation, but the maximum starch concentration was only 0.2 mg $cm^{-2}$, considerably less than that found in soybean leaves (Crookston *et al.* 1974; cf. Fig. 3.27).

Wheat leaves do accumulate starch, but to a lesser extent than soluble sugars. Azcón-Bieto (1983) found that the rate of net photosynthesis progressively declined in ambient air after the first two hours of illumination if the temperature was less than 25 °C; no such decline occurred at 30 °C unless the $CO_2$ partial pressure was raised (to 740–825 μbar). The decline in net photosynthesis at 25 °C was more dramatic at the elevated $CO_2$ partial pressure, and was further exaggerated when, in addition, the oxygen concentration was reduced to 2 per cent: in these circumstances a 35 per cent reduction in the rate of net photosynthesis was recorded after nine hours of photosynthesis (Fig. 3.28). The depression in net photosynthesis was related to the accumulation of carbohydrate within the leaf, which was greatest under those conditions favouring high rates of photosynthesis. The relative increase was the same for all carbohydrate fractions, although the concentration of soluble sugars was greater than that of starch. The effect of temperature was different in that assimilation was greater, but carbohydrate accumulation less, at 30 °C compared with 25 °C. This suggests that the export of assimilates was more effective at the higher temperature. Certainly, when export was inhibited by chilling (0–2 °C) the basal portion of a leaf, the rate of decline in net photosynthesis was increased. Stomatal closure was not responsible for any of these effects.

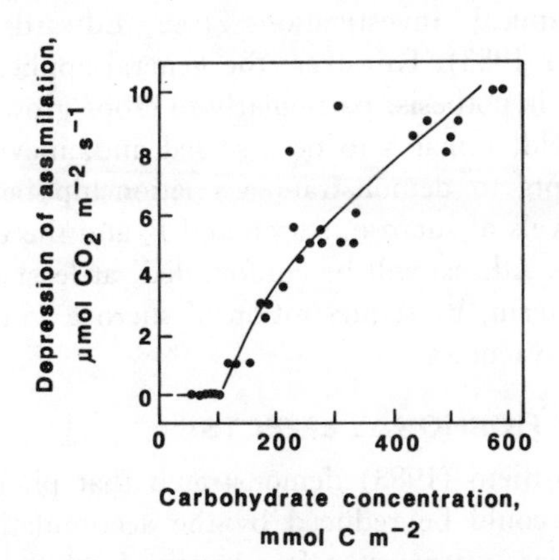


**Fig. 3.28** Relationship between the depression of net $CO_2$ assimilation and total carbohydrate concentration (except fructosans) in wheat leaves. This relationship includes data from experiments at different temperatures from 20–30 °C (from Azcón-Bieto 1983).

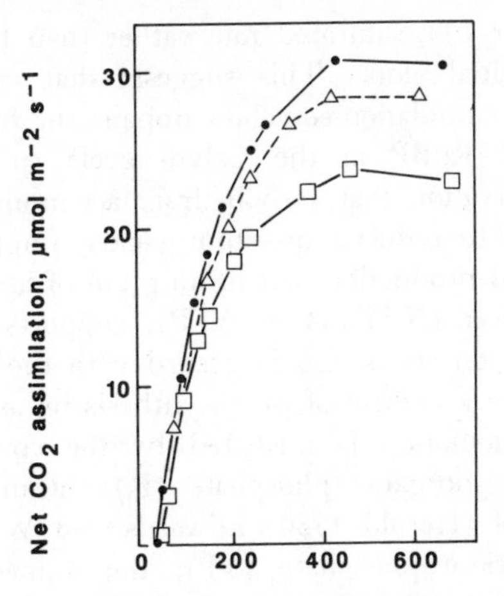


**Fig. 3.29** Effect of a period of photosynthesis on the response of net $CO_2$ assimilation to intercellular $CO_2$ partial pressure in the flag leaf of wheat. The curve was determined at the beginning of the light period (●), after 5 hours in the light at 800 μbar $CO_2$ (□), and after a further 3 hours in the dark (△). The carbohydrate concentration of these leaves increased from 12 to 260 (± 40) mmol C $m^{-2}$ after 5 hours in the light, and declined to 197 (± 11) after 3 hours in the dark (from Azcón-Bieto 1983).

### 3.4.4.2 THE IMPORTANCE OF INORGANIC PHOSPHATE

Examination of the relationship between net $CO_2$ assimilation and intercellular $CO_2$ partial pressure (Fig. 3.29) prompts speculation about the nature of the decline in photosynthesis. It was due largely to a lower $CO_2$-saturated rate rather than to an altered initial slope. This suggests that carbohydrate accumulation somehow impairs the regeneration of RuBP in the Calvin cycle (p. 48). The observation that carbohydrate accumulation leads also to reduced quantum yields, implying diminished production or consumption of assimilatory power (NADPH + ATP), supports this interpretation. It is also in accord with the idea that feedback control of photosynthesis by assimilate accumulation is mediated by the concentration of inorganic phosphate ($P_i$) within the chloroplast (Herold 1980; Edwards and Walker 1983). Triose phosphate (TP), the immediate product of the Calvin cycle, is either recycled for the regeneration of RuBP or leaves the cycle to become the precursor of either starch synthesis in the chloroplast or sucrose synthesis after export to the cytosol. This export of TP is linked with counter-movement of the $P_i$ which is released during sucrose synthesis (Fig. 3.30) Indeed, this return of $P_i$ to the chloroplast is essential for maximal operation of the Calvin cycle.

The accumulation of sucrose in the cytosol, which may result from reduced translocation out of the cell, leads to feedback inhibition of further sucrose synthesis. Less $P_i$ is therefore available to return to the chloroplast, and consequently TP export slows down, causing a greater proportion of it to be diverted towards starch synthesis. At the same time, the diminished supply of $P_i$ lowers the [ATP]/[ADP] ratio, which causes a fall in the reduction of PGA to TP (because of the requirement for ATP and NADPH) and in the regeneration of RuBP. Carbon fixation is thus slowed down, while a higher proportion of the limited supply of TP is converted into starch, the latter effect amplified by the activation of a key enzyme of starch synthesis by the high [PGA]/[$P_i$] ratio. The more starch is produced, the less TP is available for RuBP regeneration and the slower is the rate of carbon fixation. In fact, the Calvin cycle is regulated in such a way that the amount of TP recycled is strictly geared to the rate of ATP production.

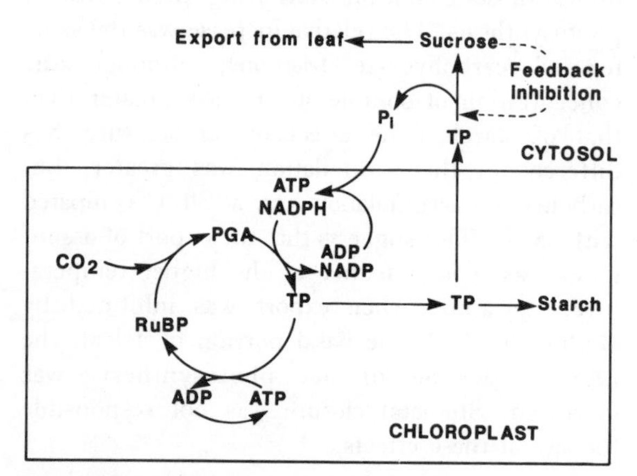


**Fig. 3.30** A mechanism for the control of photosynthesis by the availability of inorganic phosphate ($P_i$). Sucrose accumulation, as a result of a low demand for assimilates by the sinks, causes feedback inhibition of sucrose synthesis from triose phosphate and a reduction in the release of $P_i$. See text for details. Abbreviations as in Fig. 3.1 (adapted from Edwards and Walker 1983).

This hypothesis is attractive in that it provides a biochemical basis for the link between the demand for assimilate and the rate of photosynthesis. The accumulation of starch is seen to be a result of this control system rather than the cause of the reduced rates of net photosynthesis, unless of course chloroplast function does subsequently become impaired by the presence of starch granules. Various lines of evidence support the hypothesis, from observations that starch levels increase despite diminished photosynthesis in $P_i$-deficient plants (e.g. tobacco; Kakie 1969), or that the increased photosynthesis and reduced starch levels of soybean leaves under high sink demand are accompanied by a four-fold increase in $P_i$ (Thorne and Koller 1974), to more rigorous biochemical investigations (see Edwards and Walker 1983). However, the general applicability of the hypothesis, particularly to crops growing in the field, remains to be assessed and, inevitably, attempts to demonstrate a relationship between the levels of sucrose, starch and $P_i$ and the rate of photosynthesis will be confounded, at least in the short term, by sequestration of sucrose and/or $P_i$ in the vacuoles.

#### 3.4.4.3 HORMONAL EFFECTS

Azcón-Bieto (1983) demonstrated that photosynthesis could be reduced by the accumulation of assimilate after just a few hours of rapid photosynthesis; this rapid response and the avoidance of defoliation or sink removal treatments minimizes the likelihood that the effects were hormonally induced. However, hormonal effects are certainly indicated by other experiments. The removal of 70–80 per cent of the roots of young *Phaseolus* plants was followed, over a six-day period, by a reduction in net photosynthesis of the primary leaves. However, feedback inhibition of photosynthesis by assimilate accumulation was discounted in this case, since in earlier work heat-girdling of the petioles to prevent assimilate export did not result in decreased net photosynthesis, despite increased dry matter accumulation in the laminae (Carmi and Koller 1977, 1978). Removal of the growing shoot above the primary leaves had no inhibitory effect on net photosynthesis within the first four days after treatment but, perhaps unexpectedly, there were considerable increases in net photosynthesis, chlorophyll content and RuBP carboxylase activity after a further three or four days. These changes did not occur in plants in which the shoot sinks were isolated from the primary leaves by heat-girdling the stem above the primary node. Therefore, the increased rates of net photosynthesis observed in decapitated plants were not due to the removal of sinks for carbohydrate; indeed, the same effects occurred in plants in which the stem and branches were allowed to grow, but from which all trifoliate leaves were removed as soon as they started to unfold (Carmi and Koller 1979).

It was concluded that photosynthesis was influenced by the distribution of regulatory factors, such as cytokinins, carried in the transpiration stream, as suggested by Wareing *et al.* (1968). Thus, the rate of photosynthesis was depressed when the source of this influence, the roots, was partially removed; and when an increased share of the factor(s) was available, after decapitation or removal of other transpiring surfaces, the rate of photosynthesis of the remaining leaves was increased. These findings clearly have implications for the interpretation of other partial defoliation experiments and, indeed, for those in which effects attributed to the shading of certain leaves may in fact have been caused by reduced transpiration from those leaves.

Another way in which the rate of photosynthesis of a leaf may be linked to the rate of translocation is through the export from the leaf of inhibitors of photosynthesis. Thus, when export was limited in soybean, by the removal of the pods or by petiole girdling, stomatal conductance and the rate of net photosynthesis were reduced: significant differences between control and treated plants were apparent after only 30 minutes for stomatal conductance and 5 hours for net photosynthesis (Fig. 3.31). Sucrose and glucose, but not starch, accumulated in the girdled leaves, but not in the leaves of depodded plants, although in neither case was mesophyll conductance affected. Furthermore, even when depodding and petiole girdling were followed by a 24-hour period during which photosynthesis was prevented by shading of leaves or by exposure to $CO_2$-free air, the effects on stomatal conductance and net photosynthesis still occurred, and could not, therefore, be attributed to assimilate accumulation. However, free abscisic acid (ABA) levels were increased in the treated plants, and quickly became significantly different from those of the controls (Fig. 3.32): free ABA levels had risen twenty-five-fold two days after girdling and ten-fold two days after depodding (Setter *et al.* 1980a, b).

It was concluded that under normal circumstances, ABA was exported from the leaves, along with other leaf assimilates, in the phloem, but that when this export was restricted, the resultant accumulation of ABA caused stomatal closure. Increased levels of ABA in leaves and increased $r_s$ have also been observed in grape after fruit

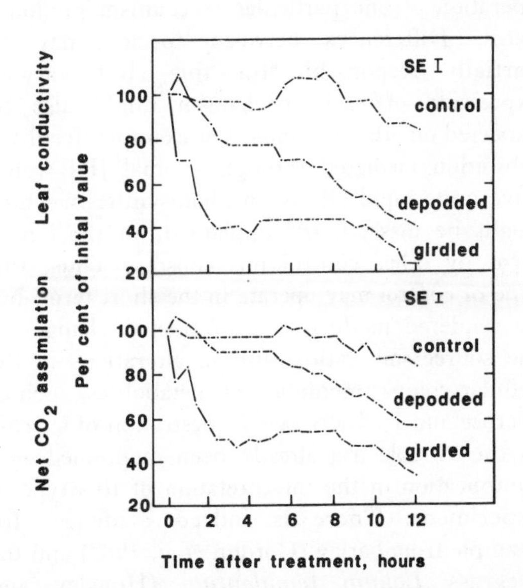


**Fig. 3.31** Time course of net $CO_2$ assimilation and leaf (primarily stomatal) diffusive conductivity (conductance) of soybean leaves after pod removal and petiole girdling (after Setter *et al.* 1980a).

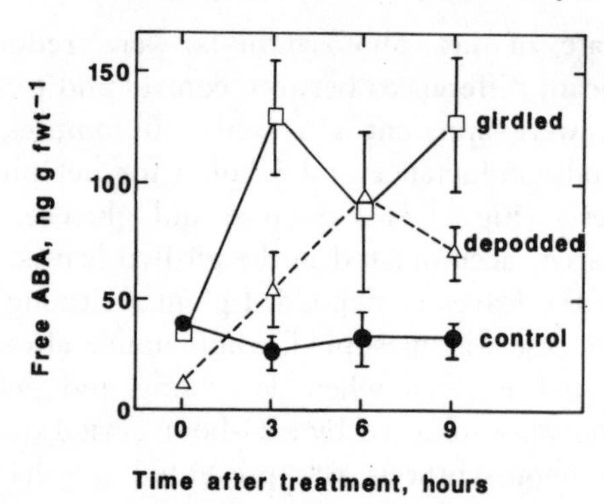


**Fig. 3.32** Time course of the effect of pod removal and petiole girdling on the levels of free ABA in soybean leaves. (●) control, (△) depodded, (□) petiole girdled (after Setter *et al.* 1980b).

removal or stem cincturing (the removal of a 2 mm collar of tissue from the stem below a node bearing fruit), and in *Capsicum* after fruit removal (Loveys and Kriedemann 1974; Kriedemann *et al.* 1976).

#### 3.4.4.4 CONCLUSIONS

Thus, there are a number of mechanisms by which the rate of photosynthesis can be regulated by the demand for assimilates. It is clear that in each of the experiments described above the operation of one particular mechanism predominates. Differences between species may be partially responsible for this, but varying expression of each mechanism might also be expected on other grounds. For example, feedback inhibition mediated through stromal [$P_i$] would have quite rapid effects on photosynthesis, which might be masked, or supplanted, in the longer term by other control mechanisms. Thus, this type of control may operate in the short term, but be rendered ineffective if protracted changes in the source/sink ratio result in alterations in the cellular compartmentation of metabolites, such as sucrose and $P_i$. Increased sequestration of sucrose in the vacuole has already been mentioned as a complication in the interpretation of this type of experiment. There is indeed evidence, for example from barley (Gordon *et al.* 1982) and the ryegrass *Lolium temulentum* (Housley and Pollock 1985), that the partitioning of sucrose between the cytosol and the vacuole allows accumulation of carbohydrate without reductions in photosynthesis. In grasses and cereals, the synthesis of fructans (fructosans), polymers of fructose units, can similarly help to regulate the cytosolic concentration of sucrose, and allow continued photosynthesis during periods of low demand by the sinks.

Controls based on hormonal effects would have longer, but nonetheless differing, response times, the direct effect of ABA retention on $r_s$ being more rapid than the effects of cytokinins on such things as chlorophyll content and RuBP carboxylase activity. It is therefore not unreasonable to expect that experiments which concentrate on the plant's behaviour at particular times after the imposition of particular treatments will each highlight a different mechanism of control. Furthermore, the operation of several such internal control systems is probably necessary to achieve fine control of photosynthesis under a range of circumstances.

### 3.4.5 ENVIRONMENTAL EFFECTS ON PHOTOSYNTHESIS

The rate of net photosynthesis is moderated by numerous environmental factors. A discussion of all these influences is beyond the scope of this book and, rather than present a superficial treatment of many factors, we have chosen to concentrate on just a few. The resistance concept applied to photosynthesis provides a useful means of analysing the effects of the environment on the various processes contributing to net $CO_2$ uptake and, from what has been said already, we can distinguish between effects on the movement of $CO_2$ as a gas, mediated largely through changes in $r_s$, and effects on some aspect of the composite intracellular resistance.

The effect of water stress is well suited to such a treatment, and will be considered in some detail. Similarly, mineral elements influence net photosynthesis in a variety of ways: nitrogen, for example, mainly affects the intracellular resistance, through leaf chlorophyll content and RuBP carboxylase activity, whereas the effect of potassium is largely on $r_s$; here we shall concentrate on the effects of nitrogen because of its profound effects on plant growth and development and its widespread use in crop production (sect. 2.3.3, 6.4, 7.5 and 8.3). We shall also consider the infection of leaves by plant pathogens, which can have a pervasive influence on all components of net photosynthesis. The responses of photosynthesis to temperature and light have already been discussed in the comparison of $C_3$ and $C_4$ plants (sect. 3.4.2), and the adaptation to reduced irradiance of leaves within a closed canopy is referred to in the discussion of crop respiration (sect. 4.1).

### 3.4.5.1 WATER STRESS

It is useful to separate the effects of water stress on net photosynthesis into stomatal and non-stomatal influences. There is much evidence (see, for example, reviews by Hsiao 1973 and Levitt 1980) that reductions in the rate of net photosynthesis of water-stressed plants are caused primarily by the closure of stomata, but that under more severe stress the stomatal effect can be accompanied by increases in the intracellular (mesophyll) resistance. This is illustrated for cotton and for *Panicum maximum* in Fig. 3.33.

Care must be taken in the interpretation of these data. In each case, $r_m$ is a residual resistance, calculated as:

$$r_m = \frac{[CO_2]_i - [CO_2]_c}{P_n} - r_a - r_s \qquad [3.11]$$

with an assumed value for $[CO_2]_c$. Troughton (1969) took this to be zero. More realistically perhaps (see p. 47), Ludlow and Ng (1976) equated $[CO_2]_c$ with the $CO_2$ compensation point ($\Gamma$), but warned that at water potentials below $-0.8$ MPa this assumption will lead to overestimates of $r_m$: this is because, although $\Gamma$ was independent of water stress above this value, it increased substantially below it. A similar increase

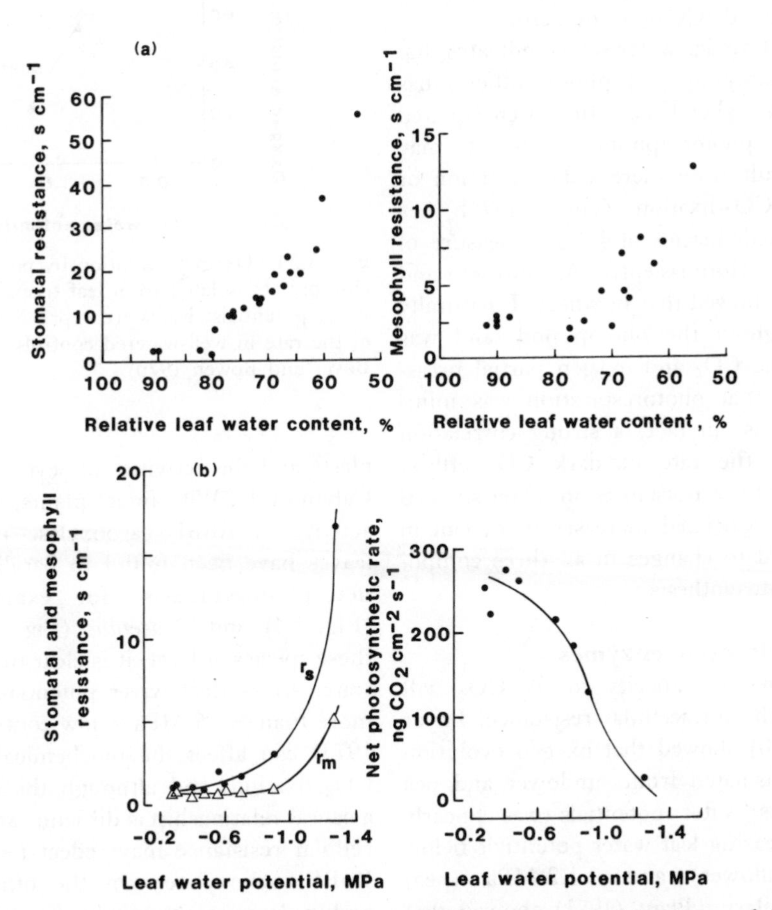


**Fig. 3.33** Effect of water stress on the stomatal and mesophyll resistances in (a) cotton and (b) *Panicum maximum*. In (b) the associated decline in the rate of net photosynthesis is also shown (after Troughton 1969 and Ludlow and Ng 1976).

**Table 3.7** Mean response of several leaf gas exchange measurements to decreasing leaf water potential in bean (*Phaseolus vulgaris*)

| $\psi_L$ | $P_n$ | $\Gamma$ | $(r_a + r_s)$ | $\Sigma_{r0}$ | $\Sigma_{r\Gamma}$ |
|---|---|---|---|---|---|
| 0.0 to −0.3 | 19.7 ± 3.0 | 61 ± 3 | 6.2 ± 1.5 | 13.7 ± 2.9 | 10.9 ± 2.3 |
| −0.31 to −0.6 | 15.1 ± 1.0 | 73 ± 4 | 8.8 ± 0.9 | 16.2 ± 1.3 | 12.2 ± 0.8 |
| −0.61 to −0.9 | 5.9 ± 1.1 | 79 ± 5 | 41.1 ± 9.5 | 74.5 ± 18.9 | 53.3 ± 12.6 |
| −0.91 to −1.2 | 0.8 ± 0.4 | 125 ± 15 | 92.4 ± 7.3 | 233.0 ± 101.0 | 142.2 ± 46.7 |

Abbreviations: $\psi_L$, leaf water potential (MPa); $P_n$, rate of net photosynthesis (mg $CO_2$ $dm^{-2}$ $h^{-1}$); $\Gamma$, $CO_2$ compensation point ($\mu l\ l^{-1}$ $CO_2$); $(r_a + r_s)$, boundary layer plus stomatal resistance to $CO_2$ diffusion (s $cm^{-1}$); $\Sigma_{r0}$ and $\Sigma_{r\Gamma}$, sum of the resistances to $CO_2$ uptake estimated on the assumption that the $CO_2$ concentration at the sites of carboxylation was 0 and $\Gamma$, respectively (s $cm^{-1}$).
(After O'Toole *et al.* 1976)

in $\Gamma$, reported for several species, has been clearly demonstrated with *Phaseolus* (Table 3.7): as a result, the residual, or intracellular, resistances calculated on the assumption that $[CO_2]_c = \Gamma$ are lower, at a given leaf water potential, than those which have assumed $[CO_2]_c$ to be zero.

An increased $\Gamma$ under water stress indicates that the plant's capacity for net photosynthesis has been reduced. A higher $\Gamma$ has often been equated with enhanced photorespiration, but it may equally well result from increased respiration or from decreased $CO_2$ fixation. Zelitch (1975b) has discussed the inadequacies of $\Gamma$ as a measure of photorespiration. More recently, Azcón-Bieto and Osmond (1983) showed that in wheat, $\Gamma$ naturally increased throughout the photoperiod, and was greatest when the $CO_2$ and oxygen partial pressures were such that photorespiration was minimized. There was, in fact, a strong correlation between $\Gamma$ and the rate of dark $CO_2$ efflux. Increased intracellular resistances in water-stressed plants, and the associated increases in $\Gamma$, can in fact be attributed to changes in all three components of net photosynthesis.

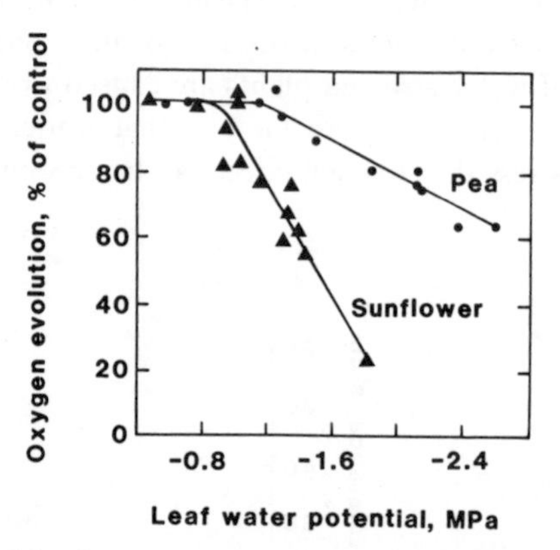


**Fig. 3.34** Oxygen evolution by pea and sunflower chloroplasts isolated from leaf tissue having various water potentials. Rates are expressed as a percentage of the rate in well-watered controls (adapted from Boyer and Bowen 1970).

### Effects on Calvin cycle enzymes

A diminution in the capacity to fix $CO_2$ will clearly increase the intracellular resistance. Boyer and Bowen (1970) showed that oxygen evolution by chloroplasts isolated from sunflower and pea leaves at various water potentials was linearly reduced by decreasing leaf water potentials below −0.8 MPa (sunflower) and −1.2 MPa (pea) (Fig. 3.34). Similarly, Plaut (1971) showed that reduced water potentials inhibited both the rate of $CO_2$ fixation by isolated intact spinach chloroplasts and the activities of several enzymes of the Calvin cycle. With intact plants, reductions in the activity of RuBP carboxylase in water-stressed leaves have been found to parallel reductions in net photosynthesis, for example in cotton (Fig. 3.4) and *Phaseolus* (Fig. 3.35). Thus, in these species at least, it is clear that even moderate water stress (leaf water potentials lowered by no more than −1.5 MPa below control values; Hsiao 1973) can affect the biochemical components of $CO_2$ fixation, and although the establishment of a causal relationship is difficult, an increased intracellular resistance may reflect reduced activity of RuBP carboxylase. On the other hand, RuBP carboxylase activity of wheat and barley leaves has been found to be much less sensitive to stress, being reduced by only about 50 per cent at leaf

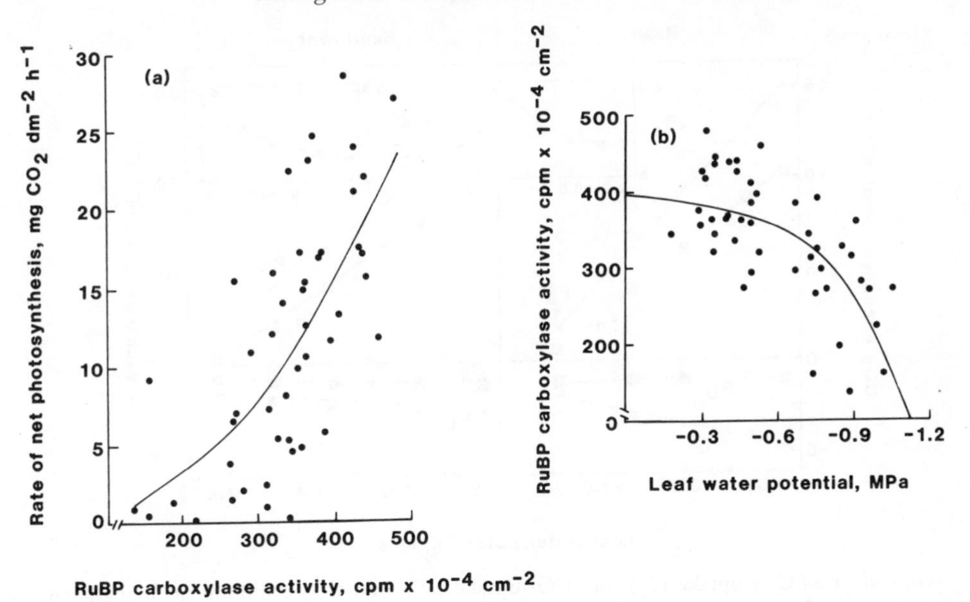


**Fig. 3.35** (a) Relationship between the rate of net photosynthesis and the activity of RuBP carboxylase in *Phaseolus vulgaris* leaves of various water potentials. (b) Relationship between RuBP carboxylase activity and water potential of leaves of *Phaseolus vulgaris* (after O'Toole *et al.* 1976).

water potentials as low as −3.3 MPa (Johnson *et al.* 1974).

**Effects on photorespiration and respiration**

Data on photorespiration and respiration in water-stressed plants are scarce, but indicate that the rates decline with increasing stress. Such effects counteract those on chloroplast function, noted above, but the important consideration is the magnitude of each effect relative to that on (gross) photosynthesis. Photorespiration and respiration both appear to be less sensitive to water stress than photosynthesis: Brix (1962) demonstrated this for the respiration of tomato leaves, and small reductions in the rate of respiration with moderate water stress have subsequently been demonstrated for *Panicum maximum* (Ludlow and Ng 1976) and soybean, sunflower and maize (Boyer 1970). Krampitz *et al.* (1984) showed that in sunflower and *Phaseolus* there was little effect of water stress on respiration, and that photorespiration was reduced less than gross photosynthesis: the ratio of photorespiration to gross photosynthesis therefore increased with reductions in leaf water potential (Fig. 3.36).

This substantiated earlier work in which the fate of accumulated $^{14}C$ was analysed immediately after feeding $^{14}CO_2$ to sunflower leaves with different water potentials. With increasing water stress ($\psi_{leaf}$ down to −1.5 MPa), less label appeared in 3-PGA, the initial product of RuBP carboxylation, and more appeared in amino acids, particularly in glycine and serine, intermediates of the photorespiratory glycollate pathway (Figs 3.1 and 3.13). The greater accumulation of label in the amino acids meant that less accumulated in the sugars and organic acids (Table 3.8). This clearly demonstrated that in water-stressed plants, photorespiration was increased relative to photosynthesis, a conclusion further strengthened by a comparison of the effects of water stress on photorespiration and photosynthesis estimated from the accumulation of $^{14}C$ by leaves in 312 vpm $CO_2$ and either 21 or 1.5 per cent oxygen. Together with inhibition of Calvin cycle enzymes, the relative increase in photorespiration contributes to a decreased efficiency of photosynthesis, which is reflected in an increased intracellular resistance.

**Parallel effects on the stomatal and non-stomatal resistances**

Lawlor and Fock (1977) concluded that the relative changes in photosynthesis and photorespira-

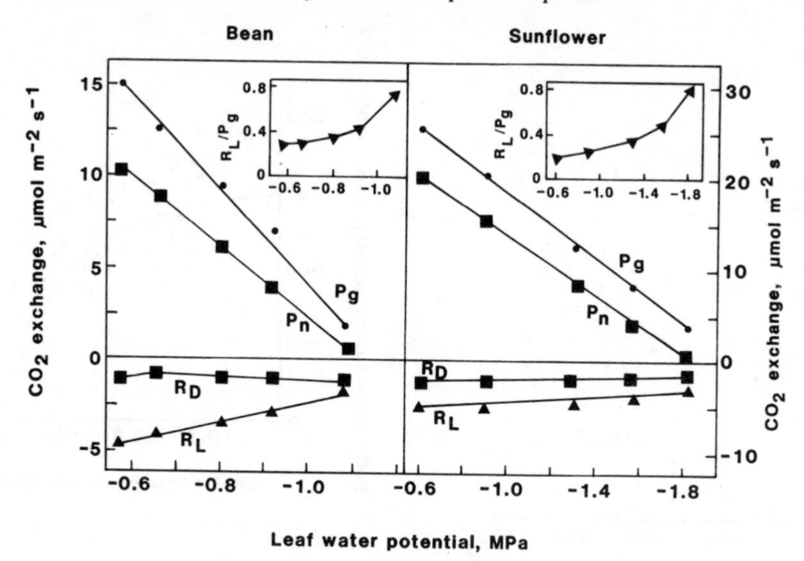


**Fig. 3.36** Rates of gross $CO_2$ uptake ($P_g$), net $CO_2$ uptake ($P_n$), and $CO_2$ evolution in the light ($R_L$) and in the dark ($R_D$) of bean (*Phaseolus*) and sunflower leaves of various water potentials. $P_g$ was calculated from $^{14}CO_2$ uptake 30 seconds after $^{14}CO_2$ was supplied to the leaf, and $P_n$ from $CO_2$ exchange measurements made with an infra-red gas analyser. $R_L$ was calculated as the difference between $P_g$ and $P_n$, and $R_D$ was determined after 20 minutes of darkness. PPFD: 1000 µmol m$^{-2}$ s$^{-1}$; temperature: 25 ± 1 °C. The insets show the $R_L/P_g$ ratios at different leaf water potentials (after Krampitz *et al.* 1984).

**Table 3.8** $^{14}C$ accumulation (dpm × $10^{-5}$) in various chemical fractions of sunflower leaves (area 26.5 cm$^2$) of different average water potential, exposed to $^{14}CO_2$ (312 vpm, 21 per cent $O_2$) for 20 minutes

| *Leaf water potential (MPa)* | *Total water-soluble extract* | *Sugars, mainly sucrose* | *Organic acids* | *3-PGA* | *Total amino acids* | *Glycine* | *Serine* |
|---|---|---|---|---|---|---|---|
| −0.5 ± 0.08 | 221.0 | 163.5 | 34.5 | 6.4 | 15.0 | 3.0 | 5.0 |
| −1.2 ± 0.14 | 170.6 | 122.8 | 20.9 | 5.5 | 22.2 | 4.1 | 11.7 |
| −1.5 ± 0.17 | 113.8 | 72.1 | 16.0 | 3.3 | 19.7 | 5.0 | 7.6 |

(After Lawlor and Fock 1977)

tion were due to increased RuBP oxygenase activity as the supply of $CO_2$ to the chloroplasts was restricted by stomatal closure. In other words, at least part of the increase in the intracellular resistance was an indirect result of the increased stomatal resistance. This interpretation, which emphasizes the stomatal response, implies a reduction in the intercellular $CO_2$ concentration ($[CO_2]_i$). However, more recent experiments have shown that $[CO_2]_i$ may be little affected by water deficits, or may increase, even though stomatal apertures and photosynthesis are reduced. It seems, therefore, that the intracellular resistance can be affected directly by water stress, rather than indirectly as a result of the stomatal response, so that the two resistances change in parallel. Indeed, there is now evidence that it is the cells' capacity for photosynthesis, and hence the intracellular resistance, which is primarily responsible for the overall photosynthetic response to water stress.

For example, the $[CO_2]_i$ was slightly higher in water-stressed, field-grown cotton and sorghum plants than in irrigated plants (Fig. 3.37). Exposure to elevated external $CO_2$ concentrations increased the rate of photosynthesis, although the stomatal conductance was reduced because of the sensitivity of stomata to $[CO_2]_i$. Significantly,

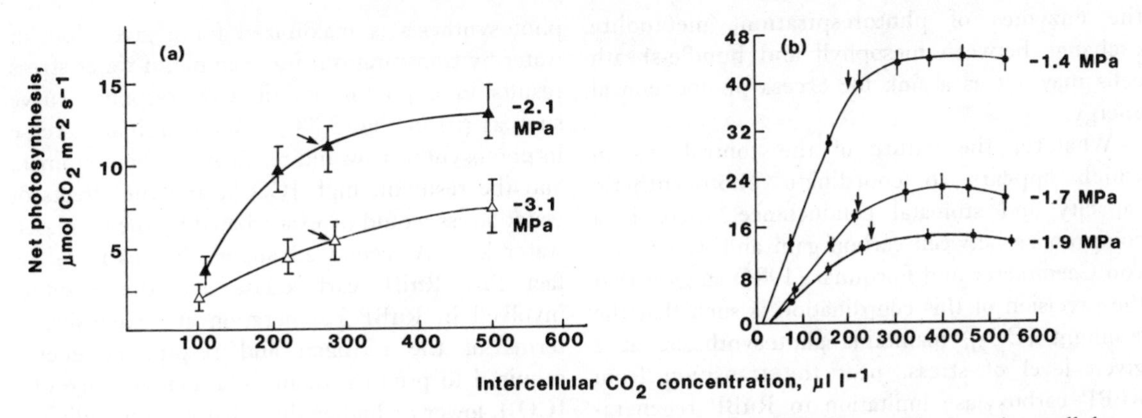


**Fig. 3.37** The influence of leaf water stress on the response of the rate of net photosynthesis to intercellular $CO_2$ concentration in (a) cotton and (b) sorghum. Measurements were made over a range of external $CO_2$ concentrations ($[CO_2]_a$), and the points obtained with a $[CO_2]_a$ of 345 μl $l^{-1}$ (cotton) and 340 μl $l^{-1}$ (sorghum) are indicated by arrows (after Hutmacher and Krieg 1983 and Krieg and Hutmacher 1986).

however, the ratio of photosynthesis to conductance was lower, at a given external $CO_2$ concentration ($[CO_2]_a$), in the water-stressed plants, indicating that photosynthesis was reduced more than conductance.

Increased leaf temperature which, in the experiment with cotton, varied over the range 24–40 °C, had a similar effect. When the temperature was more than 30 °C, photosynthesis and conductance were both reduced, but not in proportion: for example, a 41 per cent reduction in net photosynthesis was associated with an 18 per cent decline in conductance. The photosynthesis : conductance ratio therefore declined and $[CO_2]_i$ increased, both in irrigated and dryland plants. This increased non-stomatal limitation (intracellular resistance) to photosynthesis at the higher temperatures could not be attributed solely to enhanced photorespiration and respiration because gross, as well as net, photosynthesis was affected. The shape of the photosynthetic $CO_2$ response curves illustrated in Fig. 3.37 also indicates non-stomatal limitations to photosynthesis, because the carboxylation efficiency (the slope of the linear part of the curve at low $[CO_2]_i$) is clearly reduced by water deficits.

The $[CO_2]_i$ is thought to play a major part in the concerted control of the intracellular and stomatal resistances, and central to this is the closure of stomata in response to increased $[CO_2]_i$. It is suggested that a relatively constant $[CO_2]_i$ is maintained because any increase as a result of reduced photosynthetic capacity is countered by reduced stomatal aperture. Other signals may reinforce this negative feedback loop between stomata and $[CO_2]_i$. For example, abscisic acid (ABA), produced in the mesophyll chloroplasts in response to water stress, diffuses to the guard cells and causes stomatal closure, and photosynthetic metabolites have also been implicated (Walton 1980; Farquhar and Sharkey 1982). Other plant growth regulators may also be involved. Cytokinins, for example, promote stomatal opening, and may also enhance photosynthesis by stimulating chlorophyll and enzyme synthesis. Reductions in the supply of cytokinins from the roots to the leaves, as a result of water deficits, may therefore have parallel effects on both the intracellular and the stomatal resistances (Mansfield and Davies 1985).

Photorespiration has a dual significance here. Increases relative to photosynthesis would contribute to the reduced photosynthetic capacity of water-stressed plants, and hence to the tendency for $[CO_2]_i$ to increase and the resultant closure of stomata. In addition, the recycling of $CO_2$ through the photorespiratory cycle is a means of dissipating photochemical energy which is in excess of that needed for the limited operation of the Calvin cycle (p. 50; Osmond *et al.* 1980). It is possible, therefore, that if photorespiration were to be reduced in $C_3$ plants by some means, the advantage in terms of carbon gain would be outweighed by the loss of this protection. In the mesophyll cells of $C_4$ plants, which lack many of the enzymes of photorespiration, metabolite exchange between mesophyll and bundle-sheath cells may act as a sink for excess photochemical energy.

Whatever the nature of the control system which appears to coordinate photosynthetic capacity and stomatal conductance, there is a compromise between carbon gain and water loss. von Caemmerer and Farquhar (1984) suggest that the precision of this coordination is such that the resultant $[CO_2]_i$ maintains photosynthesis, at a given level of stress, near the transition from RuBP carboxylase limitation to RuBP regeneration limitation (p. 48). Thus, the $[CO_2]_i$ obtained with a normal $[CO_2]_a$ of 330–340 μbar will correspond to a rate of photosynthesis close to the point of inflection of the photosynthesis : $[CO_2]_i$ response curve, and will therefore depend on the shape of the curve. It is difficult to judge whether this is so when there is a broad inflection, as in the field-grown plants from which Fig. 3.37 was obtained. However, when water stress was induced rapidly in pot-grown plants, the more severe effect on photosynthesis revealed that the point of inflection does appear to dictate the $[CO_2]_i$ (Fig. 3.38).

A benefit of this fine control of $[CO_2]_i$ is that at the transition from limitation by RuBP carboxylase to limitation by RuBP regeneration, net photosynthesis is maximized for a given loss of water by transpiration; for example, if water stress results in a photosynthetic $CO_2$ response curve such as (ii) in Fig. 3.38, only a marginal increase in photosynthesis would result from wider stomata and the resultant high $[CO_2]_i$, and the effects of water stress would be exacerbated by the increased water loss. A second advantage derives from the fact that RuBP carboxylase and the enzymes involved in RuBP regeneration are expensive in terms of the nitrogen and respiratory energy required to produce them. The maintenance of a $[CO_2]_i$ lower or higher than that corresponding to the point of inflection on the $CO_2$ response curve makes inefficient use of either the RuBP regeneration system or the available RuBP carboxylase. In this connection, it should be remembered that the effects of water deficits may be compounded by deficiencies of nitrogen and other minerals.

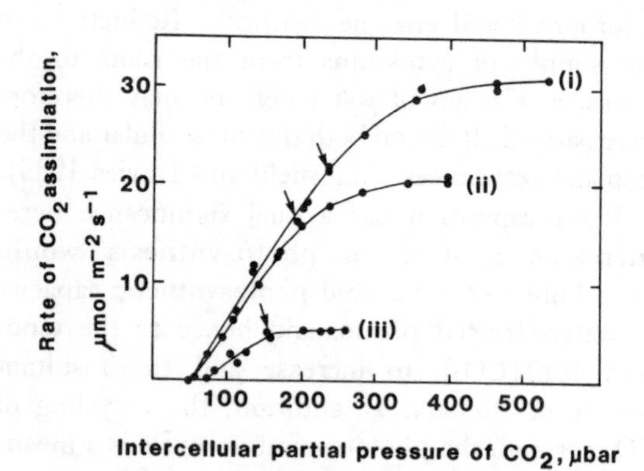


**Fig. 3.38** Responses of the rate of $CO_2$ assimilation to the intercellular partial pressure of $CO_2$ ($[CO_2]_i$, μbar) in leaves of *Phaseolus vulgaris*, measured at a PPFD of 1500 μmol $m^{-2}$ $s^{-1}$ and a leaf temperature of 28 °C. (i) before water was withheld; (ii) two days after water was withheld; (iii) four days after water was withheld. The arrows indicate the points obtained at an external $p(CO_2)$ of 330 μbar (after von Caemmerer and Farquhar 1984).

## Mechanisms by which water stress could affect stomatal and non-stomatal resistances

We have discussed the various ways in which water deficits can affect net photosynthesis. When water stress develops slowly, as normally occurs in the field, stomatal and non-stomatal resistances appear to be kept in balance, so that $[CO_2]_i$ is little changed, irrespective of the extent to which photosynthesis is reduced. Some of the ways by which this equilibrium is achieved have already been noted (p. 73), with $[CO_2]_i$ itself playing a central part. However, although it is clear that changes in photosynthetic capacity and stomatal resistance will influence $[CO_2]_i$, the means by which water deficits induce these changes are less clear. We have noted the possibility of hormonal influences, but it is difficult to explain the differential effects on photosynthetic, respiratory and photorespiratory metabolism.

It is suggested that turgor, or cell volume, rather than leaf water potential itself, is responsible for the metabolic effects (Schulze 1986), but the mechanism is not known. However, it can be seen from Figs 3.37 and 3.38 that water stress can reduce both the carboxylation efficiency (the initial slope of the net photosynthesis : $[CO_2]_i$ curve) and the $CO_2$-saturated rate of photosynthesis, indicating effects on both carbon reduction and the photochemical system. One explanation is that changes in turgor, for example, disrupt the conformation of the thylakoid membranes (cf. possible effect of starch accumulation in chloroplasts, p. 63), altering the pH gradient across the membrane, and reducing both the activities of the Calvin cycle enzymes and ATP production (Krieg and Hutmacher 1986).

The extent to which the transport component of the intracellular resistance is affected by water stress is unclear, and the mechanism of any such influence correspondingly obscure. We have already noted that in the unstressed plant the transport resistance is quantitatively less important than the biochemical (carboxylation) resistance, but there is some evidence that in water-stressed plants its significance is increased. Moorby *et al.* (1975) withheld water from potato plants for five days, reducing the mean leaf water potential from −0.23 MPa to −0.94 MPa. The rate of photosynthesis was reduced by this treatment, especially in the younger, expanding leaves (Table 3.9); associated with this were increases in the stomatal resistance and relatively larger increases in the intracellular resistance. However, the latter effect could not be attributed to reductions in the activity of RuBP carboxylase, which was unaffected by these water deficits. It was therefore concluded that the increased intracellular resistance probably reflected effects of water stress on the diffusion of $CO_2$ in the liquid phase, although the physical basis of such an effect is not clear.

Levitt (1980) speculates that loss of intercellular air could be responsible. In a fully turgid leaf, the area of contact between cells is minimized and the intercellular space maximized. As cells become less turgid, under moderate water stress, contact between them increases, air is squeezed out of the tissues and intercellular water is displaced to occupy a greater number of the reduced intercellular volumes. The path length for diffusion in the liquid phase is thus increased.

There are several ways in which water deficits, potential or actual, can influence the stomatal response, and these can usefully be interpreted in terms of first and second lines of defence (Mansfield and Davies 1985). The plant needs to respond quickly to rapidly imposed, but perhaps temporary, changes in the environment, such as increased windspeed, which could result in severe stress. The stomatal response to $[CO_2]_i$ probably serves as part of this first line of defence, because an increase in windspeed will increase the $CO_2$ supply to the leaves (p. 43). Stomata can also respond directly to the humidity of the air, an increase in the vapour pressure deficit between the leaf and the air causing stomatal closure, apparently because of reduced guard cell turgor due to

**Table 3.9** Effect of drought on $CO_2$ uptake and the gaseous ($r_l$) and intracellular ($r_r$) resistances of potato leaves

| *Treatment* | *Leaf age (days)* | $r_l$ *(s cm⁻¹)* | $r_r$ *(s cm⁻¹)* | *$CO_2$ uptake (ng cm⁻² s⁻¹)* |
|---|---|---|---|---|
| Control | 21 | 0.8 | 8.4 | 29.1 |
| | 23 | 0.9 | 9.1 | 26.6 |
| | 29 | 1.0 | 33.1 | 10.7 |
| Droughted | 21 | 7.5 | 294.6 | 0.7 |
| | 23 | 6.2 | 205.4 | 1.3 |
| | 29 | 3.7 | 67.2 | 6.1 |

Water was withheld from droughted plants for 5 days. The intracellular resistances were calculated on the basis of photosynthesis measurements derived from the accumulation of radioactivity by leaves exposed to $^{14}CO_2$ for 20 seconds; this is considered to be insufficient time for photorespiratory loss of $^{14}CO_2$, and such estimates therefore give higher rates of photosynthesis than do measurements of net $CO_2$ flux by infra-red gas analysis. Estimates of $r_r$ are therefore correspondingly reduced. The differences between treatments in the intracellular resistance are possibly exaggerated by the assumption that $\Gamma$ remains constant across the range of water deficits (see p. 69).
(From Moorby *et al.* 1975)

direct evaporation from the epidermis. This effect is independent of the bulk leaf water potential. Increased drying of the atmosphere may, of course, be independent of windspeed, but inevitably a diminished boundary layer will enhance the leaf-to-air vapour pressure deficit. Increased windspeeds may therefore bring about stomatal closure through both the $[CO_2]_i$ and the humidity response.

Stomata are thus able to sense the aerial environment, and close under conditions which might otherwise lead to water deficits. However, on their own, these measures are inadequate, because water stress can still develop as a result of the drying of the soil. Thus, when the mesophyll cells lose turgor, a second line of defence comes into play, which overrides the direct response of the stomata to the state of the atmosphere. As we have noted already (p. 73), this is mediated by ABA, perhaps interacting with other growth regulators such as cytokinins and auxin.

**Adaptation to water stress**

Finally, we draw attention to the influence of experimental procedure on the observed effects of water stress on photosynthesis. The methods of imposing stress, the age and previous environmental history of the leaves selected for measurement, and the rapidity with which stress is imposed have considerable effects on the plants' responses, and are undoubtedly responsible for much of the lack of uniformity of results in the literature. In particular, the environment to which the plant has been subjected determines the extent to which it may have become adapted to water stress. Osmoregulation, which involves the accumulation of solutes in cells and the consequent maintenance of turgor, is an important mechanism, and for further details the interested reader is referred to Turner and Kramer (1980) and Kramer (1983).

The development of water stress in the field is often more gradual than that imposed during experiments, and may thus allow time for adaptation to the changing conditions. For this reason, great care must be taken when extrapolating results obtained in the laboratory to plants growing in the field. For example, Jordan and Ritchie (1971) found that in cotton plants, grown in a greenhouse and measured in a growth chamber, the stomata closed at a leaf water potential of −1.6 MPa, whereas those of plants grown in the field were still partially open at −2.7 MPa. Similar differences have been reported for maize, sorghum and vines, and are discussed by Begg and Turner (1976).

### 3.4.5.2 NITROGEN STRESS

A deficiency of nitrogen greatly reduces crop growth, and the relief of this deficiency is most obvious in morphological and developmental responses, such as accelerated or renewed leaf expansion or tillering (sect. 2.2). The major effect of nitrogen on crop photosynthesis is, therefore, through increased light interception (sect 2.2.3, 2.3.3, 6.4, 7.5 and 8.3). However, there is much evidence that there can also be direct effects of nitrogen on the rate of photosynthesis per unit leaf area. Nitrogen is, after all, an essential component of proteins; up to 65 per cent of leaf protein occurs in the chloroplasts, of which perhaps 50 per cent is RuBisCO. In addition, it is known that nitrogen nutrition affects the size and morphology of chloroplasts (e.g. Baszynski *et al.* 1972; Hall *et al.* 1972; Natr 1972; Mader and Volfova 1984).

A deficiency of nitrogen might therefore be expected to impair photosynthesis, and in many laboratory experiments, plants given a limited supply of nitrogen conform to this expectation. For example, Nevins and Loomis (1970) showed that transferring sugar beet plants to a nitrogen-free culture solution caused a progressive reduction in the rate of net photosynthesis. Ryle and Hesketh (1969) found that the low rates of photosynthesis in nitrogen-deficient maize plants were associated with increased mesophyll and, to a lesser extent, stomatal resistances. Good correlations have been found between the nitrogen contents of leaves and their rates of net photosynthesis (Natr 1972); and the timing and amount of nitrogen supplied to young barley seedlings have been shown to affect the maximum amount of fraction I protein synthesized over a fifteen day period after sowing (Blenkinsop and Dale 1974).

However, there are dangers in extrapolating these observations to plants growing in the field. Gregory *et al.* (1981) suggest that because most agricultural soils have nitrogen contents of around $10^{-3}$ M, leaf nitrogen contents are unlikely to be reduced enough to affect photosynthesis. Furthermore, the very abundance of RuBP carboxylase and the fact that at any time part of it may be inactive suggests that photosynthesis is buffered against nitrogen deficits: the enzyme may, in fact, serve as a reserve of protein in the leaf (Huffaker and Peterson 1974). In field experiments, therefore, the rates of photosynthesis per unit leaf area often fail to respond to increasing supply of nitrogen fertilizer, as was found, for example, with spring wheat despite increased accumulation of RuBP carboxylase and chlorophyll in the leaves (Thomas and Thorne 1975). Nevertheless, nitrogen levels in the leaves may become low enough to impair photosynthesis as a result of the mobilization and export of nitrogen from them. This can occur when the uptake of nitrogen from the soil is insufficient to meet the combined demands of all the organs of the plant. Two examples of this will be used.

**Perennial ryegrass**

Robson and Parsons (1978) investigated the effects of nitrogen deficiency on photosynthesis and dry-matter partitioning in perennial ryegrass, designing the experiment so that responses to treatments could be attributed to direct effects of nitrogen on photosynthesis rather than to differences in leaf area. Small communities of plants were grown in a glasshouse at high density, so that complete light interception would be achieved while the plants were still uniform. The plants were watered with a complete nutrient solution, containing 300 ppm nitrogen, for twenty-five days, when the leaf area index of the communities was 5 and light interception was virtually complete; at this stage, each plant had five leaves on the main stem and on average five tillers. Half of the pots, the low-nitrogen (LN) communities, were then flushed through with pure water and thereafter received nutrient solution containing only 3 ppm nitrogen, while the high-nitrogen (HN) communities continued to receive 300 ppm nitrogen.

Towards the end of the experiment, when the communities had attained ceiling leaf area indices, the rates of canopy and single leaf net $CO_2$ exchange were measured over a range of irradiances and in the dark. The results are expressed on a unit ground area basis because this is the measure of net photosynthesis most closely related to the crop growth rate (Fig. 3.39a). At an irradiance of 450 W $m^{-2}$ PAR, representing the maximum irradiance on a clear summer's day in Southern England, the rate of net photosynthesis of the LN communities was almost 30 per cent lower than that of the HN communities. At low irradiances, on the other hand, the LN communities had the higher rate, the cross-over point being about 180 W $m^{-2}$. This effect was the result of a higher respiration rate in the HN communities. When gross canopy photosynthesis was estimated, by adding dark respiration to canopy net photosynthesis it was found that the HN communities had in fact fixed more carbon at all irradiances (Fig. 3.39b). It could be argued that this

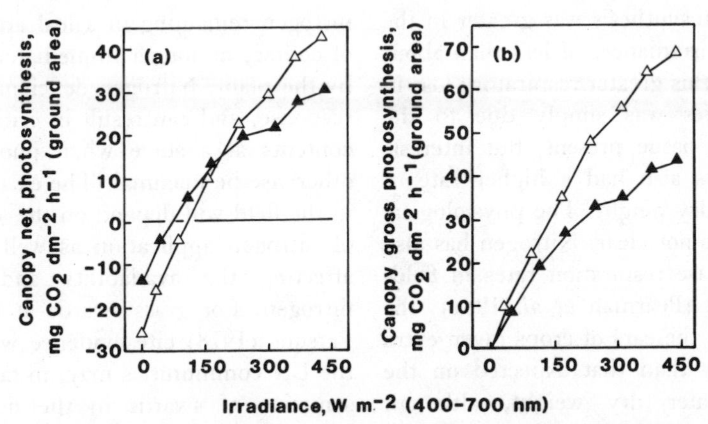


**Fig. 3.39** Light response curves of (a) canopy net photosynthesis, and (b) canopy gross photosynthesis, both on a ground area basis, of high nitrogen (△) and low nitrogen (▲) communities of S24 perennial ryegrass (from Robson and Parsons 1978).

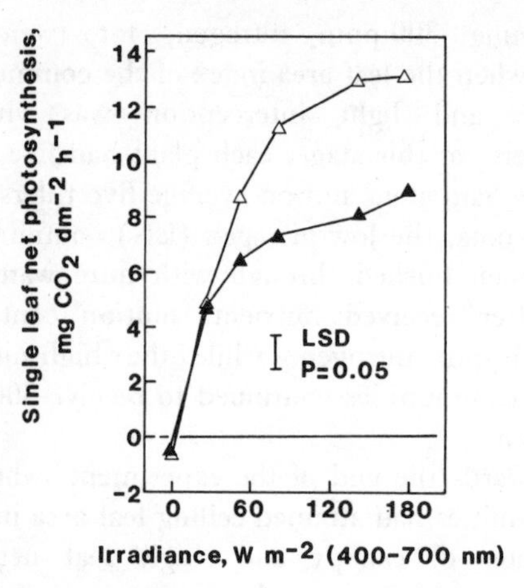


**Fig. 3.40** Light response curves of single leaf net photosynthesis of youngest fully expanded mainstem leaves from high nitrogen (△) and low nitrogen (▲) communities of S24 perennial ryegrass (from Robson and Parsons 1978).

was due simply to the larger leaf area which the HN communities developed, about twice that of the LN communities at the end of the experiment. However, the treatments were not imposed until light interception was virtually complete: a subsequent difference in leaf area should therefore have had little effect. The rates of single leaf net photosynthesis, expressed on a unit leaf area basis (Fig. 3.40), support the view that in this experiment a deficiency of nitrogen affected photosynthesis directly and not through reduced leaf area.

Thus, the HN communities used light more efficiently for photosynthesis than the LN communities, but this advantage was partially offset by higher rates of respiration, so much so that canopy net photosynthesis was greater in the LN communities at irradiances of less than about 180 W $m^{-2}$. Part of this greater respiratory loss in the HN communities was simply due to the greater mass of live tissue present, but interestingly the HN plants also had a higher rate of respiration per unit dry weight. The physiological significance of this is not clear. Nitrogen has also been found to increase respiration rates in field-grown winter wheat (Pearman *et al.* 1981): the rate of respiration in the ears of crops given extra nitrogen was greater than that expected on the basis of their greater dry weight, although increased dry weight did account for the higher respiration rates of stems and leaves.

Robson and Parsons (1978) concluded that harvestable yields could be increased by the application of nitrogen even when light was already fully intercepted, particularly as the plants receiving high nitrogen partitioned more dry matter into leaf laminae and less into roots. However, the relevance of this experiment to crops growing in the field depends on how realistic the low nitrogen treatment was. Its severity was assessed by comparing the nitrogen content of the leaf laminae with that considered to be necessary for maximum growth, accepted, from a consideration of published values, to be 4 per cent of above-ground dry weight. Nitrogen was accumulated by leaf laminae during their expansion but was subsequently lost, presumably because of remobilization back to the parent tiller during senescence (Robson and Deacon 1978). The HN treatment ensured that the peak nitrogen contents of successive laminae remained high (e.g. 5.8, 5.5 and 5.2 per cent of dry weight for leaves 5, 6 and 7, respectively), whereas the reduced availability of nitrogen in the LN communities meant that peak values declined from 5.7 per cent for leaf 5 to 3.0 per cent for leaf 7. Furthermore, the senescent decline in nitrogen content of a given lamina was more complete in the nitrogen-deficient plants, and began even before the leaves were fully expanded. For example, by the time lamina 7 of the LN plants had reached its maximum size, its nitrogen content was less than 2 per cent, whereas in HN plants it remained above 4 per cent until well after full expansion.

Grasses therefore conserve nitrogen by exporting it from a senescing leaf to the parent tiller: any nitrogen remaining in a leaf after it has fallen is, of course, no longer immediately available for use by the plant. Nitrogen deficiency accentuates this response, and can result in critically low nitrogen contents at a stage when photosynthesis would otherwise be maximal. The existence of this effect in the field will depend on the amount and timing of nitrogen application as well as all the factors affecting the availability and uptake of that nitrogen. For grass swards at least, Robson and Parsons (1978) cite evidence which suggests that the LN communities may, in fact, be more representative of swards in the field than the HN communities. For example, after the application of 84 kg N $ha^{-1}$ to a field of Italian ryegrass, the nitrogen content of the herbage fell from over 5 per cent to under 2 per cent in six weeks (Wilman 1965).

### Wheat

Mobilization of nitrogen from the leaves also occurs in cereals. In wheat, for example, the uptake of nitrogen is greatly reduced shortly after anthesis (Fig. 6.8), and nitrogen is exported to the ears to meet the demands of the growing grains. Thus, Austin *et al.* (1977b) found that only about 50 per cent of the nitrogen present in the ears of winter wheat was taken up during the grain-filling period. Gregory *et al.* (1981) confirmed this for a crop of winter wheat grown in 1977, but found that when post-anthesis uptake of nitrogen was even less during the hot, dry summer of 1976, this figure was reduced to only about 25 per cent. These differences in the weather, together with treatments which varied the amount and timing of nitrogen, and the amount of irrigation, produced varying patterns of nitrogen export from the leaves. This export of nitrogen, together with other elements, is a natural part of a leaf's senescence, but is accelerated when the rest of the plant is short of nitrogen. As the leaves aged and nitrogen was lost, the maximum (i.e. light-saturated) rate of photosynthesis ($P_{max}$) of the flag leaf decreased. However, nitrogen content *per se* was not thought to be responsible because when leaves of the same age, but different nitrogen content, were compared there was little difference in $P_{max}$ (Fig. 3.41).

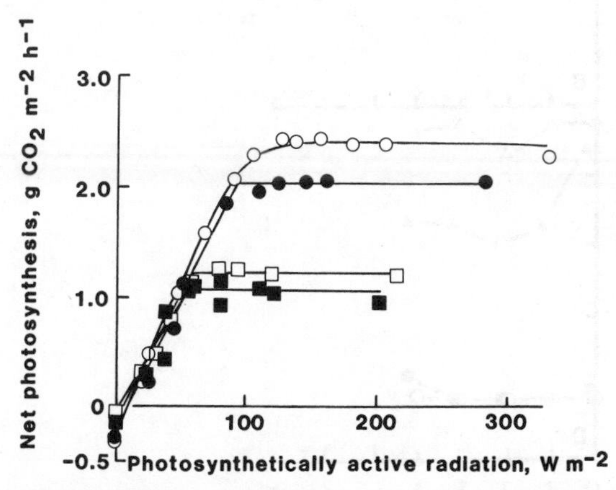


**Fig. 3.41** Photosynthetic light response curves for flag leaves of winter wheat given no nitrogen fertilizer (open symbols) and 151 kg N ha$^{-1}$ (closed symbols), measured at anthesis (circles) and about three weeks later (squares) (after Gregory *et al.* 1981).

Thus, although it was possible to present a positive correlation between the nitrogen content of the flag leaf and $P_{max}$, a causal relationship could not be inferred. However, there was also, for each season, a good inverse relationship between $P_{max}$ and the proportion of nitrogen which was lost from the leaf; furthermore, the seasonal differences disappeared when the $P_{max}$ values were normalized by expressing them as fractions of the appropriate value at anthesis (Fig 3.42). It is clear, therefore, that the decline in $P_{max}$ after anthesis was closely related to the proportion of nitrogen exported from the leaves, independent of season or treatment.

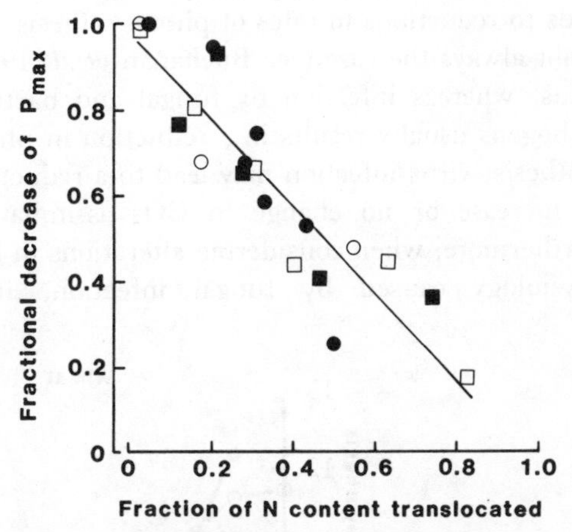


**Fig. 3.42** Relationship between the fractional decrease in the maximum rate of photosynthesis and the fraction of nitrogen content translocated from flag leaves of winter wheat in 1976 (○, no added fertilizer; ●, 151 kg N ha$^{-1}$ added) and 1977 (□, 124 kg N ha$^{-1}$ added; ■, 224 kg N ha$^{-1}$ added) (from Gregory *et al.* 1981).

Field experiments therefore suggest that the nitrogen content of most agricultural soils may be sufficient to preclude any effects of nitrogen fertilizer treatments on the rate of net photosynthesis throughout most of the crop's life. However, during the post-anthesis period, low rates of uptake, coupled with a heavy demand for nitrogen by the grains, can cause loss of nitrogen from the leaves and a reduction in photosynthesis. Gregory *et al.* (1981) further suggest that in laboratory experiments, with cereals and other crops, such a loss of nitrogen from the leaves may be precipitated at any stage of development, as a result of the imposition of sudden reductions in the nitrogen content of the growing medium. The resulting correlations between the nitrogen content of leaves and the rate of photosynthesis may therefore imply a dependence of photosynthesis on the extent of translocation of nitrogen from the leaf rather than on nitrogen content *per se*. Furthermore, although the precise effect of this loss of nitrogen is not clear, the proportion of nitrogen exported from a leaf under a particular set of conditions might be interpreted simply as a measure of the progress of senescence in that leaf.

### 3.4.5.3 PATHOGEN INFECTION

Although infection by plant pathogens usually leads to reductions in rates of photosynthesis, this is not always the case (see Buchanan *et al.* 1981). Thus, whereas infection by fungal and bacterial pathogens usually results in a reduction in photosynthesis, virus infection may lead to a reduction, an increase or no change in $CO_2$ assimilation. Furthermore, when considering alterations in host physiology caused by fungal infection, it is important to recognize that plant pathogenic fungi can be broadly split into two groups, based largely on their nutritional behaviour: biotrophic and necrotrophic fungi (Lewis 1973). Obligately biotrophic fungal pathogens, such as rusts and powdery mildews, do not kill their host plant immediately and are dependent upon viable host tissue to complete their development. Necrotrophic fungi, on the other hand, such as the blight and rotting fungi, can survive just as well on decaying or dead host tissue. Although the end result of infection by either a biotroph or a necrotroph might be to reduce rates of photosynthesis, the mechanisms responsible are different. Reductions in photosynthetic area as a result of pathogen infection have already been discussed (sect. 2.3.4).

Rates of net photosynthesis in heavily mildewed or rusted leaves are usually substantially reduced when compared with uninfected control leaves (see Walters 1985; Fig. 3.43). In general, such alterations are attributed less to stomatal behaviour than to changes in the photosynthetic machinery. Thus, only small increases in $r_s$ have

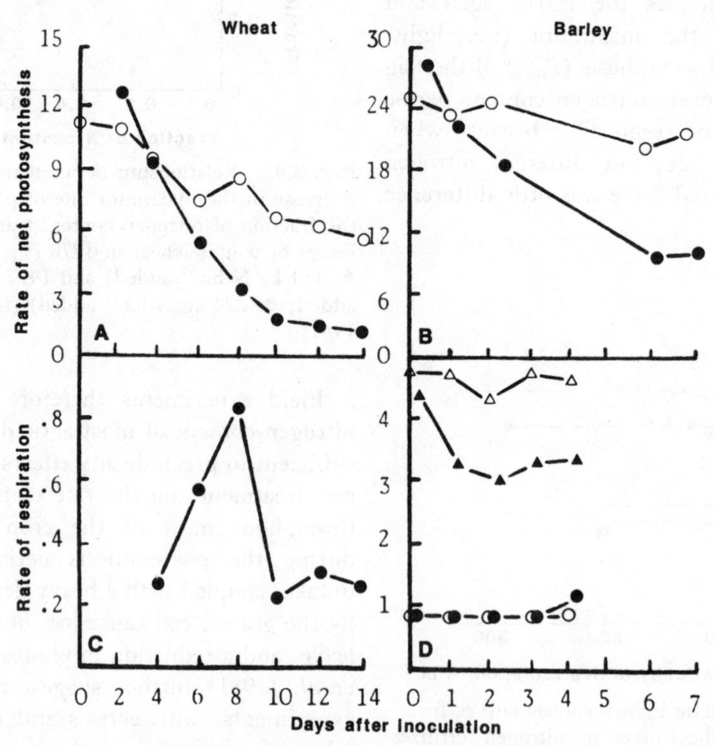


**Fig. 3.43** Effects of powdery mildew of wheat and barley on the rates of net photosynthesis (A and B: ○, healthy; ●, diseased), photorespiration (D: △, healthy; ▲, diseased), and dark respiration (C and D: ○, healthy; ●, diseased) Units: A, B, D mg $CO_2$ $dm^{-2}$ $h^{-1}$; C $mm^3$ $O_2$ $mg^{-1}$ $h^{-1}$ (after Ayres 1979).

been reported in mildewed (*Erysiphe graminis hordei*) barley (Ayres 1979) and oak (Hewitt and Ayres 1975), whereas a small decrease in $r_s$ was observed in barley infected with brown rust (*Puccinia hordei*) (Owera *et al.* 1981).

On the other hand, a characteristic symptom of mildew or rust infections is an increased chlorosis of infected leaves (e.g. Mitchell 1979), although, in general, changes in chlorophyll concentrations in diseased leaves do not correlate well with alterations in rates of photosynthesis. Nevertheless, Owera *et al.* (1981) concluded that loss of chlorophyll was a principal limiting factor for photosynthesis in barley infected with brown rust, and Ahmad *et al.* (1983) highlighted the decrease in the number of chloroplasts. More recent work (Scholes and Farrar 1986) has shown that rust infection of bluebell, caused by the fungus *Uromyces muscari*, results in reductions in chloroplast numbers per unit area, chloroplast volume, chlorophyll concentration and the ratio of chlorophyll a : b, within individual pustules. All of this tends to suggest that in rusted bluebell, chlorophyll is lost from individual chloroplasts.

Although it has been found that in some host-parasite systems, photophosphorylation was not affected (Ahmad *et al.* 1983), there is good evidence now to indicate that in some rust and mildew infections, reductions in the rate of photosynthesis are due, in part, to a parasite-induced block in the non-cyclic electron transport chain (e.g. Magyarosy *et al.* 1976). It was originally suggested that such an inhibition of non-cyclic photophosphorylation might be due to the parasite-induced formation of a compound functionally similar to the inhibitor DCMU (Montalbini and Buchanan 1974). However, subsequent work, where chloroplast membranes from rusted leaves were extensively washed, effectively ruled out that possibility (see Montalbini *et al.* 1981). It seems more likely that the inhibition of non-cyclic photophosphorylation observed in mildewed sugar beet and rusted bean is due to an alteration of certain components of the non-cyclic electron transport chain. Thus, Magyarosy and Malkin (1978) have shown that the cytochrome content of the electron transport chain was substantially reduced by mildew infection.

Scholes and Farrar (1986) have proposed another mechanism to account for the parasite-induced alterations in the electron transport chain. They suggest that for rusted bluebell, damage to the photosynthetic unit (the result of loss of chlorophylls from the light harvesting complex as well as a specific loss of chlorophyll a from the photosystem 2 light harvesting complex) would lead to a reduction in the rate of non-cyclic photophosphorylation. The authors also suggest that if such damage to photosynthetic units is a reflection of a more general decline in membrane integrity of the chloroplasts, then it is quite likely that the content of other components of the electron transport chain, such as cytochromes, will also be affected.

RuBP carboxylase is also affected. Thus, powdery mildew infection of barley has been shown to result in a substantial decrease in the activity and amount of RuBP carboxylase (Walters and Ayres 1984; Fig 3.44). A reduction in activity has also been reported in rust-infected (*Puccinia graminis tritici*) wheat (Wrigley and Webster 1966). Reductions in the amount of RuBP carboxylase in infected leaves should not be surprising in view of the multifarious changes in nucleic acid metabolism that occur in these leaves (see Chakravorty and Scott 1982). Indeed, mildew infection has been shown to result in a reduction in ribosomal RNA in chloroplasts (e.g. Dyer and

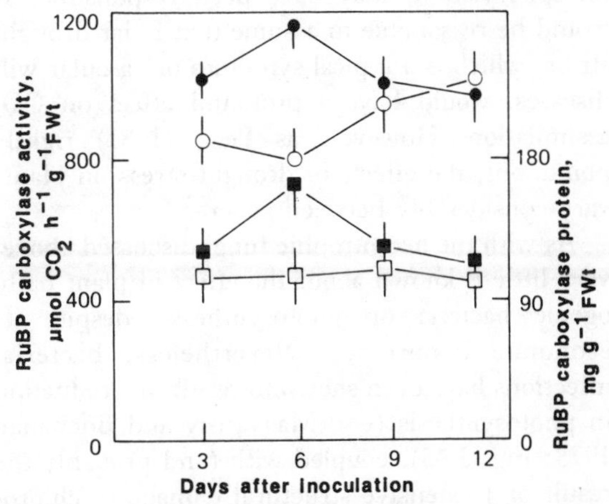


**Fig. 3.44** RuBP carboxylase activity (●) and RuBP carboxylase protein content (■) of uninfected third leaves of barley infected on the lower two leaves with powdery mildew. Healthy controls (○, □). Differences between means on day 6 significant at $P = 0.001$ for enzyme activity and at $P = 0.01$ for enzyme protein content (from Walters 1985).

Scott 1972). More recently, Higgins *et al.* (1985) have shown that mildew infection of barley results in a reduction in the amounts of mRNA coding for the two sub-units of RuBP carboxylase, a consequence of which would be a loss of chloroplast proteins and a reduction in the rate of photosynthesis.

In contrast to the effects of obligate biotrophs on $CO_2$ assimilation, those of necrotrophic fungi have been largely ignored. Obviously, leaves turned brown and necrotic, as a result of infection by such pathogens, will be photosynthetically useless. But uninfected areas of the same leaf, or even infected areas before cell death and necrosis, may also be affected. Reductions in rates of photosynthesis have been reported for infection of wheat by the glume blotch fungus, *Septoria nodorum* (e.g. Scharen and Krupinsky 1969), for infection of tomato by the wilt fungus, *Fusarium oxysporum* f.sp. *lycopersici* (Duniway and Slatyer 1971), and for infection of cotton by another vascular wilt organism, *Verticillium dahliae* (Mathre 1968). Although Duniway and Slatyer (1971) considered that reduced stomatal aperture was an important determinant of reductions in photosynthesis in wilted tomato, Mathre (1968) suggested that, in cotton leaves infected with *V. dahliae*, alterations in light and dark reactions of photosynthesis, including much reduced photophosphorylation, may have been responsible. It would be reasonable to assume that foliar drought stress, which is a typical symptom of vascular wilt diseases, would have a profound effect on $CO_2$ assimilation. However, as Pegg (1981) rightly points out, the effects of drought stress on plants vary considerably between genera.

As with the necrotrophic fungi discussed above, very little is known about the effect of plant pathogenic bacteria on photosynthesis, despite its economic importance. Nevertheless, bacterial infections have been shown to result in a reduction in photosynthesis (e.g. Magyarosy and Buchanan 1975; Fig. 3.45), coupled with (and probably the result of ) extensive structural damage to chloroplasts (e.g. Sigee and Epton 1976). A feature of many bacterial infections of plant tissues is the formation and release of toxins by the invading organism. Indeed, it has been tentatively suggested that the inhibition of photosynthesis and the structural changes in chloroplasts

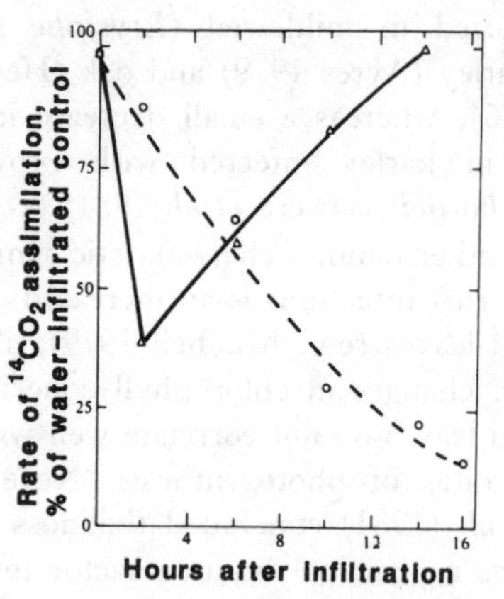


**Fig. 3.45** Effect of bacterial infiltration on the rate of photosynthetic $^{14}CO_2$ assimilation by leaf discs of bean (*Phaseolus vulgaris*). Saprophyte (△); pathogen (○) (from Buchanan *et al.* 1981).

observed in bacterial infections could be due, in part, to the release of such toxins into host tissue (Mitchell 1978). For example, tabtoxin, produced by *Pseudomonas tabaci*, has been reported to inhibit RuBP carboxylase activity in tobacco (Crosthwaite and Sheen 1979).

As mentioned earlier, different virus/host interactions may result in different photosynthetic responses. Buchanan *et al.* (1981) have suggested that depending upon the particular virus/host combination, it is possible to distinguish two effects of a virus on photosynthesis. The first is illustrated by infection of squash with squash mosaic virus and infection of piggyback (*Tolmiea menziesii*) plants with tomato bushy stunt and cucumber virus. Here, virus infection may result in a reduction in the number of chloroplasts but with very little effect on rates of photosynthesis. The second effect is illustrated by chinese cabbage infected with turnip yellow mosaic virus, where infection alters the partial reactions of photosynthesis, for example photophosphorylation, resulting in reduced plant growth. In fact, many virus infections result in dramatic reductions in photosynthesis. Naidu *et al.* (1984) have shown that the reduction in net photosynthesis caused by infection of peanut with peanut green mosaic virus was due partly to a reduction in chlorophyll a and partly to a direct inhibition of photosystem 2, mainly as a result of a smaller pool of plastoquinone.

Part of the reduction in net photosynthesis in

diseased plants could, of course, be due to increases in photorespiration and respiration. However, studies of photorespiration in diseased plants are scant, and most of the data describe changes in plants infected with obligate biotrophs. Thus, photorespiration was found to decrease in powdery mildew-infected oak and barley, due in the latter case to decreased activities of glycollate oxidase, glyoxylate reductase and RuBP oxygenase (see Walters 1985). Although reductions in photorespiration have also been observed in rusted plants (e.g. Raggi 1978), this is not always the case. Thus, whereas Mitchell (1979) could not detect any change in the rate of photorespiration in wheat with stem rust, an increase was observed in barley infected with brown rust (Owera *et al.* 1981). These differences illustrate a very important consideration in studies of host/pathogen interactions: what is true for one host/parasite combination may not be true for another, no matter how closely related the pathogens or the host plants.

More is known about the behaviour of respiration in infected plants. One of the earliest reports of respiratory changes in diseased plants is that of Allen and Goddard (1938), who showed not only that dark respiration was increased in powdery mildew-infected wheat, but also that most of the increased respiration was of host origin. In powdery mildew-infected barley, increases in dark respiration can be detected from three days after inoculation, well before the appearance of fungal growth on the leaf surface (e.g. Smedegaard-Petersen 1980; Fig. 3.46). Respiration remained high until fungal spore production stopped and leaf senescence began. Increased respiration in biotrophic infections appears to be the result of a shift from the glycolytic to the pentose phosphate pathway, the activity of which is greatly increased (Daly 1976). The pentose phosphate pathway appears to be located in the cytosol and is limited by the availability of $NADP^+$. Ryrie and Scott (1968) have suggested that the increased activity of this pathway may be due to release of $NADP^+$ into the cytosol after chloroplast breakdown in rust and mildew infections.

Whatever the cause of the increased activity of the pentose phosphate pathway, several-fold increases in the activities of enzymes of the pathway have been detected in plants infected with biotrophs (e.g. Scott 1965). Further evidence for a shift from the glycolytic pathway to the pentose phosphate pathway comes from the calculation of $C_6/C_1$ ratios. The use of this ratio in assessing the relative contributions of these two pathways in the degradation of glucose is based upon the evolution of $^{14}CO_2$ respired from glucose-1-$^{14}$C and glucose-6-$^{14}$C. During the operation of the glycolytic pathway, $^{14}CO_2$ will be produced at an equal rate from glucose-1-$^{14}$C and glucose-6-$^{14}$C, giving a $C_6/C_1$ ratio of one. However, if glucose is oxidized via the pentose phosphate pathway, the $C_1$ carbon will contribute most of the evolved $CO_2$ and the $C_6/C_1$ ratio will fall below one, as has been demonstrated several times with rusted and mildewed tissues (e.g. Shaw and Samborski 1957).

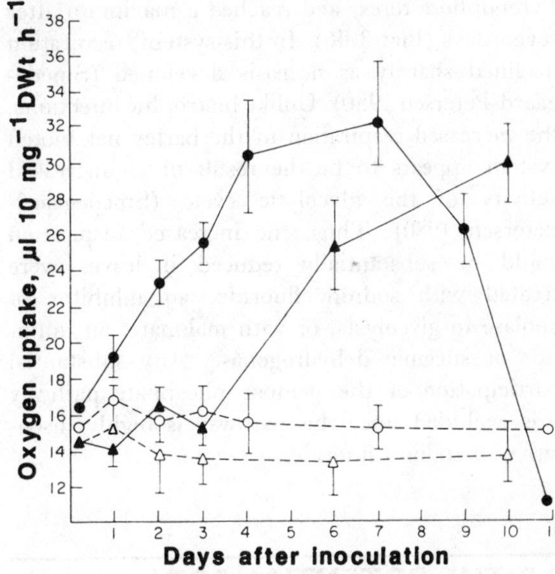


**Fig. 3.46** Time course of respiration in two susceptible barley cultivars, inoculated with either the net blotch pathogen, *Pyrenophora teres* (a necrotroph), or the powdery-mildew pathogen, *Erysiphe graminis* f.sp. *hordei* (a biotroph). ●, inoculated with *P. teres*; ○, uninoculated control. ▲, inoculated with *E. graminis*; △, uninoculated control. Vertical bars represent standard deviations (after Smedegaard-Petersen 1984).

Increased respiration in plants infected with necrotrophic pathogens follows a somewhat different pattern. Thus, in barley, increases in respiration were detected within twenty-four hours of inoculation with the net blotch fungus,

*Pyrenophora teres*, and reached a maximum after seven days (Fig. 3.46). In this system, respiration declined sharply as necrosis developed (Smedegaard-Petersen 1980). Unlike biotrophic infections, the increased respiration in the barley net blotch system appears to be the result of an increased activity of the glycolytic cycle (Smedegaard-Petersen 1980). Thus, the increased respiration could be substantially reduced if leaves were treated with sodium fluoride, an inhibitor of enolase in glycolysis, or with malonate, an inhibitor of succinic dehydrogenase. Any substantial participation of the pentose phosphate pathway was excluded since this pathway is highly insensitive to sodium fluoride.

## 3.5 THE POTENTIAL FOR INCREASING CROP PHOTOSYNTHESIS

### 3.5.1 INTRODUCTION

In this brief section we consider the ways in which the rate of crop photosynthesis might be increased. One possibility is to increase the interception of PAR by the canopy throughout the growing season, and to improve the distribution of that PAR among the various leaves by modifying canopy structure. This has been dealt with in section 2.4. The other way is to increase the efficiency of conversion of PAR into dry matter. Here we concentrate on photosynthesis and photorespiration; evidence that respiration can be reduced through breeding is examined in section 4.4.2. For a more detailed overall view, the reader is referred to Gifford and Jenkins (1982).

It should also be remembered that the absolute rate of photosynthesis may be less important than the rate relative to the consumption of some scarce resource, such as water or nitrogen. Increased water use efficiency (see p. 57), for example, is an aim of many $C_3$ crop improvement programmes. This can be achieved by increased stomatal closure, which reduces transpiration proportionately more than photosynthesis because the minimum residual (mesophyll) resistance is greater than the minimum stomatal resistance (Table 3.5). The capacity to synthesize abscisic acid (ABA) may therefore be an important characteristic, and one which is amenable to manipulation through breeding: it varies among wheat varieties, for example, and is strongly inherited (Austin *et al.* 1986).

### 3.5.2 BREEDING FOR INCREASED RATES OF NET $CO_2$ EXCHANGE PER UNIT LEAF AREA

The measure of photosynthetic performance with which economic yields are best correlated is the total productivity over the whole, or a major part, of the growing season. For example, grain yield of barley is directly proportional to the dry weight of the crop at anthesis (Dyson 1977). However, seasonal or management-induced differences in total photosynthetic production can be attributed primarily to differences in the cumulative intercepted PAR, rather than to any influence of the rate of $CO_2$ exchange per unit area of leaf (CER) (Figs 1.1, 3.3 and 7.16). In comparisons either within or between genotypes, CER does not correlate well with crop yield. This is probably largely due to the need to standardize the measure of photosynthesis, and the maximum CER, at light saturation, is commonly chosen. However, large differences in maximum CER may not be reflected in the CER of canopies, where many leaves are shaded. Developmental differences make it difficult, in fact, always to compare maximum rates, and comparisons may be further confounded by environmental adaptations in the field. The measured maximum CER is, therefore, not necessarily representative of the overall net photosynthesis of the crop throughout the season.

Historical analyses suggest that CER has been unimportant in past improvements in yield. Modern bread wheat, for example, has a lower maximum CER than some of its progenitors (Evans and Dunstone 1970), and in the twentieth century, the increased yields of cultivated wheat have been due to increases in the harvest index, the potential for total biomass production being unaltered (sect. 5.3). Improvements in yield, whether by management or through breeding, are inevitably achieved by the easiest routes. In wheat, for example, there has been scope, over several millenia, for simple selection for high grain yield because of variation in the harvest index (sect. 6.4.1). The CER has been relatively unimportant, particularly since the increased grain yields have been accompanied by increased leaf area. Indeed, the comparatively low maximum CER of *Triticum aestivum* may be an inevitable consequence of selection for high grain yield, because in wheat, and many other species, there is a negative association between leaf area and CER (e.g. Bhagsari and Brown 1986).

However, there is a limit to the extent to which improvements in harvest index and PAR interception can be relied upon to continue the trend of increased grain yields. Larger leaves, for example, bring the risk of excessive shading or water loss, and there may be little further scope for increased harvest index (sect. 5.3 and 6.4.1). Consequently, there is much interest in the reasons for variation in CER among genotypes, with the aim of identifying selection criteria and ultimately of breeding high CER genotypes. Here we consider two areas of research.

#### 3.5.2.1 ANATOMICAL CHARACTERISTICS

The trend towards larger leaves during the evolution of wheat, from diploid and tetraploid wild species to the hexaploid *Triticum aestivum*, may be seen as a consequence of the increased ploidy: it is a common observation, with many species, that cells, and consequently organs, become larger with increases in ploidy level. Attention has therefore focussed on the possibility that CER may depend on the size of the photosynthetic cells. For example, small mesophyll cells may give an advantage in terms of $CO_2$ uptake. The system of air spaces within the leaf means that the exposed surface area of mesophyll cells, across which $CO_2$ can diffuse, is much greater than the external leaf area. This ratio, internal cell surface area to external leaf area, increases as cell size is reduced, and the $CO_2$ transport resistance may consequently be lower in smaller cells.

There is some support for this idea, but it appears that any advantage of a lower transport resistance is countered by a higher biochemical (carboxylation) resistance. Thus, Evans and Seeman (1984) found that the content and specific activity of RuBP carboxylase was less in flag leaves of the diploid *T. monococcum* than in those of *T. aestivum*. However, the slope of the relationship between net photosynthesis and the intercellular $CO_2$ concentration was the same for both species, indicating that the mesophyll (residual) resistance was also the same. Clearly, the higher carboxylation resistance of the smaller cells of the diploid was offset in some way, presumably by a lower transport resistance, so that the residual resistance was the same as in the hexaploid. The photosynthetic advantage of a leaf composed of small cells cannot therefore be attributed solely to improved $CO_2$ uptake by those cells. The identification of other characteristics associated with small cells, which collectively contribute to photosynthetic superiority, is therefore being sought.

#### 3.5.2.2 PHOTORESPIRATION

Twenty years ago there was considerable optimism that selection and breeding would produce $C_3$ plants with low rates of photorespiration. However, extensive screening programmes involving several species (wheat, barley, oats, soybean, potato, tall fescue) failed to identify genotypes with low $CO_2$ compensation points. Subsequent reports of low photorespiration selections, based on various assays of photorespiration, have generally remained unconfirmed. The techniques of the search have now become more refined, with attention focussed on the kinetic properties of RuBisCO. Differences between species in the relative carboxylase and oxygenase activities have been found, and further studies on the active site chemistry may be the groundwork for subsequent alteration of the enzyme by genetic engineering.

For example, site-specific, *in vitro* mutagenesis, causing mutations at specific points in the DNA coding for RuBisCO, may bring desired changes in the catalytic properties of the enzyme. There is, however, a major problem associated with this approach as a means of manipulating RuBisCO in higher plants, because the large subunit of the enzyme, which carries the active sites, is encoded in the chloroplast, and techniques for inserting DNA into chloroplasts are in their infancy. It will be easier to alter the small subunit, which is encoded in the nucleus, and it is therefore important to learn the role of this component, which at present is not known. For further details of the genetic manipulation of RuBisCO, see Ellis and Gatenby (1984) and Somerville (1986).

### 3.5.3 CHEMICAL CONTROL OF PHOTORESPIRATION

Inhibitors of various steps in the metabolism of glycollate have tended to reduce net photosynthesis: there is no end-product inhibition of glycollate synthesis, and the carbon lost in this way from the Calvin cycle is not returned (sect. 3.4.1). On the other hand, differential inhibition of the oxygenase function, preventing the synthesis of glycollate, is likely to be beneficial. Two inhibitors, glycidate and hydroxylamine, are reported to have this effect, but the results remain equivocal. It seems, therefore, that chemical treatment offers less scope for the control of photorespiration than does the genetic approach.

# CHAPTER 4 PHOTOSYNTHETIC EFFICIENCY: RESPIRATION

*To the physiologist or biochemist, respiration is a process by which materials a, b and c are converted into materials x, y and z. To the systems analyst, it is a leak in a "black box". It is time we attempted to reconcile these two views.*

(K J McCree 1970.)

## 4.1 VARIATION IN THE RATE OF RESPIRATION

It was not until attempts were made to estimate the potential productivity of agricultural land in various parts of the world, or to construct mathematical models of crop growth, that the deficiencies in our knowledge of the magnitude of respiratory losses in whole plants or plant communities became apparent. Crop respiration was simply assumed to be a constant rate on a unit plant weight (or leaf area or photosynthesis) basis. For example, in his estimates of potential productivity, de Wit (1959) assumed that respiration was 20 per cent of gross photosynthesis, while Loomis and Williams (1963) used a figure of 33 per cent of net photosynthesis. Duncan *et al.* (1967) developed a model for simulating community photosynthesis based on estimates of solar radiation absorbed by successive layers of leaves. When a single light response curve for net photosynthesis/respiration was used to convert the absorbed radiation of each layer into dry-matter production, the model increasingly underestimated the crop growth rates at values of $L$ greater than 3. A light response curve for the photosynthesis of individual leaves had been used, and the failure of the model was due to the high respiration rates thereby assigned to heavily shaded leaves towards the base of the canopy (sect. 2.4.2). By modifying the low light portion of the curve to give a lower light compensation point and lower rates of respiration, the agreement between simulated and measured crop growth rates was greatly improved (Loomis *et al.* 1967). Clearly, leaves growing under low light, and therefore having low rates of photosynthesis, had lower rates of respiration than more highly illuminated leaves.

Ludwig *et al.* (1965) had, in fact, already clearly demonstrated that the rate of respiration of crop communities was not merely a function of biomass or leaf area, but was also related to the light environment in which each individual leaf had developed. Cotton plants, which had been grown at a wide spacing, were transferred to high population densities in a growth cabinet to give leaf area indices of up to 12. The rates of net photosynthesis per unit ground area were measured over the range of $L$ values at various times after transfer to the cabinet (Fig. 4.1).

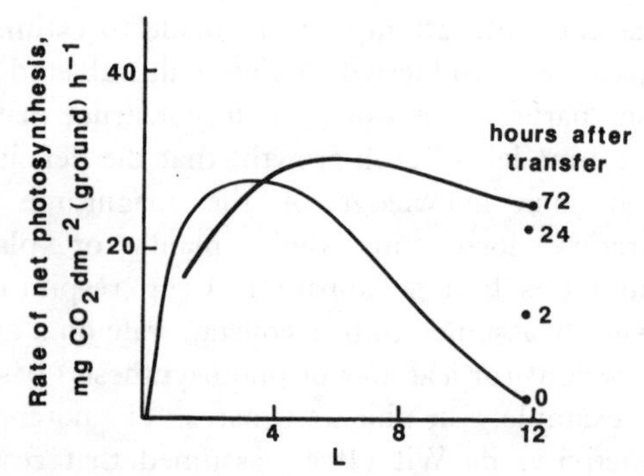


**Fig. 4.1** Changes with time in the relationship between net photosynthesis and leaf area index of communities of cotton plants in a controlled environment chamber. The plants were grown initially at wide spacing, before being crowded together in the chamber to give a range of values of $L$. Measurements were made immediately after transferring the plants to the chamber (0 h) and subsequently over several days (data of Ludwig *et al.*, from Loomis *et al.* 1967).

The plants initially had high rates of respiration adapted to the high light environments in which they had developed. These rates were maintained immediately on transferring the plants to the high densities in the cabinet, where the intense mutual shading of leaves resulted in low rates of photosynthesis per unit leaf area. Consequently, net photosynthesis at the highest density, directly after transfer, was zero, and there was an optimum $L$ at which net photosynthesis was at its highest value. Within twenty-four hours, however, the shaded leaves had adapted to the new light environments by decreasing their rates of respiration, so that there was no longer a dramatic decline in the rate of net photosynthesis as the values of $L$ increased.

The distinct responses of net photosynthesis to increasing $L$ seen immediately after crowding the plants and twenty-four hours later can therefore be interpreted in terms of differing respiratory responses to increased $L$ (or biomass). In fact, they represent two extreme views about the behaviour of respiration in crop communities: the classical concept of $L_{opt}$ based on the assumption that the rate of respiration was proportional to biomass, which held sway until the mid-1960s, and the more recent realization that it is more closely related to the rate of photosynthesis (Fig. 4.2). The latter point of view was substantiated by McCree and Troughton (1966a, b), who showed first that individual young white clover plants adapted their rates of respiration to changes in the incident light level and, secondly, that the same adaptation occurred in communities, so that the respiration rate of the whole community was proportional to its rate of photosynthesis. Thus, a physiological explanation had been given for the response of net photosynthesis (or dry-matter production) to increasing leaf area index, which marked an important step in our understanding of crop respiration, and which has subsequently been confirmed for a variety of crops (e.g. King and Evans 1967).

The extent of the variation in the rate of respiration associated with differences in the rate of photosynthesis (or incident radiation) was demonstrated by Ludwig *et al.* (1965): the night respiration rate of a cotton leaf which had been exposed to high irradiance during the preceding day was up to 130 per cent higher, depending on temperature, than that of a leaf previously in darkness. Similar observations have since been made with other crops. For example, Fig. 4.3 shows results from a field experiment with potato. Daytime respiration measurements, taken on sunny days (daily mean irradiance 30.7 MJ $m^{-2}$) and cloudy days (daily mean irradiance 12.2 MJ $m^{-2}$), were made by excluding light from the assimilation chamber just before measurement; similar measurements were made during the following nights. Respiration rates at night

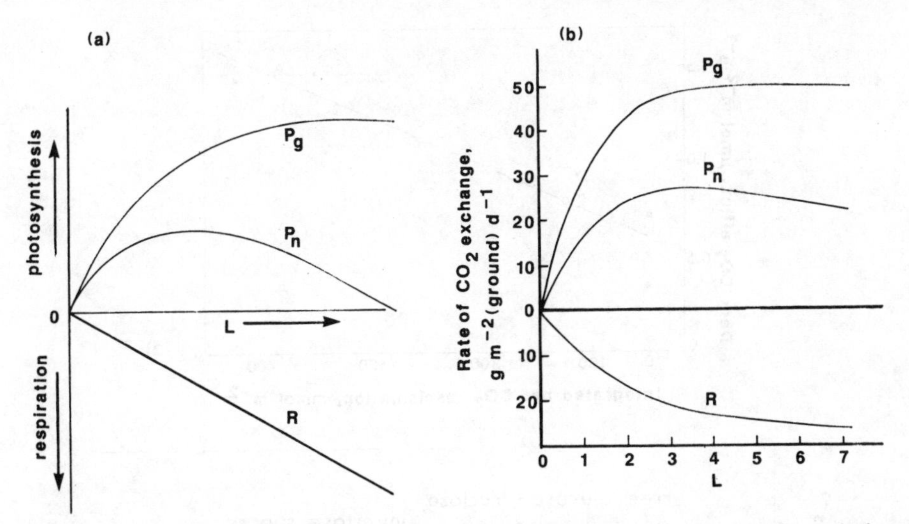


**Fig. 4.2** Relationship between the rate of $CO_2$ exchange of crop communities and leaf area index. (a) The theoretical linear response of respiration to increasing $L$ (or biomass) results in a reduction in net photosynthesis at values of $L$ above an optimum. (b) Results obtained from communities of white clover plants, in which the rate of respiration is related to that of gross photosynthesis so that there is little reduction in net photosynthesis above a critical value of $L$ (after McCree and Troughton 1966b).

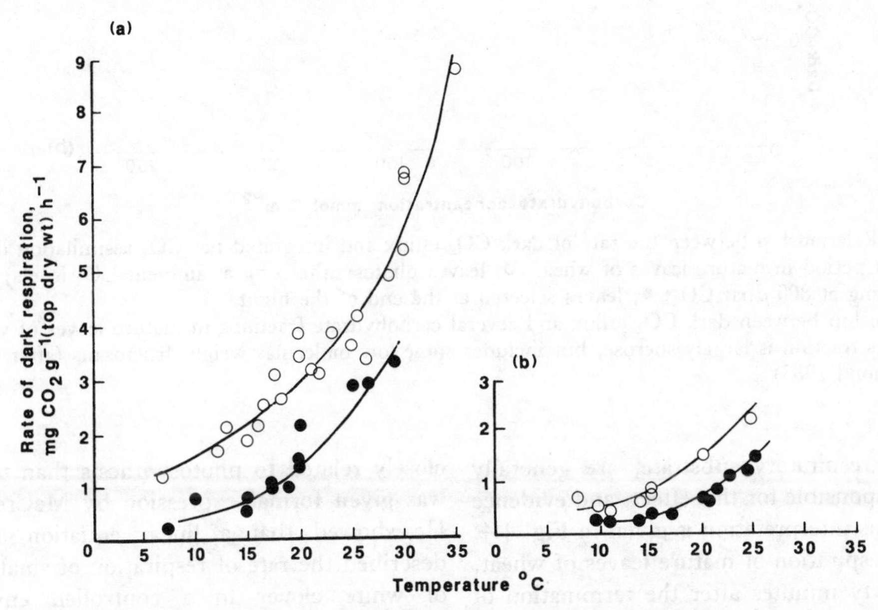


**Fig. 4.3** Relationship between temperature and the rate of dark respiration of the above-ground parts of potato plants, measured during the day (open symbols) or during the following night (closed symbols) on (a) four sunny days, and (b) two cloudy days (after Sale 1974).

were only 30–75 per cent of those during the day at the same temperature, and during a sunny day the rate of respiration was markedly higher than during a cloudy day.

In fact, a direct relationship between the respiration rate in the dark and the preceding rate of net photosynthesis has been demonstrated in maize (Heichel 1970), tomato (Ludwig *et al.* 1975) and wheat (Azcón-Bieto and Osmond 1983). Variations in the amount of soluble carbo-

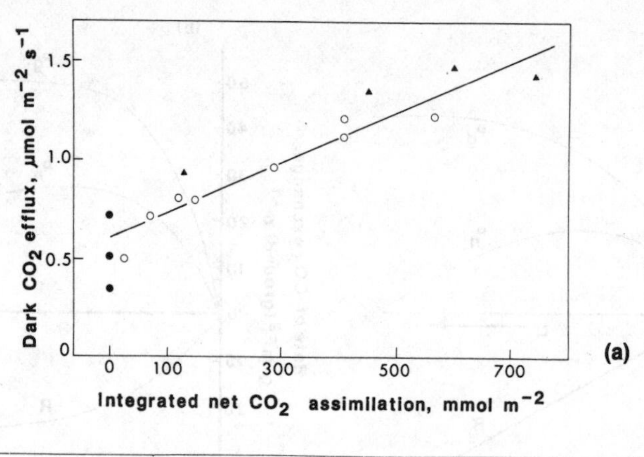


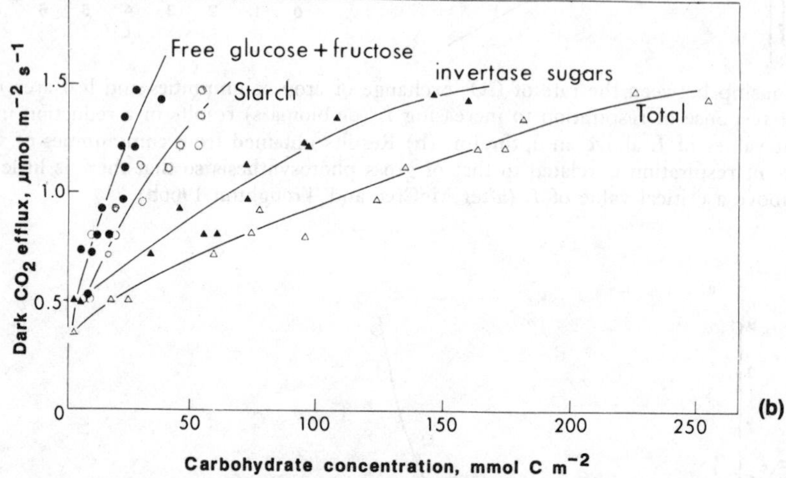


**Fig. 4.4** (a) Relationship between the rate of dark $CO_2$ efflux and integrated net $CO_2$ assimilation during the preceding light period in mature leaves of wheat. ○, leaves photosynthesizing at ambient $CO_2$ levels; ▲, leaves photosynthesizing at 800 μbar $CO_2$; ●, leaves selected at the end of the night.

(b) Relationship between dark $CO_2$ efflux and several carbohydrate fractions in mature leaves of wheat. The invertase sugars fraction is largely sucrose, but includes some low molecular weight fructosans (after Azcón-Bieto and Osmond 1983).

hydrate, the respiratory substrate, are generally held to be responsible for this effect, and evidence supporting this interpretation is given in Fig. 4.4. The rate of respiration of mature leaves of wheat, measured thirty minutes after the termination of the light period, was proportional to the total net photosynthesis during that period, and correlated with several leaf carbohydrate fractions.

## 4.2 GROWTH AND MAINTENANCE RESPIRATION IN CONTROLLED ENVIRONMENTS

The recognition that respiration could be more closely related to photosynthesis than to biomass was given formal expression by McCree (1970). He showed that a linear equation adequately described the rate of respiration of small canopies of white clover in a controlled environment (20 °C, 100 W m$^{-2}$ PAR):

$$R = aP_g + bW \qquad [4.1]$$

where $R$ and $P_g$ are the rates of respiration and gross photosynthesis, respectively (g $CO_2$ m$^{-2}$ ground area d$^{-1}$), and $W$ is the crop dry matter (g $CO_2$ equivalents of the dry matter). McCree found that the constants of the equation, $a$ and $b$, were 0.25 (dimensionless, see p. 92) and 0.015 d$^{-1}$. In other words, the daily respiratory loss of $CO_2$ from this community of plants was equal to 25 per cent of the daily gross photosynthesis plus 1.5 per cent of the existing dry weight.

McCree interpreted the two components of eqn [4.1] in terms of the crop's need to support both growth and maintenance through the provision of high-energy compounds and respiratory intermediates. Growth respiration supports the synthesis of new biomass, and is proportional to the rate of gross photosynthesis: it represents the metabolic cost of converting the products of photosynthesis into the various compounds required for growth. Maintenance respiration supports the turnover and replacement of existing cell components, and is proportional to plant or crop dry weight.

McCree cautioned that 'it would be foolish to claim that the equation has any general validity'. However, a similar equation, developed independently by Hesketh, Baker and Duncan (1971), described the rate of respiration of cotton leaves and gave values of 0.54 and 0.026 $d^{-1}$ for the coefficients $a$ and $b$, respectively. Furthermore, considerable support for the concept came from the theoretical work of Penning de Vries (1972, 1975a), who calculated the minimum amounts of carbon lost through respiration during the conversion of photosynthate into biomass of given chemical composition. These calculated values for growth respiration corresponded closely with measured values.

Unfortunately, the rate of maintenance respiration cannot yet be derived theoretically with as much confidence because of inadequate knowledge about the materials being replaced and the rates of their degradation. However, Penning de Vries (1975b) has used measured rates of protein and lipid turnover in plants, together with estimates of the energy required for the maintenance of ion gradients, to calculate that maintenance requires the consumption of 15–25 mg glucose $g^{-1}$ dry weight $d^{-1}$. This figure agrees with at least some of the measured rates of maintenance respiration, although rates of up to 80 mg glucose $g^{-1}$ dry weight $d^{-1}$ have been reported. This discrepancy is not surprising: first, because of differences in the conditions of growth between the plants used for measuring protein turnover rates and those used for the measurements of maintenance respiration; and secondly, because of uncertainties about some of the assumptions which have to be made in the calculation of maintenance requirements. Nevertheless, this 'quantitative biochemistry' has provided a valuable theoretical basis for the interpretation of $CO_2$ exchange data from whole-plant and crop experiments in terms of growth and maintenance processes.

McCree's method of estimating the growth and maintenance components of respiration involved subjecting the plants to a series of different light levels. The daytime influx of $CO_2$ therefore varied, and the night-time efflux was linearly related to this (cf. Fig. 4.4a). The maintenance component was taken to be the efflux at zero influx; the additional efflux at night when daytime influx was increased by enhanced light was the growth component. This approach was subsequently simplified (McCree 1974): maintenance respiration was estimated as the steady-state efflux of $CO_2$ after more than forty-eight hours in the dark at a constant temperature. Growth respiration was then the difference between this value and the total efflux during a normal night period.

The $^{14}C$-labelling technique used by Ryle *et al.* (1976) was in principle the same, but had the advantage that extended nights were not necessary. Uniculm barley and maize plants in a growth cabinet were exposed to $^{14}CO_2$. Immediately after labelling, the $^{14}C$ contents of samples of plants were measured to determine the amount of $^{14}C$ assimilated, and thereafter the $^{14}CO_2$ efflux from the remaining plants was monitored. Analysis of the efflux curves revealed a rapid loss of $^{14}CO_2$ for about twenty-four hours accompanied by a much slower efflux which then continued for many days: the former was equated with growth respiration and the latter with maintenance. In this way, growth respiration was found to consume 25–35 per cent of each day's assimilate, while maintenance respiration accounted for 2–3 per cent of the current dry weight per day, depending on the particular light and temperature conditions imposed.

McCree could perhaps have been more sanguine: the $^{12}CO_2$ exchange and $^{14}CO_2$ efflux methods have produced comparable results for a variety of crops in controlled environments, and are supported by the theoretical estimates of growth and maintenance respiration. However, before going on to look at the validity of the growth and maintenance concept to crops growing in the field, it is necessary to examine the characteristics of growth and maintenance respiration in more detail.

It must be emphasized that the distinction between growth and maintenance respiration is conceptual, or operational: no biochemical differences are implied. This approach to a quantitative understanding of crop respiration simply shows how the products of a single process are allocated to two broad purposes. However, it is invariably found that growth and maintenance respiration differ in two important respects: growth respiration appears to be independent of temperature but affected by limitations in substrate supply, whereas the reverse applies to maintenance respiration. This distinction has caused some confusion (see, for example, the discussion of the papers by McCree and Van Bavel 1977 and Penning de Vries and Van Laar 1977), but it is due to a fundamental difference in the nature of the constants *a* and *b* in eqn [4.1] rather than to differences in metabolism.

This dilemma can be resolved by following Thornley's (1970, 1976) logic. In the following scheme, it is assumed that substrate produced by a plant in a given time is used completely in that time for either maintenance or growth. The symbols are explained in the text, and all quantities are in $CO_2$ equivalents.

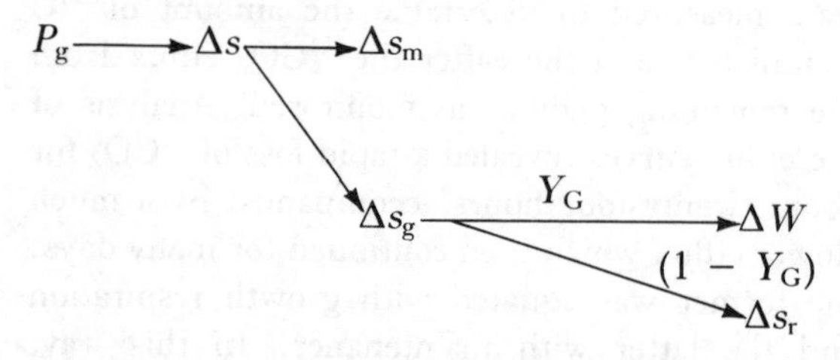


Thus, daily gross photosynthesis ($P_g$) produces a certain amount of substrate ($\Delta s$), some of which, depending on the dry weight of the plant, is allocated for maintenance, i.e. it is respired to provide the energy and carbon skeletons necessary to maintain the *status quo* (therefore $\Delta s_m = bW$). The rest of the substrate ($\Delta s_g$) is allocated to growth processes.

Maintenance is assumed to take precedence over growth, because it is found that maintenance respiration is not normally limited by substrate supply and is dependent on the existing dry weight to be maintained. *b* is the rate of substrate use per unit weight of material to be maintained, and will have units of, for example, $g\ g^{-1}\ d^{-1}$ (or simply $d^{-1}$). Because it is a rate, it will be affected by temperature, and has been found to have a $Q_{10}$ of 1.8–2.0 (but see p. 92).

The amount of substrate which is incorporated into new plant dry weight ($\Delta W$) depends on the amount of substrate allocated for growth ($\Delta s_g$: the amount remaining after the requirements of maintenance respiration have been met), and the efficiency with which that substrate is converted to the various chemical constituents of the plant ($Y_G$). Thus:

$$\Delta W = Y_G \cdot \Delta s_g \text{ or } Y_G = \Delta W/\Delta s_g \qquad [4.2]$$

and

$$Y_G = \Delta W/\Delta W + \Delta s_r \qquad [4.3]$$

This conversion efficiency ($Y_G$) is therefore the ratio of the rate of production of dry weight to the rate of production of the substrate available for growth. A conversion efficiency of 0.75 means that 75 per cent of $\Delta s_g$ is converted into $\Delta W$; the other 25 per cent represents the metabolic cost of these conversions, i.e. growth respiration ($\Delta s_r$). Thus:

$$\Delta s_r = (1 - Y_G)\ \Delta s_g \qquad [4.4]$$

Since $Y_G$ is a ratio of two rates, each of which is affected more or less equally by temperature, it is itself unaffected by temperature. As it is a ratio, it has no units (i.e. it is dimensionless). $Y_G$ values of 0.75 are typical for vegetative growth, but will vary according to the overall chemical composition of the plant. The importance of plant composition becomes clear when conversion efficiencies are calculated simply in terms of the number of grams of product produced from one gram of substrate. (However, it should be noted that the numbers calculated in this way are only approximations of $Y_G$, which is defined in terms of $CO_2$ equivalents.) Examples of the efficiencies of synthesis of individual biochemical components, which Penning de Vries has called 'production values', are given in Table 4.1.

The production values of glucose for carbohydrates are clearly much higher than those for nucleic acids, amino acids and proteins, which in turn are higher than the production values for lipids. These differences are reflected in the efficiencies with which substrate is converted into plant biomass of differing chemical composition

**Table 4.1** Some characteristic values for the conversion of 1.0 g of glucose into various plant components, using the most efficient biochemical pathways

| *End product* | *Production value* ($g\ g^{-1}$) |
|---|---|
| Sucrose | 0.92 |
| Cellulose and starch | 0.86 |
| Nucleic acids* | 0.57 |
| Amino acids* | 0.54 |
| Proteins* | 0.45 |
| Lipids | 0.36 |

* N and S supplied as $NO_3^-$ and $SO_4^{2-}$.
(After Penning de Vries 1972)

(Table 4.2). Thus, 1 g of glucose can produce 0.75 g of a high carbohydrate seed such as barley or rice but only 0.43 g of a high lipid, high protein seed such as rape. The high metabolic cost of synthesizing rape seed therefore imposes a limitation on yield which is independent of photosynthetic abilities.

Growth respiration ($aP_g$) is therefore largely independent of temperature because $a$ is a ratio of two rates which are more or less equally affected by temperature. It is reduced by limitations in substrate supply because the maintenance of existing biomass takes precedence over growth. It is also strongly dependent on the chemical composition of the plant material being synthesized. On the other hand, maintenance respiration ($bW$) is strongly affected by temperature because $b$ is a true rate. It is independent of substrate supply because reserve carbohydrates are normally available should supplies of recent assimilate be exhausted; even if this pool of reserve carbohydrates is depleted, as might happen with high night temperatures, maintenance respiration can continue at the expense of net degradation of protein.

The overall effect of temperature on plant or crop respiration will therefore depend on the relative contributions of growth and maintenance respiration, which will vary during the life-history of the crop (see p. 96): accordingly, the $Q_{10}$ will vary. Artificial manipulation of the growth and maintenance requirements of field bean (*Vicia faba* L.) plants has been shown to cause considerable variation in $Q_{10}$ values, and at the same time has lent support to the accepted ideas about the effects of temperature on growth and maintenance respiration (Breeze and Elston 1978). The substrate contents of plants were varied by imposing light pre-treatments, after which the $CO_2$ efflux rates in the dark were measured at 10, 20 and 30 °C (Table 4.3). This experiment clearly generated considerable variation (1.5–4.0) in the $Q_{10}$ for respiration, which can be attributed to differences in the soluble carbohydrate contents of the plants. Over both temperature ranges, the $Q_{10}$ or, in other words, the sensitivity to temperature decreased as the substrate contents increased. This is to be expected because the maintenance component of respiration would be

**Table 4.2** Yield of seed from 1.0 g of photosynthate in relation to seed composition

| *Crop* | *Seed composition (% D Wt)* | | | | *Biomass productivity*[†] |
|---|---|---|---|---|---|
| | *Carbohydrate** | *Protein* | *Lipid* | *Ash* | ($g\ g^{-1}$ *glc*) |
| Barley (*Hordeum vulgare*) | 80 | 9 | 1 | 4 | 0.75 |
| Rice (*Oryza sativa*) | 88 | 8 | 2 | 2 | 0.75 |
| Wheat (*Triticum aestivum*) | 82 | 14 | 2 | 2 | 0.71 |
| Oats (*Avena sativa*) | 77 | 13 | 5 | 5 | 0.70 |
| Pea (*Pisum sativum*) | 68 | 27 | 2 | 3 | 0.65 |
| Soybean (*Glycine max*) | 38 | 38 | 20 | 4 | 0.50 |
| Rape (*Brassica napus*) | 25 | 23 | 48 | 4 | 0.43 |

* Mainly polysaccharides
† Estimates based on the composition of the seed and the appropriate production values for the various biochemical components.
Abbreviation: glc, glucose.
(After Sinclair and de Wit 1975)

**Table 4.3** $Q_{10}$ values for the respiration rates of field bean (*Vicia faba*) plants with differing amounts of respiratory substrate

| *Growing period* | *Soluble carbohydrate content* | *Temperature range (° C)* | |
|---|---|---|---|
| | | *10–20* | *20–30* |
| Winter | Low | 4.0 | 2.9 |
| | High | 2.5 | 2.1 |
| Summer | Low | 2.1 | 2.1 |
| | High | 1.7 | 1.5 |

Light (PAR) pre-treatments were as follows:
winter-grown plants (glasshouse) 10 W $m^{-2}$ ('low'), 100 W $m^{-2}$ ('high');
summer-grown plants (outside) 20 W $m^{-2}$ ('low'), 200 W $m^{-2}$ ('high').
In each case the duration of treatment was 18 h light, 6 h dark, 18 h light.
(After Breeze and Elston 1978)

dominant in plants with low substrate contents, and a marked temperature dependence would result. The potential errors which could accompany the common assumption of a $Q_{10}$ of 2.0 for respiration are highlighted.

We have so far considered maintenance respiration as a function of plant or crop dry weight. This relationship is often adequate, but in many cases maintenance respiration has been found to be more closely related to protein content. This is in keeping with the proposition that protein (enzyme) turnover, the repair and replacement of unstable cell components, and the maintenance of gradients of ions and metabolites are largely responsible for maintenance respiration. Penning de Vries (1975b) has also included as maintenance the physiological adaptations that ensure cell function in a changing environment: a change in the enzyme complement of cells would be an example of such adaptation. Thus, some of the $CO_2$ production which we attribute to maintenance respiration is due to a requirement for energy for cellular work; some results from a need for carbon skeletons; and some is 'non-respiratory' in the sense that it arises directly from protein breakdown. Maintenance, particularly that which involves enzyme and membrane turnover, clearly involves some synthesis. The distinction made between maintenance respiration and growth (or synthetic) respiration is therefore an arbitrary one, maintenance respiration simply referring to processes which result in no net synthesis of dry matter. Barnes and Hole (1978) have stressed the synthetic character of maintenance by pointing out that turnover is merely the resynthesis of degradable tissue.

This appreciation of the nature of the distinction between growth and maintenance processes is important because it underlines the basic unity of respiration as a biochemical process. There is also some evidence that growth respiration and maintenance respiration are not as independent of each other as McCree's two-component equation suggests. Penning de Vries' (1975b) survey of measured maintenance respiration rates suggests that, at a given temperature, leaves with high assimilation rates during the days preceding the measurements have higher maintenance respiration rates than those with low assimilation rates. Some recent experiments by McCree (1982) support this possibility. White clover plants were grown at 25 °C and a high irradiance to give a rapid growth rate (25 g C $m^{-2}$ $d^{-1}$). The irradiance was then reduced until the growth rate was zero, although the plants adapted somewhat to this change so that the growth rate gradually increased over the next twenty days to about 3 g C $m^{-2}$ $d^{-1}$. This change from a very high growth rate to a low (or zero) growth rate was accompanied by a reduction in the maintenance coefficient ($b$ in eqn [4.1]) of about 40 per cent. Penning de Vries has surmised that protein turnover may in fact be related to overall metabolic activity, which reflects gross assimilation, whereas the active ion transport component of maintenance is a function of biomass.

## 4.3 GROWTH AND MAINTENANCE RESPIRATION IN THE FIELD

Controlled environment and theoretical studies have therefore indicated that young plants use 25–35 per cent of their daily assimilate to support growth and 1.5–3.0 per cent of their dry weight ($CO_2$ equivalents) for maintenance processes. These results imply a system of physiological control over the magnitude of plant respiration. If this is so, McCree's formula should be applicable to crops growing in the field, although a number of limitations to its use might be envisaged. For example, the steady-state conditions of controlled environments do not exist in the field, and over the life of a crop there will be considerable changes in the plants' chemical composition.

Biscoe *et al.* (1975c) compared respiration rates derived from measured $CO_2$ fluxes above a barley crop with estimates based on McCree's formula. At night, the measured flux of $CO_2$ from the soil-crop system ($R_a$) is the sum of respiration by the canopy ($R_c$), by the roots ($R_r$) and by soil organisms ($R_s$). The net loss of $CO_2$ by plant respiration in the dark ($R_d$) can therefore be calculated as:

$$R_d = (R_a - R_s) \quad [4.5]$$

Total loss of $CO_2$ from the soil ($R_r + R_s$) was estimated by cutting the foliage at ground level within an area of 500 $cm^2$, placing a weighed dish of soda-lime just above the soil surface, and covering the area with a tin box. Temperature differences between the air and the soil inside and outside the box were minimized by covering the box with aluminium foil. The increase in the weight of the soda-lime over a period of 72–100 hours was attributed to the absorption of $CO_2$ released from the soil surface. Root respiration ($R_r$) over the same period was then calculated on the basis of measured changes in root dry weight and assumptions about the chemical composition of the roots, allowing $R_s$ to be estimated by difference.

The amount of $CO_2$ respired during the day ($R_l$) was estimated as the night-time loss corrected for temperature differences by a $Q_{10}$ value:

$$R_l = R_d\, Q_{10}{}^{(T_l - T_d)/10} \quad [4.6]$$

where $T_l$ and $T_d$ are the mean air temperatures in the canopy at the height of maximum foliage density during the day and night, respectively. A $Q_{10}$ of 2 and a 14-hour day were assumed, allowing respiration during the light to be estimated as:

$$R_l = (14/10)\, R_d 2^{(T_l - T_d)/10} \quad [4.7]$$

Weekly amounts of total crop respiration ($R_l + R_d$), and corresponding estimates of respiration in terms of the growth and maintenance requirements of the crop, are shown in Fig. 4.5. The estimates were made using McCree's formula (eqn [4.1]) and the values of *a* and *b* found for white clover. Thus, growth respiration was calculated as $0.25P_g$ and assumed to be independent of temperature, and the weekly maintenance respiration was calculated as:

$$7 \times 0.015 \times 2^{(\bar{T} - 20)/10} \times \bar{W}$$

where $\bar{T}$ is the mean air temperature during the week and $\bar{W}$ the standing crop dry weight ($CO_2$ equivalents), taken to be the mean value of successive weekly harvests. Weekly amounts of $P_g$ were calculated as the sum of net photosynthesis, derived from $CO_2$ fluxes above the canopy, and corresponding totals of ($R_l + R_d$).

There was good agreement between the respir-

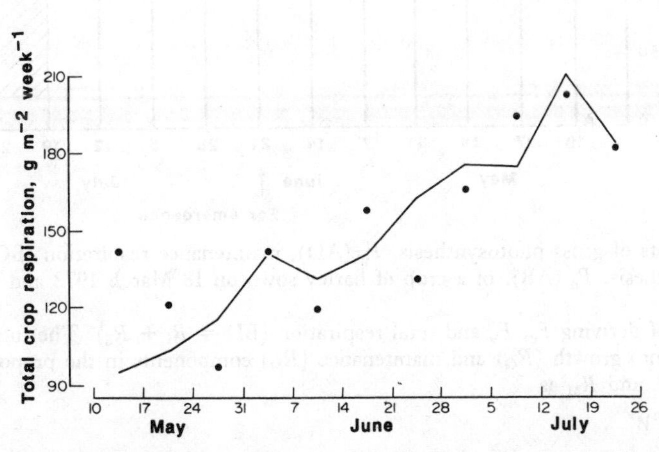


**Fig. 4.5** Comparison of the total respiration ($R_l + R_d$) of a barley crop calculated from $CO_2$ fluxes (●) with that estimated from McCree's formula (——) (from Biscoe *et al.* 1975c).

atory losses calculated from $CO_2$ fluxes and those derived from McCree's formula. The actual values of *a* and *b* for the barley crop were estimated by correcting the measured rates of respiration to a standard temperature of 20 °C, and dividing eqn [4.1] through by $P_g$ to give $R/P_g = a + bW/P_g$. *a* and *b* were then given as the intercept and slope when $R/P_g$ was plotted against $W/P_g$. The values obtained, $a = 0.34$ and $b = 0.012$, are consistent with those found in controlled environment experiments, and very close to those found for field-grown cotton ($a = 0.37, b = 0.013$) (Baker *et al.* 1972).

The application of eqn [4.1] over an extended period to crops growing in the field therefore allows the course of respiration to be considered in terms of its growth and maintenance components. This is important because it has helped to relate variations in respiration rate to the prevailing weather, and hence has aided our understanding of the effects of weather on dry-matter production. Respiration should not be considered in isolation, however, because other factors, notably the size and activity of the photosynthetic system and the receipt of solar radiation, operate simultaneously to determine dry-matter production. Biscoe *et al.* (1975c) have shown how the seasonal pattern of net $CO_2$ uptake by the barley crop results from the interplay of these factors (Fig. 4.6).

During the first five weeks, total respiration changed little. Net photosynthesis therefore followed changes in gross photosynthesis, which depended on the amount of radiation intercepted. *L* was still increasing during this period, so that the proportion of incident radiation intercepted by the canopy increased from 70 to 95 per cent. Increased respiration was associated with the development of the ear during the week before emergence, and resulted in a reduction in net $CO_2$ uptake from 57 per cent of gross photosynthesis to 39 per cent. By this time, maintenance respiration had increased from 11 per cent of total

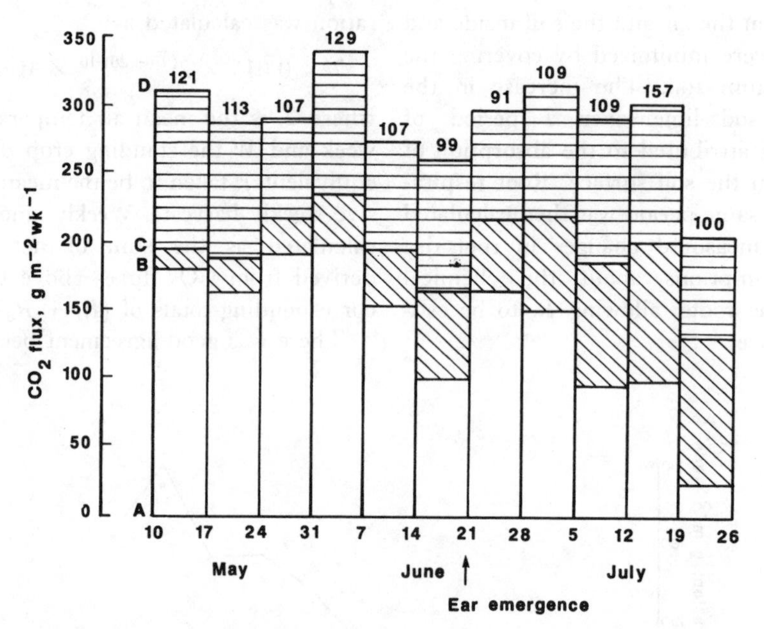


**Fig. 4.6** Weekly amounts of gross photosynthesis, $P_g$ (AD), maintenance respiration (BC), growth respiration (CD), and net photosynthesis, $P_n$ (AB), of a crop of barley sown on 18 March 1972 and harvested on 21 August.

See text for method of deriving $P_g$, $P_n$ and total respiration (BD = $R_l + R_d$). The total measured respiration was divided into growth ($R_G$) and maintenance ($R_M$) components in the proportions derived from estimates of $R_G$ as $0.34P_g$ and $R_M$ as

$$7 \times 0.012 \times 2^{(\bar{T}-20)/10}\bar{W}$$

The numbers above the histograms are the weekly totals of incident solar radiation (MJ m$^{-2}$) (adapted from Biscoe *et al.* 1975c).

respiration to 40 per cent, in response to a fourfold increase in standing crop dry weight and an increase in mean daily temperature from 9.1 to 12.4 °C. In the week after ear emergence, increased gross photosynthesis, attributed to assimilation by the ears and peduncles, reinstated net $CO_2$ uptake to its former level. However, over the next four weeks there was a dramatic decline in net assimilation, from 150 to 27 g $CO_2$ $m^{-2}$ $week^{-1}$, caused by an increasing respiratory load superimposed upon declining amounts of gross photosynthesis. This reduction in gross photosynthesis reflected the decreasing size and activity of the photosynthetic system associated with the senescence of the canopy (Fig. 4.7) and, to some extent, with water stress. The increased respiratory losses were due to the continuing high growth respiration (30–35 per cent of gross photosynthesis throughout the first half of July) and the steadily increasing maintenance component: in the final week, respiration accounted for 88 per cent of gross photosynthesis, and 60 per cent of this respiration was required for maintenance.

We have already seen that net photosynthesis, and therefore dry-matter production, during the early part of the growing season is directly related to intercepted radiation (Fig. 1.1), and this is supported by Fig. 4.6 almost until ear emergence. Thereafter, solar radiation becomes a less important factor, as $L$ declines (perhaps exacerbated by water stress) and respiration increases. The extent of the respiratory losses will be increasingly dependent on temperature as the maintenance component becomes more important. Thus, a 3 °C rise in temperature during the week 17–24 May would reduce dry-matter production by only 3 per cent, but two months later the same rise in temperature would reduce dry-matter production by 27 per cent. Biscoe and Gallagher (1977) suggest that the increase in maintenance respiration and the consequent reduction in crop growth rate with increasing temperature may explain an association between cool summers and improved crop growth.

Jones *et al.* (1978) have also demonstrated that the concepts of growth and maintenance respiration developed from controlled environment and theoretical studies can usefully represent the respiration of a crop growing in the field, but by using different methods to those of Biscoe *et al.* (1975c). The $CO_2$ exchange of that part of a perennial ryegrass sward within a transparent enclosure (50 × 50 cm area) was measured. Growth and maintenance respiration (of the above-ground parts only) were estimated by measuring the $CO_2$ efflux during a 46-hour period of darkness (see p. 91): the steady rate of efflux

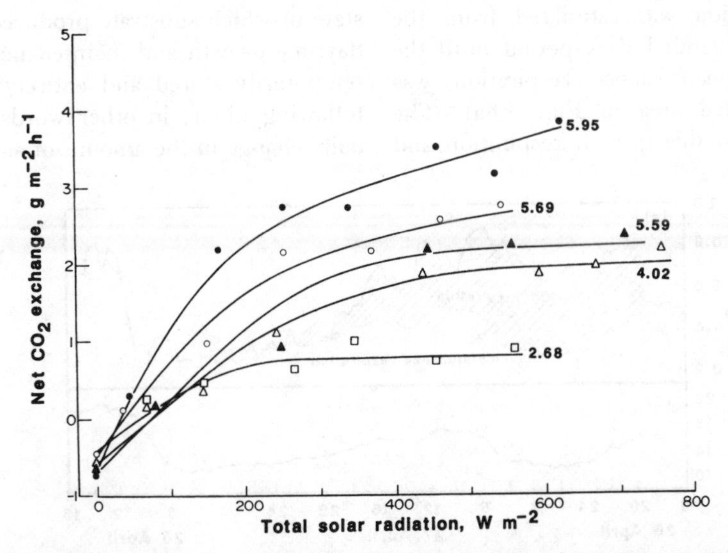


**Fig. 4.7** Relationship between the net $CO_2$ fixation of a barley crop and irradiance at weekly intervals after anthesis. The green leaf area index on each occasion is shown. The curves were derived from hourly averages of solar radiation and the corresponding $CO_2$ fluxes measured throughout the day (after Biscoe *et al.* 1975c).

established after 40 hours of darkness was taken to represent maintenance respiration, and the difference between this and the total $CO_2$ efflux during the preceding period of darkness was the growth component. As an example, Fig. 4.8 shows the respiratory $CO_2$ efflux of a reproductive sward during the period of its maximum rate of growth. In Fig. 4.8a, the course of maintenance respiration is shown as the steady rate of $CO_2$ efflux after 40 hours of darkness adjusted for the effect of temperature. A $Q_{10}$ of 2 was used, after establishing that this was an appropriate temperature coefficient for plants subjected to prolonged darkness; the air temperature in the enclosure is shown in Fig. 4.8b. Measurements were made during both reproductive (between late March and early June) and vegetative (in July and August) growth periods. Variations in the rate of photosynthesis were achieved by shading part of the sward so that the irradiance was reduced by about half.

Reproductive growth resulted in a large increase in dry weight but, in contrast with the work already discussed, estimates of maintenance respiration were poorly correlated with crop dry weight. However, maintenance respiration was closely correlated with sward protein content (Fig. 4.9a), supporting the proposition that protein turnover is largely responsible for maintenance respiration (p. 94). The total $CO_2$ efflux attributed to growth respiration was estimated from the beginning of the extended dark period until the constant rate of maintenance respiration was achieved (the shaded area in Fig. 4.8a). The relationship between this growth respiration and the total net $CO_2$ fixed in the preceding photosynthesis is shown in Fig. 4.9b: clearly, growth respiration increased as net photosynthesis increased, but the constant proportional relationship predicted by the McCree formula and demonstrated in controlled environment studies was not found. Growth respiration in fact accounted for between 15 and 35 per cent of the $CO_2$ fixed.

This variation in the relationship between growth respiration and $CO_2$ fixation is not unexpected. Growth respiration depends on the efficiency of conversion of substrate into new dry weight ($Y_G$), and this will vary with changes in the chemical composition of the plant: McCree and Kresovich (1978), in fact, accurately predicted that changes in $Y_G$ would occur in long-term experiments with flowering plants. There may also be apparent changes in $Y_G$ due to varying degrees of utilization of stored substrate, which might be expected in the fluctuating environment of long-term experiments in the field. For example, translocation and utilization of substrate is likely to be maintained at the expense of stored material when $CO_2$ fixation is reduced (e.g. in soybean (Thorne and Koller 1974) and tomato (Ho 1976a,b)), weakening any subsequent relationship between fixation and growth respiration. The constant relationship found in controlled environment studies implies a steady state in which substrate produced in excess of the daytime growth and maintenance requirements is temporarily stored and entirely used during the following night; in other words, there is no net daily change in the amount of stored substrate. In

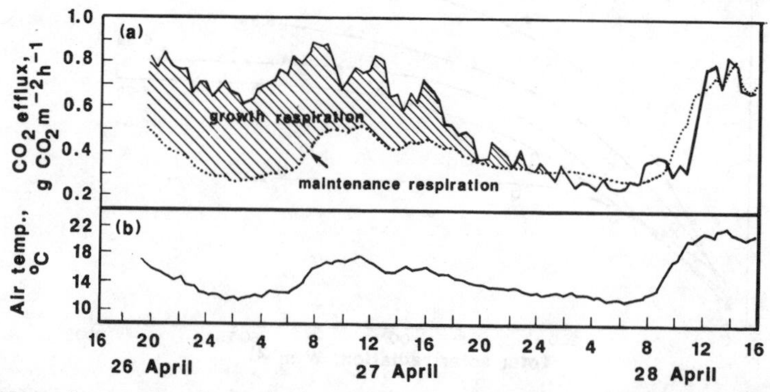


**Fig. 4.8** $CO_2$ efflux of an enclosed sward of perennial ryegrass during reproductive growth, at ambient temperature, during two days of continuous darkness. (a) ——, continuous record of $CO_2$ efflux; . . . . . . . ., maintenance respiration, calculated as the mean rate of $CO_2$ efflux after 40–46 hours of darkness, adjusted for the effect of temperature. (b) continuous record of air temperature in enclosure (after Jones *et al.* 1978).

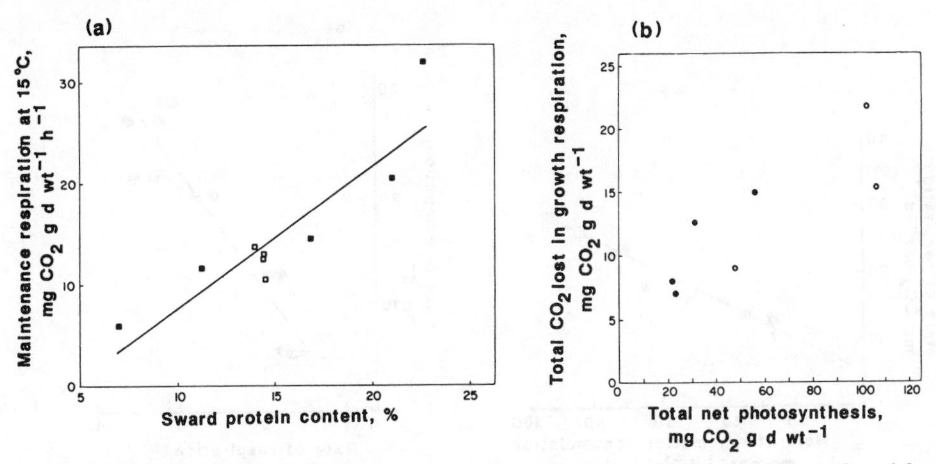


**Fig. 4.9** (a) Relationship between maintenance respiration, adjusted to 15 °C assuming a $Q_{10}$ of 2, and protein content of reproductive (■) and vegetative (□) swards of perennial ryegrass.
(b) Relationship between total $CO_2$ lost as a result of growth respiration during an extended dark period and total net $CO_2$ fixed in the preceding light period, for unshaded (○) and shaded (●) reproductive swards of perennial ryegrass (after Jones *et al.* 1978).

the present experiment, the continuation of growth respiration for 24 hours after the start of the extended dark period (Fig. 4.8a) clearly shows that assimilate can be carried over from one photoperiod to the next, although this may not necessarily occur in plants experiencing normal day/night cycles.

The field-scale experiments discussed above show that measurements of the respiration of crops growing in the field can be analysed in the same way as those of plants in controlled environments. The description of respiration by the two-component McCree formula, while less precise, remains adequate, despite variation in plant composition, non steady-state carbon partitioning in a changing environment, and the difficulties of measuring root respiration. However, attempts to apply the McCree equation to younger plants (e.g. 18-day old barley; Farrar 1980) have been unsuccessful.

## 4.4 WASTEFUL RESPIRATION: THE ALTERNATIVE PATHWAY AND WASTEFUL MAINTENANCE RESPIRATION

### 4.4.1 THE ALTERNATIVE (CYANIDE-RESISTANT) PATHWAY

Respiration is an essential process. If it is assumed that 'a high rate of evolution of $CO_2$ indicates a rapid synthesis of new material' (McCree 1974), and there are many examples of a direct relationship between the rates of respiration and of growth (e.g. Fig. 4.10), then the prospects of increasing dry-matter productivity through reduced respiration might be thought to be limited. In particular, growth respiration is, by definition, useful, and can be beneficially reduced only by increasing the efficiency of conversion of substrate to new material. The good agreement between measured values of $Y_G$ and those determined theoretically (p. 91) suggests that there is little scope for improvement. However, Lambers *et al.* (1983) suggest that this close agreement may be fortuitous, the result of conversion efficiencies in the leaves which exceed the calculated maximum efficiencies and much lower conversion efficiencies in the roots.

There may be several reasons for this difference between shoots and roots. First, respiratory energy may be supplemented in the leaves by energy derived from photosynthetic phosphorylation (sect. 3.1.2). Thus, estimates of $Y_G$ may be of little value when considering the efficiency of use of respiratory ATP. Secondly, the growth of the shoots benefits from energy expended by the roots in ion uptake, and this would reduce the conversion efficiencies of roots relative to those of shoots. Hansen (1978) suggested this as an explanation for the reduction in $Y_G$ in roots of Italian ryegrass after defoliation: $Y_G$ declined from 0.54

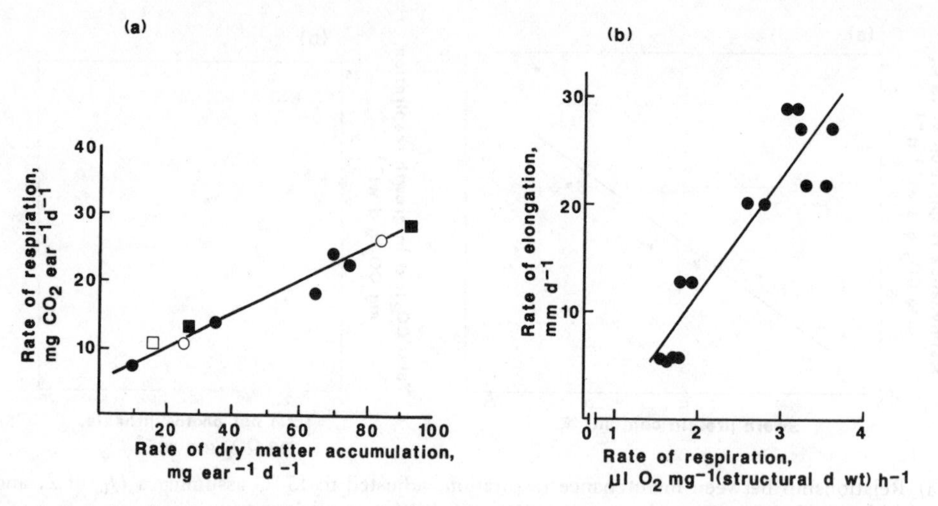


**Fig. 4.10** (a) Relationship between the rate of dry-matter accumulation by ears of wheat and respiration rate, as influenced by temperature. Ear temperature was 16 °C throughout the grain-filling period (●) or was increased to 22 °C at 6 (○), 17 (■) or 24 (□) days after anthesis (after Vos 1979).

(b) Relationship between the rate of leaf elongation and the rate of apical meristem respiration in tall fescue (*Festuca arundinacea*) plants, kept in the dark for up to 16 days (after Moser *et al.* 1982).

before defoliation to 0.36 during regrowth, possibly because root growth is then low while ion uptake remains unchanged. It is worth noting here that Hansen's data substantiate the point made by Lambers *et al.* that estimates of $Y_G$ for whole plants may mask considerable variation in conversion efficiencies between roots and shoots. Thus, $Y_G$ values for shoots before and after defoliation were 0.83 and 0.82, respectively, and the corresponding values for whole plants were 0.78 and 0.73. The effect of defoliation on the conversion efficiency of the whole plants was therefore slight compared with that of the roots, probably because shoot growth relative to root growth was higher during the regrowth period than it was before defoliation. Thirdly, the efficiency of respiration may be lower in the roots because of the greater involvement there of uncoupled respiration associated with the cyanide-resistant, or alternative, pathway. This pathway can account for over 50 per cent of the respiration in roots of young plants (de Visser and Lambers 1983) and, because it is largely a non-phosphorylating branch of the mitochondrial electron transport pathway, ATP production is reduced. It is detected as oxygen uptake in the presence of cyanide.

One function of the alternative pathway appears to be the oxidation of NADH produced by the TCA cycle in excess of the requirements for subsequent ATP generation by the cytochrome electron transport pathway (Lambers 1980; de Visser and Lambers 1983). Two main reasons for the production of excess NADH have been suggested. First, the import of carbohydrate may exceed its consumption in respiration, growth and storage, resulting in the surplus being oxidized in the TCA cycle (Fig. 4.11). For example, the alternative pathway comprised about 50 per cent of total respiration in young roots of carrot and sugar beet which had little capacity to store carbohydrate, but only 25 per cent in older plants which had developed tap roots (Lambers 1980). Alternatively, carbohydrate import is more precisely geared to the requirements of the root which include a high carbon flow through the TCA cycle in order to produce intermediates, such as α-ketoglutarate, for biosynthetic reactions. If this is associated with a low requirement for ATP, excess NADH is an inevitable result.

If the alternative pathway is the inevitable consequence of the root's requirement for TCA cycle intermediates, there is no reason to consider it wasteful, assuming of course that the syntheses sustained by the withdrawal of intermediates are necessary. On the other hand, respiration solely as a means of disposal of surplus carbohydrate is

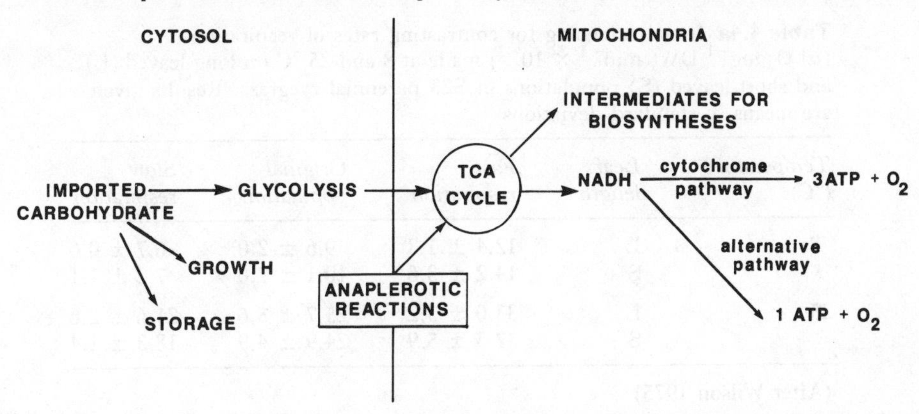


**Fig. 4.11** Outline scheme showing the relationship between glycolysis, the TCA cycle and the cytochrome and alternative pathways of electron transport.

a waste of resources. Its existence calls into question the assumption of a fine control between a sink's requirement for assimilate and the amount imported (sect. 5.2), although this import may simply be satisfying the poorly regulated glycolytic demands of the root. Variation among species in the extent to which the rate of glycolysis is geared to the root's respiratory demands has been reported: in bean (*Phaseolus*), glycolysis appears to be regulated so that the supply of substrate to the mitochondria does not exceed the capacity of the cytochrome system, and the alternative pathway does not operate; in spinach and maize, there is some control of glycolysis, but not enough to prevent saturation of the cytochrome pathway and the consequent necessity for the alternative pathway; in wheat, the rate of glycolysis was sufficient to permit full expression of both the cytochrome and the alternative pathway (Day and Lambers 1983). It appears, therefore, that when the alternative pathway is truly wasteful, this is a consequence of restricted (controlled) operation of the cytochrome pathway associated with excessive glycolysis.

However, while the 'energy overflow' theory of a non-phosphorylating alternative pathway, outlined above, has received experimental support (see Siedow and Berthold 1986), recent work with pea suggests that ATP synthesis may, in fact, be involved (de Visser *et al.* 1986). The pathway is envisaged not so much as an alternative to the cytochrome pathway, but more as a supplementary (but less efficient) route to ATP production, helping to meet the energy requirements of the cell when the capacity of the cytochrome pathway is saturated. Clearly, much remains to be learned about the functions of the alternative pathway and its regulation, before an assessment can be made of its significance in the metabolism and carbon economy of crops. It does appear to be more widespread among the tissues of a plant than was formerly thought, having been found in leaves as well as in roots (Lambers *et al.* 1983; Azcón-Bieto *et al.* 1983).

### 4.4.2 'WASTEFUL' MAINTENANCE RESPIRATION

Maintenance processes may include an element of respiration which is wasteful, not in the sense that it is uncoupled and therefore energetically wasteful, but because it supports unnecessarily rapid rates of protein turnover. The function of turnover in plants is discussed briefly by Penning de Vries (1975b). Protein turnover may account for the ability of the biochemical machinery to adjust to environmental changes, such as the formation of iso-enzymes in response to a change in temperature. Such adaptation may improve the plant's competitiveness, but is at the expense of protein turnover. Moreover, much of this degradation and resynthesis may be superfluous in modern agriculture where interspecific competition can be effectively controlled by the farmer. Reductions in the rate of protein turnover could increase productivity: thus, Penning de Vries has calculated that complete removal of protein turn-

**Table 4.4a** Initial screening for contrasting rates of respiration ($\mu l\ O_2\ mg^{-1}\ DWt\ min^{-1} \times 10^{-3}$) made at 8 and 25 °C on long-leaved (L) and short-leaved (S) populations of S23 perennial ryegrass. Results given are means ± standard deviations

| *Temperature (°C)* | *Leaf length* | *Fast respiration* | *Original population* | *Slow respiration* |
|---|---|---|---|---|
| 8 | L | 12.4 ± 1.3 | 9.6 ± 2.0 | 6.7 ± 0.6 |
| | S | 14.2 ± 3.6 | 10.1 ± 2.6 | 7.2 ± 1.1 |
| 25 | L | 33.0 ± 3.2 | 25.7 ± 5.6 | 21.6 ± 2.6 |
| | S | 32.3 ± 5.9 | 24.9 ± 4.9 | 18.3 ± 1.4 |

(After Wilson 1975)

**Table 4.4b** Mean crop growth rates ($g\ DWt\ m^{-2}\ d^{-1}$), at two successive harvest intervals in the growth room at 25 °C, of simulated swards of plants with slow or fast dark respiration. Each value is the mean of eight replicates

| *Harvest interval* | *Selection group* | | *5% LSD* |
|---|---|---|---|
| | *Fast respiration* | *Slow respiration* | |
| 1 (4 wks) | 5.78 | 9.14 | 2.03 |
| 2 (5 wks) | 4.97 | 8.19 | 1.80 |

(From Wilson 1975)

over could reduce the cost of maintenance by 10–40 kg carbohydrate $ha^{-1}\ d^{-1}$. It is therefore conceivable that increases in dry matter production could be made by chemical inhibition of protein turnover or by plant breeding.

The latter approach requires that there be sufficient intraspecific genetic variation in the amount of maintenance respiration to allow selection and subsequent breeding of slow respiration lines which would have higher rates of net photosynthesis, and therefore of dry-matter production, than fast respiration lines. Such variation was identified by Wilson (1975) in young, fully expanded leaves ('mature tissue') of S23 perennial ryegrass, originally selected for either long or short leaves. The plants were grown at either 8 or 25 °C, and respiration measurements were made in the dark at the growing temperature (Table 4.4a).

There were considerable differences in respiration rate (up to two-fold) between plants from within the original long-leaved or short-leaved populations. These differences were reflected in the rates of growth of simulated swards, comprising either fast-respiring or slow-respiring groups, in a growth room at 25 °C (Table 4.4b). Differences in dry-matter production were also found when simulated swards were grown from September to April in a glasshouse. These differences were significant in the autumn and spring but not in winter, presumably because the winter temperatures prevented full expression of the differences in respiration rate (Fig. 4.12a). These results encouraged the continuation of the selection programme, and the growth of small plots of clonally replicated genotypes, selected for slow and fast respiration, was measured during June 1974 to October 1975 (Wilson 1976; Fig. 4.12b). Dry-matter production of the slow respiration plots was greater than that of the fast at all harvests in both seasons.

A more detailed controlled environment study was made by Robson (1982) to answer the questions: (i) is the selection for single mature leaf respiration reflected in differences in $CO_2$ efflux of corresponding swards, and (ii) is this factor alone sufficient to explain differences in dry-matter production? Sixteen simulated swards of each of two lines of S23 perennial ryegrass, which differed in the rate of mature tissue respiration, were established in growth rooms which provided day/night temperatures of 20/15 °C and a photoperiod of 16 hours at 107 W $m^{-2}$ (PAR). Once dense, closed canopies were achieved (>90 per cent light interception), the swards were cut to a 5 cm stubble, and the two swards of each selection line which had produced the greatest herbage dry weight, and the two which had produced the least,

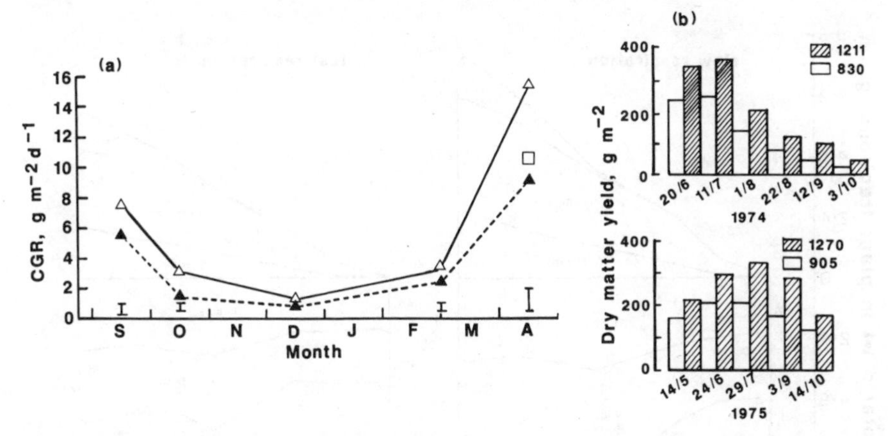


**Fig. 4.12** (a) Mean crop growth rates in the glasshouse of simulated swards of S23 perennial ryegrass selected for slow (△) and fast (▲) respiration. The CGR of the original, unselected population in April is also shown (□). Vertical lines represent 5 per cent LSD.

(b) Accumulation of dry matter between sequential cuts of simulated swards established from tillers of genotypes selected from S23 perennial ryegrass for either slow (▨) or fast (□) respiration. Total annual production (g m$^{-2}$), excluding winter growth, is given as the numbers above the histograms (from Wilson 1975, 1976).

were discarded. The remaining twenty-four swards were allowed to regrow and were cut again for four successive regrowth periods, the average length of which was 39 days. The slow respiration line outyielded the fast line by an amount which increased with each successive regrowth period from 13 per cent to 28 per cent (Table 4.5).

During the fourth regrowth period the swards were studied in detail as they grew from bare stubble to communities with leaf area indices of 10 or 14. The greater weight of herbage of the slow line was not due to greater growth of the shoot at the expense of root and stubble: the harvest indices (percentage of dry weight above the 5 cm cut) of the two lines were the same, and the slow line clearly outyielded the fast 'below ground' (root and stubble) as well as above (Fig. 4.13). The rate of respiration per unit dry weight of the slow line was between 22 and 34 per cent lower than that of the fast line (Fig. 4.14a). However, the rate of respiration per unit ground area was almost identical in the two lines because of the greater biomass of the slow line (Fig. 4.14b and c). Thus, the slow and fast swards lost the same total amounts of carbon through respiration, and the higher dry weight of the slow line throughout the regrowth period must have been due to higher rates of carbon fixation. Measurements confirmed that canopy net photosynthesis per unit ground area was indeed greater in the slow line, but only during the first half of the regrowth period.

**Table 4.5** Dry weight of herbage (above a 5 cm cut) of lines with 'slow' and 'fast' respiration, and the difference between them, at the end of each of four successive regrowth periods

| *Length of regrowth period (days)* | *Dry weight of herbage (g m$^{-2}$)* | | *Difference between lines (%)* |
|---|---|---|---|
| | *'Slow'* | *'Fast'* | |
| 46 | 5.80 | 5.08 | 13 |
| 30 | 3.50 | 2.95 | 17 |
| 26 | 3.10 | 2.40 | 25 |
| 53 | 8.69 | 6.57 | 28 |

(From Robson 1982)

Associated with this higher rate of carbon fixation was a greater number of tillers (114 dm$^{-2}$ ground area in the slow line compared with 80 dm$^{-2}$ in the fast line), and consequently a more rapid rate of canopy leaf area expansion (Fig. 4.15c and a). Extension growth per tiller was unaffected (Fig. 4.15b). Thus, the greater number of tillers caused the slow line to expand its leaf area more rapidly than the fast line, leading to earlier canopy closure and hence to the greater carbon fixation. This interpretation is supported

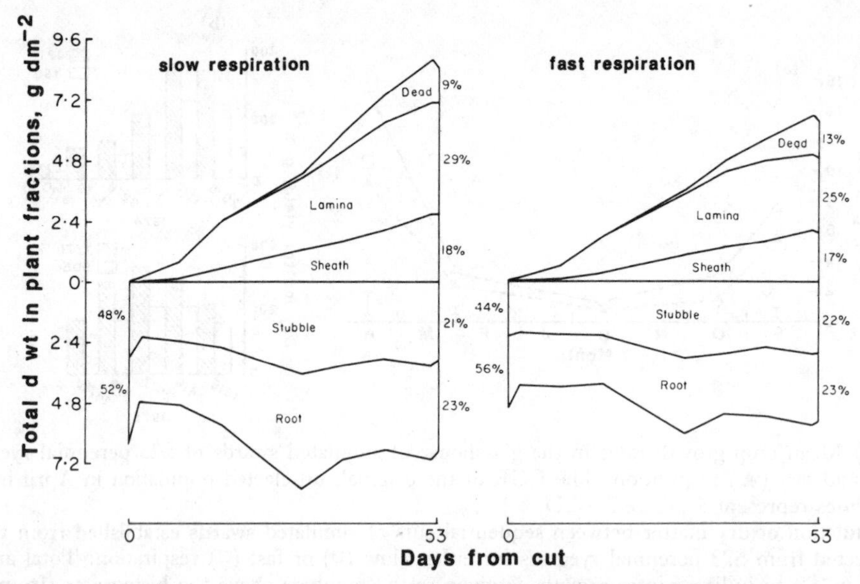


**Fig. 4.13** Dry weight of herbage and its components (sheath, lamina and dead leaf), and of root and stubble, of simulated swards of lines of S23 perennial ryegrass selected for slow or fast respiration, during the fourth period of regrowth, in a growth cabinet. Percentages represent the initial weight of root and stubble, and the final weight of all sward components, relative to the total weight of crop (from Robson 1982).

by the coincidence between the attainment of complete light interception by the two lines and the disappearance of the difference in their rates of canopy net photosynthesis. This chain of events applies equally to the earlier regrowth periods. Indeed, during any period of regrowth the more efficient use of fixed carbon by the slow line (i.e. the lower rate of respiration per unit dry weight) reinforces, and capitalizes upon, the greater dry weight (root + stubble) resulting from the previous period of growth. This is the reason for the increasing superiority of the slow line over the successive regrowth periods. For example, the advantage of the slow line over the fast increased from 1.2 g $dm^{-2}$ at the start of the fourth regrowth period to 2.96 g $dm^{-2}$ at the end of that period.

The superiority of the slow line can therefore be attributed to (i) a direct effect of the lower rate of respiration per unit dry weight, which allows a greater biomass to be supported for the same overall cost in terms of respiratory loss of carbon per unit ground area; and (ii) the greater number of tillers which results directly from the more abundant supply of assimilates in the slowly respiring plants. It should be borne in mind, however, that the 28 per cent advantage in herbage yield of the slow line was achieved relative

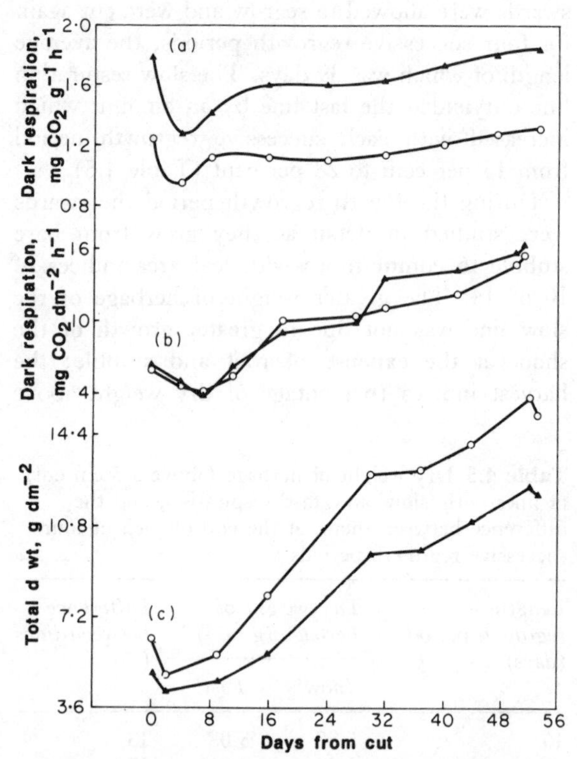


**Fig. 4.14** Rate of respiration, at 15 °C, per unit dry weight (a) and per unit ground area (b), together with the crop dry weight (c), of lines of S23 perennial ryegrass selected for slow (○) and fast (▲) respiration, during the fourth regrowth period (from Robson 1982).

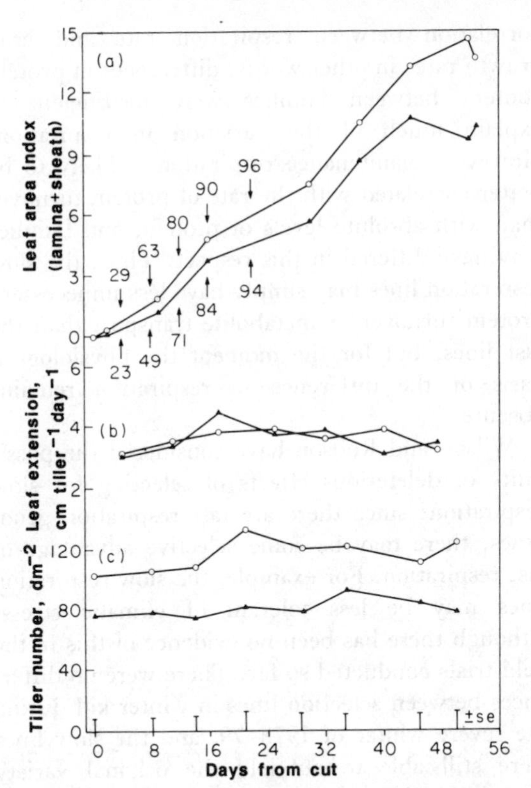


**Fig. 4.15** Leaf area index (lamina + sheath) (a), daily rate of leaf extension per tiller (b), and tiller number per unit ground area (c) of lines of S23 perennial ryegrass with slow (○) and fast (▲) respiration during the fourth regrowth period. The figures indicate the percentage of incoming radiation intercepted by the crop canopies on the days arrowed. The standard errors refer to the tiller numbers (from Robson 1982).

to the fast line, whereas the important comparison would be between the slow line and the parent variety. Field trials, using 2 $m^2$ plots, have indeed shown that two promising slow lines (code-named GL83 and GL112) can perform well in comparison not only with the parent cultivar S23, but also with the more recently introduced cultivars Perma and Melle (Wilson and Robson 1981; Wilson and Jones 1982). GL 83 and GL 112 yielded 11 and 13 per cent more dry matter than S23 under simulated grazing and 5 per cent more under cutting for conservation (Table 4.6).

It is reasonable to assume that it is the maintenance component of respiration which has been reduced by selection of the slow lines. The relative importance of maintenance respiration increases as the biomass of the crop increases, and the progressively increasing differences between the dry weights of the slow and fast lines, observed in these experiments, might therefore be expected on the basis of this difference in respiration alone. However, it should be noted that this increasing disparity between the rates of growth of the slow and fast lines cannot continue indefinitely. At a certain crop biomass, the respiratory $CO_2$ efflux of the slow line on a ground area basis will be greater than that of the fast line, despite the slow line's lower respiration rate per unit dry weight. Thus, an equilibrium is established, with the slow line's increasing advantage due to greater carbon fixation in the first half of the regrowth period counterbalanced by its increasing respiratory

**Table 4.6** Annual herbage dry-matter yields of perennial ryegrass populations selected from S23 for reduced mature tissue dark respiration (GL83, GL112) compared with S23 and two high-yielding varieties (Perma, Melle)

| *Population* | *Conservation** | | *Simulated grazing†* | |
|---|---|---|---|---|
| | *DM yield (kg ha⁻¹)* | *Yield relative to S23* | *DM yield (kg ha⁻¹)* | *Yield relative to S23* |
| S23 | 13,983 | 100 | 9,757 | 100 |
| Perma | 14,039 | 100 | 10,397 | 106 |
| Melle | 14,261 | 102 | 10,675 | 109 |
| GL83 | 14,678 | 105 | 10,814 | 111 |
| GL112 | 14,650 | 105 | 11,009 | 113 |
| 5% LSD | 704 | | 608 | |

Plots were sown May 1979 and harvested during April–October 1980.
* 5 cuts (9 June to 30 October).
† 9 cuts (10 April to 30 October).
(After Wilson and Robson 1981)

losses (per unit ground area) during the second half of the regrowth period: in a growth room experiment (Wilson 1982), the slow line's advantage increased only over the first four of seven successive regrowth periods. The apparent effect of temperature, implied by seasonal variations in the yield advantage of the slow lines (e.g. Fig. 4.12a), would also be expected if selection were for differences in maintenance respiration.

The alternative, cyanide-resistant pathway has been found to have substantial capacity in mature leaf tissue from both fast and slow respiration selections, but it did not contribute to respiration in the absence of cyanide. In both the fast and slow lines, respiration is apparently limited by glycolysis, and the alternative pathway is not engaged (Day *et al.* 1985). It is interesting to note here that in leaves, as in roots, there is considerable variation between species in the degree of coupling between glycolysis and the cytochrome pathway. In leaves of wheat, for example, glycolysis is less tightly controlled and, with high leaf sugar levels, can supply an excess of substrate to the cytochrome chain, which brings the alternative pathway into operation (Azcón-Bieto *et al.* 1983).

The relationship between protein content and maintenance respiration has already been discussed. It might be expected that selection for reduced maintenance respiration would result in reduced herbage protein levels, a possible inherent disadvantage of such selection. Wilson (1982) found a weak correlation between protein concentration and respiration rate in twelve F1 perennial ryegrass families selected from S23 for either slow or fast dark respiration. However, even when the variation in protein content was taken into account, by expressing respiration rates per unit protein weight, there was still a good negative correlation between respiration rate and crop growth rate; in other words, differences in protein content between families were insufficient to explain much of the variation in respiration. However, maintenance respiration is likely to be better correlated with the rate of protein turnover than with absolute levels of protein, and families may have differed in this respect. Thus, the slow respiration lines may simply have less unnecessary protein turnover or metabolite transport than the fast lines, but for the moment the physiological basis of the difference in respiration remains obscure.

Wilson and Robson have considered the possibility of deleterious effects of selecting for slow respiration: since there are fast respiration genotypes, there may be some selective advantage of fast respiration. For example, the slow respiration lines may be less tolerant of climatic stress, although there has been no evidence of this in the field trials conducted so far. There were no differences between selection lines in winter kill during the severe winter of 1978–79, and the slow lines were still able to outyield the original variety during the drought of 1976. Why have the slow respiration genotypes not come to dominate existing swards? The competitive advantage which must accrue from greater dry-matter production is perhaps associated with an unidentified disadvantageous feature, although in mixtures the superiority of the slow line is enhanced. However, the difference in respiration, and hence in dry-matter production, is not apparent when nutrients are in short supply, and under such conditions the (natural) selection pressure on the fast respiration genotypes would be considerably reduced (Robson *et al.* 1983).

# CHAPTER 5 DRY-MATTER PARTITIONING

*The sap having been elaborated and conveyed to the different parts of the plant, is applied to the nourishment and development of its various organs. But of the manner in which this assimilation of the nutritious fluid takes place, it appears that nothing is known with precision.*

(W MacGillivray 1840)

*. . . it is one of the important tasks before crop physiology to analyse the basis of the competing power of an organ for assimilates, because it is on empirical selection for this that past increases in yield have largely been based.*

(L T Evans 1975)

## 5.1 INTRODUCTION

### 5.1.1 THE RATE AND VELOCITY OF TRANSLOCATION

The yield of the economically important parts of most crops depends on the translocation of assimilates, in most species largely sucrose, from the leaves or other photosynthetic tissues. This is most obvious when the economic yield comprises completely non-photosynthetic tissue such as the underground tubers of potato. The tubers are morphologically distinct sinks for assimilate, dependent on the above-ground sources of assimilate production, and linked to them by a transport pathway of phloem sieve elements. Of course, the tubers are not the only sinks on the plant. The roots, stems, young leaves and fruits also rely on the assimilates produced by the mature leaves. The assimilate is thus partitioned among the various sinks in a coordinated way according to their changing requirements throughout the life of the plant. It is the control of this partitioning which we shall explore in this chapter.

The obvious dependence of potato tubers on a supply of assimilate from a distant source, and the translocation of that assimilate to each tuber through a slender stolon, made this species a suitable subject for one of the earliest quantitative assessments of assimilate transport. Dixon and Ball (1922) found that a tuber accumulated 50 g of carbohydrate in 100 days, and that the cross-sectional area of phloem in the stolon was 0.422 mm$^2$. This now classic set of data is instructive, allowing translocation to be quantified in two ways. Dixon and Ball's reasonable assumption that the carbohydrate had moved as a 10 per cent sucrose solution allows the velocity of translocation to be estimated as 49 cm $h^{-1}$. Subsequently, estimates of velocity have been derived for many species, in more recent years by timing the movement of a front of radioactivity between two detectors after allowing the plant to assimilate $^{14}CO_2$. However, there are limitations to the usefulness of such velocity measurements on their own, because unless the concentration of sucrose in the phloem is known, the actual rate of mass transfer of sucrose (or dry matter) into the sink cannot be inferred. The rate of mass transfer has units of, for example, g $h^{-1}$, and the size of the transport pathway supplying the sink can be taken into account by deriving the specific mass transfer rate (or flux density). This has units of g $cm^{-2}$ (cross-sectional area of phloem) $h^{-1}$, and can be thought of as the product of the mean velocity and the mean concentration of sucrose in the phloem. Thus:

$$\underset{(\mathrm{g\ cm^{-2}\ h^{-1}})}{\text{specific mass transfer rate}} = \underset{(\mathrm{cm\ h^{-1}})}{\text{velocity}} \times \underset{(\mathrm{g\ cm^{-3}})}{\text{concentration}} \qquad [5.1]$$

Dixon and Ball's data therefore give a specific mass transfer rate of $50/0.00422 \times 100 \times 24 = 4.94$ g $cm^{-2}$ $h^{-1}$, and it is clear from eqn [5.1] that this term conveys more information than velocity on its own.

However, it is important to remember that this value has been derived on the basis of the amount of carbohydrate which accumulated in the tuber. This is an underestimate of the amount of assimilate which was actually imported by the tuber, because a certain proportion would have been lost by respiration. It is also an underestimate of the flux of assimilate through the sieve tubes because it is based on total phloem cross-sectional area: the sieve elements themselves may comprise less than 30 per cent of that area (Grange and Peel 1975) and, furthermore, not all the sieve elements will be involved in translocation at a given time.

We can view the overall rate of translocation as the result of component processes operating at the source, the transport pathway and the sink. Simple measurements of the rate of dry-weight accumulation by sinks, such as those made by Dixon and Ball, provide a reference to which more detailed studies of the component processes must ultimately be related: clearly, an explanation of these processes must be able to account for the measured rates or velocities of transport. Dixon and Ball, for example, were unable to reconcile their estimated velocity of carbohydrate movement with the structure of the phloem. They argued that movement through a conduit of phloem cells, which had many cross-walls and viscous contents, could only be by diffusion, but realized that diffusion was grossly inadequate to account for the velocity of movement along the potato stolon. Attention was thus focussed on the problems of reconciling the structure of the phloem with its function, setting the scene for a preoccupation with the mechanism of translocation through the sieve tubes, which persisted for fifty years (see, for example, Weatherley and Johnson 1968).

The question of the mechanism of transport through the phloem cannot be resolved, however, until the precise structure of the functioning sieve tube is known. This knowledge eludes us because the functioning sieve element, with its high sugar content and therefore high turgor pressure, is easily disrupted when the tissue is prepared for microscopy. Cutting, or loss of membrane semipermeability during chemical fixation, releases the internal pressure, causing the displacement of cytoplasmic components: the resulting micrographs are inevitably ambiguous. The mechanism of phloem translocation therefore remains an intriguing problem. However, it is now generally agreed that longitudinal transport within the phloem does not normally restrict the movement of assimilate, which must therefore be controlled by processes operating at the source and sink (e.g. Milthorpe and Moorby 1969). The bulk of this chapter will therefore be concerned with the relative importance of source and sink limitations to yield and the factors responsible for these constraints. First, however, it will be useful to consider in more detail the terms source and sink.

### 5.1.2 SOURCES AND SINKS

Young leaves are sinks because they cannot fix enough carbon to support their growth. However, the status of the leaf progressively changes from sink to source as the leaf's photosynthetic capacity increases during its ontogeny (e.g. Fig. 3.20). An organ is defined as a source or sink according to the direction of net transport associated with it: this is important in the case of a developing leaf because it matures basipetally, the basal region continuing to import assimilates while the mature tip exports. This maturation encompasses not only an increased photosynthetic activity, but also the development of mature sieve elements in the minor veins and their associated capacity for the accumulation of sugar. The functional maturation of transfer cells has also been reported to coincide with the onset of this capacity to export (sect. 5.4.3). In squash (*Cucurbita pepo*), the tip of the leaf lamina stops importing and begins exporting when the leaf is only 10 per cent of its final area. Export from this region initially supplements the supply of assimilate imported from older leaves by the still immature leaf base. However, by the time the leaf is 35 per cent of its final area it has become a net exporter of assimilate, although simultaneous import continues until the leaf is 45 per cent expanded (Turgeon and Webb 1973, 1975). Similar basipetal progression in the sink to source transition within a leaf has been observed in other species, such as sugar beet (Fellows and Geiger 1974) and tomato (Ho and Shaw 1977), and in each case there is again a period during which the leaf is importing and exporting assimilate simultaneously.

Mature leaves are the primary sources of assimilate, from current photosynthesis, but this can be supplemented by the mobilization of stored reserves. This is well illustrated by cereals, such as wheat and barley. For the first 15–20 days after anthesis, net assimilation exceeds the requirements of the growing grains, and soluble carbohydrate accumulates in the stems and leaf sheaths, reaching a maximum of up to 2 t $ha^{-1}$ in British crops. Subsequently, as net assimilation declines with the senescence of the canopy, mobilization of these reserves maintains the translocation of assimilate to the grains. This accumulation and depletion of reserves helps maintain a constant rate of grain growth over a large part of the grain-filling period (Fig. 5.1), and is reflected in the changes in dry weight of the stems after anthesis, when stem elongation has ceased (Fig. 5.2). The storage of assimilate produced during the early

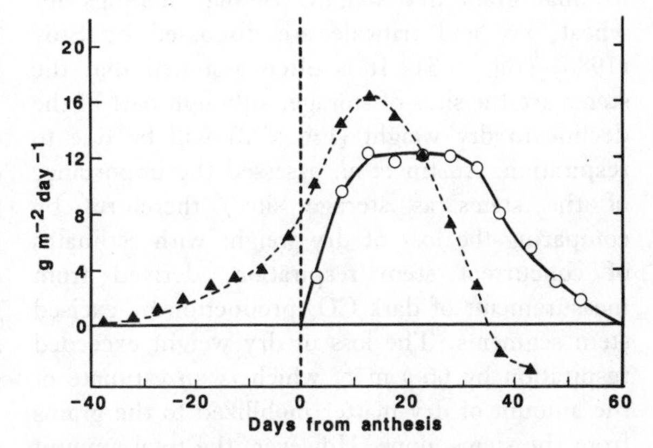


**Fig. 5.1** Comparison between the rate of grain growth (○) and the calculated rate of fixation of carbon recovered from the grains (▲) of winter wheat cv. Skandia IIIB (after Stoy 1980).

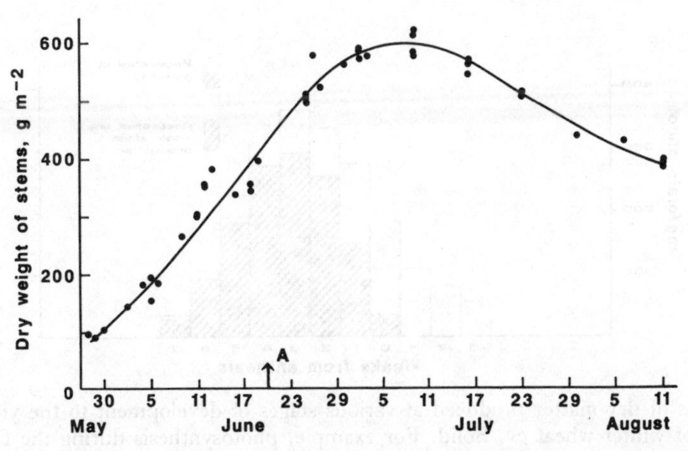


**Fig. 5.2** Time course of change in the dry weight of stems of winter wheat cv. Huntsman. A indicates the time of anthesis (after Austin *et al.* 1977a).

part of the grain-filling period clearly buffers the growing grains against the later shortfall in current assimilate production (sect. 6.6). There is, in addition, a further contribution to grain yield, from reserves accumulated before anthesis. This becomes particularly important when current photosynthesis cannot meet the demands of the growing grains, as a result, perhaps, of adverse weather or pest or disease damage to leaves.

For example, Austin *et al.* (1977a) found that although 48 per cent of the grain yield of winter wheat was derived from assimilation during the first sixteen days after anthesis, half of this was temporarily stored before being transported to the grains later during the grain-filling period. Assimilation before anthesis contributed about 8 per cent to final grain dry weight. Similar findings for wheat, rye and triticale are discussed by Stoy (1980) (Fig. 5.3). It is often assumed that the stems are the sites of storage, although part of the decline in dry weight (Fig. 5.2) will be due to respiration. Austin *et al.* assessed the importance of the stems as storage sites, therefore, by comparing the loss of dry weight with estimates of concurrent stem respiration, derived from measurements of dark $CO_2$ production by excised stem segments. The loss of dry weight exceeded respiration by 66 g $m^{-2}$, which is an estimate of the amount of dry matter mobilized to the grains from the stems alone. However, the total amount of stored dry matter, derived from assimilation during the pre- and post-anthesis periods, amounted to 197 g $m^{-2}$. Since only 66 g $m^{-2}$ was stored in the stems, the remainder, 131 g $m^{-2}$, must have been stored elsewhere, presumably in the leaf laminae and sheaths.

The increased significance of pre-anthesis assimilation to grain yield when current assimilation cannot meet the demands of the growing grains was revealed in a companion study with barley (Austin *et al.* 1980a). Measurements were made during 1976, when the growing season in the UK was much hotter and drier than average, and 1977, when it was cooler and wetter than average. Reflecting these differences in the weather, the grain yields in 1976 were less than half those in 1977. Even so, the effects of drought and high temperatures on yields in 1976 were mitigated considerably by a much greater contribution to grain growth from assimilation before, and during the first few days after, anthesis (Table 5.1). The genotypic variation in this contribution is clearly much lower than the variation between seasons. A *pro rata* reduction in the mean values of 59 per cent (1976) and 14 per cent (1977) to exclude the products of assimilation during the five days after anthesis gives pre-anthesis contributions to grain yield of 44 per cent and 11 per cent. The latter value is close to the 8 per cent found for wheat in 1975, and these estimates probably represent the minimum contribution to grain yield of pre-anthesis assimilation.

It should be noted here that nitrogenous compounds, such as amino acids and amides, as well as soluble carbohydrates are mobilized and transferred to the grains. Indeed, up to 75 per cent of the nitrogen in the grains at maturity is derived from that taken up by the plant before

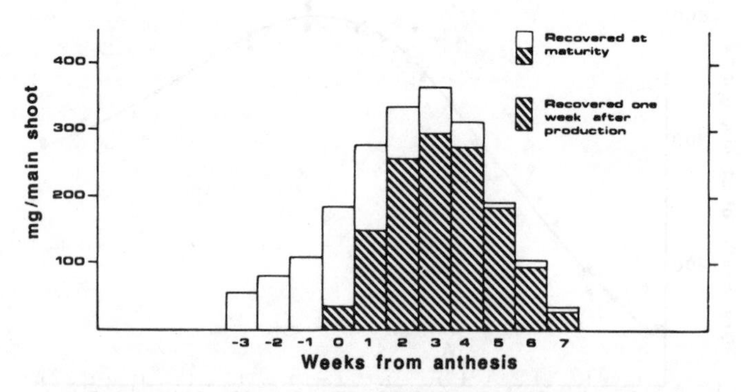


**Fig. 5.3** Contributions of dry matter produced at various stages of development to the yield of grain on the ear of the main shoot of winter wheat cv. Solid. For example, photosynthesis during the first week after anthesis contributed 280 mg to grain yield, although only 150 mg of this had reached the grains one week after its production (from Stoy 1980).

**Table 5.1** Grain yields of spring barley genotypes classified according to height, and the calculated contributions to grain yields from assimilation up to five days after anthesis

| *Year* | *Height to base of ear (cm)* | *Grain yield* ($g\ m^{-2}$) | *Contribution to grain yield from assimilation during the period 18 days before anthesis to 5 days after anthesis* | |
|---|---|---|---|---|
| | | | ($g\ m^{-2}$) | (%) |
| 1976 | 45 | 318 | 173 | 54 |
| | 36 | 296 | 180 | 61 |
| | 30 | 291 | 178 | 61 |
| 1977 | 45 | 715 | 107 | 15 |
| | 36 | 654 | 89 | 14 |
| | 30 | 659 | 91 | 14 |

(After Austin *et al.* 1980a)

anthesis (sect. 3.4.5.2). Calculations suggest, in fact, that the minimum contributions estimated in these experiments, in unstressed plants, largely represent the amount of dry matter needed to supply the grains with the required quantity of nitrogen. Of course, in stressed plants, in which photosynthesis during grain-filling is restricted, the greater dependence on pre-anthesis assimilation is an expression of the grains' requirement for a supplementary supply of carbohydrate.

The mature leaves, as the primary sources, therefore support the growth of sinks by maintaining a supply of assimilate from current photosynthesis and stored reserves. This can be supplemented by longer term storage in other organs such as the stems, which may be thought of as secondary sources. In grasses, regrowth after cutting or grazing may be substantially dependent on reserves stored in the roots, and the success of many perennial weeds can be attributed to rapid regeneration sustained by reserves in their rhizomes or stolons. Storage is also important, of course, in the shorter term, the regular accumulation of reserves during the day and their mobilization at night ensuring continuity of export from the leaf (sect. 5.4.2).

At the other end of the scale, the potato tuber, whose status as a classic sink has already been noted, clearly becomes a source when, after a period of dormancy, it supports the growth of the sprouts which give rise to a new plant (sect. 7.1 and 7.2). In fact, the developing shoots remain largely dependent on the mother tuber until a leaf area of 200–400 $cm^2$ has been produced (Moorby, 1978). As in cereals, reserves stored in the stem subsequently help to buffer the growth of the crop, especially that of the new generation of tubers, against short-term deficiencies in current photosynthesis.

True seeds resemble 'seed' tubers, and other organs of vegetative reproduction such as bulbs and corms, in that the reserves accumulated in one generation support the initial growth of the next. Seeds of different species, however, show a variety of small-scale source–sink relationships, which may change in character during the course of germination and early seedling growth. In most dicotyledonous seeds the cotyledons act initially as a source of reserves, but are subsequently raised above ground, when they assume a photosynthetic function which supports further growth, until the first true leaves are photosynthetically competent.

In some dicots, however, such as the castor oil plant (*Ricinus communis*), the storage function is taken over by the endosperm, from which mobilized reserves are absorbed by the cotyledons, and thence translocated to the embryonic axis. Sucrose derived from the endosperm is rapidly accumulated by the phloem cells of the cotyledons, in a process considered to be analogous to that in mature photosynthesizing leaves. However, uptake by the cotyledons, and the subsequent transport, is much more rapid than in mature leaves, and in many ways more amenable to experimentation. Study of this source–sink system has therefore given valuable insights into the mechanism by which sucrose is loaded into the phloem. The sucrose exported from the cotyledons largely sustains growth of the hypocotyl and roots. Elongation of the hypocotyl then raises the cotyledons, still partially surrounded by the diminishing endosperm, above ground so that, when the reserves are finally depleted, photosynthesis can take over. Germinating monocotyledonous seeds, such as those of grasses and cereals, function similarly, in that sucrose derived from reserves in the endosperm is absorbed and exported by the cotyledon (scutellum); in this case, however, the cotyledon remains below ground and does not have a photosynthetic function.

It is clear from what has been said above that, irrespective of the type of source, export must be preceded by the loading of sucrose into the phloem sieve elements. Furthermore, this loading is merely the final step in a system of partitioning within the source, by which a proportion of assimilate is allocated for export. The control of these processes, which are not only fundamental to the transport of assimilate within the plant, but also relevant to the uptake and distribution of alien substances, such as herbicides, will be considered further in section 5.4.

Sinks are net importers of assimilate. Thus, all parts of the plant, other than the mature leaves or sites of reserve mobilization and export, can be considered sinks; even those organs or tissues not increasing in dry weight must be importing enough assimilate to maintain the *status quo*. Clearly there is considerable structural, and presumably metabolic, diversity among different sinks. However, each can be broadly classified as a meristematic (cell division), elongation (cell growth) or storage sink, although within a given tissue or organ these characteristics are not mutually exclusive. Thus, in many sinks, such as developing leaves, an initial phase of meristematic growth is largely supplanted by expansion or elongation due to cell enlargement, with cell division restricted to localized areas (sect. 2.1, Figs 2.3–2.5); roots are similarly differentiated into zones of cell division and of cell expansion. The growth of storage sinks, such as potato tubers or cereal grains, obviously involves cell division and enlargement as well as the accumulation of food reserves, and again these different components may be separated in time. In the grains of wheat and barley, for example, the meristematic phase of endosperm growth is completed by two or three weeks after anthesis.

Different types of sink may therefore be recognized, but they all share the common characteristic of a high metabolic activity. This is particularly clear when storage sinks are considered, especially in those species for which we have analyses of both the imported assimilate and the stored products. In white lupin (*Lupinus albus*), for example, in which the phloem supplies 98 per cent of the fruit's carbon and 89 per cent of its nitrogen, the dry matter imported by the fruit is 82 per cent sucrose, with the remainder largely amino acids and amides. Within the seeds, these assimilates are rapidly processed into protein (63 per cent of the dry weight), oil (20 per cent) and polysaccharides (17 per cent) (Pate *et al.* 1977). The composition of phloem sap in white lupin is, in fact, typical of most species. In some, however, notably in the cucurbit family (squash, cucumber), varying quantities of raffinose, stachyose and verbascose (in which the sucrose molecule has one, two or three galactose moieties added to it) replace sucrose itself as the major translocated sugar. In certain tree species (e.g. ash and apple), sugar alcohols such as mannitol and sorbitol are dominant. However, in most higher plants it is established that sucrose is the major vehicle by which carbon is translocated from source to sink.

The ability to metabolize the imported sucrose is a characteristic feature of sinks and, indeed, there is growing evidence that certain aspects of this metabolism are involved in the overall regulation of sucrose translocation. However, although the final composition of the seed, or of any other storage sink, shows that in the long term there may well have been extensive reorganization of the carbon skeletons imported as sucrose, this does not indicate which steps in that metabolism might be involved in the control of import. It will also be recognized that the sucrose must first move out of the phloem (unloading) and, either as sucrose or as the initial products of its metabolism, must enter the cells of the sink tissue. These dual aspects of sink physiology, physical transfer and biochemical conversion, vary considerably among the different types of sink, both in anatomical and physiological detail and in their relative importance in the control of import: examples of diverse sinks will be examined in detail in section 5.5.

## 5.2 SOURCE AND SINK LIMITATIONS IN THE DETERMINATION OF ECONOMIC YIELD: A QUESTION OF BALANCE

The realization that the overall rate of translocation is controlled by processes operating at the source or sink led to many investigations of whether the growth of a particular economic sink was limited primarily by the supply of assimilates (source-limited) or by the capacity of the sink to import and use those assimilates (sink-limited). In many cases it is found that if photosynthesis is allowed full expression, sink limitation prevails. Control of translocation by the sink is also implied by the pronounced effects on translocation when sink function is perturbed by local treatments, such as exposure to low or high temperatures or anoxia. However, although such treatments may be valuable in attempts to elucidate the processes within the sink which can effect control of translocation, they rarely tell us the extent to which these processes assume control in plants growing in the field.

Nevertheless, sink growth can be largely unaffected by treatments which reduce source capacity, such as partial defoliation. Similarly, constant rates of growth of a population of potato tubers or of the grains on a cereal ear, despite pronounced variation in solar radiation, suggest excess source capacity (but see sect. 7.7). We have already noted that the rate of photosynthesis may indeed be geared to the sink's demand for assimilate (sect. 3.4.4), and that current photosynthesis may be supplemented by mobilized reserves at times of peak demand.

However, the value of such considerations is limited when they are applied only to the final storage phase of sink growth. The question of whether the economic yield of a crop is source-limited or sink-limited is more complicated because during the development and growth of the sink, the relationship between source and sink inevitably changes. Thus, in many crops, final economic yield can be attributed to a series of sequentially determined yield components, each of which is the result of source–sink relationships operating during successive phases of plant development. Crops such as cereals and legumes, in which the economic yield depends on the development and growth of fruits, exemplify yield component analysis. In cereals, for example, grain yield can be viewed as the product of the number of grains per unit area (which is itself the product of earlier formed yield components) and the mean individual grain weight (sect. 6.1). The first component sets the potential for grain yield, which will be realized if enough assimilate becomes available after anthesis for maximum growth of all the grains. We shall therefore use cereals as an example to examine the source–sink relationships which underlie the determination of yield components, and to assess the extent to which those relationships are dominated by either source or sink limitations.

Evans and Wardlaw (1976) have suggested that for productive cultivars growing in the environments to which they are adapted, one might expect source and sink to be in balance. Gifford *et al.* (1973) similarly concluded, from comparisons of the changes in total crop dry weight and grain dry weight following shading or $CO_2$ enrichment, that source and sink imposed partial, co-existing limitations to grain-filling in wheat. Certainly, feedback control of photosynthesis by sink demand, and the partial ability of later-formed yield components to compensate for deficiencies in those determined earlier (but see sect. 6.6), would favour such a balance. However, the extent to which either source or sink limitations predominate at any particular stage is an indication of how far the system is out of balance. This can be equated with loss of yield potential, and is therefore worthy of further consideration.

### 5.2.1 PROBLEMS OF INTERPRETATION

Various treatments, such as shading, partial defoliation, carbon dioxide enrichment and selective grain removal, have been imposed in order to investigate the effects of variations in assimilate supply on grain growth. The results from shading experiments are inconsistent: in some cases, grain yield is reduced, while in others there is little effect. Similarly, increased grain yield has not always resulted from $CO_2$ enrichment. Clearly, a reduction in grain yield as a result of shading suggests that under these conditions the supply of assimilate limited grain growth, but this is not to say that under natural field conditions such reductions in photosynthesis would occur. Similarly, lower grain yields after partial defoliation could be attributed to nitrogen deprivation as well as to a shortage of assimilate. On the other hand, a lack of yield response to shading, or to partial defoliation, does suggest sink limitation. In these circumstances the apparent spare source capacity no doubt depends partly on mobilization of reserves or enhanced photosynthesis in the leaves remaining after partial defoliation. The success of such compensation will depend on the severity and duration of the treatments imposed, as well as on the extent of the reserves and the plants' inherent ability to adjust the rate of photosynthesis to sink demand. It is thus not unreasonable to expect inconsistencies between experiments.

The timing of shading treatments may also be significant. It is now established that the potential weight of the individual grain is set by the number of cells formed during the endosperm's meristematic period, which in wheat lasts for 16–20 days after anthesis (but longer in barley: see Cochrane and Duffus 1981). A reduction in assimilate supply during this period, by thus limiting the storage capacity of the grain, may therefore have a greater effect than similar later treatments. Experiments have indeed shown that variations in irradiance during the meristematic phase have parallel effects on grain dry weight and endosperm cell number (Fig. 5.4).

The same interpretation can be applied to experiments in which some of the grains are removed from an ear. This usually results in an increase in the final weights of the remaining grains, suggesting that in the intact ear the grains compete for a limited supply of assimilate. It is possible, however, that in the intact ear the grains do receive sufficient assimilate for their needs, and the effect of selective grain removal is to enhance the sink capacity of the grains through an increase in endosperm cell numbers. Thus, in each case, grain growth may be limited by the existing sink capacity, although this attribute is itself regulated by assimilate supply. Increased grain weights associated with greater numbers of endosperm cells have been observed in experiments involving selective removal of grains (Brocklehurst 1977; Radley 1978).

The interpretation of grain growth in terms of source–sink relationships is therefore made easier if individual grain weight is itself considered as the product of two yield components, the number of endosperm cells and the mean weight per cell. Even so, to envisage these relationships solely in terms of assimilate supply and demand is undoubtedly simplistic: hormonal influences must also be considered, particularly in grain removal experiments, in which the normal hormonal status of the ear may be altered. For example, cytokinin activity of the grain rises to a peak shortly after anthesis (Wheeler 1972), and it is tempting to assume a causal relationship between this event and the meristematic activity of the endosperm. Certainly, varietal differences in grain size in barley have been associated with significant differences in cytokinin activity (Michael and Seiler-Kelbitsch 1972), and treatment of intact ears of wheat with cytokinins during the first sixteen days after anthesis has resulted in increased grain weights (Herzog 1982).

The source of the cytokinins active in the grain is not known precisely: some may originate in the ears and some certainly appears to be carried from the roots in the transpiration stream. Thus, barley grains from de-awned ears, in which transpira-

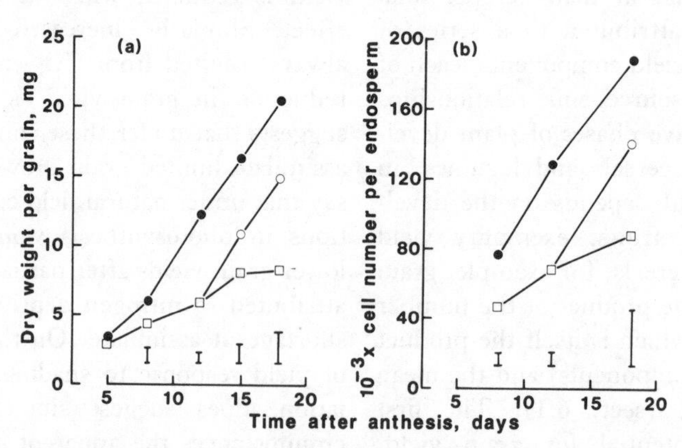


**Fig. 5.4** Grain dry weight (a) and endosperm cell number (b) of the *a* grain taken from wheat cv. Sonora plants growing continuously at high PPFD (●, 560 $\mu$mol m$^{-2}$ s$^{-1}$), or transferred 2 days after anthesis to low PPFD (□, 55 $\mu$mol m$^{-2}$ s$^{-1}$) until the grains had matured. Another set of plants (○) was returned to high PPFD on day 12. Vertical bars represent LSD ($P < 0.05$) between means for times (from Singh and Jenner 1984).

tional losses were reduced, had lower cytokinin activities than those from intact ears. Furthermore, a reduction in cytokinin production by the roots, caused by waterlogging, was reflected in lower cytokinin activity in the grains and an associated reduction in grain growth (Michael and Seiler-Kelbitsch 1972). It is possible, therefore, that grain removal results in increased availability of cytokinins for the remaining grains, an interpretation which is reinforced by the existence, at least in wheat, of competition for cytokinins between leaves and grains (Herzog 1982).

This digression indicates some of the problems in the interpretation of experiments which modify normal source–sink relationships during grain-filling. However, even if a less equivocal picture were forthcoming, it would, as we have already noted, be inappropriate to consider grain-filling in isolation from the earlier source–sink relationships which determine the potential yield. The number of grains per unit area is itself the product of other yield components: the number of fertile (ear-bearing) tillers per unit area, the number of spikelets per tiller, the number of florets per spikelet and the number of florets which set grain. These components are determined during the period of reproductive development, from ear initiation until a few days after anthesis (Fig. 6.5; sect. 6.1). They are dependent on the availability of assimilate, and competition between the developing ear and several already established sinks suggests that the determination of potential yield is source-limited.

### 5.2.2 COMPETITION FOR ASSIMILATES BETWEEN THE DEVELOPING EAR AND OTHER SINKS

Stem growth begins towards the end of spikelet differentiation and accelerates during the period of floret development. At the same time, growth of the upper leaves, particularly the flag leaf, continues (Fig. 5.5). Competition for assimilate is suggested by the many field and controlled environment experiments in which assimilate supply has been altered by pre-anthesis shading or $CO_2$ enrichment. Shading, for example, reduces both tiller survival and floret formation, as well as increasing floret death, so that the crop has fewer ears per unit area and fewer grains per ear. That competition for assimilate is responsible for these effects has been assumed rather than proved (see Gallagher and Biscoe 1978 for the circumstantial evidence), although the assumption is reinforced by the experiments of Fischer and Stockman (1980), who measured carbohydrate contents of ears as well as yield components. They found that shading treatments of about 10 days centred 10–13 days before anthesis were most effective, through reductions in the number of grains per spikelet. Not only was the number of florets reduced but there was an increase in the number which did not progress to full anther development ('non-competent' florets); shading did not affect the number of competent florets which subsequently set grain. The total dry weight increment of the plants was reduced by shading, although the proportion allocated to the developing ear, about 30 per cent, was unaffected.

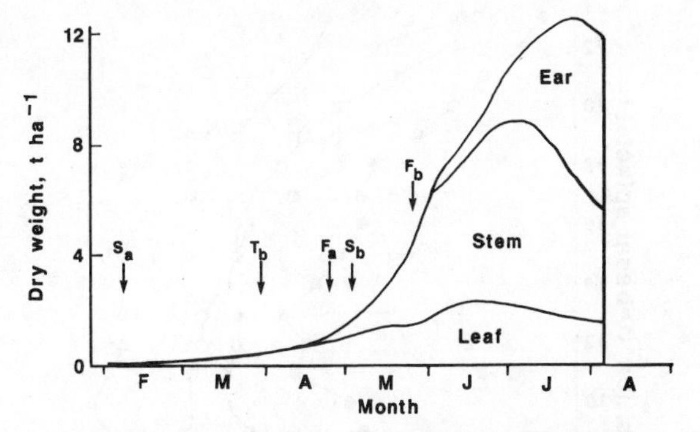


**Fig. 5.5** Time course of the changes in total, leaf, stem and ear dry weight for a crop of winter wheat cv. Huntsman, sown in October 1974. The times when tillering ended ($T_b$), spikelet production started ($S_a$) and stopped ($S_b$), and floret production started $(F)_a$ and stopped ($F_b$), are indicated (from Biscoe and Gallagher 1978).

It appears, therefore, that even at this relatively advanced stage of development, when it is doubling its weight every 3–4 days, the ear does not take precedence over the other major sink, the stem, when there is a shortage of assimilate. The decreased growth rate of the ears of shaded plants was preceded by reduced water-soluble carbohydrate (WSC) levels in the ears, and associated with reductions in the starch content of carpels. Further support for the idea of florets in competition for assimilate comes from subsequent work in which increased light was given for 8 days,

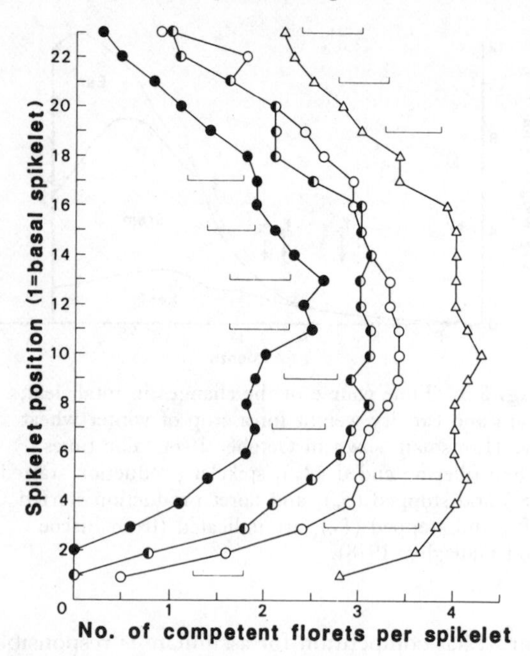


**Fig. 5.6** Competent floret profiles at anthesis for the Mexican semidwarf wheat, Yecora 70. Plants were grown in a controlled environment cabinet with a 14 h photoperiod during which the PPFD at the level of the uppermost leaves was 700 $\mu$mol m$^{-2}$ s$^{-1}$. At flag leaf emergence, the following treatments were imposed for 8 days: △, additional light–plants raised in the cabinet to give 15 per cent more light, and spaced further apart; ○, control; ◐, shaded to 40 per cent below control level; ●, shaded to 70 per cent below control level (from Stockman *et al.* 1983).

starting at flag leaf emergence (14 days before anthesis): this allowed greater floret survival, just as shading increased floret mortality (Fig. 5.6).

Full floret development appears to be particularly sensitive to low WSC levels, and even in plants not subjected to experimental shading a substantial proportion of florets dies prematurely (Fig. 6.5). The question of floret death is oversimplified, however, if it is represented solely in terms of competition for carbohydrate. Other limiting substrates, such as reduced nitrogen, are undoubtedly also involved, and we have already noted the possibility of intraplant competition for growth substances, such as cytokinins (p. 115).

$CO_2$ enrichment experiments support the implication that the developing ear competes for assimilate with the growing stem and flag leaf. Grain yield is increased most when extra $CO_2$ is given during the late reproductive period, between jointing (the appearance of the first node, signalling the start of stem extension) and anthesis (decimal code 3.1 to 6; Table 6.1), largely due to an increase in the number of fertile tillers per unit area and in some cases to a greater number of grains per ear. Havelka *et al.* (1984) extended these studies by showing that the higher rates of net photosynthesis, and the greater number of fertile tillers, in $CO_2$-enriched wheat plants were indeed associated with increased levels of sucrose and starch in the flag leaves. That this extra carbohydrate was available for mobilization and export was suggested by its rapid depletion during the first two weeks after anthesis.

### 5.2.3 THE MEANING OF 'SOURCE LIMITATION'

We have seen that during its development before anthesis and during the early part of grain-filling, the small size of the ear and its inability to develop fully in competition with other sinks clearly suggests source limitation of potential yield. Any such source limitation will, of course, be exacerbated when plants are growing in environments not wholly favourable for their photosynthesis, e.g. maize in Northern Europe or temperate cereals in prolonged cloudy weather. The term 'source limitation' requires clarification because it conveys information not only about the availability of assimilate but also about the ability of the economic sink to attract that assimilate in competition with other sinks. For example, the ear may be source-limited during its pre-anthesis development because it is unable to compete adequately for the relatively small quantity of assimilate which it requires. Later, during grain-filling, the degree of source limitation may be lessened, despite reduced supplies of assimilate from the senescing canopy, because the grains are now established as strong sinks with little competition.

The same reasoning explains why early grain growth might be limited by a shortage of assimilate at a time when 'surplus' assimilate is being stored in the stems (Fig. 5.1): the stems at this time remain more powerful sinks than the young grains, able to monopolize the assimilate supply. The grains constitute a much smaller sink, and are further disadvantaged by the obvious requirement for leaf assimilate to pass through at least the upper internode of the stem *en route* to the ear. Because increased assimilate supply at this stage results in higher final grain weight, grain growth is, by definition, source-limited, despite the fact that it is the grains' inability to attract sufficient assimilate, surplus to the rest of the plant's requirement for growth, which is responsible for their being unable to grow to full potential. Only if the growth of the grains does not respond to increased assimilate supply is it said to be sink-limited, implying saturation of the grains' requirement for assimilate and the attainment of potential grain weight (but see p. 120).

### 5.2.4 SOURCE OR SINK LIMITATIONS AFTER ANTHESIS

Difficulties in the interpretation of experiments investigating post-anthesis limitations to grain growth, particularly concerning the possible regulation of endosperm cell number by assimilate supply, have already been discussed. However, if such regulation exists, it follows that this yield component at least is source-limited. Furthermore, it would provide another means of keeping source and sink in balance, because the potential size of the individual grain would be geared to the likely subsequent assimilate supply. Of course, the system swings out of balance, more towards either source limitation or sink limitation, if the assimilate supply during the rest of the grain-filling period is less or more than the 'likely' amount. Source limitation at some stage during grain-filling is therefore a possibility, and most likely during the early stages. The amount of available assimilate and the capacity of the grains to store that assimilate determines whether source limitation continues to predominate or whether, in the absence of competing sinks, the storage sites become saturated. The many factors regulating assimilate supply alone make generalizations futile.

Furthermore, the storage capacity of the collective grain sink depends not only on the number of storage sites (numbers of grains and of endosperm cells), but also on the innate ability of those sites to accumulate assimilate. Physical resistances between the phloem within the grain and the endosperm impose an upper limit to the transport of sucrose, and certainly towards the end of grain-filling the ability of the grains to convert sucrose into starch declines. These and other controls operating within the grains themselves will be discussed in section 5.5.

Environmental influences on the extent to which source or sink limitations prevail during the post-anthesis period are therefore inevitable, and we might also expect differences between species and cultivars. Fischer and HilleRisLambers (1978) assessed the degree of source limitation to grain-filling in cultivars of *Triticum* species and *Triticale*, grown under irrigation and high fertility during five successive seasons in Mexico. Potential grain weight, attained with unlimited assimilate supply, was assumed to be that which resulted from a substantial (*c.* 80 per cent) reduction in grain number per ear at anthesis. The degree of source limitation (SR) was then taken to be the percentage increase of potential grain weight over control grain weight (no reduction in grain number). That assimilate supply was the prime cause of this response was confirmed by calculating a similar index of source limitation (ST) from the increased grain weight which resulted from thinning a plot at anthesis to reduce shading.

Further validation came from comparisons with the responses of grain weight to post-anthesis shading or $CO_2$ enrichment and to the removal of the flag leaf lamina at anthesis. Differences in SR were consistent with the presumed variations in

**Table 5.2** The effect of flag leaf removal at anthesis on the extent to which grain weight was source-limited (SR) in plants of wheat cv. Yecora 70 growing in three post-anthesis environments

| *Shoot treatment* | *Grain weight (mg)* | | |
|---|---|---|---|
| | *Unmodified crop* | *Thinned crop* | *Shaded crop* |
| No flag leaf removal: | | | |
| Control | 44.7 | 57.6 | 38.7 |
| Potential grain wt. | 55.5 | 55.7 | 55.8 |
| SR | 24% | −3% | 44% |
| Flag leaf removal: | | | |
| Control | 39.7 | 52.7 | 28.5 |
| Potential grain wt. | 54.9 | 55.1 | 51.7 |
| SR | 38% | 5% | 81% |

(After Fischer and HilleRisLambers 1978)

assimilate supply resulting, for example, from shading or thinning treatments (Table 5.2). These estimates of source limitation apply, of course, over the whole post-anthesis period, and do not discriminate between effects on endosperm cell number and subsequent effects on grain-filling. However, the response of grain weight declined markedly when treatments were delayed up to 20 days after anthesis (Fig. 5.7).

There were marked seasonal effects on SR and ST: SR, for example, ranged from 7 to 35 per cent across thirteen crops of the cultivar Yecora 70, and was positively related to mean post-anthesis temperature, which varied from 15.7 to 20.4 °C. There was apparently a direct effect of temperature on grain growth because the potential grain weight was reduced at higher temperatures, but there was an additional depressing effect on the grains of the control crop. This may have been indirect, the result of a reduced supply of assimilate, perhaps due to greater respiratory losses or hastened senescence. Whatever the cause, the increasing difference between potential grain weight and control grain weight indicated an

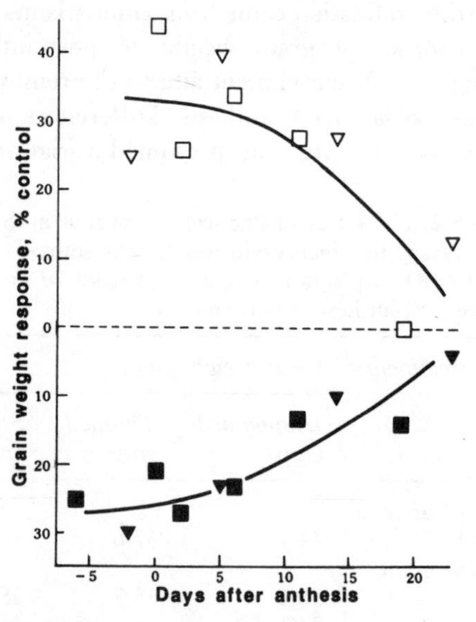


**Fig. 5.7** Effect of timing of treatment upon the response of grain weight to grain number reduction (open symbols) and to flag leaf removal (closed symbols). Yecora 70 in 1972 (□ ■) and 1974 (▽ ▼). 20 days after anthesis was about half-way through the grain-filling period, when grain weight was still less than 50 per cent of its final value (from Fischer and HilleRisLambers 1978).

increase in the degree of source limitation as the temperature rose. There was no significant effect of solar radiation on source limitation, perhaps because the range in radiation values was small (18.2–25.3 MJ $m^{-2}$ $day^{-1}$); certainly, artificial shading increased SR (Table 5.2).

Old, low-yielding bread wheats and *Triticales* showed the least source limitation, with SR and ST usually less than 10 per cent. Modern, shorter wheats, whose higher yields were primarily associated with greater grain numbers, had values of SR and ST averaging 20 per cent. This more pronounced source limitation was not entirely due to the higher grain numbers, however, because potential grain weight was also higher in the modern cultivars. This trend towards more severe source limitation with higher yields did not apply to a comparison of modern wheats on their own. It seems that within this group the increases in grain number have been accompanied by comparable increases in post-anthesis photosynthetic capacity, which have prevented the establishment of excessive source limitation, and thus reduced the risk of poorly-filled grains. Natural and artificial selection in the past will, however, have favoured plump grains, and, thereby, the mechanism by which a reduction in grain number under adverse conditions guarantees subsequent grain-filling. It follows that selection for higher grain yields through increased grain number must also have involved a reduction in the source limitation which drives this mechanism. The way in which this has been achieved in British wheats is discussed in section 5.3.

Of course, plump grains could be assured if their numbers were reduced sufficiently before anthesis for their subsequent growth to be sink-limited, but this could well mean a loss of yield potential through unused assimilate. Nevertheless, we have already noted that sink limitation may exist during the latter part of the grain-filling period when the demands of competing sinks have subsided. The inability of grains to respond to enhanced assimilate supply during the second half of their period of growth (Fig. 5.7) supports this view. Furthermore, the balance established by the prevailing supply of assimilate, between the potential yield (number of grains, potential grain weight) and subsequent grain growth, which prevents extreme source or sink limitation, can be upset. Fischer and HilleRisLambers found, for example, that the newer Mexican cultivars of *Triticale*, selected specifically for high potential grain weight, were severely source-limited (SR and ST averaging 36 per cent), suggesting that photosynthetic capacity had not kept pace with the improvements in potential grain weight.

Source limitation of grain-filling would also be promoted if an abundant supply of assimilate before anthesis, allowing the production of many grains, was curtailed after anthesis, by drought, high temperatures or pest or disease attack, for example. Conversely, the effect of pre-anthesis shading may be such that after anthesis there is a surfeit of assimilate available for the fewer grains produced, so that their growth is consequently sink-limited, particularly if the weather becomes sunnier after anthesis. For example, Stockman *et al.* (1983) found that when wheat plants were grown in a controlled environment at a PPFD of 700 $\mu$mol m$^{-2}$ s$^{-1}$ (PAR), the mean final weight of the fifty-eight grains per ear was, at 53 mg, close to the potential weight for that variety, given unlimited assimilate. However, when plants received 15 per cent extra PAR for 8 days, beginning 14 days before anthesis, grain number was increased to seventy and the mean grain weight reduced to 48 mg. This indicates that with this number of grains the balance had been tipped towards source limitation, because individual grain weight was now significantly less than the potential. It also verifies that maximum grain weight had been reached in the control plants without exhausting the assimilate supply, because in the controls total grain weight per ear was only 3074 mg (fifty-eight grains at 53 mg each), 286 mg less, for the same post-anthesis PPFD, than in the plants which had received additional pre-anthesis light.

This sink limitation was exaggerated in plants which had been shaded, to 70 per cent below the control level, for the same 8-day period before anthesis: grain number per ear was reduced to twenty-eight, but the grains were no heavier than those of the control plants. Admittedly, this shading treatment, giving light levels of about one-tenth full sunlight, was very severe. However, the adverse effects of shading on reproductive development can set an upper limit to the increases in the yield of cereals and legumes which result from increases in plant population density. Furthermore, other environmental stresses can have similar effects. Floret development is particularly sensitive to water stress, for example.

Whether a particular environmental stress promotes a predominance of source limitation or sink limitation may well depend not only on the timing of the stress, which is clearly central to its influence on reproductive development, but also on its severity. For example, we have already noted that high temperatures can have a direct effect on potential grain weight (p. 118), probably because the duration of grain-filling is reduced as temperature increases (Sofield *et al.* 1977); and that in the temperature range studied (15–20°C) there were, apparently, greater effects on assimilate availability, so that the degree of source limitation was more pronounced at the higher temperatures. However, still higher temperatures, allied perhaps with water stress, have been found to exaggerate the direct effect(s) on the sink, such that grain-filling is more dramatically cut short, and final grain weight correspondingly low (Gallagher *et al.* 1976). The reason for this premature termination of grain-filling, which imposes an overall limitation to grain growth, is not clear. In this particular case, the supply of assimilate did not appear to be limiting, and effects of high temperature and water stress on the grains' ability to synthesize starch may have been responsible.

## 5.3 THE IMPORTANCE OF THE HARVEST INDEX

In modern cultivars of cereals, such as wheat, it would appear that empirical selection for grain yield has produced plants in which yield under favourable conditions is not restricted disproportionately by either source or sink limitations. However, as we have seen, this balance is itself established by the pre-anthesis source limitation of grain number, because this militates against subsequent excessive source limitation during the grain-filling period. The central role of this constraint, in determining the yield components which comprise potential yield, inevitably means that source limitation is a more dominant influence than the sink limitation which operates during the later phase of grain-filling.

This is illustrated by comparisons of yield within a cultivar (over a number of seasons, for example), which show that yield variation is associated mainly with differences in grain number per unit area, and that individual grain weight is a comparatively conservative quantity (e.g. Gallagher and Biscoe 1978) (sect. 6.6). A wide range of individual grain weights is unlikely, first, because low values are avoided by reductions in grain number, and a shortage of current assimilate can be supplemented by the mobilization of stored reserves; and secondly, because high values are restricted either by assimilate supply or by the inherent capacity of the grain to accumulate starch.

The point has already been made (p. 114), but bears repeating, that grain growth can be unresponsive to increased assimilate supply, and therefore sink-limited, even though its final weight does not reach the potential for that cultivar. This is because the final weight of a grain, though less than the potential, may be the maximum possible within the confines set by its complement of endosperm cells. The potential grain weight, however, is that attained by grains which have received an unlimited supply of assimilate from anthesis throughout grain-filling, and in which, therefore, the division of endosperm cells has not been restricted. Thus, an overall source limitation may prevail over the growth of grains whose final weight is less than the potential, but which are unable to respond to increased supplies of assimilate over perhaps half of the grain-filling period, during which growth is therefore sink-limited.

There is more variation in individual grain weight between cultivars, however. A comparison of varieties of wheat introduced in Britain from 1908 until 1978 showed that the greater yields of the modern varieties could not be attributed primarily to any one yield component: high yields clearly resulted from various combinations of yield components (Austin *et al.* 1980b; Table 5.3). Past increases in yield must have been due to reductions in the source and sink limitations operating during the development of these components, and it is instructive to see how this has been achieved.

One way of relieving source limitation would be to increase net carbon fixation and, therefore, the amount of dry matter available to the various competing sinks. The existing pattern of assimilate allocation among the sinks may not be altered, as Havelka *et al.* (1984) found in their $CO_2$-enrichment experiment (p. 116). Thus, although the increased assimilate production allowed a

**Table 5.3** Grain yield, and its components, of eleven varieties of winter wheat introduced in the UK this century. The data are from high-fertility plots intensively managed against weeds and diseases, and in which lodging was prevented by allowing the plants to grow through netting

| *Variety* | *Height to base of ear (cm)* | *Grain yield, 0% mc ($t\ ha^{-1}$)* | *Ears ($m^{-2}$)* | *Grains ($ear^{-1}$)* | *Mean grain weight (mg)* |
|---|---|---|---|---|---|
| Little Joss (1908) | 142 | 5.22 | 366 | 36.8 | 38.8 |
| Holdfast (1935) | 126 | 4.96 | 468 | 32.5 | 32.6 |
| Cappelle-Desprez (1953) | 110 | 5.86 | 435 | 30.1 | 44.8 |
| M. Widgeon (1964) | 127 | 5.68 | 415 | 30.3 | 45.1 |
| M. Huntsman (1972) | 106 | 6.54 | 379 | 36.2 | 47.7 |
| Hobbit* (1977) | 80 | 7.30 | 429 | 46.1 | 36.9 |
| Mardler* (1978) | 77 | 6.21 | 386 | 42.1 | 38.2 |
| 370/500* | 78 | 6.34 | 385 | 38.1 | 43.2 |
| 730/3637* | 78 | 7.07 | 424 | 39.8 | 41.9 |
| 989/10* (1981) | 84 | 7.57 | 366 | 43.6 | 47.5 |
| Armada (1978) | 97 | 6.86 | 535 | 33.9 | 37.8 |

The varieties identified by code numbers refer to advanced breeding lines, one of which, 989/10, was subsequently released as Norman.
* Homozygous for the dwarfing gene Rht2.
(After Austin *et al.* 1980b)

greater number of ear-bearing tillers to be supported and, consequently, an increased grain yield, there was a proportionate increase also in non-grain weight: the harvest index (HI) was therefore unchanged. However, comparisons of old and new varieties of cereals and legumes (e.g. Donald and Hamblin 1976; Gifford *et al.* 1984) reveal that increases in economic yield potential have been due to improved harvest index, and that total (above-ground) yield has remained largely unchanged (sect. 6.4.1).

This is well illustrated by the survey of twentieth-century British winter wheat varieties referred to above (Fig. 5.8). The correspondence between grain yield and HI is striking, although improved HI itself was not a selection criterion, but rather a secondary result of direct selection for grain yield or for shorter, stiffer straw to withstand lodging. However, the benefits of this inadvertent selection for increased HI, in the absence of any lodging, were clearly revealed in this experiment. Furthermore, the yield advantage of the newer varieties did not depend on high soil fertility, because the ranking of varieties in terms of grain yield and HI was the same on both high and low fertility sites.

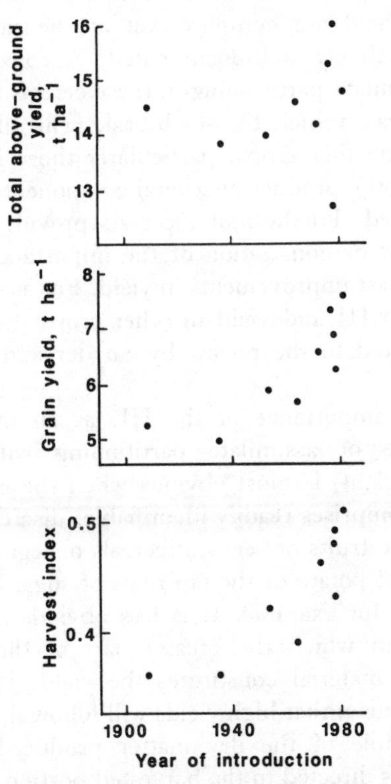


**Fig. 5.8** Total above-ground yield, grain yield and harvest index of the named varieties of winter wheat in Table 5.3 plotted against the year of introduction (after Gifford *et al.* 1984 from data of Austin *et al.* 1980b).

The higher grain yields and harvest indices in the shorter, modern varieties were at the expense of stem weight per unit area: the investment of assimilate in leaf production was similar in the different varieties. We have already examined the evidence for competition for assimilate between the developing ear and other sinks, notably the stem. The reduction in competition, which would presumably be a consequence of reduced stem weight and height, would therefore allow less restricted development of the ear. The way in which this apparent relief of source limitation is expressed differs among the varieties, however. For example, the yield advantage of Armada over Little Joss can be attributed to its ability to produce more ears m$^{-2}$, whereas Norman benefits by having larger ears and heavier grains. Nevertheless, there is some consistency in that Norman and the other varieties possessing the Norin 10 dwarfing gene, Gai/Rht2 (for 'gibberellic acid insensitive' and 'reduced height'), have more grains per ear than the taller varieties (gai/rht2).

Brooking and Kirby (1981) confirmed this characteristic, and showed that it was due not to any difference in the numbers of spikelet and floret primordia initiated, but to improved floret survival (cf. Fischer and Stockman 1980; Fig. 6.5). This was associated with the partitioning of a greater proportion of total growth to the ear in Gai/Rht2 lines compared with gai/rht2 lines and, indeed, there was a direct relationship between the dry weight of the ear at anthesis and the number of grains per ear at maturity. However, this increased partitioning of dry matter to the ear, and the resultant higher HI, was associated not with reduced stem height and weight *per se*, but with the possession of the Gai/Rht2 gene (Table 5.4: compare the genotypes ST (gai/rht2) and TD (Gai/Rht2) which have identical stem weights and similar heights). The mechanism of this influence is unclear.

If past improvements in yield, other than those due to decreased susceptibility to lodging and disease, have resulted from higher harvest indices,

**Table 5.4** Grain yield and its components at maturity of six genotypes of winter wheat. All data per main shoot, except total grain yield

| *Genotype* | *gai/rht2* | | | *Gai/Rht2* | | | *LSD* ($P = 0.05$) |
|---|---|---|---|---|---|---|---|
| | *CD* | *TT* | *ST* | *HS* | *SD* | *TD* | |
| Number of grains | 36.8 | 42.2 | 39.1 | 48.8 | 52.2 | 44.3 | 2.35 |
| Number of grains per spikelet | 1.93 | 2.30 | 2.15 | 2.81 | 2.82 | 2.61 | 0.14 |
| Grain weight (mg) | 53.3 | 52.9 | 51.0 | 47.6 | 48.9 | 53.4 | 1.2 |
| Grain yield (mg) | 1961 | 2228 | 1992 | 2323 | 2553 | 2361 | 127 |
| Stem weight (mg) | 1211 | 1260 | 1065 | 791 | 919 | 1065 | 51.9 |
| Stem length (mm) | 949 | 932 | 775 | 683 | 674 | 825 | |
| Harvest index (%) | 46.6 | 47.7 | 47.0 | 55.4 | 53.9 | 51.7 | 1.0 |
| Total grain yield (g $m^{-2}$) | 565 | 558 | 568 | 664 | 719 | 636 | 86 |

HS: Hobbit 's' (sister line of Hobbit).
CD: Cappelle Desprez.
The other genotypes were progeny of a cross between HS and CD, classified according to height and dwarfing gene constitution: TT, tall tall; ST, short tall; SD, short dwarf; TD, tall dwarf.
(After Brooking and Kirby 1981)

to what extent can this trend be further exploited? Austin *et al.* (1980b) calculated that a reduction in stem and leaf sheath dry matter to half current average values and a reallocation of this dry matter to the ear could raise the HI from about 0.5 to 0.62, assuming a constant biomass yield (sect. 6.4.1). Whether such a dramatic change is possible, or even desirable in view of the stem's role in supporting the ear and maintaining an effective display of leaves for light interception, is doubtful. Increases in total dry-matter production, while maintaining the current highest harvest indices, therefore provide the most likely route to further increases in yield (sect. 6.6). This would be most effectively achieved by higher rates of photosynthesis per unit leaf area, since in many circumstances higher water requirements would make a larger leaf surface counterproductive.

We would expect increased rates of net assimilation to support the development of more grains per unit area, but full exploitation of greater availability of assimilate may also require an increase in the capacity of individual grains to accumulate dry matter. We have already noted differences between varieties in mean grain weight, but there is also variation in the potential grain weight within a variety, according to the position of the grain on the ear. Similar intraplant variation in the potential size of the economic sinks occurs also in other crops, such as tomato, and indicates the need to understand the properties of the sink which determine its potential for growth (sect. 5.5).

We have concentrated on wheat and barley in the preceding sections because the source–sink relationships involved in the determination of grain yield are complex, yet at the same time comparatively well-documented. A consideration of assimilate partitioning in these cereals therefore provides a vehicle by which basic principles applicable to other crops, particularly those in which yield is the product of several components, can be presented. Furthermore, cereals provide perhaps the best demonstration of the importance of the HI to past improvements in yield. For associations between HI and yield in other crops, the reader is referred to the review by Snyder and Carlson (1984).

The importance of the HI, as an integrated measure of assimilate partitioning within the whole plant, is most obvious when the economic yield comprises readily identifiable, discrete sinks, as in the fruits or seeds of cereals or legumes, the tubers of potato or the tap roots of sugar beet and carrots, for example. It is less obvious in forage crops, in which the greater part of the above-ground material constitutes the yield. However, it is a truism that high yields will follow if as much as possible of the dry matter produced in the season is directed to the harvested portion, so long as the roots are not starved to the extent that their functioning limits productivity. The highest yields will not be achieved, though, merely by the maximum allocation of dry matter to the growing shoots: the way in which this material is used is also important. In order to make full use of the available solar radiation, crops must reach the $L_{crit}$ as soon as possible, and this is as important for the rapid regrowth of forage crops after grazing or cutting as it is for seedling growth (Fig. 2.18; sect. 2.4.2 and 8.2). High yields would therefore be promoted when the allocation of assimilate to the shoots is expressed as rapid leaf area expansion rather than merely as increased dry weight.

For example, Potter and Jones (1977) showed that the relative growth rates of various crop and weed seedlings, over a range of temperatures, were dependent on the rate of leaf area production as a proportion of the concurrent dry weight increase, a function which they termed the leaf area partition coefficient. Similarly, Wilhelm and Nelson (1978) compared the regrowth after cutting of four genotypes of tall fescue (*Festuca arundinacea*), selected for high or low net photosynthetic rates and high or low yields. Regardless of their rating for net assimilation, the more rapid growth exhibited by the high-yielding genotypes was due to their more rapid leaf area expansion, achieved by the partitioning of a greater proportion of newly-assimilated dry matter into leaf area, and by the possession of larger carbohydrate reserves to initiate new leaf growth.

The picture is more complicated in perennial crops, in which there must be a compromise between a high HI in one season and the need to accumulate the reserves necessary to sustain the plants during the winter and for the resumption of growth in the next season. For example, the ability of white clover cultivars to resume growth in the spring in Britain is positively correlated with the weight of stolon present at the start of the winter. Clovers compete with grasses in a mixed sward, and cultivars with a high HI, expressed as long petioles and large leaves, are more competitive in their establishment year than shorter, smaller-leaved cultivars (sect. 8.4.1). However, the greater allocation of dry matter to petioles and leaves is at the expense of stolon biomass, so that regrowth and competitive ability in the second season is comparatively poor. The greater investment of dry matter in the stolons of smaller, naturalized or indigenous British cultivars compared with that of the larger cultivars, which are derived from South European or New Zealand plants, is apparently an adaptive feature which ensures satisfactory overwintering and regrowth. As well as providing reserves, the stolons are themselves able to photosynthesize. On the other hand, in the warmer winters of New Zealand, for example, the provision of reserves from large stolons is less important because new leaves are produced throughout the year (Harris *et al.* 1983).

## 5.4 CARBON PARTITIONING WITHIN THE LEAF, AND PHLOEM LOADING

### 5.4.1 INTRODUCTION

So far, we have discussed dry-matter partitioning in terms of the distribution of available assimilate among competing sinks, involving assimilate movement over comparatively long distances. However, partitioning occurs also on a smaller scale, within the source leaf, and is responsible in part for determining the amount of assimilate available for subsequent distribution. We have already seen that the availability of assimilate can be an important determinant of yield, particularly during the early, formative stages of economic sink development, when more powerful, alternative sinks receive the lion's share. It is important, therefore, to understand the partitioning processes which determine how much of the carbon fixed by net photosynthesis becomes available for export, largely as sucrose.

The term partitioning encompasses the biochemical and spatial compartmentation of fixed carbon within the photosynthesizing cells as well as the subsequent movement of sucrose to the phloem of the minor veins, where it is loaded into the sieve elements. In this sense, partitioning begins with the divergence of photosynthetic metabolism at the very point of carbon fixation, as a result of the oxygenation of RuBP (and the consequent photorespiratory cycle), as well as its carboxylation. The extent of carbon flow through the photorespiratory diversion is just one of the factors influencing net photosynthesis. Here, however, we are concerned with the partitioning of the products of net photosynthesis, the most obvious manifestation of which, recognized over a century ago, is the formation of sucrose and starch in the photosynthesizing cells. Triose phosphate which leaves the Calvin cycle is converted into starch within the chloroplasts, or is exported to the cytosol to form sucrose. The sucrose represents assimilate potentially available for export, whereas starch serves as a reserve, subject to mobilization into sucrose to maintain continuity of export at night or whenever photosynthesis is too low to meet the assimilate requirements of the rest of the plant.

The importance of sucrose synthesis to the status of the leaf as a source is illustrated by several studies which show that the well-documented, and not unexpected, correspondence between the rates of translocation and of photosynthesis over a range of irradiances is largely due to the concentration of sucrose in the source leaf. This has been demonstrated in several crops, including tomato (Ho 1976a, 1979), barley (Gordon *et al.* 1980) and sugar beet, in which the close coupling of the rates of translocation and of photosynthesis (Fig. 5.9) has been studied in some detail. The possibility that the rate of translocation responded directly to the irradiance was discounted by manipulating the $CO_2$ and $O_2$ concentrations so that the rate of net photosynthesis (and, therefore, of translocation) remained nearly equal under different irradiances (Servaites and Geiger 1974). Subsequently, both the rate and velocity of translocation (eqn 5.1) were found to respond to changes in the sucrose concentration in the source leaf (e.g. Fig. 5.10). However, it can be seen from Fig. 5.10 that the direct relationship between the translocation velocity and the sucrose concentration held only over a limited range of low sucrose concentrations. The lack of correlation at concentrations over about 15 g $cm^{-2}$ was attributed to compartmentation of sucrose into a non-transport, or storage, pool, as well as into the transport pool upon which, it was surmised, the velocity and rate of translocation depended.

In this experiment, the existence of the storage pool was betrayed by subjecting the plants to non-steady-state conditions. Treatments 25 and 27, for example, involved maintaining the plants first under a high (1100 $\mu l\ l^{-1}$) $CO_2$ concentration in the light for about 24 hours, and then lowering the $CO_2$ concentration to 405 $\mu l\ l^{-1}$ for the measurements of sucrose concentration and the velocity of translocation. Sucrose which accumulated as a storage pool during the high $CO_2$ treatment was relatively unavailable for export from the leaf when the $CO_2$ concentration was lowered. The velocity of translocation was thus related not to the total sucrose concentration, but to the amount of currently-formed sucrose entering the transport pool as a result of photosynthesis under the lower $CO_2$ concentration. Furthermore, sucrose

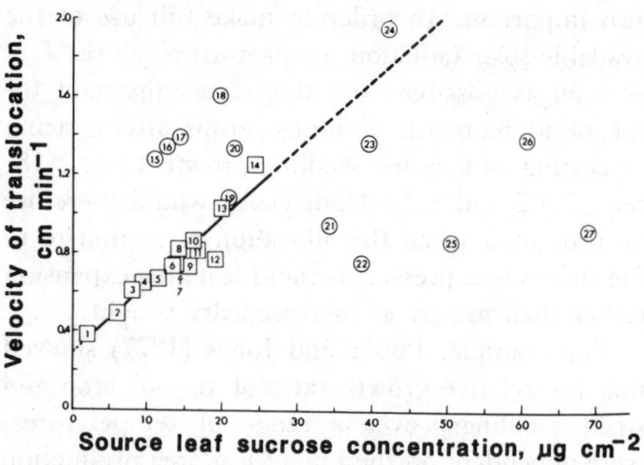


**Fig. 5.10** Relationship between the velocity of translocation and the concentration of sucrose in the source leaf of sugar beet plants. Data were obtained from $^{14}CO_2$ pulse-labelling experiments, and the numbers refer to different treatments (duration of light and dark before labelling, and $CO_2$ concentrations before and after labelling) designed to cause variations in photosynthetic rates and sucrose concentrations.

Pulse-labelling refers to the introduction of a limited amount of radioactive tracer so that the movement of the labelled 'pulse' can subsequently be monitored (from Christy and Swanson 1976).

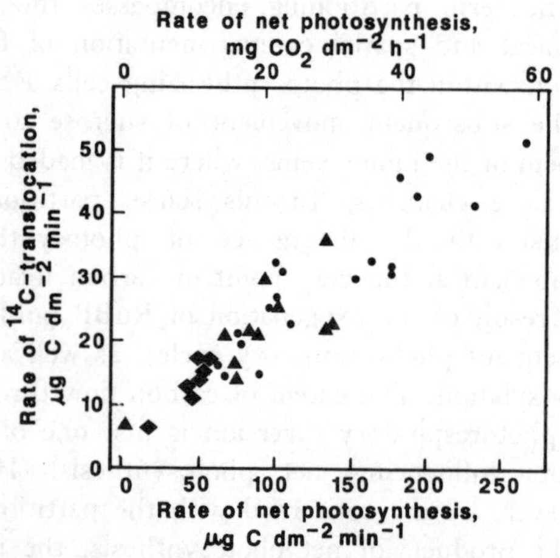


**Fig. 5.9** Relationship between the rate of translocation and the rate of photosynthesis in sugar beet plants. The symbols represent three different irradiances, and further variation in photosynthesis was achieved by varying the $CO_2$ concentration at the source leaf (after Servaites and Geiger 1974).

accumulation was exaggerated in this experiment because the sink tissue was restricted to a single young leaf, translocation to the root having been prevented by heat-girdling the hypocotyl. In a normal plant, however, photosynthesizing under steady-state conditions, the storage pool would be less apparent because it would be not only smaller but also in equilibrium with the transport pool.

Other evidence of two distinct sucrose pools, in sugar beet and other species, comes from the time-course of $^{14}C$ depletion in a leaf following steady-state labelling with $^{14}CO_2$. Thus, an initial, rapid, exponential loss of $^{14}C$-sucrose implies a pool of sucrose readily available for export and rapidly replenished by newly synthesized, unlabelled sucrose. This is followed by a slower exponential loss caused by the movement of labelled sucrose out of a second, storage pool, which clearly has a lower rate of turnover than the first, transport pool (see, for example, Geiger *et al.* 1983; cf. Fig. 5.19).

Clear evidence of histological compartmentation of sucrose has been demonstrated in leaflets of *Vicia faba* (Outlaw *et al.* 1975). The kinetics of $^{14}C$-sucrose specific activity (sucrose radioactivity per unit weight of sucrose) were followed in the palisade parenchyma, the lower layer of spongy parenchyma, and the upper layer of spongy parenchyma which contained the veins, after leaflets had been labelled with a 1-minute pulse of $^{14}CO_2$ (Fig. 5.11). The specific activity increased to a maximum as newly-synthesized sucrose became labelled and, thereafter, declined as labelled sucrose moved out of the leaf, while that remaining was diluted by more recently synthesized sucrose and by the transfer of radioactivity to other compounds. The palisade parenchyma had the highest rates of net photosynthesis, and this is reflected in the higher maximum specific activity of sucrose in this tissue compared with that of the lower spongy parenchyma. However, the highest specific activities were found in the parenchyma containing the veins, implying preferential transport of newly-synthesized sucrose, as a discrete pool, from the sites of $^{14}CO_2$ fixation to this tissue. The lower specific activity attained by the palisade parenchyma was presumably due to dilution of labelled sucrose in the larger, storage pool, assumed to be in the vacuole.

This picture of intracellular compartmentation of sucrose was substantiated by direct measurements of the specific activity of sucrose in samples of palisade parenchyma containing either mostly cytoplasm or mostly vacuole (Fisher and Outlaw 1979). The samples were obtained from leaf discs, taken at various times after pulse-labelling with $^{14}CO_2$, and centrifuged so that most of the cytoplasm collected at the centrifugal end of the cells. Cytoplasmic and vacuolar samples had distinctly different kinetics, confirming that a separate sucrose pool was associated with each compartment: the transport pool in the cytoplasm (strictly speaking, the extra-chloroplastic cytoplasm), where it is available for rapid, preferential movement to the minor veins, and the storage pool in the vacuole.

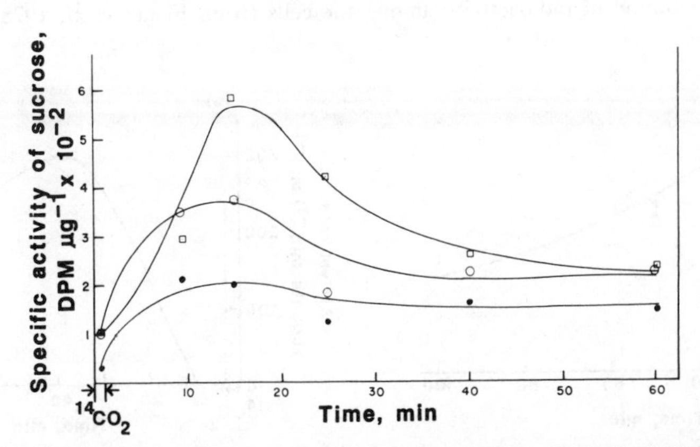


**Fig. 5.11** Kinetics of $^{14}C$-sucrose specific activity in different tissues of a leaf of *Vicia faba*, pulse-labelled with $^{14}CO_2$ for 1 minute: palisade parenchyma (○), lower layer of spongy parenchyma (●), upper layer of spongy parenchyma which contained veins (□) (from Outlaw *et al.* 1975).

The sucrose available for export at any given time can therefore be identified with that comprising the transport pool, and not necessarily with the total pool of sucrose in the leaf. Although the transport pool of an actively photosynthesizing leaf comprises primarily newly-synthesized sucrose, it can, when required, be supplemented by sucrose from the storage pool in the vacuole, or by the mobilization of starch. In any event, the bulk of the transport pool is considered to be sited in the sieve element–companion cell complex of the minor veins. Here, an active loading process is responsible for the accumulation of sucrose to higher concentrations than in the surrounding mesophyll cells. The accumulation of radioactivity in the minor veins of leaves fed $^{14}CO_2$ has been demonstrated, by autoradiography, in numerous species. The refinement of the technique, as quantitative microautoradiography, allows the kinetics of labelling in individual cells to be followed (Fig. 5.12). We would expect that in a leaf labelled with a brief pulse of $^{14}CO_2$, the site of the transport pool would be revealed as the cells in which the kinetics of labelling most closely matched the kinetics of arrival of radioactivity, after an appropriate time-lag, in the transport pathway or in a nearby sink. In this way, the companion cells of the minor veins of soybean leaves have been clearly identified as the site of a major part of the transport pool (Fig. 5.13).

Thus, the availability of sucrose for export from a mature leaf involves more than the parti-

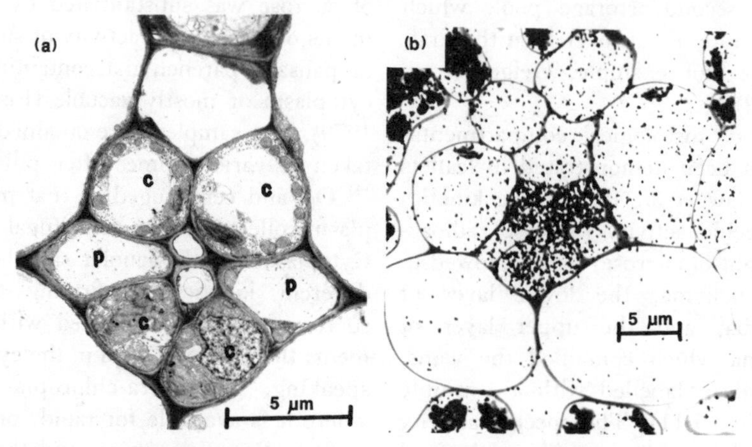


**Fig. 5.12** (a) Electron micrograph of minor vein of soybean leaf. c, companion cell; p, phloem parenchyma. (b) Autoradiograph of a soybean minor vein with cell types corresponding to those in (a). The black specks are the silver grains produced by the exposure of the autoradiograph film to radioactivity. Their density therefore indicates the distribution of radioactivity among the cells (from Fisher *et al.* 1978).

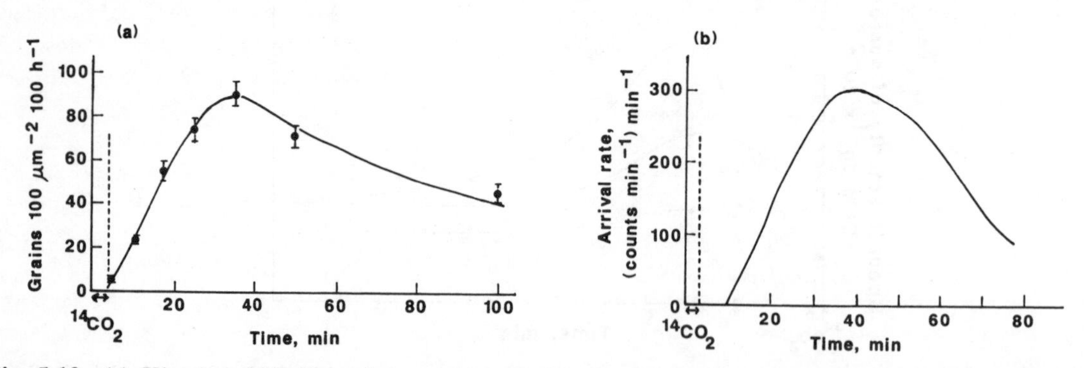


**Fig. 5.13** (a) Kinetics of labelling of companion cells in minor veins of soybean leaf after pulse-labelling with $^{14}CO_2$. Vertical bars represent SD for silver grain counts. (b) Arrival kinetics for $^{14}C$-photosynthate in a sink leaf of soybean 8 cm from pulse-labelled leaf (from Fisher *et al.* 1978).

tioning of fixed carbon between sucrose synthesis and starch synthesis: it also depends on the subsequent allocation of sucrose to transport and storage pools, which involves intracellular compartmentation, intercellular transfer, and finally the loading process which delivers the sucrose into the sieve elements. We shall now consider these component processes in more detail.

## 5.4.2 BIOCHEMICAL PARTITIONING

### 5.4.2.1 SUCROSE AND STARCH SYNTHESIS

The biochemical partitioning of triose phosphate (TP) between starch synthesis and sucrose synthesis (Fig. 3.30) also involves spatial partitioning, because TP (largely in the form of dihydroxyacetone phosphate) must be exported from the chloroplast for sucrose synthesis in the cytosol. This export of TP is an obligatory exchange for inorganic phosphate ($P_i$), which is released during sucrose synthesis and required inside the chloroplast for ATP synthesis. The exchange is facilitated by the phosphate translocator, a protein carrier situated in the inner envelope of the chloroplast. A consequence of this counter-exchange is that if sucrose synthesis is restricted, by feedback inhibition when sink demand is low, the reduction in $P_i$ availability inhibits TP export: more TP is thus available within the chloroplast for starch synthesis, which is further promoted by the high $[PGA]/[P_i]$ ratio (sect. 3.4.4). Although starch synthesis releases $P_i$ within the chloroplast, this is insufficient to sustain the maximum running of the Calvin cycle, which therefore relies on the import of $P_i$ from the cytosol. This dependence means that sucrose can exert a feedback control on the rate of photosynthesis.

However, there must also be feed-forward control of sucrose synthesis, to prevent the export of TP from becoming excessive, in the sense that it jeopardizes the regeneration of RuBP in the Calvin cycle (Fig. 3.30). The need for such control is demonstrated in experiments with isolated chloroplasts: rapid photosynthesis is impossible without a suitable concentration of $P_i$ in the bathing medium, but if too much is added the rapid withdrawal of TP causes a depletion of Calvin cycle intermediates and an inhibition of carbon fixation. The pathway of sucrose synthesis is shown in Fig. 5.14. An important regulatory factor is thought to be the level of the cytosolic metabolite, fructose-2,6-bisphosphate (F-2,6-$P_2$) (see Stitt 1985 for details and references). This compound inhibits fructose-1,6-bisphosphatase (FBPase), the enzyme catalysing the first irreversible step in the synthesis of sucrose. Its own synthesis is inhibited by high TP concentrations. Thus, when the rate of photosynthesis is low, the resultant low level of TP allows an increase in the synthesis of F-2,6-$P_2$, which in turn inhibits FBPase. Sucrose synthesis and the release of $P_i$ are therefore blocked, and TP is retained within the chloroplast. The effect of higher levels of TP in promoting sucrose synthesis is two-fold: first, the inhibition of FBPase is relieved as the level of F-2,6-$P_2$ declines, and secondly the concentration of the substrate for FBPase, F-1,6-$P_2$, is increased.

Changes in the ratio of substrate : inhibitor therefore represent a subtle control system by which the level of TP in effect regulates the activity of FBPase. There is evidence that this control allows for a threshold level of TP below which FBPase is inactive, and sucrose synthesis is therefore switched off. On the other hand, it effects large increases in FBPase activity, and rapid sucrose synthesis, for small increases in TP above the threshold, thus ensuring that potentially high rates of net photosynthesis are not limited by the availability of $P_i$ or the accumulation of Calvin cycle intermediates. The feed-forward control is further amplified by the stimulation of sucrose phosphate synthase activity by increased levels of glucose-6-P.

As we have seen already, TP is diverted towards starch synthesis once sucrose is produced in excess of the leaf's ability to export or store it. The feedback control of sucrose synthesis instigates this change in partitioning, and here also F-2,6-$P_2$ plays a part, through the inhibition of FBPase. The accumulation of sucrose inhibits sucrose phosphate phosphatase and sucrose phosphate synthase, causing a rise in fructose-6-P, which both stimulates the synthesis and restricts the breakdown of F-2,6-$P_2$. The increased level of F-2,6-$P_2$ then lowers the activity of FBPase, so reducing the conversion of TP into sucrose and thereby favouring starch synthesis.

FBPase is therefore a key enzyme in source leaf

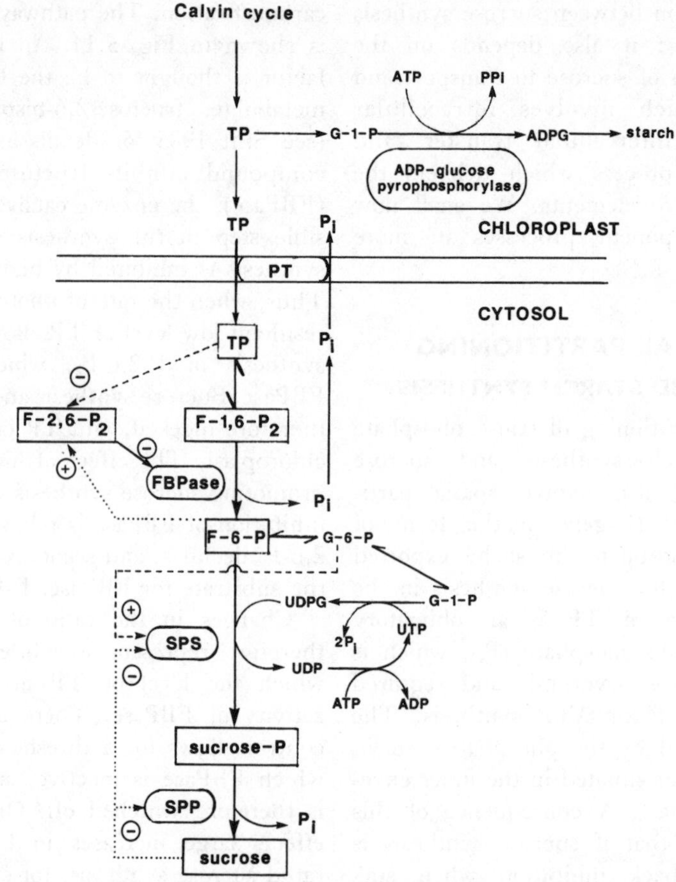


**Fig. 5.14** Pathway of sucrose synthesis and its control. Abbreviations: as Fig. 3.1, and F-1,6-$P_2$, fructose-1,6-bisphosphate; F-2,6-$P_2$, fructose-2,6-bisphosphate; FBPase, fructose-1,6-bisphosphatase; PT, phosphate translocator; SPP, sucrose phosphate phosphatase; SPS, sucrose phosphate synthase.

metabolism, its activity subtly modulated so that the rate of sucrose synthesis is geared both to the rate of net photosynthesis and to sink demand. The feed-forward control, of which it is part, allows sucrose synthesis to respond rapidly to increased rates of photosynthesis, but in many species only up to an upper limit set by the activity of sucrose phosphate synthase (SPS). A leaf's capacity for sucrose synthesis and export may thus be correlated with the activity of this enzyme. Certainly, the transition from import to export in leaves of several species, including sugar beet and soybean, is marked by the appearance of SPS activity (see Giaquinta 1980). Furthermore, we have already seen that partitioning into sucrose is favoured by the presence of active sinks, and that conversely starch synthesis becomes more important when the rate of photosynthesis is high relative to sink demand. If SPS activity can limit the leaf's capacity for sucrose synthesis, and in so doing regulate partitioning between sucrose and starch, we might therefore expect it to be influenced by changes in the source : sink ratio.

Several experiments, notably with soybean which can accumulate ample storage reserves of starch, have confirmed this expectation. For example, the presence of root nodules imposes an added demand for assimilates, which is met by increased partitioning of carbon into sucrose, associated with enhanced SPS activity (Huber and Israel 1982). Similarly, the increased sink demand made on an individual source leaf, when other mature leaves were removed, resulted in the cessation of starch accumulation. This was associated with increased activities of SPS and FBPase, and changes in sucrose content which

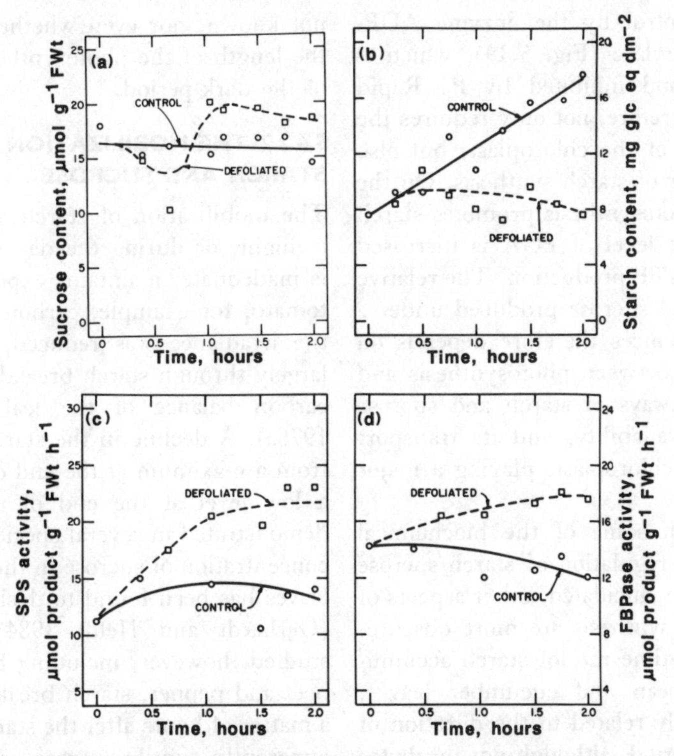


**Fig. 5.15** Changes in (a) sucrose, (b) starch, (c) SPS activity, and (d) FBPase activity in the first trifoliate leaf of Ransom soybean plants after defoliation of all other source leaves. The first trifoliate leaves of other, intact, plants were used as controls. Abbreviation: glc eq, glucose equivalent (from Rufty and Huber 1983).

were consistent with the re-equilibration of sucrose pools after a rapid increase in sucrose export and enhanced sucrose synthesis (Fig. 5.15).

The rate of net photosynthesis was unchanged over the 2-hour span of this experiment, in contrast with the pronounced increase which has been found to develop over a period of days after partial defoliation (Fig. 3.26). The change in partitioning, away from starch and in favour of sucrose synthesis, may therefore be an immediate response to increased sink demand, with photosynthesis increasing more slowly. Certainly, in both pepper (*Capsicum annuum*) and cucumber (*Cucumis sativus*), photosynthesis was found to increase during the transition from vegetative to reproductive development (Pharr *et al.* 1985).

Species such as wheat and barley, which accumulate little starch in their leaves, have higher SPS activities than those which form large amounts of starch, such as soybean, red beet, peanut and tobacco. This negative correlation between partitioning into starch and SPS activity has also been found in comparisons of varieties within a species, including wheat and peanut (see Huber 1983). It appears, however, that in species having a high SPS activity, partitioning towards sucrose is further enhanced by a lower sensitivity of SPS to feedback inhibition by sucrose: apparently, different forms of SPS exist in different species. In those species in which SPS activity is comparatively insensitive to sucrose level, an alternative means of initiating feedback control of sucrose synthesis seems likely. Rufty and Huber (1983) have suggested that ABA may be involved in the regulation of SPS, and the report of non-stomatal effects of ABA (Cornic and Miginiac 1983) warrants further investigation.

Starch synthesis clearly depends on the amount of TP which remains within the chloroplast, and is therefore indirectly controlled by the rate of sucrose synthesis. However, the reciprocal relationship between starch synthesis and sucrose synthesis is not due solely to their dependence on a common precursor. Starch synthesis is also subject to direct control by the enzyme ADP-glucose pyrophosphorylase (Fig. 5.14), which is activated by PGA and inhibited by $P_i$. Rapid sucrose synthesis, therefore, not only requires the diversion of TP out of the chloroplast, but also generates an inhibitor of starch synthesis. On the other hand, rapid photosynthesis promotes starch synthesis because the level of PGA is increased and $P_i$ is used up in ATP production. The relative amounts of starch and sucrose produced under a given set of circumstances therefore depends on complex interactions between photosynthesis and the biochemical pathways of starch and sucrose synthesis, with $P_i$ availability, and its transport between cytosol and chloroplast, playing a major coordinating role.

However, although some of the biochemical links involved in the regulation of starch/sucrose partitioning have been elucidated, other aspects of the control of starch synthesis are more obscure. For example, the daytime rate of starch accumulation, in both soybean and cucumber leaves, appears to be inversely related to the duration of the photosynthetic period, although not to photoperiod *per se* (Fig. 5.16). The greater rate of starch accumulation under short days, which was at the expense of current export and residual (structural) dry weight, may be seen as an adaptation to provide for the increased respiratory and export requirements during the correspondingly longer nights. The mechanism of this control is not known, nor even whether it is a response to the length of the photosynthetic period or to that of the dark period.

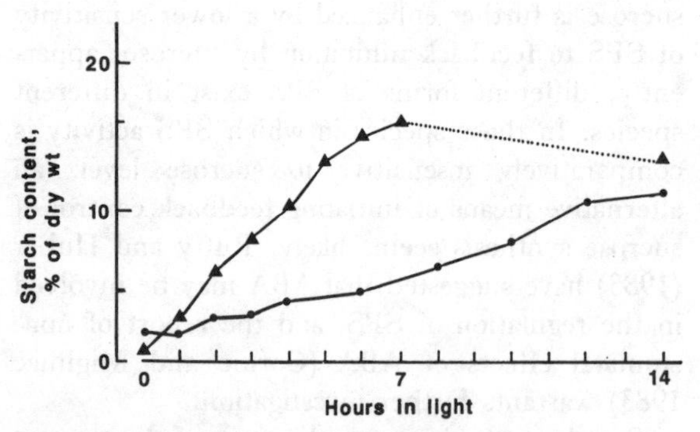


**Fig. 5.16** Changes in starch content of soybean leaves grown for 28 days in a growth cabinet under either a 14-hour (●) or a 7-hour (▲) photosynthetic period (640 $\mu$mol m$^{-2}$ s$^{-1}$). The photoperiod was the same in each treatment because the 7-hour photosynthetic period was followed by 7 hours of low incandescent light (10 $\mu$mol m$^{-2}$ s$^{-1}$) (from Chatterton and Silvius 1979).

### 5.4.2.2 THE MOBILIZATION OF STORED STARCH AND SUCROSE

The mobilization of starch and vacuolar sucrose at night, or during the day when photosynthesis is inadequate, maintains export from the leaf. In tomato, for example, carbon export decreased as the irradiance was reduced, but was continued largely through starch breakdown, even when the carbon balance of the leaf was negative (Ho 1976a). A decline in the starch content of leaves, from a maximum at the end of the light period to a low level at the end of the night, has been demonstrated in several species and, similarly, the concentration of sucrose in the vacuoles of spinach leaves has been found to decline during the night (Gerhardt and Heldt 1984). In most species studied, however, including barley, cotton, sugar beet and pepper, starch breakdown is delayed for a matter of hours after the start of the dark period, apparently until sucrose reserves have been exhausted. Little starch or sucrose appears to be carried over from one photoperiod to the next (see Geiger and Giaquinta 1982), presumably because, as noted above, the rate of starch accumulation is adjusted according to the demands imposed on the leaf by the duration of the dark period.

Details of the control of starch mobilization are, however, poorly understood. Starch granules are made up of two components, each a polymer of glucose units: amylose is a long, straight-chain molecule of around 1000–2000 glucose units, linked by (1→4) glycosidic bonds, and arranged as a helix; amylopectin, the major component, is a branched polymer of about 100,000 units, each branch chain linked by an (1→6) glycosidic bond (Fig. 5.17). How the two components are arranged within the granule is not clear.

Starch degradation in the chloroplasts is controlled by the activities of four main enzymes: α-amylase, β-amylase, debranching enzyme [α(1→6)glucosidase], and phosphorylase (Fig. 5.17). The products are maltose and glucose, together with glucose-1-phosphate which is further metabolized, via glycolysis and the pentose phosphate pathway, to give TP and PGA. Sucrose for export can then be synthesized after

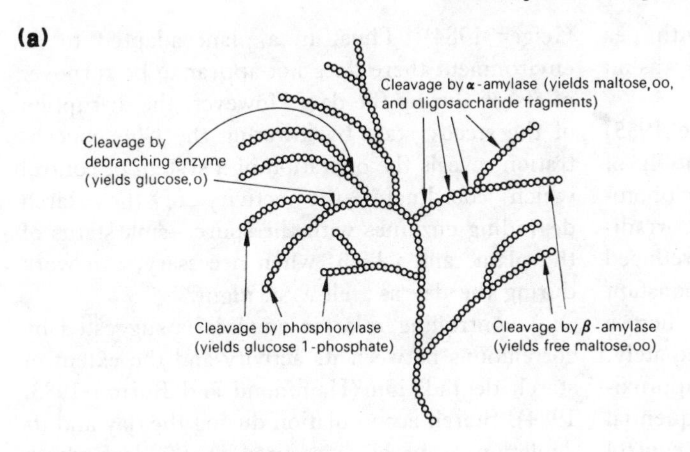


(b)

$CH_2OH$ $CH_2$ (glucose)$_n$

⋯→ cleaved by β-amylase to give maltose
→ cleaved by α-amylase to give maltose
cleaved by α(1→6) glucosidase to give glucose

O—(glucose)$_n$ + $HPO_3^-$

⋯→ cleaved by phosphorylase to give α-D-glucose-1-phosphate

**Fig. 5.17** (a) Enzymic breakdown of amylopectin. Each glucose molecule is represented by a 'bead'. α-amylase attacks the molecule internally and externally; β-amylase and phosphorylase work inwards from the non-reducing ends of the exposed chains; and the α (1→6) glucosidase hydrolyses the α(1→6) branch points.
(b) Cleavage of amylopectin: by α- and β-amylases and α(1→6) glucosidase, and by phosphorylase (from Hall *et al.* 1974).

export of these products to the cytosol. The activities of these enzymes, and thus the mode of degradation, differ among species: for example, pea chloroplasts, unlike those of spinach, do not contain significant amylase activities, and starch breakdown here is therefore predominantly phosphorolytic.

What controls the activities of these enzymes is not clear. At the simplest level, the switch from starch accumulation during the day to degradation at night may result merely from the cessation of synthesis, the rate of degradation being constant throughout. Simultaneous synthesis and degradation of starch implies turnover of the starch pool in the light, and this has been investigated in work with pepper (*Capsicum annuum*) plants (Grange 1984). Leaves were labelled with $^{14}CO_2$ of constant specific activity for between four and seven hours from the start of the photoperiod, to produce a uniformly-labelled pool of starch. For the remainder of the photoperiod the irradiance was reduced to a level which caused no net change in the starch content of the leaves, and the supply of $^{14}CO_2$ was removed. Further photosynthesis therefore fixed only unlabelled carbon. If there were turnover of starch, the specific activity of the starch pool would decline as labelled starch was degraded and replaced by new, unlabelled starch.

In this, and in a similar experiment with pea leaves (Kruger *et al.* 1983), however, there was no such evidence of turnover.

Subsequent work with pepper (Grange 1985) has supported this conclusion. Mobilization of reserves during the latter part of a 10-hour photoperiod was brought about by lowering the irradiance. Even when net photosynthesis was reduced by 90 per cent, export was maintained constant throughout the photoperiod, but although sucrose and the hexoses were mobilized immediately, starch degradation did not begin until approximately 1 hour later. This evidence for sequential mobilization of reserves suggests that the control of starch degradation involves more than merely the absence of its synthesis. In sugar beet leaves there was no evidence of simultaneous synthesis and degradation of starch at normal (340 $\mu l\ l^{-1}$) $CO_2$ concentrations, the rate of accumulation of starch equalling the rate of its synthesis. However, lowering the $CO_2$ concentration, and therefore the rate of current photosynthesis, caused starch to be degraded, although synthesis of new starch continued: at a $CO_2$ concentration of 120 $\mu l\ l^{-1}$, synthesis and degradation were equal (Fox and Geiger 1984). Thus, in a plant adapted to its environment there does not appear to be turnover of starch during the day. However, the disruption of the steady state by lowering the $CO_2$ concentration reveals the operation of a system of control which coordinates the activity of the starch degrading enzymes with the source–sink status of the plant, and which, when necessary, can work during the day as well as at night.

A controlling role for amylase is suggested by correlations between its activity and the extent of starch degradation (Hammond and Burton 1983, 1984). Starch accumulation during the day and its depletion at night (measured as the leaf starch contents an hour after dusk ('evening') and an hour after dawn ('morning'), respectively) increased markedly between the appearance of the first flower buds and early fruit development (Fig. 5.18a). A steady rise in the rate of photosynthesis during this period was sufficient to support this enhanced starch accumulation and the consequent higher export rates at night. Amylase activity in the evening increased in parallel with the higher evening starch levels which accompanied reproductive development

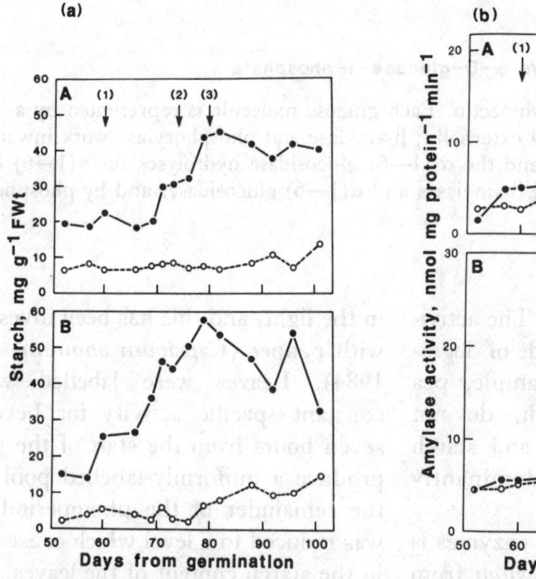


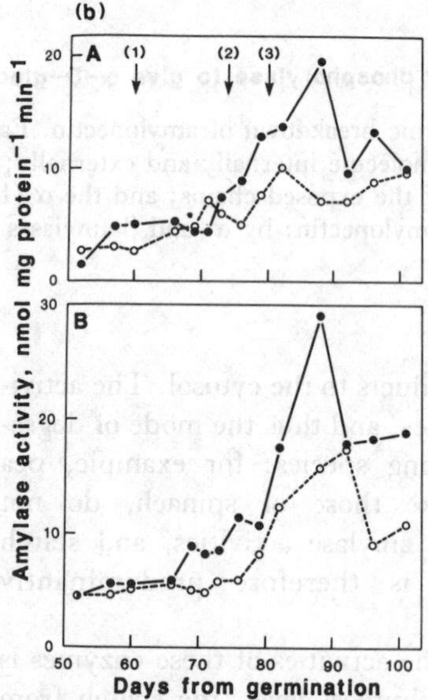


**Fig. 5.18** (a) Leaf starch content an hour after dawn (○) and an hour after dusk (●) in: A, one-fifth expanded leaves; B, two-thirds expanded leaves of pepper (*Capsicum anuum.*).
(b) Amylase activity in leaf extracts an hour after dawn (○) and an hour after dusk (●) from: A, one-fifth expanded leaves; B, two-thirds expanded leaves.
Reproductive development is indicated by: (1) appearance of first flower bud; (2) anthesis; (3) first fruit development (>10 mm diameter) (from Hammond and Burton 1983).

(Fig. 5.18b); maximum amylase activity occurred 6–8 hours after dusk. The response was the same, and occurred simultaneously, in both young and old leaves, suggesting that the regulation of amylase activity, which in turn controlled starch breakdown, was coordinated between leaves, perhaps through hormonal or sink effects.

However, morning amylase activities, although generally lower than in the evening, and particularly so after anthesis, also increased during fruit development. This is difficult to explain if we accept that under normal conditions there is no starch turnover during the day. It is possible that at least some of this amylase activity is due to the presence of the enzyme in the cytosol, where there is no starch and where it could therefore have no direct effect on starch degradation. Similarly, the lack of a direct relationship in this experiment between phosphorylase activity and starch breakdown could be attributed to the location of phosphorylase in the cytosol as well as in the chloroplasts. The function of extra-chloroplastic enzymes of starch breakdown is unknown (see Preiss 1982).

There is clearly much to be learned about the control of starch mobilization, even in terms of the extent to which different enzymes are involved. In the experiment referred to above, starch degradation in pepper leaves appeared to be controlled by amylase activity, although the complications of interpretation caused by the presence of starch-degrading enzymes in the cytosol as well as in the chloroplasts means that a similar role for phosphorylase cannot be discounted. As we have already noted, phosphorylase activity predominates in some species, and it may be that the balance between amylase and phosphorylase activities varies with plant development according to the metabolic requirement for either non-phosphorylated or phosphorylated sugars (Hammond and Burton 1983). Even less is known about the underlying biochemical controls which coordinate the activities of the starch-degrading enzymes with the requirements of the plant for respiratory substrate and sucrose for export (see Stitt 1984).

As we have already seen, sucrose may be stored, in the vacuoles, even in the leaves of those species such as *Vicia* and sugar beet which have starch as the major storage carbohydrate. Species which store little starch in the leaves, such as wheat, barley and many grasses, rely more heavily on vacuolar sucrose; fructans, oligosaccharides of varying degrees of polymerization and for the most part water-soluble, can be accumulated in leaf blades, sheaths and stems. Evidence for the important role of vacuolar sucrose comes from a series of experiments on the fluxes of carbon in mature leaves of young barley plants (Farrar and Farrar 1985a, b; Sicher *et al.* 1984). Sucrose was the dominant carbohydrate, more abundant throughout most of a 16-hour photoperiod than the rest of the carbohydrates (hexoses, fructans and starch) put together. It was also more heavily labelled than the other carbohydrates after $^{14}CO_2$ labelling, and accounted for most of the diurnal changes in leaf carbohydrate content. Loss of label from leaves fed $^{14}CO_2$ for 30 minutes took the form of double exponential curves (Fig. 5.19), suggesting rapid loss of label from a readily available transport pool, and slower efflux from a less accessible storage pool, assumed to be in the vacuole (cf. p. 125).

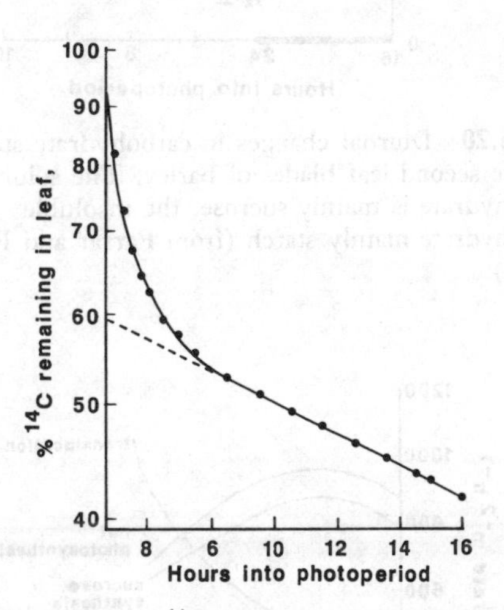


**Fig. 5.19** Efflux of $^{14}C$ from an intact leaf of barley after a 30-minute period of labelling with $^{14}CO_2$ (from Farrar and Farrar 1985b).

Soluble carbohydrate, of which 60–90 per cent was sucrose, accumulated rapidly during the photoperiod (Fig. 5.20), and calculations indicated that only between 15 and 25 per cent of that sucrose was in the transport pool. It is not surprising, therefore, to find that the rate of export was largely independent of the rate of sucrose synthesis. Thus, for the first 4 hours of the photoperiod, the rate of carbohydrate accumulation exceeded that of export, as the vacuolar sucrose pools, depleted during the night, were replenished. Thereafter, the relationship was reversed because, as the rate of accumulation declined, an increasing rate of export could be sustained by diminishing rates of sucrose synthesis. Indeed, towards the end of the photoperiod, mobilization of carbohydrate allowed the rate of export to exceed the rate of sucrose synthesis (Fig. 5.21).

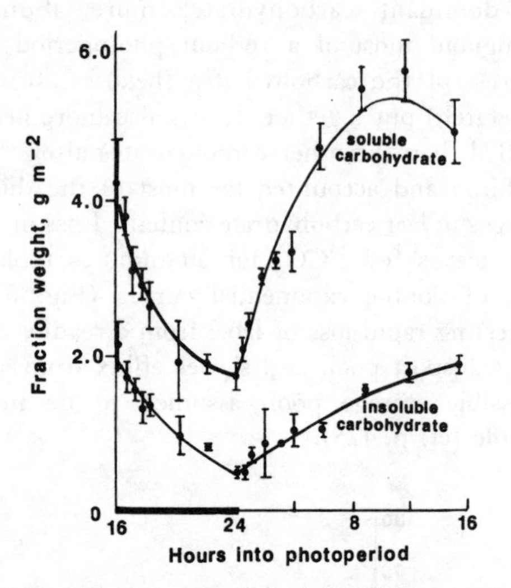


**Fig. 5.20** Diurnal changes in carbohydrate storage in mature second leaf blades of barley. The soluble carbohydrate is mainly sucrose, the insoluble carbohydrate mainly starch (from Farrar and Farrar 1985b).

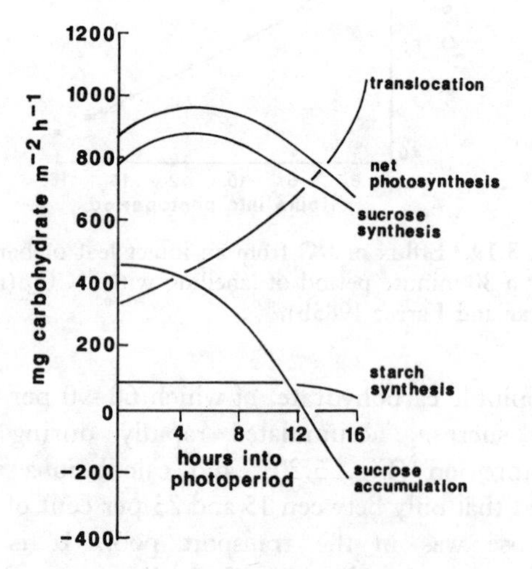


**Fig. 5.21** Carbon fluxes in mature second leaf blades of barley (from Farrar and Farrar 1985b).

Exchange of sucrose between the vacuole and the cytosol clearly plays a key role in determining the amount of sucrose present in the cytosol and, therefore, constituting the transport pool. The vacuole may, in a sense, compete with the phloem for cytosolic sucrose, with a considerable measure of control invested in the tonoplast. This control must also extend to cover the release of sucrose at night and, when necessary, during the day. Experiments with vacuoles isolated from mesophyll protoplasts of barley have, in fact, demonstrated that photosynthetic products are rapidly transferred into the vacuoles, apparently by diffusion facilitated by a tonoplast carrier (Kaiser and Heber 1984). This uptake was not energy-dependent, and could be reversed in response to a loss of sucrose from the cytosol. In pea, on the other hand, the tonoplast carrier appears to be energy-dependent (Guy *et al.* 1979).

The removal of soluble photosynthates from the cytosol may also prevent the disruption of metabolism which could otherwise result from a lowering of the cytosolic water potential and the dilution of metabolites caused by the consequent influx of water. Photosynthesis, for example, may be inhibited by the dilution of stromal metabolites. Such a buffering function would be less important in species which were able to lessen the effects of carbohydrate production by extensive starch accumulation, and may help to explain why movement of sucrose into vacuoles is much slower in spinach than in barley (Asani *et al.* 1985).

### 5.4.3 MOVEMENT OF SUCROSE TO THE MINOR VEINS, AND PHLOEM LOADING

The pathway from the mesophyll to the minor veins could be symplastic, apoplastic or a combination of both. Work with several species suggests that symplastic movement, down a concentration gradient and through plasmodesmata, prevails. However, close to the phloem, sucrose is released across the plasmalemma into the apoplast, from where it is loaded into the sieve element–companion cell (se–cc) complex.

Such a scheme, involving phloem loading from the apoplast, is in accordance with several undisputed features of phloem loading, confirmed for various species, but studied in most detail in sugar beet by Geiger and his colleagues (see Geiger 1975; Geiger and Giaquinta 1982; Giaquinta 1983). First, it is a selective process: sucrose is the major constituent of the phloem sap of most species, with glucose and fructose virtually absent, despite their presence, together with sucrose, in the apoplast. The high amino acid content of phloem saps suggests that they too may be preferentially loaded. Secondly, the concentration of sucrose in the se–cc complex is markedly higher than in the surrounding phloem parenchyma and mesophyll cells. For example, in sugar beet leaves, solute potentials of −1.3 MPa (mesophyll), −0.8 MPa (phloem parenchyma) and −3.0 MPa (se–cc complex) have been measured. A concentrating mechanism therefore exists, apparently situated at the plasmalemma of the companion cell. We have already noted that the companion cells have been identified as the site of a major part of the transport pool of sucrose. Any subsequent resistance to the transfer of sucrose from companion cell to sieve element is thought to be negligible because the cells have identical solute potentials, the equilibrium accomplished by communication through large plasmodesmata. Thirdly, loading is an energy-requiring process, stimulated by ATP and inhibited by dinitrophenol, which uncouples ATP synthesis; ATP is present in the phloem sap and has a high rate of turnover.

A selective, active accumulation of sucrose can readily be explained on the basis of a membrane carrier process, and there is strong evidence to suggest that such a process operates at the plasmalemma of the se–cc complex. There is, however, no evidence to suggest that plasmodesmata are able to discriminate between solutes or to actively pump them into the se–cc complex, as would be required of a solely symplastic pathway.

Further circumstantial evidence for loading from the apoplast comes from the intimate association between transfer cells and minor veins in some species. Transfer cells develop from phloem parenchyma and companion cells, and are characterized by ingrowths of the cell walls, and the consequent amplification of membrane surface area. It is thought that the greater number of potential carrier sites provided by the increased membrane surface facilitates the accumulation of sucrose from the apoplast. Certainly, the development of transfer cells has been found to coincide with, or slightly precede, the onset of export from the leaf, when the membranes also acquire ATPase activity. Not all species have transfer cells, but those without, such as sugar beet, squash and soybean, may have compensatory advantages. For example, ATPase activity has been found in the plasmalemmas of both companion cells and sieve elements of squash, but it is absent from the plasmalemmas of sieve elements associated with transfer cells. For further details of transfer cell structure and function see, for example, Pate and Gunning (1972) and Gunning *et al.* (1974).

Evidence for symplastic movement through the mesophyll towards the minor veins has been less forthcoming. A symplastic pathway would, however, keep sucrose movement separate from the transpiration stream moving in the opposite direction in the apoplast. Moreover, in many species the apoplastic pathway is blocked by a barrier of impermeable cell walls: in wheat, for example, the veins are surrounded by a suberized cell layer called the mestome sheath, and in maize the bundle sheath cells are suberized, dictating symplastic movement, at least at that point. Certainly, extensive plasmodesmatal connections between mesophyll cells and veins, well-documented in $C_4$ species, support the idea of symplastic movement. Similarly, in certain $C_3$ species, symplastic continuity has been observed. In soybean, this is particularly clear in a layer of cells in the centre of the leaf between the veins, and it is suggested that this paraveinal mesophyll serves to channel assimilates from the mesophyll to the phloem. To what extent such a specialization might compensate for large interveinal distances, and the associated time-lag between the fixation of carbon and its movement out of the leaf, is not clear. Correlations have been found, however, between the rates of export and the distances which assimilate must travel within the leaf before it reaches a minor vein. For example, in a comparison of $C_3$ and $C_4$ grasses, veins were on average 270 $\mu$m apart in the $C_3$ species but only 110 $\mu$m apart in the $C_4$ species, which had higher rates of export. Similar differences have been found between *Lolium temulentum* ($C_3$; 330 $\mu$m) and *Sorghum sudanense* ($C_4$; 100 $\mu$m) (see Wardlaw 1976).

The extent to which the veins branch may therefore influence the rate of export by determining the distance which assimilate must travel through the symplast by diffusion. Efflux into the apoplast is another potential point of control, because this could limit the amount of sucrose available for loading into the phloem. The mechanism of this control is unknown, but its potential influence has been revealed by the finding that enhanced export from leaves of sugar beet sprayed with KCl solutions could be attributed to a stimulation of sucrose entry into the apoplast rather than effects on loading itself. Photosynthesis was unaffected (Doman and Geiger 1979).

In a discipline remarkable for its history of discord, albeit largely over the structure of the functioning sieve tube and the mechanism of movement within it, the study of phloem loading has proved to be exceptional: there is general agreement about the basic mechanism. Moreover, this agreement has come from investigations of diverse experimental subjects, notably leaves of sugar beet, *Vicia, Ricinus* and maize, as well as the cotyledons of the developing *Ricinus* seedling. It is accepted that the driving force for sucrose uptake is an electrochemical potential difference established across the plasmalemma of the se–cc complex by the extrusion of protons into the apoplast. This implies a proton pump, assumed to be a membrane ATPase, and would account for the pH of the phloem sap (8.0–8.5) being higher than that of the apoplast (5–6). A membrane carrier then binds selectively with both protons and sucrose, and the resultant ternary complex (carrier-sucrose-$H^+$) moves across the membrane down the gradient of electrochemical potential of protons (Fig. 5.22). Sucrose, an uncharged molecule, does not itself respond to the electrical polarity of the membrane, but owes its transport, across the membrane against its gradient of chemical potential, to its ability to bind to the membrane carrier. The 'uphill' movement of sucrose is thus considered to be a secondary consequence of the 'downhill' movement of protons, and this scheme for phloem loading is therefore referred to as sucrose–proton co-transport (or symport).

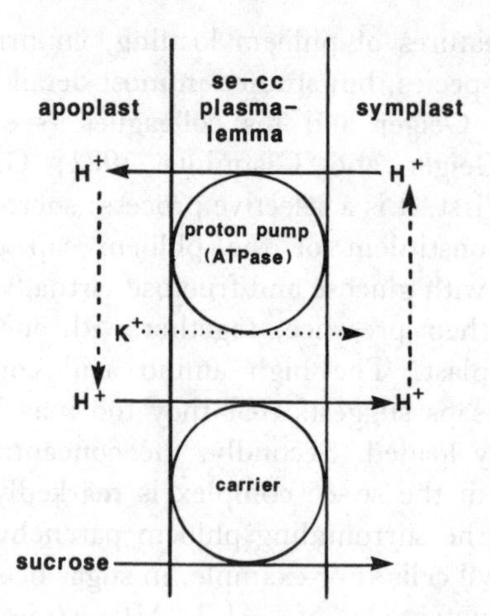


**Fig. 5.22** A scheme for phloem loading of sucrose from the apoplast by a sucrose–proton co-transport system. See text for details.

The two essential features of this mechanism, an ATP-driven extrusion of protons and coupled sucrose–proton uptake, are supported by a considerable body of evidence. We have noted already the ATPase activity of phloem membranes, the rapid turnover of ATP in the phloem sap, and the pH gradient across the membrane. If this gradient is reduced, by raising the pH of the apoplast with alkaline buffers unable to penetrate the membrane, loading is inhibited. On the other hand, the enhanced export which results from ATP application to leaves may be due to a stimulation of ATPase activity and a consequently steeper pH gradient; this may be accompanied by enhanced efflux of sucrose from the mesophyll. Coupled sucrose–proton uptake would cause a temporary depolarization of the membrane, and this effect is implied by the transient rise in the pH of the apoplast which accompanies sucrose uptake, transient because the proton efflux pump quickly re-establishes the proton gradient. Inhibition of this pump by *para*-chloromercuribenzene-sulphonic acid (PCMBS), observed *in vitro* and implied by a reduction in proton efflux, is accompanied by reductions in sucrose loading, although this may in part be caused by direct effects of PCMBS on the membrane carrier. Stimulation of proton efflux by the phytotoxin fusicoccin enhances loading. There may be some exchange of potassium for protons pumped out of the phloem cells.

Although the mechanism of phloem loading has been satisfactorily resolved, details of its regulation are much less clear. Little is known about the nature of the carrier, nor about the structural characteristics of molecules which affect their ease of entry into the phloem, despite the relevance of this knowledge to the efficacy of systemic pesticides. The exclusion of sugars other than sucrose from the phloem sap of most species is attributed to carrier-specificity, but it is not known whether the loading of several sugars, in species such as squash, involves a single, less specific carrier, or several. Regulation of the rate of loading, particularly in relation to sink demand, is also obscure. However, the capacity of the loading system is thought not to limit the overall rate of translocation, the controls being sufficiently flexible to accommodate changes in sink demand (see, for example, Geiger and Fondy 1980). When source limitation of assimilate partitioning is identified, it can be attributed to a shortfall in the amount of assimilate available for loading rather than to inadequacies of loading itself. We shall therefore consider the loading process no further, and the reader is referred to recent reviews (Geiger and Giaquinta 1982; Giaquinta 1983; Ho and Baker 1982; Baker 1985; Delrot and Bonnemain 1985) for discussions of its regulation.

## 5.5 UNLOADING AND SINK METABOLISM

### 5.5.1 INTRODUCTION

It is clear from our discussion of source and sink limitations to yield (sect. 5.2) that the capacity to attract assimilate (sink strength) varies among the different sinks of a plant, and changes during the development and growth of an individual sink. The size of a sink is a contributory factor, as is its position relative to other sinks and to the sources of assimilate. However, other factors, loosely and collectively referred to as sink activity, must be invoked in order to account for differences in import rate which cannot be explained by size or position, and which have often been found to be related to sink metabolism, or to the transfer of assimilate out of the phloem within the sink and into the surrounding cells. In many cases it has proved useful to consider translocation into sinks as a response to a concentration gradient: loading at the source maintains a comparatively high sucrose concentration in the phloem, while in the destination cells, sucrose is efficiently metabolized or sequestered in the vacuoles. The rate of transport to the sink is then determined by the steepness of the gradient and by any resistances to movement offered by the pathway.

The recognition that the import and utilization of sucrose involves both physical transfer and metabolic conversion raises a number of questions. For example, does sucrose move out of the se–cc complex into the apoplast, or is there symplastic transfer, through plasmodesmata, directly into the cytosol of the sink cells? If the former, is sucrose then taken up by the cells unaltered or is prior metabolism, such as hydrolysis to glucose and fructose, a prerequisite for uptake? Is active transport involved, at the plasmalemma of the se–cc complex, or at the plasmalemma or tonoplast of the destination cell? How important are physical resistances to carbon movement within the sink? And what is the nature of any link between carbon transfer and metabolism? Such questions cannot be answered in general terms because the diverse nature of different sinks means that different controls operate. In this section we shall consider some of these questions, although in no one sink are the answers to all the relevant questions known. We shall examine in detail the wheat grain, the legume seed, the tomato fruit and the diseased plant, which illustrate in different ways some of the factors which contribute to sink strength.

### 5.5.2 THE WHEAT GRAIN

We have discussed at length the evidence for source and sink limitations to grain yield in cereals (sect 5.2): source limitation during ear development and the early part of grain-filling gives way to sink limitation as the grains become the dominant sinks. Consequently, much work has been done to investigate the factors limiting the uptake of sucrose by grains during this sink-limited phase of growth.

#### 5.5.2.1 VARIATION IN GRAIN WEIGHT WITHIN THE EAR

Sink limitation implies that the process of assimilate uptake by the grains is saturated, and that individual grains attain their potential weights. However, within an ear there is considerable variation in grain weight at maturity, depending on the position of the grain within the spikelet and of the spikelet on the ear (Fig. 5.23). Assuming that there is adequate assimilate, the final weight of an individual grain will reflect its inherent capacity to accumulate dry matter and the resistance to assimilate movement. It is therefore pertinent to consider which of these attributes is responsible for the apparent natural variation in sink limitation suggested by the pattern of grain weights shown in Fig. 5.23. The grain's capacity for growth may depend on such things as endosperm cell number and the activities of starch- and protein-synthesizing enzymes. The resistance component includes that offered to movement within the phloem of the rachis and rachilla, as well as that associated with movement out of the phloem and through the tissues of the grain. The phloem of the peduncle, which delivers assimilate to the rachis, is thought not to limit translocation: half of the vascular system here can be severed without reducing the rate of import of assimilate by the ear (Wardlaw and Moncur 1976).

Bremner and Rawson (1978) selectively removed grains as a means of increasing assimilate supply to those remaining, which were therefore freed from any source limitation. Under these circumstances the weights of most grains were increased but, since the treatment was imposed 9 days after anthesis, it is possible that the remaining grains' growth potential was altered, perhaps through effects on endosperm cell number (sect. 5.2). Nevertheless, spikelets were more or less equally affected by grain removal, so that the pattern of grain weight along the ear remained largely the same as in intact ears. Similarly, there was a generally uniform reduction in grain weight in all spikelets when assimilate supply was reduced by shading. These results suggest that resistance to assimilate movement along the rachis was not responsible for differences in grain weight between spikelets.

Within spikelets, however, shading affected distal grains proportionately more than those nearer the base of the spikelet (Fig. 5.24). This pattern of response (grains *a* and *b* affected equally but an increasingly pronounced effect on grains further along the rachilla) is consistent with what is known about the vascular connections

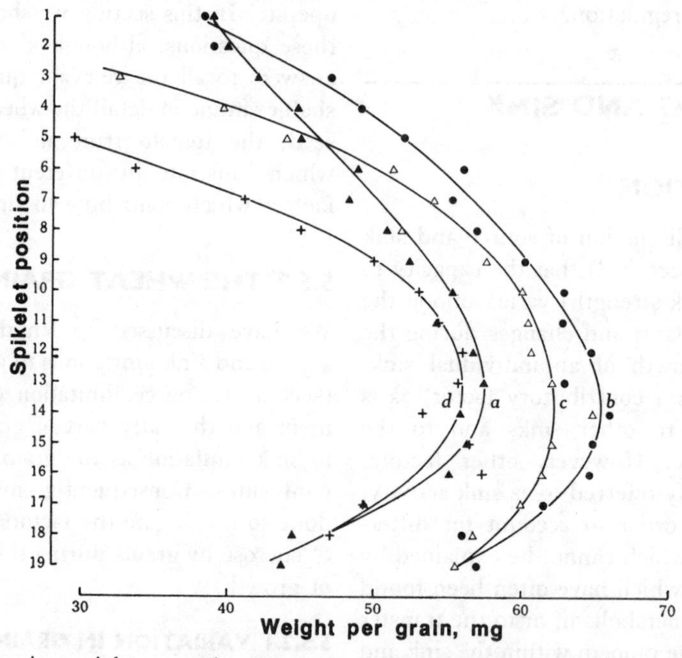


**Fig. 5.23** Profiles showing weight per grain at maturity for grains *a*, *b*, *c* and *d* in each spikelet position (numbering from the ear apex) on intact ears of the Mexican wheat cv. WW15. Plants were grown in a glasshouse at day/night temperatures of 15/10 °C (from Bremner and Rawson 1978).

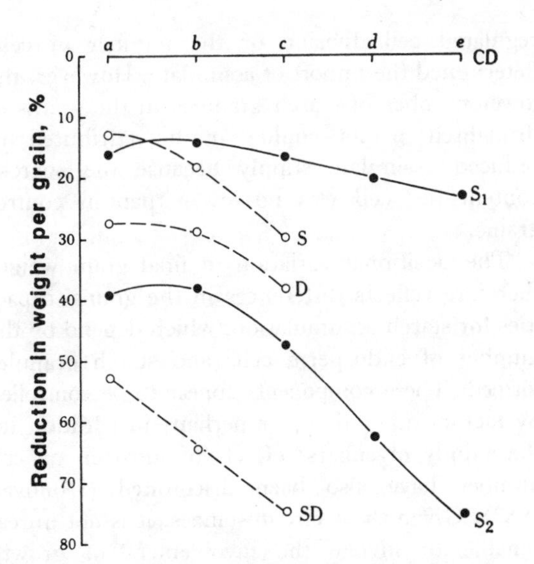


**Fig. 5.24** Percentage reduction in weight of grains *a*, *b*, *c*, *d* and *e* from the central spikelets (11–15) of wheat when shoots were subjected to 45 per cent ($S_1$) or 70 per cent ($S_2$) shade, as compared with control grains (CD). The broken lines are from an earlier experiment, in which plants were either shaded (S) or defoliated (D), or shaded and defoliated (SD) (from Bremner and Rawson 1978).

within the spikelet. The three basal grains, *a*, *b* and *c*, have independent vascular strands connecting them *via* a plexus of transfer cells to the vascular system in the rachis; grain *d* is served by a branch originating near the base of grain *c*, grain *e* by a branch from the base of grain *d*, and so on. Grains *a*, *b* and *c* are thus connected 'in parallel' to the source of assimilate, and might therefore be expected to respond similarly to a reduction in assimilate supply. However, the vascular connection to grain *c* is longer than that serving grains *a* and *b*, and the effect of the consequently greater resistance becomes increasingly obvious as the supply of assimilate is reduced (Fig. 5.24). Grain *c* is further disadvantaged by its connection 'in series' with grains *d*, *e*, etc., and such a linkage could account for the response of the more distal grains to a shortage of assimilate.

This interpretation implies that growth of the distal grains in the spikelet is restricted by the limited amount of assimilate which reaches them. However, although the influence of vascular anatomy, and hence resistance, may manifest itself when the supply of assimilate is reduced by shading, under favourable conditions it seems to be outweighed by the higher concentration gradient resulting from a greater supply of assimilate. We have already seen that the resistance of the pathway to grains *a* and *b* appears to be similar, and yet in practically all spikelets grain *b* is heavier; indeed, in many spikelets grain *a* weighs less than grain *c* (Fig. 5.23). Furthermore, the concentration of sucrose and other soluble sugars has been found to be no less, and in some cases more, in grain *d* than in the proximal grains of a central spikelet, at least during the first 15 days after anthesis. Similarly, the concentrations of sugars were either unrelated, or inversely related, to grain weight in comparisons between spikelets (Singh and Jenner 1982). The pattern of grain weight within and between spikelets cannot, therefore, be attributed predominantly to the differing resistances along the various assimilate pathways to the grains: control must reside within the grains themselves. We must therefore consider both the capacity of the endosperm for starch synthesis and the resistance to assimilate movement through the grain.

Both these properties appear to be important. Differences in grain weight between and within spikelets on intact ears have been shown to be well correlated with differences in the number of endosperm cells, and therefore with the capacity for starch accumulation (Table 5.5). Genotypic differences in grain weight have similarly been attributed to endosperm cell number (Gleadow *et al.* 1982). However, the causes of such variation are not clear. Although assimilate supply during the crucial period of cell division can be important as, for example, when grain number is reduced or when plants are shaded (p. 114), other factors may also be involved.

The lack of a correlation between sucrose concentration in the grain and final grain weight (and therefore endosperm cell number) suggests at least that sucrose supply is not the main limiting factor in plants which have remained free from direct experimental manipulation of assimilate supply. Some supporting evidence comes from plants subjected to environmental stress. Drought during the period of endosperm cell division can reduce final grain weight by reducing the numbers not only of cells but also of starch

**Table 5.5** Dry weight, cell number and volume of grains taken from different positions within the ear of wheat

| *Comparison* | *Grain dry weight (mg)* | *$10^{-3}$ × endosperm cell no.** | *Grain volume (μl)* |
|---|---|---|---|
| Within spikelets: | | | |
| Floret *a* | 51.6 | 158.9 | 43.3 |
| Floret *b* | 55.4 | 167.6 | 45.0 |
| Floret *c* | 46.3 | 124.2 | 37.6 |
| Floret *d* | 27.6 | 78.0 | 22.7 |
| LSD† | 1.6 | 15.8 | 2.5 |
| Between spikelets (floret *a*): | | | |
| Lower | 40.6 | 116.6 | 36.1 |
| Middle | 51.6 | 158.9 | 43.3 |
| Upper | 41.3 | 127.1 | 34.3 |
| LSD† | 2.1 | 9.4 | 2.0 |

* Cell number was estimated in the endosperm of 18-day-old grains (after anthesis), whereas dry weight and volume data are for ripe grains.

† Level of probability is $<0.05$.

(After Singh and Jenner 1982)

granules within cells. Nicolas *et al.* (1985) found a correlation between the sucrose contents of grains and cell division, but such a correlation does not establish whether the sucrose level regulated cell division or the number of cells determined the import of assimilate. However, the lower number of starch granules in the grains of droughted plants could not be attributed to reduced assimilate supply because the sucrose content per cell was no lower than in control grains.

The positional variation in final grain weight therefore reflects differences in the grains' capacities for starch accumulation, which depend on the number of endosperm cells and starch granules formed. These components appear to be controlled by factors other than, or perhaps in addition to, the supply of sugars; effects of nitrogen on cell number have also been discounted (Donovan 1983). Given these circumstances, it is not unreasonable to invoke the involvement of growth regulators. Some circumstantial evidence for cytokinin-mediated control of endosperm cell number has already been presented (p. 114). There is also evidence that IAA is involved in the pattern of grain weight within the ear. The concentration of IAA in wheat grains shows a characteristic rapid increase from about 8 days after anthesis, reaching a maximum 15–20 days later, and thereafter dramatically declining. Positive correlations have been found between the increase in IAA

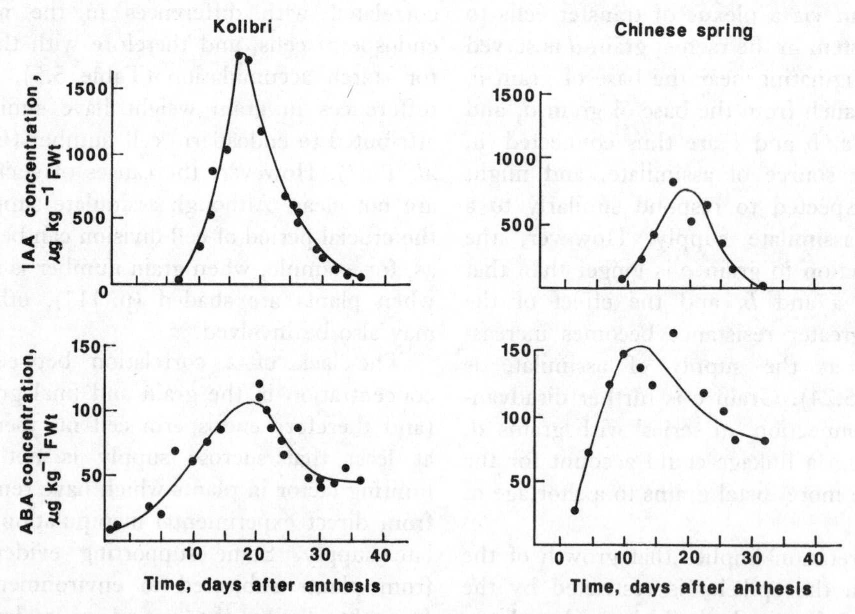


**Fig. 5.25** Changes in the concentrations of IAA and ABA during the growth of grains of the spring wheat cultivars Kolibri (mean final grain dry weight 38 mg) and Chinese Spring (mean final grain dry weight 25 mg) (after Rademacher and Graebe 1984).

concentration and the rate of dry-weight accumulation of grains between and within spikelets (Bangerth *et al.* 1985). Similarly, Rademacher and Graebe (1984) found that the more rapid accumulation of starch in a large-grained cultivar compared with a smaller-grained cultivar was paralleled by differences in the IAA content of the grains. In contrast, the maximum abscisic acid concentration was higher in the smaller-grained cultivar (Fig. 5.25).

The periodicity of growth regulator concentration is suggestive, but causal relationships between the presence or activity of growth regulators and the development and growth of grains cannot be assumed. However, speculation that IAA may have a promotive role in starch accumulation is reinforced by suggestions of similar means of control in other sinks (e.g. tomato fruits, p. 152). In addition, IAA may be involved at an earlier stage of development, in the correlative inhibition of younger grains by those developing earlier. Such effects could be manifested as differences in endosperm cell number, but there is also evidence that these differences are themselves superimposed upon an already existing pattern of variation between florets, established before anthesis. Thus, in both wheat (Singh and Jenner 1982) and barley (Scott *et al.* 1983) close correlations have been found between ovary weight at anthesis and final grain weight at various positions on the ear. Indeed, differences in potential grain weight may be established even earlier in development, since a good correlation has been found between spikelet width at the double ridge stage and final grain weight (Cottrell and Dale 1984).

#### 5.5.2.2 A PHYSICAL RESISTANCE TO SUCROSE MOVEMENT WITHIN THE GRAIN

Whatever the factors responsible for endosperm cell division, the greater the number of endosperm cells the greater is the grain's capacity for sucrose consumption, primarily for starch synthesis. The flow of sucrose into and within the grain, down the resultant concentration gradient, keeps pace with this consumption because, during the sink-limited phase of grain growth, the sucrose supply in the vascular tissue serving the grain is, by definition, more than adequate. The effect of varying the concentration gradient has been studied by culturing detached ears, darkened to prevent photosynthesis, in sucrose solutions of a range of concentrations. The intracellular concentration of sucrose in the endosperm, and the rate of starch synthesis, were responsive to increases in the concentration of sucrose in the medium up to about 25 g $l^{-1}$ (Fig. 5.26). Higher concentrations produced no further increase, which means that neither the sucrose concentration nor the rate of starch accumulation in the endosperm could be induced to exceed the levels found in the grains of well-illuminated intact plants. However, in all tissues of the ears except the grains, the sucrose concentration increased in direct proportion to that supplied to the cut peduncle. Something therefore imposes an upper limit on the movement of sucrose into the grain, in contrast to the apparently unrestricted access to closely associated floral organs, such as the glumes and paleas.

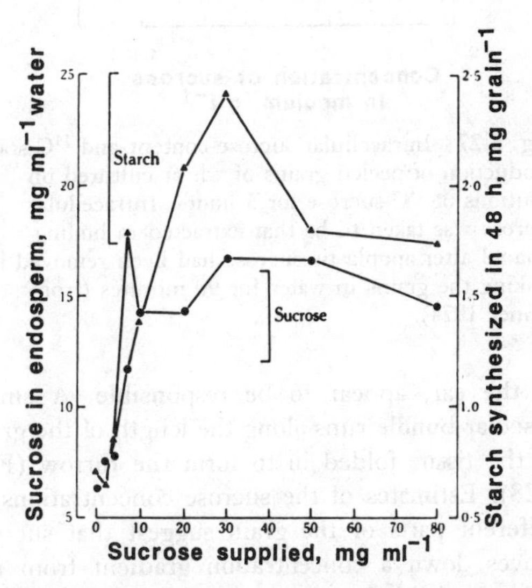


**Fig. 5.26** Final concentration of sucrose (●) and the amounts of starch (▲) produced in 48 hours in grains from detached ears of wheat cultured on solutions of sucrose. Vertical bars represent LSD ($P = 0.05$) (from Jenner 1970).

This restriction is not a property of the endosperm itself because in detached grains, peeled to expose the endosperm and cultured directly in solutions of sucrose, the endosperm cells accumulated sucrose and starch in direct proportion to the external concentrations, up to 50 g $l^{-1}$ (Fig. 5.27). Physical resistances to movement within the grain, and perhaps within the phloem

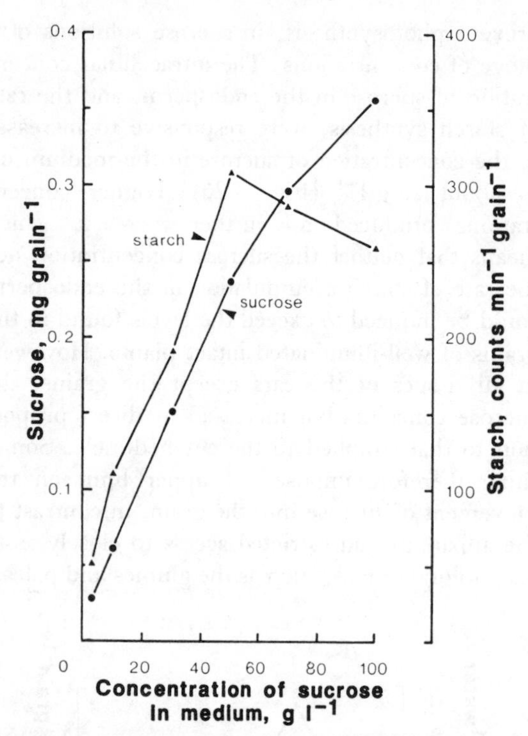


**Fig. 5.27** Intracellular sucrose content and $^{14}$C-starch production of peeled grains of wheat cultured on solutions of $^{14}$C-sucrose for 5 hours. Intracellular sucrose was taken to be that extracted in boiling ethanol after apoplastic sucrose had been removed by soaking the grains in water for 90 minutes (from Jenner 1974).

of the ear, appear to be responsible. A single vascular bundle runs along the length of the grain in the tissue folded in to form the furrow (Fig. 5.28). Estimates of the sucrose concentrations in different parts of the grain suggest that sucrose moves down a concentration gradient from the phloem of this bundle to the endosperm (Jenner 1974). It is thus assumed to be unloaded passively from the phloem, and then appears to move symplastically through the chalaza and the cells of the nucellar projection. Further movement must be apoplastic, however, because the endosperm is separated from the nucellar projection by a liquid-filled space, the endosperm cavity. From here, sucrose spreads out through the apoplast of the endosperm and is finally absorbed by the cells.

We have already noted that when sucrose was given direct access to the endosperm cavity, as in the peeled grains cultured in sucrose solutions, a major limitation to sucrose uptake by the endo-

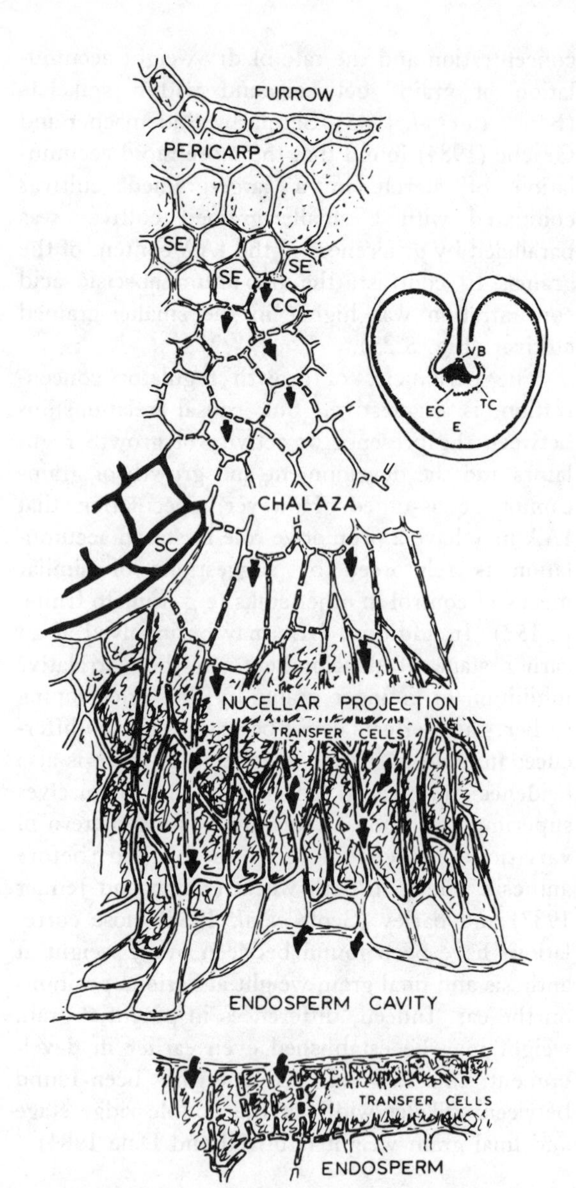


**Fig. 5.28** Diagrammatic representation of the pathway of assimilate movement from the vascular bundle to the endosperm of the wheat grain.

Arrows within cells indicate probable symplastic transport *via* plasmodesmata; those within cell walls indicate apoplastic transport. Curved arrows represent possible facilitated transport.

Abbreviations: SE, sieve element; CC, companion cell; SC, seed coat.

Inset: median cross-section through a wheat grain, indicating the position of the furrow (crease) vascular bundle (VB), and the dark-staining pigment strand which contains the chalaza and nucellar projection, in relation to the endosperm cavity (EC), aleurone transfer cells (TC), and endosperm (E) (from Thorne 1985).

sperm was relieved. The upper limit to sucrose movement within the grain can therefore be attributed, at least partly, to a resistance to movement from the phloem to the endosperm cavity. The precise cause of this resistance is unknown, but it could, for example, be related to the capacity of unloading from the phloem or to the functioning of the transfer cells in the nucellar projection. Other observations have confirmed that the resistance to diffusion in this part of the pathway is relatively high. Thus, in grains taken from intact plants, Jenner (1974) found that the gradient of sucrose concentration between the vascular bundle and the endosperm cavity was steeper than that between the cavity and the endosperm itself, implying a correspondingly greater resistance. Similarly, when detached ears were cultured in $^{14}C$-sucrose solutions, radioactivity was detected in the grain within 5 minutes, and yet even after 20 minutes 86 per cent of this activity was still confined to the tissue of the furrow (comprising a strip of pericarp, the vascular bundle, chalaza, and nucellar projection) (Donovan *et al.* 1983).

#### 5.5.2.3 SUCROSE UPTAKE BY THE ENDOSPERM CELLS

Thus, for a grain of given sink capacity, as set by its complement of endosperm cells and starch granules, the rate of sucrose uptake increases with increases in assimilate supply until an upper limit is imposed by restrictions on the movement of assimilate through the outer (maternal) tissues of the grain. This restriction presumably accounts for the limited response of grains, whose endosperm cell number has been fixed, to increased supplies of assimilate (Fig. 5.7, p. 118). Peeled grains show that endosperm cells themselves are capable of much higher rates of sucrose uptake and starch synthesis, apparently because the diffusion of sucrose into the endosperm cells is facilitated by a membrane carrier system, which has properties similar to those of phloem loading.

This is suggested, first, by the effect of PCMBS on the time-course of $^{14}C$-sucrose uptake by detached grains, peeled to expose the aleurone lining of the endosperm, and incubated in sucrose solution (Fig. 5.29). It is assumed that the early curvilinear part of the $^{14}C$-sucrose uptake curve represents movement into the apoplast, while the linear part was due to accumulation within the

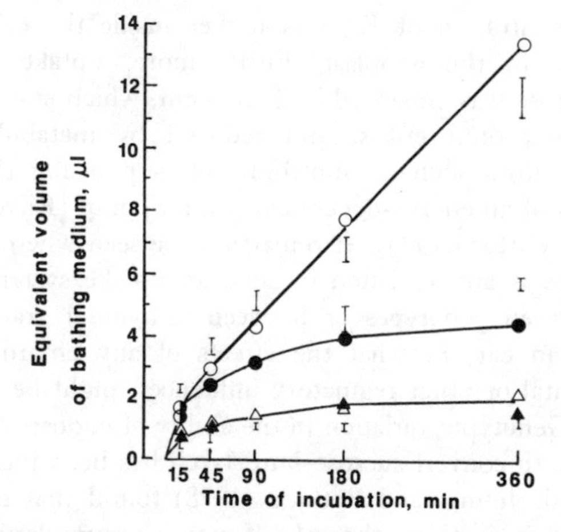


**Fig. 5.29** Uptake of $^{14}C$-sucrose (circle) and $^{3}H$-PEG900 (triangle), with PCMBS (closed symbols) or without PCMBS (open symbols), over a period of 6 hours.

Grains of wheat cv. Cleveland (20 days after anthesis) were pre-incubated in a bathing medium containing 20 mg $ml^{-1}$ sucrose with or without 0.55 mM PCMBS for 90 minutes before the radioactive markers were added. The amount of label taken into a grain at each time was converted to the equivalent volume of bathing medium, using the known radioactivity per unit volume of medium. PCMBS, *p*-chloromercuribenzenesulfonic acid (after Ho and Gifford 1984).

cells. When PCMBS, which is known to inhibit transport by membrane carriers, was included in the bathing medium, $^{14}C$-sucrose uptake over a period of 6 hours was considerably reduced. PCMBS had no effect on the uptake of $^{3}H$-polyethylene glycol from the bathing medium, however, because this large molecule cannot permeate the plasmalemma. The uptake curve for $^{3}H$-PEG therefore represents movement into the apoplast only. The difference between this curve and that of $^{14}C$-sucrose uptake in the presence of PCMBS presumably represents unaided diffusion of sucrose into the cells, reminding us that the carrier system facilitates sucrose movement down, and not against, a concentration gradient.

Circumstantial evidence suggests sucrose–proton co-transport. The apoplastic sap, obtained by centrifugation, had a comparatively low pH (6.3–6.6), possibly the result of proton extrusion from the endosperm cells. As in phloem loading, protons can be exchanged for potassium ions, which would account for the observation that the concentration of $K^+$ was higher inside the cells than in the apoplast. Furthermore, uptake of sucrose was enhanced by fusicoccin, which stimulates proton efflux, and reduced by metabolic inhibitors such as dinitrophenol, supporting the idea of an energy-dependent proton pump (Rijven and Gifford 1983). It remains to be seen whether there is any variation in such an uptake system, between genotypes or between individual grains on an ear, or what the effects of any environmental or other regulatory influences might be.

Genotypic variation in the ability of endosperm cells to convert sucrose into starch has been identified. Jenner and Rathjen (1978) found that for a given cultivar, the rate of starch accumulation was directly related to the concentration of sucrose in the endosperm cells, which was varied by culturing detached ears in different concentrations of sucrose solution. However, a given amount of intracellular sucrose produced different rates of starch accumulation in different cultivars, as illustrated by the slope of the lines in Fig. 5.30. There is thus genetic variation in the kinetics of the conversion of sucrose to starch or, put another way, variation in the intracellular resistance to the conversion of sucrose to starch. This resistance can be envisaged in terms of a transport and a biochemical component, and is analogous to the intracellular resistance to $CO_2$ uptake by the chloroplasts in photosynthesizing leaves.

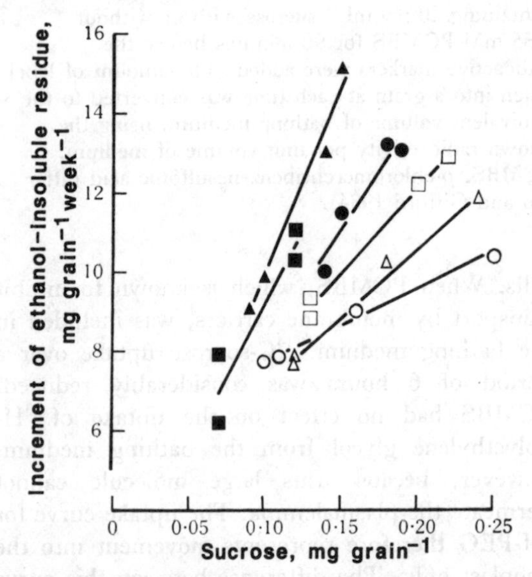


**Fig. 5.30** Varietal differences in the dependence of the rate of production of ethanol-insoluble matter (mainly starch) on intracellular amounts of sucrose in the endosperm of wheat grains.

Detached ears were cultured for 7 days on sucrose solutions at concentrations of 4, 10, 25 or 40 g $l^{-1}$. Sugars in the free space (apoplast) were rinsed out, and the endosperm was extracted in hot ethanol so that intracellular sucrose could be estimated.

Slopes were calculated by linear regression analysis and values of the regression coefficients are Pitic (○) 18, Gabo (△) 35, Nainari (□) 44, Warimba (●) 61, Warimba-sib (▲) 95, and WW15 (■) 73; units are mg $grain^{-1}$ $week^{-1}$ $mg^{-1}$ sucrose (from Jenner and Rathjen 1978).

The detached ear technique allowed the varietal differences to be revealed because a range of intracellular sucrose concentrations and rates of starch accumulation was generated for each cultivar. Measurements on grains from intact plants were consistent with these differences, however. Thus, for cultivars flowering at the same time (e.g. Pitic, Gabo and Nainari), a lower value for the slope (i.e. a higher resistance to starch synthesis) was associated with a higher steady-state level of sucrose in the endosperm cells. Whether differences in this resistance among the grains of a single ear are responsible for any of the positional variation in grain growth rate is not known.

Several points of control over the movement of sucrose into and within the grain have therefore been identified, although the precise nature of these controls and their relative importance during the linear phase of grain growth remain obscure. However, the termination of grain growth appears to be due to a decline in the capacity of the endosperm to synthesize starch, rather than to a reduction in the supply of assimilate. Thus, the ability of dissected endosperms, incubated in $^{14}C$-sucrose solutions, to synthesize $^{14}C$-starch declined with age: endosperms taken at 38 days after anthesis produced less than one-third of the amount of starch produced 10 days earlier, although no less sucrose was absorbed. Furthermore, the reduction in the rate of accumulation of starch over this period, in grains growing normally in the field, was accompanied by the build-up of sucrose (Jenner and Rathjen 1975).

Blockage of the transport pathway within the grain, previously thought to be responsible for the termination of grain-filling, is now discounted. Lignification of the cell walls of the chalaza, which began half-way through the grain-filling period, did not affect the rate of grain-filling, and the collapse of cells in the phloem and chalaza happened after linear dry weight accumulation had ended (Lingle and Chevalier 1985). However, anatomical changes in the vascular and chalazal region are thought to be involved in the control of both water entry to the endosperm during the linear phase of grain growth, and the loss of water towards the end of grain-filling (Cochrane 1983).

### 5.5.3 THE LEGUME SEED

Insights have been gained into various aspects of the unloading of sucrose and the accumulation of starch in the wheat grain by considering the resistances to sucrose movement between the phloem and the endosperm cells. One such resistance is associated with the efflux of sucrose from the phloem and/or its passage through the chalaza and nucellar projection to the endosperm cavity. The importance of this constraint has been recognized, but any investigation of the delivery of sucrose to the cavity has been hindered by the inaccessibility of the tissues involved.

In legume seeds, sucrose and other assimilates are also delivered to an apoplast which separates the maternal tissues from those of the next generation, but here the separation is more complete. One or two vascular bundles serve the seed coat and, in *Phaseolus* and soybean, these reticulate to form a network of veins, from which imported assimilate is rapidly distributed throughout the seed coat (Fig. 5.31). In *Vicia* and pea, however, there is no such network, and it is possible that these anatomical differences between species are associated with differences in the way in which import is controlled. Radial movement inwards, after unloading, is initially symplastic, but at some point assimilate is transferred to the apoplast for the final stage of its journey through the maternal tissues. It then enters an apoplastic transfer zone which completely surrounds the two cotyledons and the embryonic axis (Fig. 5.32).

In essence, this is what happens in the wheat

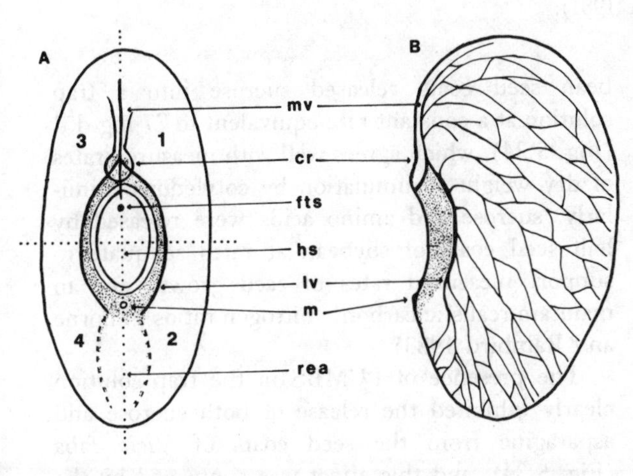


Time course of $^{14}$C-labelled photosynthate distribution within bean seed coats.
A leaflet was exposed to $^{14}CO_2$. Ovules were harvested sequentially, at the intervals shown, through windows cut in the pod wall. Values of radioactivity are expressed as counts $min^{-1}$ per plant portion.

| Time after exposure to $^{14}CO_2$ (h) | Seed coat sectors 1 | 2 | 3 | 4 |
|---|---|---|---|---|
| 1.0 | 18 | 8 | 12 | 13 |
| 1.5 | 174 | 190 | 108 | 121 |
| 2.0 | 3 370 | 2 550 | 3 620 | 3 100 |
| 3.0 | 3 450 | 2 610 | 3 695 | 4 325 |
| 4.0 | 6 940 | 3 845 | 4 840 | 3 370 |

**Fig. 5.31** Vasculature of *Phaseolus* seed coat.
A, surface features after removal of the funicle (the tissue attaching the seed to the pod), with the position of the veins superimposed. The numbers designate the sectors of the seed coat referred to in the inset table.
B, the vascular network which proliferates from the median and one lateral vein. Abbreviations: mv, median vein; fts, funicular trace scar; hs, hilum scar; lv, lateral vein; m, micropyle; rea, region above embyronic axis.
Inset: table showing distribution of imported $^{14}$C-labelled assimilate throughout the seed coat at various times after feeding $^{14}CO_2$ to a leaflet (after Offler and Patrick 1984, Patrick and McDonald 1980).

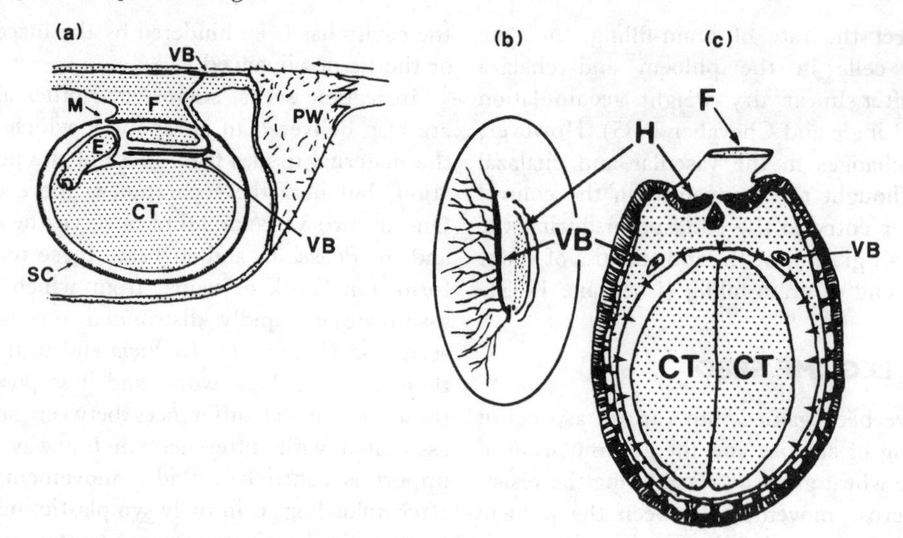


**Fig. 5.32** (a) Median sagittal section of soybean seed attached to the pod wall (PW) by the funicle (F). Two large vascular bundles, one of which is shown (VB), supply assimilate to the pod walls and seeds. The pod walls are extensively vascularized by branches from these main bundles, but only a single branch supplies each seed. Within the seed coat (SC), it joins two parallel bundles which extend the length of hilum near the inner surface of the seed coat. M, micropyle; E, embryonic axis; CT, cotyledon.
(b) Surface view of seed coat showing one of the two vascular bundles and its initial branching to form a reticulate venation, and its position in relation to the hilum (H).
(c) Transverse section of a seed attached to the funicle, showing the approximate position of the two main vascular bundles. The small arrows indicate the movement of assimilate from the vein branches in the seed coat into the space surrounding the cotyledons (after Thorne 1981).

grain, but here the much more limited vascularization reflects the fact that the apoplast providing the zone of assimilate transfer between the maternal tissues and the endosperm is restricted to the endosperm cavity. The advantage of the legume seed as a system for the study of unloading (and here it seems sensible to broaden the meaning of the term from the strict definition of assimilate exit from the sieve tubes to include subsequent transfer to the apoplast) is that the embryo can be removed with a minimum of surgery and, in particular, without damage to the phloem. Part of the pod wall is cut away, exposing one or more seeds, from which the embryos are removed, either through rectangular 'windows' cut from the seed coats, or after excising the distal half of the seed (Fig. 5.33). Samples taken at intervals from a trap solution, or from agar solidified *in situ*, allow unloading from the 'empty' seed coat to be monitored.

Removal of the embryo does not affect the rate of unloading from the seed coat, at least over periods of up to 6 hours. Thus, 'empty' broad bean seed coats released sucrose into a trap solution at a constant rate equivalent to 27 mg $d^{-1}$ (Fig. 5.34), which agrees well with measured rates of dry weight accumulation by cotyledons. Similarly, sucrose and amino acids were released by half seed coats of soybean at rates adequate to support measured rates of seed growth and to maintain realistic carbon : nitrogen ratios (Thorne and Rainbird 1983).

The presence of PCMBS in the trap solution clearly inhibited the release of both sucrose and asparagine from the seed coats of *Vicia faba* (Fig. 5.34), and this effect was confirmed by the accumulation of radioactivity in the trap solution after feeding labelled solutes to a cut petiole (Fig. 5.35). This inhibition suggests that membrane carriers and/or an energy-driven pump (cf. p. 143) facilitate the membrane transfer step of unloading. Similar inhibition, when the trap solution contained sodium azide ($NaN_3$) or carbonyl cyanide *m*-chlorophenylhydrazone (CCCP), or when it was at 0°C, indicates some measure of metabolic control.

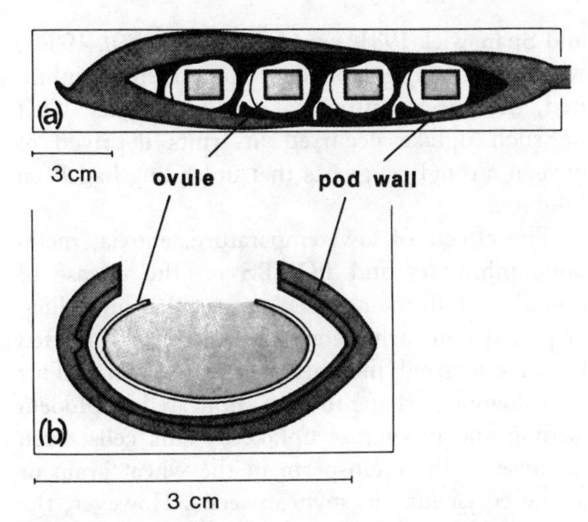


**Fig. 5.33** Diagrammatic representation of the procedure for obtaining 'empty' seed coats of *Vicia*, from which the release of sugar and amino acids could be measured.

(a) The pod was placed in a metal trough, and part of the pod wall was cut away to expose the ovules. Windows were cut in the seed coats of four ovules, and the embryos removed.

(b) Cross-section of one 'empty' ovule, filled with a buffered solution, within the pod wall.

A high humidity was maintained within the trough by lining the bottom with damp tissue paper and fitting an aluminium foil lid (from Wolswinkel and Ammerlaan 1983).

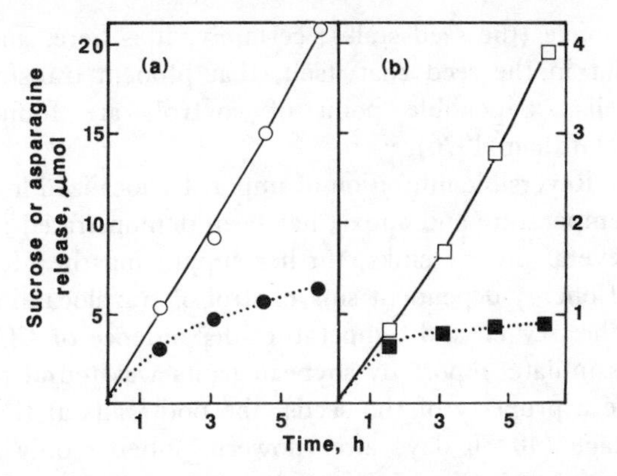


**Fig. 5.34** Effect of PCMBS on the release of sucrose (○, ●) and asparagine (□, ■) by attached seed coats of *Vicia* over a period of 6 hours. Data for control ovules (open symbols) and PCMBS-treated ovules (closed symbols) at a given time were obtained from two pairs of 'empty' seed coats in a single pod (after Wolswinkel and Ammerlaan 1983).

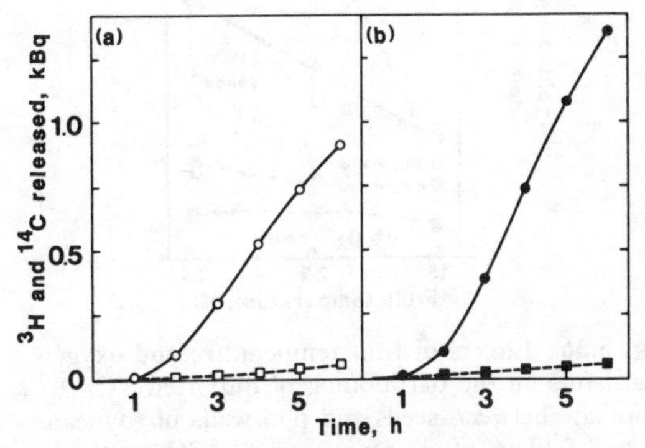


**Fig. 5.35** Effect of PCMBS on the release of labelled solutes by attached seed coats of *Vicia*, after pulse-labelling with an isotope mixture, containing 74 kBq $^3$H-valine and 37 kBq $^{14}$C-asparagine, administered to leaf 13.

(a) $^3$H-solutes: ○, control; □, PCMBS-treated.

(b) $^{14}$C-solutes: ●, control; ■, PCMBS-treated (from Wolswinkel and Ammerlaan 1983).

More direct evidence for a respiratory requirement of unloading has come from pulse-labelling experiments in which KCN in the trap solution inhibited the release of labelled assimilates from empty seed coats of pea (*Pisum sativum*). Similar inhibition of solute release, this time from excised seed coats in a bathing solution containing KCN, ruled out the possibility that the inhibitor acted on the phloem transport of assimilate into the seed coat (Wolswinkel *et al.* 1983).

However, the precise effects of PCMBS, and the respiratory inhibitors CCCP and DNP, depend on the timing of their introduction into the trap solution relative to that of pulse-labelling. When inhibitors were added within 20 minutes of labelling, before any radioactivity had reached the seed coat, none was subsequently released into the trap solution. On the other hand, if labelled assimilates were already being released when the inhibitor was added, the response was delayed (Minchin and McNaughton 1986). This delay, which was greater than could be accounted for by the washing out of label already unloaded into the apoplast, indicates perhaps that the inhibitor's effect was not on unloading from the seed coat *per se*, but on a membrane-transport process deeper in the maternal tissue. Thus, labelled assimilate already in the seed coat would be unloaded, but further phloem transport into the seed coat would be stopped. Minchin and McNaughton suggested that the site of inhibition could be within the funicle (the seed stalk); certainly, it is here, and not in the seed coat itself, that phloem transfer cells, a possible point of control, are found (Hardham 1976).

Reversible inhibition of import by localized low temperature and anoxia has been demonstrated in several diverse sinks, further supporting the idea of energy-dependent sink control of translocation. The oxygen and temperature dependence of $^{14}C$-assimilate import by soybean fruits was found to be a property of the seeds, the pod walls at this stage (40–50 days after flowering) being only a minor sink (Fig. 5.36).

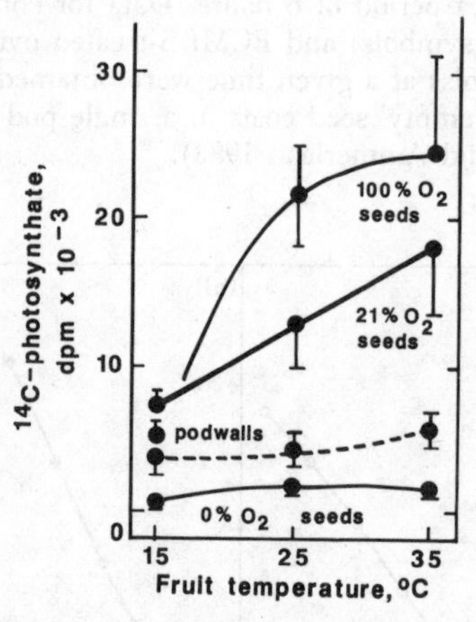


**Fig. 5.36** Effects of fruit temperature and oxygen treatments on the partitioning of imported $^{14}C$-assimilate between seeds and pod walls of soybean fruits 2.5 hours after application of $^{14}CO_2$ to the leaves. The data for pod walls represent all treatments since none was significantly different (from Thorne 1982).

Nevertheless, the limited amount of $^{14}C$ which was imported by fruits completely deprived of oxygen was restricted to the pod walls (75 per cent) and the seed coats (25 per cent): none was found in the cotyledons, implying that either the release of assimilate by the seed coat or its uptake by the cotyledons, or both, was prevented by the lack of oxygen. However, isolated embryos (largely cotyledons), incubated under anaerobic conditions, took up sucrose at rates only 30 per cent less than aerobic embryos. This is consistent with experiments which suggest that a sucrose–proton co-transport system facilitates sucrose uptake by soybean cotyledons (Lichtner and Spanswick 1981; see also Schmitt *et al.* 1984): when this energy-dependent component is inhibited, the purely diffusional uptake remains. That no such uptake occurred in fruits deprived of oxygen strongly suggests that unloading had been inhibited.

The effects of low temperature, anoxia, metabolic inhibitors and PCMBS on the release of assimilates into the apoplast suggest that unloading, or perhaps an earlier stage of import, is facilitated by an energized membrane-carrier system. This is analogous perhaps to that implicated in phloem loading and in sucrose uptake by sink cells, such as those of the endosperm in the wheat grain or of the cotyledons in soybean seeds. However, the major difference is that unloading involves assimilate transfer from symplast to apoplast, rather than in the opposite direction, and the mechanism remains unclear.

Furthermore, unloading from legume seed coats undoubtedly involves controls other than those operating directly at the plasmalemma. For example, steady-state unloading (for periods of 7–12 hours) from attached, empty seed coats of pea was influenced by the osmolarity of the trap solution. When the concentration of sucrose or mannitol was 350 mM, the import of radiolabelled assimilate was maximized, at rates comparable with transport into intact ovules, but at lower concentrations import was reduced (Wolswinkel and Ammerlaan 1984). The reason for this is not known, but it could be related to the osmotic efflux of water from the sieve tubes and, therefore, a reduction in sieve tube turgor, caused by adding osmotica to the trap solution. The propagation of this reduced turgor back along the phloem pathway to the leaves would then stimulate phloem loading, which has been shown to be sensitive to sieve tube turgor, resulting in increased pressure flow of assimilate to the seeds. In this simple form, such an explanation is clearly too gross to account for observed differences in import between seeds of the same fruit. The mechanism, and indeed the significance, of turgor-sensitive transport therefore remain to be established (see Wolswinkel 1985).

Growth regulators have often been implicated in the control of assimilate partitioning by sinks. For example, correlations have often been observed between the growth rates of sinks and the activities of their endogenous hormones (see p. 140), although proof of causal relationships is usually elusive. Effects could be mediated through sink size, as seems to be the case with cytokinins and endosperm cell number, but in addition there has long been evidence for influences in the short term, acting independently of growth on some aspect(s) of sink activity. A consideration of all the evidence for the involvement of hormones in assimilate partitioning is beyond the scope of this book (for recent reviews, see Patrick and Wareing 1980; Patrick 1982). Here we shall examine recent evidence for the influence of ABA on the unloading of assimilates, and their subsequent uptake by the cotyledons, in legume seeds.

ABA is difficult to classify because, although it is best known as a growth inhibitor (see Walton 1980), positive correlations have been found between the rates of growth and ABA content of several sinks, including the cereal grain, grape berry and legume seed. Large-seeded pea genotypes have been found to accumulate ABA more rapidly than smaller-seeded types, although as we have seen (p. 140), the opposite was true for two cultivars of wheat. Similarly, there have been reports of both stimulatory and inhibitory effects on assimilate transfer as a result of ABA application to ears of wheat (for references to the above work, see Schussler *et al.* 1984). However, it is not surprising to find these conflicting reports in view of the various roles which have been attributed to ABA in developing seeds, including the suppression of precocious germination and the control of the duration of seed filling, as well as possible effects on the rate of import.

As we have seen, the legume seed provides an ideal system for the study of this last role, because unloading by the seed coat and uptake by the cotyledons can be isolated from each other. Furthermore, ABA accumulates rapidly in the developing seed, to comparatively high levels (Fig. 5.37, cf. Fig. 5.25). The close correspondence between ABA concentration in the soybean seed and the seed growth rate conforms with previous results suggesting ABA involvement in seed growth. Alternatively, it could be merely the inevitable result of the simultaneous import of assimilate and ABA, although the changes in ABA concentration do appear to anticipate those of growth rate by about four days. Throughout seed development, the concentration of ABA was about four times higher in the seed coat and cotyledons than in the embryonic axis. Comparison between genotypes having different sized seeds (because of differences in growth rate) showed that the seed coats of large seeds always had higher concentrations of ABA than those of smaller seeds, supporting the view that ABA enhances unloading. A rapid, stimulating effect of ABA (and of the synthetic cytokinin, 6-benzylaminopurine) on unloading from excised seed coats of *Phaseolus* has, in fact, been demonstrated (Clifford *et al.* 1986).

Vreugdenhil (1983) has suggested that ABA's inhibitory effect on phloem loading could be due to enhanced leakage of sucrose back out of the phloem, and that unloading could therefore be stimulated by ABA if the same passive leakage occurred in the sink. Alternatively, as Tanner (1980) has proposed, ABA may stimulate sucrose–proton symport into the apoplast. Experiments on assimilate release from empty seed coats may help to resolve these differences.

ABA may also stimulate the uptake of assimilate by the cotyledons. Sucrose uptake by embryos excised from plants raised in a growth chamber was stimulated by ABA added to the incubation

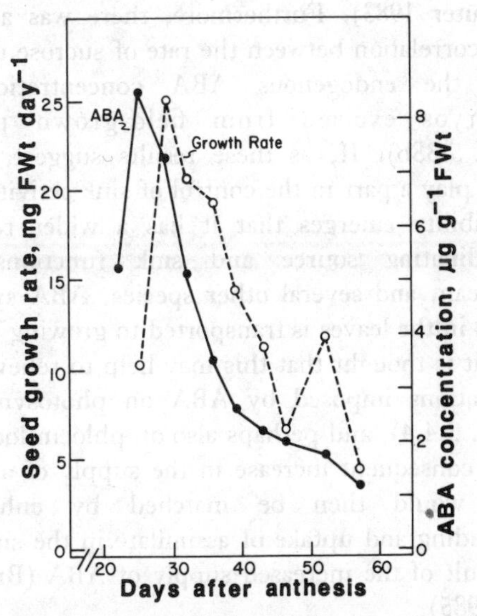


**Fig. 5.37** Changes with time in the rate of seed growth (○) and the concentration of ABA (●) in seeds of field-grown soybean plants. The lines represent pooled data from three soybean genotypes differing in seed size (from Schussler *et al.* 1984).

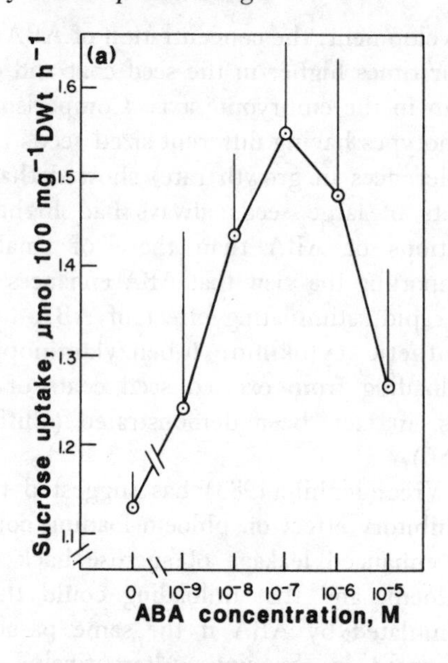


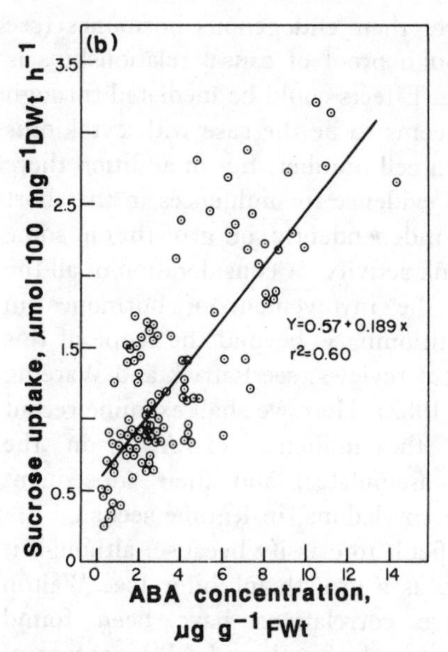


**Fig. 5.38** (a) Effect of exogenous ABA on *in vitro* rate of sucrose uptake by excised soybean embryos from plants grown in a controlled environment chamber.

(b) Relationship between *in vitro* rate of sucrose uptake and ABA concentration of excised field-grown soybean embryos. Data were collected from embryos of three genotypes, which differed in the rate of seed growth, throughout the seed-filling period (from Schussler *et al.* 1984).

medium (Fig. 5.38a). Similar effects have been found on sucrose uptake by discs of sugar beet root (Saftner and Wyse 1984) and on sorbitol uptake by slices of tissue from apple fruits (Berüter 1983). Furthermore, there was a positive correlation between the rate of sucrose uptake and the endogenous ABA concentration of embryos excised from field-grown plants (Fig. 5.38b). If, as these results suggest, ABA does play a part in the control of sink activity, the possibility emerges that it has a wider role in coordinating source and sink functions. In soybean, and several other species, ABA synthesized in the leaves is transported to growing seeds, and it is thought that this may help to relieve any limitations imposed by ABA on photosynthesis (sect. 3.4.4), and perhaps also on phloem loading. The consequent increase in the supply of assimilate would then be matched by enhanced unloading and uptake of assimilate in the sink, as a result of the increased supply of ABA (Brun *et al.* 1985).

## 5.5.4 THE TOMATO FRUIT

The wheat grain and legume seed have yielded information about different aspects of sink function. In the wheat grain we have seen the importance of physical resistances to assimilate movement from the phloem to the endosperm, and are beginning to learn something about assimilate uptake by the endosperm cells, although nothing is known about the energetics of unloading. On the other hand, the size and anatomy of the legume seed have made the study of both unloading and assimilate uptake comparatively easy, with the minimum of structural or physiological disturbance. In both cases, uptake of assimilate from the apoplast by sink cells appears to be facilitated by an energy-dependent membrane process similar to that responsible for phloem loading from the apoplast. One general difference is that in the sink cells rapid metabolism of imported assimilate seems to ensure that movement continues down a concentration gradient, whereas loading proceeds uphill, and is therefore considered an active process. Unloading from legume seed coats has similar energy-dependent characteristics, but here movement is from symplast to apoplast, and details of the mechanism of assimilate transfer are lacking.

One aspect of sink function which we have not yet touched upon is the biochemical conversion of imported assimilate, clearly important in determining the chemical composition of the sink, and perhaps significant in sink growth because it helps maintain the gradient of assimilate concentration between source and sink. However, the importance of assimilate conversion in comparison with other aspects of sink function, such as unloading, is, for most sinks, unclear. Certainly, in the wheat grain the ability of the endosperm to convert sucrose into starch becomes a limiting factor only towards the end of the grain-filling period.

In tomato, however, there is evidence that the rate of assimilate import by a young fruit is inversely related to the sucrose content of the fruit (Fig. 5.39), implying that the rate of transport depends on the steepness of the concentration gradient between source and sink. Sucrose is the major mobile assimilate, but little is found in the young fruit, where glucose and fructose are the major sugars. Lowering the fruit temperature to 5 °C reduced the import rate, or in some cases caused export, and was associated with increased rates of sucrose accumulation and lower rates of accumulation, or depletion, of the hexoses, starch and insoluble residue. Conversely, fruits maintained at 35 °C had higher rates of import, associated with less accumulation of sucrose and more of the hexoses, compared with controls at 25 °C. Since import rates were related only to the concentration of sucrose in the fruit, and not to that of any of the other major carbon metabolites, it was suggested that import was regulated by the rate of conversion of imported sucrose into glucose and fructose, probably by acid invertase, high activities of which have been identified in tomato fruits (Walker *et al.* 1978).

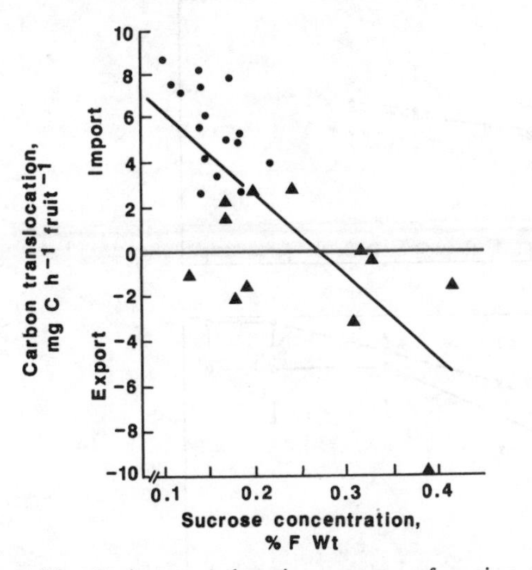


**Fig. 5.39** Carbon translocation rate as a function of the mean sucrose concentration over a 48-hour period in young tomato fruits maintained at either 25 °C (●) or 5 °C (▲). Translocation rates were measured as the sum of the change in carbon content and the respiratory losses of carbon over a 48-hour period (from Walker and Ho 1977).

Support for this idea has come from experiments with detached fruits importing $^{14}$C-sucrose from an agar medium into which the pedicels were inserted. Temperature treatments, giving a range of sucrose concentrations, confirmed the inverse relationship between sucrose uptake and fruit sucrose concentration (Dinar and Stevens 1982). Furthermore, genotypic variation in the rate of fruit growth, due to a difference in the proportion of assimilate partitioned to the fruits at the expense of the stem, appeared to be associated with the ability of the fruits to hydrolyse sucrose. Thus, detached fruits of the genotype with the higher rate of dry-weight accumulation took up $^{14}$C-sucrose more rapidly, but had lower sucrose contents than those of the slower growing genotype (Hewitt *et al.* 1982).

Such circumstantial evidence for the involvement of sucrose hydrolysis in the control of import complements the common observation that many sink tissues have high activities of invertase and, indeed, that the growth of sinks has often been directly correlated with invertase activity (see Morris 1982). The fruits of wild species of tomato, for example, are smaller than those of the cultivated species, and have lower activities of acid invertase (pH optimum 4.0–4.5) and correspondingly high sucrose contents (Manning and Maw 1975).

Similarly, the ability of the developing inflorescence of tomato to compete for assimilate with the vegetative apex has been attributed to its acid invertase activity. Reductions in assimilate supply, as a result of the removal of source leaves or by mutual shading of crowded plants, caused the rate of inflorescence growth and development to be reduced, such that flower buds aborted. This was associated with low reducing sugar (hexoses) concentrations, low reducing sugar/sucrose ratios and low invertase activities. In contrast, the removal of leaves at the shoot apex allowed increased inflorescence growth, associated with a high activity of invertase (Russell and Morris 1982).

High temperatures (> 30/20 °C, day/night), as well as low irradiance, can cause flower abortion, as a result of a reduction in the proportion of exported assimilate which reaches the inflorescence. Again, this reduced ability to import assimilate is attributed to impaired hydrolysis of sucrose by the young flowers. Significantly, hydrolysis was less susceptible to high temperatures in a heat-tolerant cultivar than in one more sensitive (Dinar and Rudich 1985).

Study of the causes of natural variation in fruit size may help to explain the control of import. For example, there is considerable variation in the final size of fruits at different positions on a truss, the earlier-developing proximal fruits being larger than the more distal ones. This is partly because the proximal fruits are the closest to the source of assimilate, and it is therefore not surprising that they gain more than distal fruits when assimilate supply is increased, yet are least affected by reductions in supply. However, the pattern of growth seen in the intact truss persists even when competition for assimilates is eliminated by fruit thinning: the distal fruits are unable to make use of the assimilate which is available to them (Fig. 5.40) and, therefore, have a lower potential for growth.

Part of this difference in potential appears to be established before fruit set, although the natural sequence of floral development and fruit set inevitably means that the later-developed distal fruits are disadvantaged by having to compete for assimilates with the larger proximal fruits. By emasculating flowers and then inducing parthenocarpic fruit set, Bangerth and Ho (1984) found that distal fruits induced first grew more than proximal fruits induced five days later. However, the dry weight advantage of distal fruits induced first was less than that of proximal fruits induced first. This suggests an inherent superiority of proximal fruits, accentuated perhaps by their positional advantage. We have already seen that the potential weight of wheat grains depends on the number of endosperm cells formed, and it is possible that cell number similarly limits the potential growth of tomato fruits. Thus, it has been observed that the ovaries of proximal flowers, before anthesis, have significantly more cells than those of distal ones (see Bangerth and Ho 1984). Any advantage which this confers could be compounded during the first one or two weeks after fertilization, when the final cell number of the fruits is determined, with IAA apparently having a controlling influence. The greater IAA

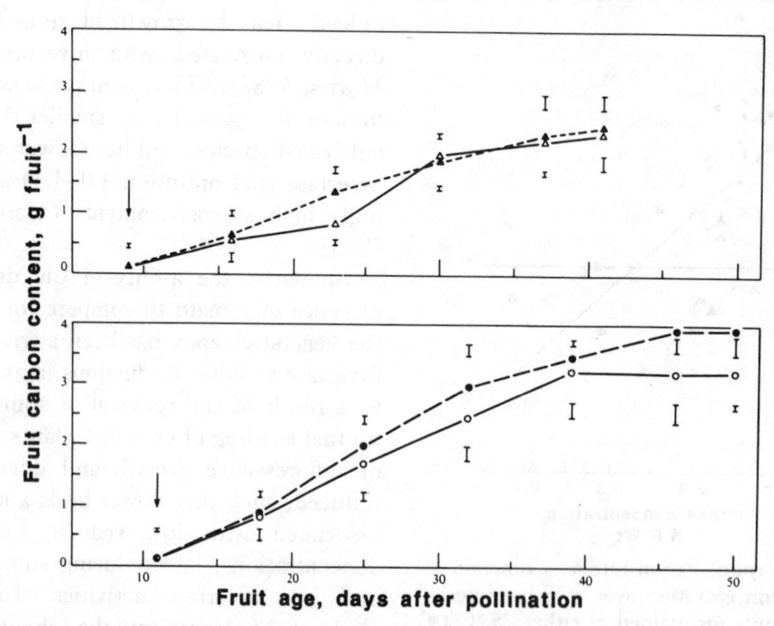


**Fig. 5.40** Effect of fruit thinning on carbon contents of tomato fruits at proximal (○, ●) and distal (△, ▲) positions. Open symbols, control (unthinned) plants; closed symbols, treated plants. Arrows indicate the time of fruit thinning.) The number of fruits was reduced to 3 at either the proximal or distal position when the average age of proximal and distal fruits was 11 and 9 days, respectively (from Ho *et al.* 1983).

content observed in proximal fruits may therefore be significant (Ho *et al.* 1983). However, we do not at present know whether, or by how much, final cell numbers differ between fruits on a truss.

IAA may play a part in determining the sink activity of fruits, in addition to its possible effects on sink size. For example, there is considerable evidence that invertase activity is regulated by IAA and by other growth regulators, particularly in elongating stem sinks (see Morris 1982, 1983), but also in fruits (e.g. strawberry; Poovaiah and Veluthambi 1985). However, with little other than circumstantial evidence for invertase having a major role in the control of sink activity in tomato fruits, it is pointless to speculate further on the possible involvement of IAA. However, a consistent picture does emerge from the available evidence, with the larger proximal fruits having higher IAA contents, echoing the association between growth and IAA content found in grains of wheat (p. 140).

### 5.5.5 THE DISEASED PLANT

Alterations in the pattern of assimilate translocation as a result of infection by fungi, bacteria or viruses have been known for some time. Here we concentrate on the effects of fungi because, in many cases, changes in the distribution of assimilate seem to be due to the establishment of sinks for assimilate at the sites of infection. Whether bacterial and viral infections have similar effects is unclear. The incomplete picture for fungal infection is complicated by the differing nutritional behaviour of the fungi (biotrophic or necrotrophic), and by specific interactions between host and fungus. Biotrophic fungi derive their nutrients from living host cells and thus use either current or recently mobilized assimilates. Necrotrophs, on the other hand, derive their nutrients mainly from tissue killed before colonization. This is inevitably an oversimplification and, as Whipps and Lewis (1981) point out, the nutritional modes of fungi lie somewhere along the continuum biotrophy–necrotrophy–saprotrophy (see Lewis 1973 and Cooke and Whipps 1980 for fuller accounts of fungal nutrition). According to Cooke and Whipps (1980), fungi can be split into two broad nutritional groups, biotroph–hemibiotroph and hemibiotroph–necrotroph, where the former are characterized by a long biotrophic phase and the latter by a long necrotrophic one.

Infection by fungi of both nutritional groups commonly results in a reduction in the export of assimilates from infected leaves. In addition, increased import of photosynthetic products by infected tissues, and the accumulation of carbohydrates, has been demonstrated in some plants infected with fungi belonging to the biotroph–hemibiotroph grouping. Livne and Daly (1966) showed that when unifoliate leaves of the french bean (*Phaseolus vulgaris*) were infected with the rust, *Uromyces appendiculatus*, there was a substantial reduction in the amount of newly-fixed carbon exported from that leaf. Import from the next trifoliate leaf increased, but this diversion of assimilates to the rust was at the expense of the roots and newly-emerging leaves (Fig. 5.41). Prolonged assimilate import into infected leaves was also demonstrated by Thrower and Thrower (1966), working on broad bean infected with the rust, *Uromyces fabae*.

However, the situation for monocots is not straightforward. For example, whereas single leaves of wheat infected with the rust, *Puccinia striiformis*, did not attract assimilates from other leaves (Doodson *et al.* 1965), later experiments using completely infected plants showed that the proportion of labelled assimilate moving to the infected leaves was increased while that going to roots was greatly decreased (Siddiqui and Manners 1971). However, Owera *et al.* (1981, 1983) have suggested that in barley infected with the rust, *Puccinia hordei*, there is enough carbon left in infected leaves to account for fungal growth and starch synthesis. Thus, when $^{14}CO_2$ was fed to a non-infected leaf, very little label appeared in the infected leaf (Farrar 1984).

Necrotrophs can also affect translocation, although to a lesser extent than biotrophs. The extent of infection and the age of the plant have a large effect on assimilate distribution patterns. Thus, in the early stages of infection of tomato with *Alternaria solani*, there was increased retention of assimilate in the single infected leaf (Coffey *et al.* 1970). During later stages, when there was extensive chlorosis and necrosis, export from diseased leaves was usually increased. Furthermore, although uninfected leaves were not at first affected by disease development, during later stages of infection they made an increased contri-

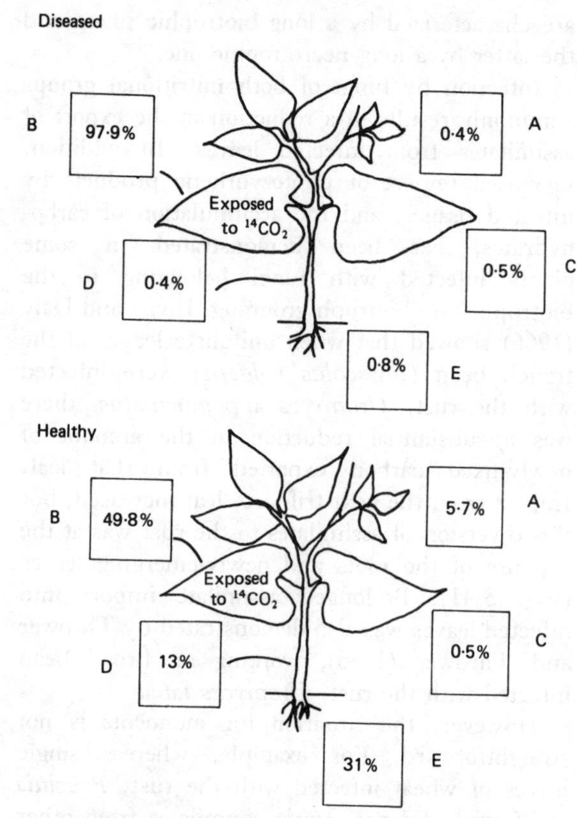


**Fig. 5.41** Distribution of radioactivity in bean (*Phaseolus*) plants 5 hours after exposure of diseased (rust) or healthy unifoliate leaves to $^{14}CO_2$. Boxes present the amounts of radioactivity in different organs as percentages of total activity: A, both trifoliate leaves; B, unifoliate leaf supplied with $^{14}CO_2$; C, opposite unifoliate leaf; D, stem below trifoliate leaves; E, roots (from Walters 1985, adapted from Livne and Daly 1966).

bution to actively-growing regions of the plant. In contrast, there was no evidence for increased export of assimilates from uninfected leaves to those which were infected.

Little is known about the causes of these effects on the distribution of assimilate, although some progress has been made in understanding changes in carbohydrate metabolism at the sites of infection. We have seen that during infection by biotrophic pathogens, patterns of translocation are altered so that sites of infection accumulate photosynthetic products at the expense of other regions of the plant (see also Scott 1972; Lewis 1973). Most studies show a pronounced initial accumulation of carbohydrates, which in some cases may subsequently decline (see Long *et al.* 1975). A model system for such work has been the rust fungus, *Puccinia poarum*. It has two hosts, one a grass, *Poa pratensis*, and the other a composite, *Tussilago farfara* (coltsfoot). Holligan *et al.* (1973) and Fung (1975) have shown that at infection sites in either host there is an accumulation of sucrose, glucose, free fructose and fructose polymers (Table 5.6).

However, although in *Poa pratensis* there is also an increase in starch deposition in host cells around fungal pustules, in coltsfoot no such starch accumulation occurs. Instead, increased fructan levels have been detected, with the greatest accumulation occurring within pustules, and a mere fraction of that concentration occurring in regions immediately surrounding pustules. At the same time, the fungus accumulates mannitol, arabitol and glycogen (Holligan *et al.* 1974). These metabolites are not readily utilized by host tissues and, in this way, their synthesis from host sugars absorbed by the fungi maintains a chemical concentration gradient which may promote further uptake of the substrates (Smith *et al.* 1969).

There is evidence to show that the carbon source used by rust fungi is sucrose, which is hydrolysed before or during transfer to the fungus. Some of this evidence comes from an examination of invertase activity in rust-infected leaves. Increases in acid invertase activity have been found in several biotrophic infections (see Whipps and Lewis 1981). Thus, in coltsfoot infected with *P. poarum*, the increase in invertase activity is more than ten-fold at the pustule centres with activity declining rapidly with increasing distance from the pustule. It has been suggested that the enhanced invertase activity in infected leaves is primarily due to a fungal enzyme, although a stimulation of host invertase is also thought to make some contribution to the overall increase (see Whipps and Lewis 1981 for further details). Regarding the site of the increased invertase activity, it is known that sucrose may be hydrolysed by enzymes in the cell wall (apoplast) or cytoplasm. On the other hand, Greenland (1979) has suggested that it may be hydrolysed later as it crosses the walls of intercellular hyphae.

In leaves infected with biotrophic fungi, increased concentrations of hexoses are present,

**Table 5.6** Increase in concentrations of soluble carbohydrates at infection sites in leaves of *Tussilago farfara* and *Poa pratensis* infected by *Puccinia poarum* at different stages in its lifecycle (values as the ratio, diseased : healthy)

| *Disease stage* | *Fructose* | *Glucose* | *Sucrose* | *Fructans* |
|---|---|---|---|---|
| On *Tussilago farfara* | | | | |
| Early | 2.1 | 7.0 | 0.6 | 1.2 |
| Late | 3.7 | 14.0 | 1.6 | 4.0 |
| On *Poa pratensis* | | | | |
| Uredial | 7.4 | 7.6 | 2.1 | 15.0 |

(After Lewis 1976, from data of Holligan *et al.* 1973 and Fung 1975)

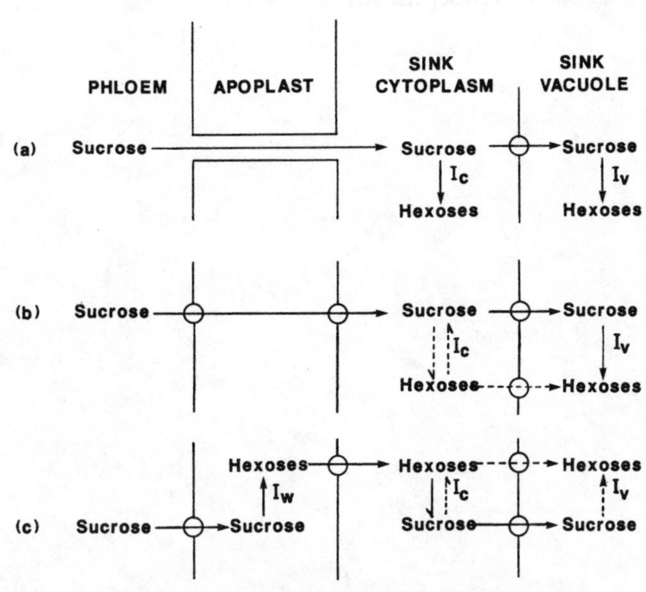


**Fig. 5.42** A simplified representation of possible pathways of sucrose transfer across the sieve element/sink cell boundary, and the subsequent use of sugars for storage or metabolism.

(a) Symplastic transfer *via* plasmodesmata, e.g. in roots of pea and maize, apical meristems, young leaves.

(b) Transfer of intact sucrose through the apoplast, e.g. in grains of wheat and barley and seeds of legumes, in which sucrose is largely metabolized in the cytoplasm; and in sugar beet roots, in which sucrose is stored in the vacuole.

(c) Sucrose is unloaded into the apoplast, where it is hydrolysed by a wall-bound invertase. The resultant hexoses are taken up by the sink cells and largely reconverted into sucrose before transport into the vacuole, e.g. in sugar cane stems.

$I_w$, $I_c$ and $I_v$: wall, cytoplasmic and vacuolar invertases.

—⊖→ : movement across the plasmalemma or tonoplast, which may be passive or energized 'downhill' transport, or active transport against a concentration gradient (after Morris 1983).

due to the increased invertase activity. These hexoses can readily cross membranes, and their carbon skeletons, after entering the glycolytic cycle, can be converted to triose phosphates. These triose phosphates can enter chloroplasts and be utilized in starch synthesis (Walker 1974), and since many chloroplasts within pustules are destroyed, starch synthesis occurs in the first available chloroplasts, in cells surrounding fungal pustules. Fungal sequestration of inorganic phosphate (Bennett and Scott 1971) may also be partly responsible for the enhanced starch synthesis in infected leaves. A reduction in $P_i$ in host cells would encourage retention of triose phosphates in the chloroplast and discourage starch breakdown (Chen-She *et al.* 1975). It should be noted, however, that this hypothesis is speculative and requires critical examination (cf. sect. 3.4.4 and 5.4.2).

The high activity of invertase in the diseased leaf, and in numerous other diverse sinks, is consistent with the idea, developed for the tomato fruit (sect. 5.5.4), that the import of sucrose is controlled by its hydrolysis in the sink. However, this simple picture is complicated by the presence, in many sinks, of more than one invertase, usually having different pH optima: acid invertase is usually either bound to the cell walls (apoplastic) or vacuolar, whereas neutral or alkaline invertases are cytoplasmic. The relative importance of these invertases in the diseased leaf or the tomato fruit is unknown, because the route and mechanism of sucrose transfer from the phloem of the sink into the surrounding cells is unknown. There are several possibilities, however, based on what is known about other sinks, and their depiction in Fig. 5.42 can serve to illustrate some of the diversity of function displayed by different types of sink.

# PART 2 CROP CASE HISTORIES

# CHAPTER 6 TEMPERATE CEREALS

*There is plenty of evidence to show that these components (of grain yield) are interdependent to a greater or lesser degree; that for example a greater number of heads per acre is counteracted by a smaller number of grains per head. If yields of current varieties are to be substantially increased this compensatory mechanism must be suppressed.*

(Evans 1977)

*No single yield component predominated in determining yield. However, combining ears/m² with grains per ear showed that many grains per unit field area were correlated with large yields. (The data) do not support the idea that yield components tend to be mutually compensating.*

(Gallagher and Biscoe 1978)

## 6.1 INTRODUCTION – CEREAL PLANT DEVELOPMENT AND GRAIN YIELD COMPONENTS

The individual plants of a cereal crop (barley, oats, rye, wheat) progress through a series of well-defined developmental stages (generally, and incorrectly, called growth stages) from germination and establishment, through tiller production, stem extension and ear emergence, to grain filling and maturity. Because management practices such as the application of crop protection chemicals must be carried out at particular stages of plant development, there has been developed a series of simple keys for the recognition and definition of crop growth stages in the field. The Feekes system (Fig. 6.1) was the first to be widely used, but it has since been superseded by the more comprehensive Zadoks, or Decimal Code system (Table 6.1, as illustrated by Tottman *et al.* (1979).

These keys are very useful, but they do not provide a sufficiently detailed description of plant development for crop physiological work, or even for crop management in some cases, because much of the recent interest of physiologists has concentrated upon the development of the stem apex; in particular, in studies of the influence of management or environment upon ear and grain size, it is important to know at what stages of overall plant development the following stages of apex development (Figs 6.2 and 6.5) occur:

(a) the end of leaf primordium production and the start of production of primordia which

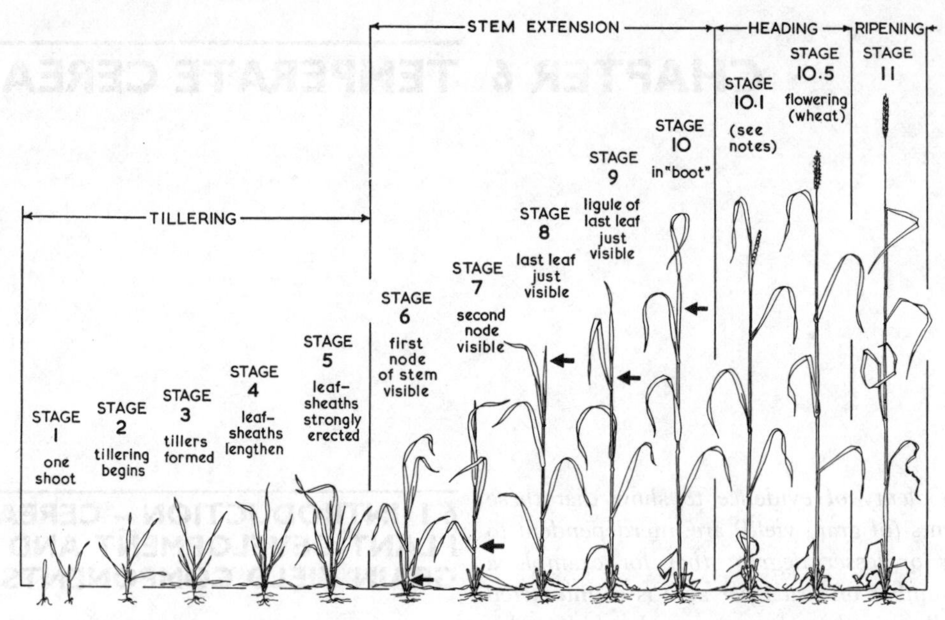


**Fig. 6.1** The 'Feekes' Scale of developmental (or growth) stages of wheat (from Large 1954). Compare with Table 6.1.

will ultimately become spikelets (first spikelet stage, allowing one primordium for the collar before spikelet initiation can begin);

(b) the start of spikelet differentiation, i.e. the first unequivocal signs of the switch from vegetative to reproductive development (double ridge stage, occurring when 50–80 per cent of the total number of spikelets has been initiated); and

(c) the end of spikelet initiation (terminal spikelet stage in wheat, maximum number of primordia in barley).

Unfortunately, there are few consistent relationships between crop growth stage and either the first spikelet or double ridge stages (Kirby and Appleyard 1984b), although there is evidence that double ridge is related to leaf sheath development in wheat (Hay 1986). Consequently, the timing of (b) is normally determined by dissection or estimated by means of a crop simulation model (Ch. 9), whereas (a) can be determined only in retrospect by calculation when the total number of leaves and the maximum number of primordia are known. However, it is now clear that for wheat mainstems, the terminal spikelet stage occurs shortly after the start of true stem elongation (not pseudostem or sheath elongation) (Kirby and Appleyard 1984b; Hay 1986); consequently, the 'ear at 1 cm' stage (above the crown, measured easily in the field), which signals the start of stem internode elongation, is a suitable field index of terminal spikelet. Because stem extension is so rapid (Hay 1978), the appearance of the first visible stem node (Decimal code 31, Feekes stage 6) can also be used, as a (retrospective) indicator of the achievement of terminal spikelet. The 'ear at 1 cm' stage also tends to be associated with the start of the rapid decline in tiller number in wheat (Fig. 6.11), but the limited evidence which is available for barley suggests that the interrelationship between maximum number of primordia (awn initiation), stem extension and tiller mortality is slightly different (Hay 1986).

The grain yield of a cereal crop can be split up into three major components:

grain yield = ear population density (no. of ears/unit area) × ear size or length (no. of grains/ear) × individual grain weight

each of which can, to a certain extent, vary independently of the others, as shown in the following sections. Their magnitudes are determined at

**Table 6.1** Primary and secondary growth stages used in the Decimal Code for the description of the development of cereal crops

**0 Germination**
- 01 Water absorption
- .
- .
- .
- 07–09 coleoptile above ground

**1 Seedling growth**
- 10 first leaf through coleoptile
- 11 first leaf emerged*
- 12 2 leaves emerged
- .
- .
- .
- 18 8 leaves emerged
- 19 9 or more leaves emerged

**2 Tillering**
- 20 main shoot only
- 21 main shoot and 1 tiller
- .
- .
- .
- 29 main shoot and 9 or more tillers

**3 Stem extension**
- 30 pseudostem (leaf sheath) extension (= 5 cm)
- 31 first node detectable (above ground)
- .
- .
- .
- 36 sixth node detectable
- 37–39 flag leaf emergence

**4 Ear or panicle in 'boot'**
- 49 first awns visible (barley)

**5 Ear or panicle emergence**

**6 Anthesis**

**7 Milk development**

**8 Dough development**

**9 Ripening**

* A leaf is classed as emerged (or unfolded) when the ligule is visible.
(Adapted from Tottman *et al.* 1979)

different stages of crop development. Ear population density depends primarily upon the number of tillers formed at Decimal code 2, Feekes stage 2–3, but conditions around the time of terminal spikelet (especially nitrogen supply, sect. 6.4.2; Fig. 6.11) determine the proportion of these tillers which survives to carry ears. The second component is determined primarily during the period of spikelet initiation (the product of rate and duration of spikelet initiation) but unfavourable conditions during floret development (up to the appearance of the flag leaf, Baker and Gallagher 1983) can lead to extensive spikelet mortality Figs 5.5 and 6.5). In general, the duration of spikelet initiation will be shorter for tiller ears than for mainstem ears because tiller development begins later, but ear emergence is essentially synchronous for all the fertile stems of a crop. In barley, three spikelets, each containing a single floret, are initiated at each node of the rachis, but in the more common two-row varieties, only one of these spikelets is fertile. The determination of ear size is slightly more complex in wheat because each spikelet can accommodate up to ten florets (potential grains), although each normally holds only two or three. Typically, about 15 of the 40 spikelets initiated by a spring barley apex will die before ear emergence, whereas the mortality of wheat florets is generally much higher (Kirby and Appleyard 1984b). Finally, mean grain weight is determined primarily by the quantity of assimilate available for transport to the ear between anthesis and maturity; this, in turn, depends upon green leaf area duration after anthesis and the photosynthetic activity of the ear, as well as source/sink relationships. However, in seasons when leaf area duration is low because of drought or leaf diseases, for example, substantial amounts of assimilate stored in the stem and leaf sheaths before anthesis can become available for grain-filling. The extent of these stores will depend upon environmental conditions before ear emergence. The factors influencing the partitioning of dry matter to the developing cereal grain are reviewed in detail in Chapter 5.

In temperate zones, the weather is rarely unfavourable throughout crop development; consequently, the fact that these three components are determined at different stages of crop development means that the crop is buffered against very low grain yield. For example, in an unusually dry season, such as 1976 in the UK, the curtailment of leaf area duration after anthesis by water stress tends to lead to poorly-filled grain. However, in addition to the possibility of enhanced translocation of assimilate from reserves (sect. 5.12), low grain weights will be at least partly compensated for by the magnitudes of the other two components (ear number and size) which were determined at earlier stages of development before water supply became limiting. There is also the possibility that similar grain yields in different years can result from different combinations of components (e.g. the yields of spring barley in 1969 and 1970, Table 6.2). This protects the farmer against disastrously low yield, although there can be financial penalties to pay for shrivelled grain (Taylor and Blackett 1982), but it

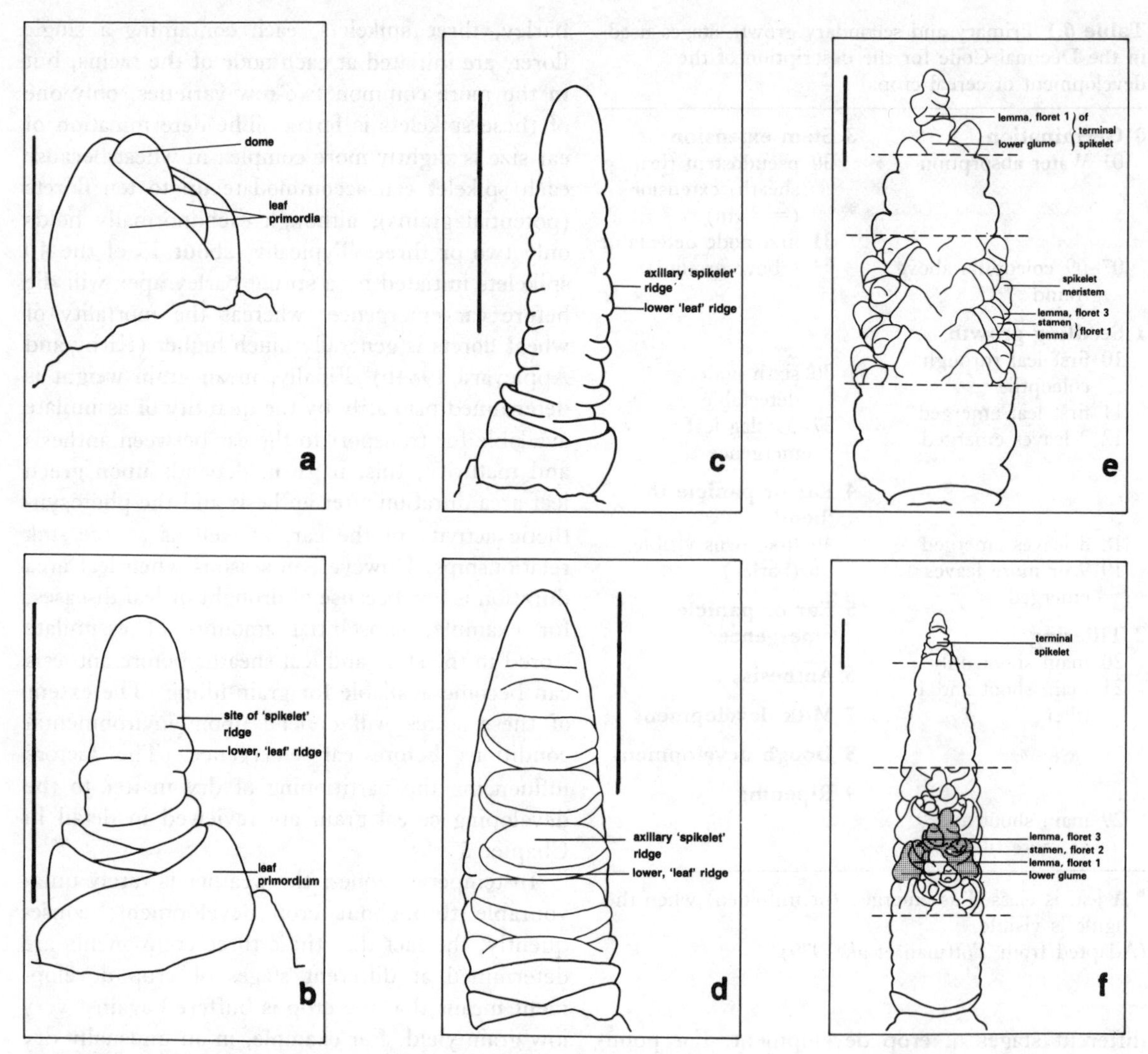


**Fig. 6.2** Developmental stages of the wheat stem apex: a, early vegetative; b, late vegetative/apex elongation (during this stage the first spikelet primordia are normally formed – see text); c, double ridge, profile view; d, double ridge, face view; e, terminal spikelet, profile view; f, terminal spikelet, face view. The vertical bars indicate 0.5 mm (adapted from Kirby and Appleyard 1984a).

certainly does not rule out the possibility of above-average yields when the growing season is uniformly favourable (e.g. 1972, Table 6.2).

In several studies, it has proved useful to combine the first two of these components of grain yield to give a single composite component, the number of grains per unit area, which is an expression of the capacity of the crop sink for assimilate after anthesis. The fact that grain yield tends to be more closely related to grain population density than to grain weight (Willey and Holliday 1971; Gallagher *et al.* 1975; Gales 1983) has led several authors to propose that cereal grain yields are sink-limited and that grain weight is the most stable of the three basic yield components, except in extreme seasons such as the drought years of 1975 and 1976 (see above and sect. 5.2 and 6.6).

This chapter is concerned mainly with the interactions between varying management practices (especially seed rate, sowing date, nitrogen fertilization and water supply) and the environment in the determination of grain yield and its components. However, work since 1970 has

**Table 6.2** The components of grain yield of crops of spring barley cv. Zephyr grown under uniform conditions (conventional cultivation and 100 kg N $ha^{-1}$) at the East of Scotland College of Agriculture, UK, over five seasons 1968–72

| | *1968* | *1969* | *1970* | *1971* | *1972* | *Mean* |
|---|---|---|---|---|---|---|
| Grain yield, t $ha^{-1}$ at 15% mc | 3.5 | 4.8 | 4.5 | 3.1 | 5.7 | 4.32 |
| Ear population density, no. $m^{-3}$ | 533 | 517 | 811 | 544 | 617 | 604.4 |
| Ear size, grains $ear^{-1}$ | 20 | 20 | 13 | 20 | 22 | 19.0 |
| Individual grain weight, mg | 34 | 43 | 36 | 28 | 38 | 35.8 |

(Hay 1982)

shown that grain yield is determined primarily by the biomass of the crop which, in turn, is determined by the quantity of radiation intercepted by the crop canopy (e.g. McLaren 1981; sect. 3.2). It is, therefore, one of the tasks of this chapter to attempt to reconcile the traditional yield components approach with the more recent biomass approach (see sect. 6.6). Throughout the chapter, for simplicity, crop/weed interactions are ignored, although other aspects of crop protection are considered where relevant.

## 6.2 PLANT POPULATION DENSITY AND YIELD

The population density of mature plants in a cereal crop at harvest depends upon a series of factors and processes, including the initial seed rate, seedling mortality and predation before crop emergence, and plant losses resulting from environmental stresses, pests and diseases. Such losses will be greatest at high seed rates owing to intense intraspecific competition (e.g. Puckridge and Donald 1967). As a result, a given seed rate can give widely differing plant densities in years of contrasting weather or pest incidence (e.g. Cowton 1982).

Investigations of the influence of population density upon crop dry matter yield have generally shown increases up to a plateau value at moderate densities and a significant reduction in production only at very high densities (Holliday 1960; Donald 1963)(Fig. 6.3). The level of the plateau yield and the population density at which it is achieved depend, of course, on other factors, particularly the nitrogen supply (sect. 6.4.1 and 6.6). This pattern, in which the yield plateau is the result of decreasing plant size (here principally number of tillers per plant) with increasing plant numbers, is similar to that shown by a range of seed-bearing annual species (Harper 1977). Grain yields tend to respond similarly to population density; for example, in the wheat crops studied by Puckridge and Donald (1967) (Fig. 6.3), grain yield was relatively unaffected by seed rates above 35 seeds $m^{-2}$, falling by only 25 per cent at the exceptionally high rate of 1078 seeds $m^{-2}$ (equivalent to a spacing of only 6 mm between seeds sown in rows 15 cm apart). Similarly, work in the Netherlands has shown a plateau of yield in winter wheat grown at plant population densities varying between 100 plants $m^{-2}$ (8.4 t $ha^{-1}$) and 800 plants $m^{-2}$ (8.9 t $ha^{-1}$) (Darwinkel 1978). However, in some experiments, especially using older varieties, substantial grain yield depressions were recorded at more modest population densi-

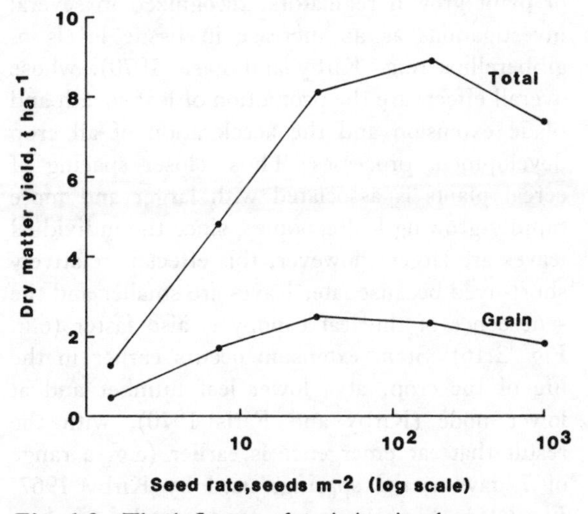


**Fig. 6.3** The influence of variation in plant population density on the total dry-matter and grain yields of a wheat crop cv. Insignia 49 grown in a Mediterranean environment, Adelaide, South Australia (from data of Puckridge and Donald 1967).

ties (Holliday 1960). This effect is almost invariably related to the incidence of lodging which is more prevalent in dense crops and in older varieties, in both cases due to weaker stems (sect. 6.4.1).

### 6.2.1. POPULATION DENSITY AND THE COMPONENTS OF GRAIN YIELD

Responses to variation in population density such as that shown in Fig. 6.3 demonstrate that the grain yield of a temperate cereal crop can be buffered against variation over a very wide range of seed rates, within a given growing season. However, this is not simply the result of alterations in the number of fertile tillers to compensate for changes in plant numbers; increased plant density is associated with a wide range of changes in the rate and pattern of development of cereal plants which can cause changes in the magnitude of each of the three components of grain yield (especially ear population density and ear size) which tend to be mutually compensating.

The primary effect of increasing plant population density is to increase competition between adjacent plants; the resulting shading of plant tissues (alteration of both the quantity and spectral composition of radiation incident upon shaded leaves) has a profound influence upon the balance of plant growth regulators, recognized in several investigations as an increase in tissue levels of gibberellins (e.g. Kirby and Faris 1970), whose overall effects are the promotion of leaf sheath and blade extension and the acceleration of all crop development processes. Thus, closer spacing of cereal plants is associated with larger and more rapidly-growing leaf canopies, since the individual leaves are larger; however, this effect is relatively short-lived because later leaves are smaller and the senescence of the leaf canopy is also faster (e.g. Fig. 2.16). Stem extension occurs earlier in the life of the crop, at a lower leaf number and at lower node (Kirby and Faris 1970), with the result that ear emergence is earlier (e.g. a range of 7 days at the spacings used by Kirby 1967, Fig 2.16), but the stem tends to be weaker, leading to increased incidence of lodging at high seed rates, in spite of the fact that the straw is generally shorter.

#### 6.2.1.1 NUMBER OF EARS PER UNIT AREA

In both spring and winter cereals, increase in plant population density is almost invariably associated with a continuous increase in ear population density across a very wide range of seed rates from the very low, through the optimal range (200–500 seeds $m^{-2}$) to high rates (600–800 seeds $m^{-2}$), beyond which the crop contains only mainstems (e.g. Fig. 6.4; Table 6.3; Kirby and Faris 1972, where the effect, in spring barley, was still apparent at 1600 seeds $m^{-2}$). In stands where the plants are very widely spaced and inter-plant competition is slight, low ear numbers are the consequence of the inability of each plant to produce enough fertile tillers to compensate fully for lower plant numbers. For example, in the case of Table 6.3, each plant would need to carry 160 ears to give an ear population density of 800 ears $m^{-2}$. At higher densi-

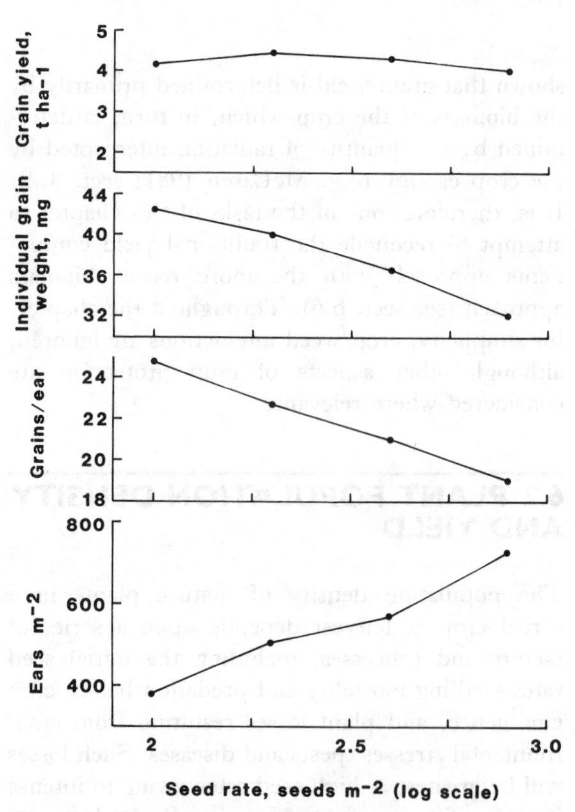


**Fig. 6.4** The influence of variation in plant population density on the components of grain yield of spring barley (averaged over several varieties) grown at Cambridge, UK, in 1963 (adapted from Kirby 1967).

**Table 6.3** The influence of plant population density on the components of grain yield of a crop of winter wheat cv. Lely grown in the Netherlands

| *Plants*($m^{-2}$) | *Ears* ($m^{-2}$) | *Ears/plant* | *Maximum no. of tillers/plant* | *Tiller mortality* (%) |
|---|---|---|---|---|
| 5 | 118 | 23.6 | 29.0 | 18.6 |
| 25 | 272 | 10.9 | 19.7 | 44.7 |
| 50 | 322 | 6.4 | 13.8 | 53.6 |
| 100 | 430 | 4.3 | 9.8 | 56.1 |
| 200 | 490 | 2.5 | 4.7 | 46.8 |
| 400 | 582 | 1.5 | 3.1 | 51.6 |
| 800 | 777 | 1.0 | 2.2 | 54.5 |

(Adapted from Darwinkel 1978)

ties, the number of ears per plant decreases progressively for two reasons: first, because of the acceleration of plant development at higher densities, the duration of the period of tiller appearance (between the first and terminal spikelet stages of the mainstem apex of wheat, Kirby and Appleyard 1982) is progressively reduced, with the result that the maximum number of tillers formed is also progressively reduced. Secondly, the intensified competition among stems for solar radiation and nutrients, especially after stem extension has begun (Fig. 6.8), can lead to progressive increases in tiller mortality (Table 6.3).

### 6.2.1.2 NUMBER OF GRAINS PER EAR

Increase in plant population density is, therefore, associated with a progressive decrease in the number of fertile tillers per plant up to populations of 600–800 $m^{-2}$, at which all the plants are uniculm (e.g. Table 6.3). Consequently, since, in a given crop stand, tiller ears tend to be smaller than mainstem ears (Darwinkel 1980a), it might be predicted that the number of grains per ear would increase with increasing density. However, as illustrated by Fig. 6.4, the converse is true, so that there are opposing trends in the number and size of ears with changes in plant population density.

In general, variations in management or in the environment can affect final ear size by their influence upon one or more of the following (Fig. 6.5):

(a) the rate of spikelet initiation (commonly expressed as the number of spikelets per day or per unit of accumulated temperature, °C day);

(b) the duration of spikelet initiation (first to terminal spikelet in wheat, or anther initiation in barley, although floret initiation in wheat can continue after terminal spikelet up to flag leaf appearance, Figs. 5.5 and 6.5);

(c) spikelet/floret mortality between terminal spikelet and anthesis; and

(d) the loss of potential grains due to failure of

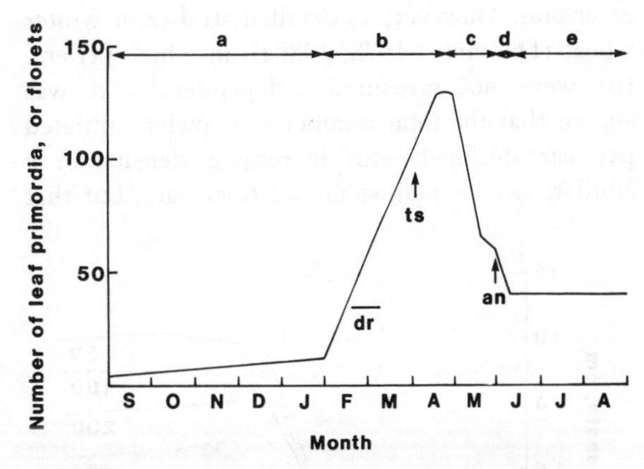


**Fig. 6.5** Schematic diagram of mainstem apical development in wheat, divided into five phases: a, leaf initiation; b, spikelet and floret initiation; c, floret death before anthesis; d, loss of florets during pollination; e, grain (fertilized floret) filling. The primordium formed at the intersection of a and b will differentiate into the collar of the future ear. dr, ts and an indicate the timing of the double ridge and terminal spikelet stages, and anthesis, respectively. Out of a total of 118 initiated florets, 40 survived as grains in the mature ear (adapted from Thorne and Wood 1982). Note that several investigations, particularly with winter varieties, have shown that the switch from phase a to b need not coincide with the point of inflection, as shown here.

pollination at anthesis (normally small, and can be neglected in most experiments).

The results from the few critical experiments which have investigated this phenomenon (e.g. Kirby and Faris 1970) indicate that: (a) is relatively unaffected by changes in spacing (although the process does begin slightly earlier in more closely-spaced crops) and that, in barley at least, the predominant effect is the reduction of (b), the duration of spikelet initiation, with increasing density (Fig. 6.6). Most of the relevant information comes from studies of mainstems only; although it is known that the duration of spikelet initiation is shorter in tiller ears, there are no reports of the effect of population density on the rate of initiation in tiller ears. The factors determining floret mortality are also little understood although it is thought to be related to floret age (i.e. the most recently-initiated are the most vulnerable, Whingwiri and Stern 1982) and enhanced by environmental stresses (drought, low temperature, and especially carbohydrate and nitrogen supply – sect. 5.2 and 6.4) during stem extension. However, in detailed studies of winter wheat (Darwinkel 1978, 1980a), in which (a) and (b) were not measured independently, it was shown that the total number of spikelets initiated per ear declined with increasing density in a similar way for mainstem and tiller ears, but that this trend was accompanied by increased spikelet mortality. Furthermore, at a given population density, spikelet mortality was greater in tiller ears than mainstems and, in general, floret mortality within fertile spikelets increased with increasing density, giving a decline in number of grains per fertile spikelet from 3.5 (at 5 plants $m^{-2}$) to less than 2 (800 plants $m^{-2}$). Overall, it is clear that the ears carried by tillers in less dense crops can be substantially larger than the mainstem ears in very dense, uniculm, crops.

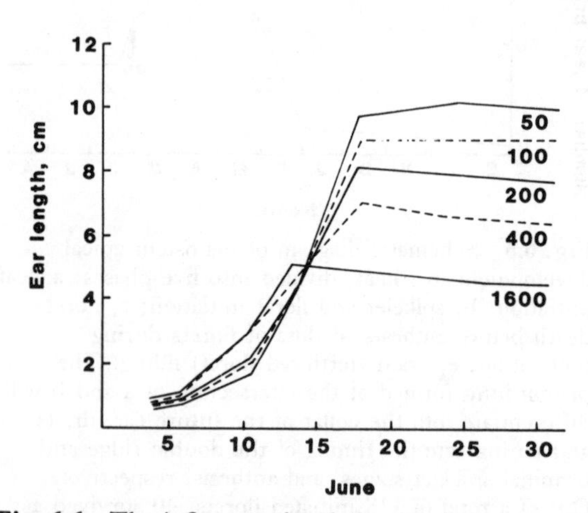


**Fig. 6.6** The influence of variation in plant population density from 50 to 1600 plants $m^{-2}$ on the length of the mainstem ear of barley plants cv. Proctor grown at Cambridge, UK, in 1969 (adapted from Kirby and Faris 1970).

#### 6.2.1.3 INDIVIDUAL GRAIN WEIGHT

Using older, taller varieties of spring barley such as Proctor, Kirby (1967) showed that individual grain weight declined sharply with increasing plant population density (Fig. 6.4). This effect was probably caused mainly by increased lodging since later studies with more modern cultivars have failed to confirm the response, suggesting that grain weight is relatively insensitive to plant density (e.g. Evans 1977). The pattern of response in winter wheat appears to be distinctly different, with modest, but statistically significant, increases in mean grain weight with density (Evans 1977) being explained by the observation that mainstem grains are heavier than those from tillers, even at relatively high population densities (McLaren 1981). There are, presumably, important differences between the responses of different cultivars to density, which may explain the finding of Darwinkel (1978) that individual grain weight in cv. Lely reached a maximum at a relatively low population density (44.6 mg at 50 plants $m^{-2}$), falling by more than 10 per cent to 39.0 mg when the stand was uniculm (800 plants $m^{-2}$). In this investigation, mainstem grains were heavier than tiller grains only at low population densities; the effect was reversed above 50 plants $m^{-2}$. A very similar trend was observed in a subsequent field experiment (Darwinkel 1980a).

In conclusion, it appears that the buffering of grain yield of barley and wheat, grown under northern European conditions, against variation in population density is the result of changes in ear population and ear size which are mutually compensatory and that changes in grain weight are of less importance, provided that the crop is not subject to lodging.

### 6.2.2 THE INFLUENCE OF SPATIAL ARRANGEMENT

Most of the results presented in the previous section were derived from experiments in which the plants were grown in small plots in precise, normally square, arrangements. In contrast, most cereal crops are drilled in parallel rows, 11–20 cm apart, within which the seeds are spaced in a random (Poisson) arrangement; the resulting clumping of seeds is demonstrated graphically by Soetono and Puckridge (1982). A series of theoretical analyses of plant stands (e.g. Pant 1979) has indicated that crop dry-matter yield can be maximized by employing uniformly-spaced, hexagonal or square planting patterns. However, in view of the insensitivity of the dry-matter and grain yields of temperate cereals to variation in population density over a wide range of values (e.g. Fig. 6.3), it would be predicted that such effects of planting pattern would be expressed only at low densities, below those generally achieved using commercial seed rates. This prediction has been confirmed in a number of small-scale investigations (e.g. Soetono and Puckridge 1982; Auld *et al.* 1983) although, as indicated by the latter authors, the effect of planting pattern could become significant in dry areas, where lower seed rates are more common.

The importance of more uniform spacing of cereal plants, permitting each plant to fulfil its tillering potential, has also been investigated on a field scale by comparing the yields of crops which have been precision sown (giving a regular rectangular arrangement of seeds, typically in rows 10 cm apart, 2.5 cm between seeds), broadcast (giving a more random, equally-spaced crop) or drilled in a conventional way. The results of a substantial number of experiments in the UK over twenty years (reviewed by Graham and Ellis 1980) are equivocal; in general, broadcast crops yielded as well as conventionally drilled crops grown under identical conditions, whereas precision sowing, a slower and more expensive process which requires ideal seedbed conditions, had a negligible influence on winter wheat yields but did result in yield increases of up to 9 per cent in high-yielding spring barley crops.

As shown by Roebuck and Trenerry (1978), conventional sowing practice gives a wide range of sowing depths, as well as spacings between seeds; for example, in a barley crop whose target sowing depth was 2–3 cm, 25 per cent of the seeds were deeper than 4 cm and 7 per cent were more than 5 cm from the settled soil surface. The range of depths will tend to be greater with unfavourable or variable soil conditions and uneven soil surfaces. In contrast, in precision-sown crops and in small-scale experiments, sowing depth is much more closely controlled. Experiments carried out in the East of Scotland have indicated that the yield penalty is normally modest even if an entire stand of spring barley is sown to a depth of 5–7.5 cm (Rodger 1982). However, where a crop is unevenly sown, so that emergence is spread over several days, there is evidence from studies of Australian wheat crops to suggest that the plants emerging first establish a dominance such that, making some allowance for spacing, later-emerging plants do not achieve their full yield potential (Soetono and Donald 1980; Knight 1983). In view of the prediction that spacing should not affect yield at higher population densities (Kemp *et al.* 1983), this seed depth effect could explain the higher yields obtained from some precision-sown barley crops.

### 6.2.3 OPTIMUM PLANT POPULATION DENSITY IN PRACTICE

Since cereal crops achieve maximum grain yield at relatively low plant population densities, in some cases below 200 plants $m^{-2}$, it is clear that any increase in seed rate above that necessary to reach this yield plateau serves only to increase production costs without increasing yield. On the other hand, the established plant population density which results from a given seed rate can vary considerably among seasons, soil types, fields, and even within a single field (Hubbard 1976; Sim 1982). Given a high level of crop management, it is reasonable to expect an 80 per cent establishment of sown seed but, even in experimental work, it is not unknown for establishment to be less than 25 per cent when soil conditions are very unfavourable or pests and diseases unusually prevalent (e.g. Cowton 1982). Consequently, in drawing up recommended seed rates for cereal growers, it is necessary to make an appropriate allowance for establishment losses to ensure that the resulting crop achieves the yield plateau. For example, seed rates of 400 $m^{-2}$ are usually adequate to achieve the optimum plant population for winter wheat grown in northern Europe (normally within the range 200–300 plants $m^{-2}$), although allowance must also be made for factors such as sowing date (Darwinkel 1980a; Barling 1982; Sim 1982; sect. 6.3). Finally, in spite of the increased interest in minimum plant population densities for maximum yield (sect. 6.6), recommended seed rates are still widely quoted in terms of kg $ha^{-1}$. In view of the fact that mean grain weight can vary between crops (Fig. 6.7), cultivars and seasons, a given recommendation expressed in this way can result in differences in true seed rate (seeds $m^{-2}$) of up to 100 per cent (Rodger 1982).

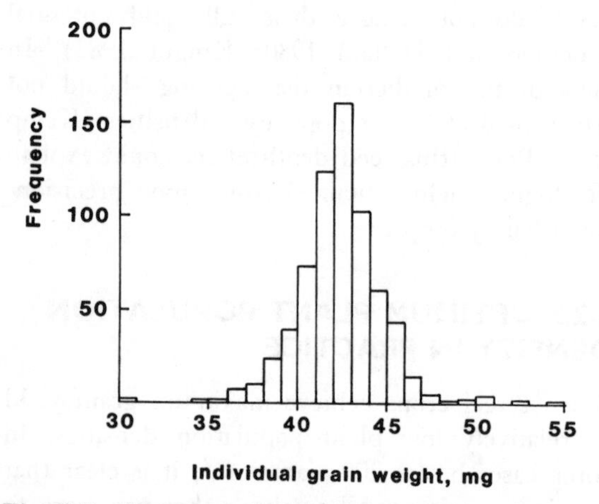


**Fig. 6.7** Frequency distribution of mean individual grain weights of samples of (potential) seed barley cv. Golden Promise submitted for testing to the Department of Agriculture and Fisheries for Scotland, season 1978. Mean 42.3 mg (from data provided by DAFS Agricultural Scientific Services, Edinburgh).

## 6.3 SOWING DATE AND YIELD

Of all the management aspects of growing a cereal crop (cultivar selection, seed rate, amount and timing of fertilizer, etc.), sowing date is probably the most subject to variation because of the very great differences in weather at sowing time between seasons even within the range of climates highly suited to arable agriculture. In principle, delay in sowing beyond a given date results in a progressive reduction in the potential yield of the crop because an increasing proportion of the available solar radiation will not be intercepted by the crop canopy. In practice, yield does normally decline with delay in sowing (e.g. yield penalties for winter cereals of the order of 20 per cent within the normal range of sowing dates, Green *et al.* 1985; Knop 1985) but, as discussed fully in section 6.3.3, the results of sowing date experiments can be highly inconsistent between seasons and sites; for example, it is not unusual for a relatively late sown crop to outyield the 'control' crop sown within what would be considered to be the optimum period.

There are several reasons for such inconsistencies and unexpected results. First, the soil conditions at different sowing dates will inevitably be different; in northern Europe, unfavourable conditions (excess or deficiency of soil moisture, serious incidence of disease etc.) can occur at almost any point during the normal range of sowing date for winter or spring crops. Consequently, the observed differences in the performance of cereal crops sown on different dates are commonly a reflection of differences in established plant population density. Secondly, crops sown at different dates pass through each developmental stage at slightly different times and, therefore, under different environmental conditions (especially photoperiod and temperature); thus any one of the developmental stages which determine the components of yield could conceivably occur under more (or less) favourable conditions in a later-sown crop. For these reasons, it is not easy to carry out a critical comparison of the grain yields (and their components) of the different crops in a sowing date experiment.

The interaction between sowing date and environmental factors is further complicated by the following phenomena:

(a) some cereal crops (in particular, most, but by no means all, of the winter wheat and barley cultivars originating at higher latitudes, e.g. northern Europe and Canada) have a specific requirement for a period of low temperature (vernalization) for floral initiation, followed by exposure to long daylengths, for full reproductive development; temperate spring wheats tend to require only long days for normal ear development, whereas most temperate spring barleys (Knop 1985), and cereal cultivars from lower latitudes, e.g. Mexico, India, Australia, require neither vernalization nor long days for the initiation of reproductive development.

(b) in addition to promoting reproductive development, lengthening days in spring are associated with the acceleration of all phases of plant development, at least in temperate cultivars (e.g. leaf appearance and extension, sect. 2.2). The resulting telescoping of development means that crops sown over a long period tend to achieve maturity within a few days or weeks (Table 6.4).

The quantitative aspects of the control of flowering by vernalization and photoperiod in winter cereals have been studied in greatest detail for winter rye cv. Petkus (Purvis 1961), and it is generally assumed that other cultivars and species behave in a broadly similar manner. This assumption may well be unjustified, but there are few comparable data for wheat and barley. Petkus rye plants can be vernalized most effectively by exposure to temperatures within the range 0 to 5 °C for periods of about 6 weeks, although considerably longer periods at temperatures as low as −4.5 °C and as high as 10 to 13 °C can also induce floral initiation (Hänsel 1953). Since there is no juvenile phase and the low-temperature treatment is perceived by the stem apex, vernalization can begin in the imbibed seed before germination is complete. Once vernalized, Petkus rye becomes a quantitative long-day plant, with no absolute requirement for long days, but flowering sooner the longer the daylength. However, in all cases, a minimum of eight leaves must be initiated before the apex can form spikelets whereas, in short days, a maximum of twenty-two leaves can be produced before the apex becomes reproductive.

In the simulation of the growth, phenology and yield of the winter wheat crop described in full in Chapter 9, it is proposed (on the basis of limited evidence) that temperatures between 3 and 10 °C are fully, and equally, effective in bringing about vernalization, and formulae are derived to compute the (lower) effectiveness of periods during which the temperature falls within the ranges −4 to 3 °C and 10 to 17 °C. The progress of vernalization can be followed by accumulating the number of days of vernalizing temperatures (vernal days) which the plant has experienced from the beginning of germination, taking into account the fact that high temperatures (>30 °C) result in devernalization. Weir *et al.* (1984) propose that 33 vernal days (the equivalent of 33 days continuously within the temperature range 3 to 10 °C) are required for the completion of vernalization, but that there is a threshold of 8 vernal days before the process can begin.

The overall effects of the requirement for vernalization, and the promotion of flowering by long days, on the phenology of winter wheat crops of different sowing date can be illustrated by the results presented in Table 6.4. In the case of the first two sowings (9 Sept., 5 Oct.), at least, vernalization would be achieved before the end of the year but, because of low temperatures (slowing down all the processes of development) and short photoperiods, double ridge did not occur until February or April (note that it is unusual even for crops sown in September in the south of England to achieve double ridge before January; Porter *et al.* 1987). For sowings 3 (21 Oct.) to 6 (14 Jan.), vernalization would be complete in time for the plants of each sowing to respond similarly to lengthening days, achieving double ridge within a period of 5 days, although the sowing dates were up to 3 months apart. This telescoping of development is even more marked for sowings 7 (7 Feb.) and 8 (9 Mar.) where the temperatures early in the life of the crop were low enough to give satisfactory vernalization, but development was accelerated by long days such that the dates of double ridge were only a few days later in May than that of the third sowing. To a certain extent, this 'catching up' by later sowings is achieved at the expense of leaf initiation (Kirby and Appleyard 1984b); the number of leaves per mainstem declined from fourteen at the first sowing to ten for the December sowing but there was no further penalty for later sowing, indicating that, under normal combinations of temperature and daylength, the minimum leaf number for the cultivar used was ten (Hay 1986).

The telescoping of the development of later sowings, under the control of photoperiod,

**Table 6.4** The phenology of winter wheat crops (cv. Maris Hustler) sown on nine dates in 1982–3*

| *Sowing date* | *Emergence* | *Double ridge* | | *Terminal spikelet* | | *Flag leaf* | | *Ear emergence* | *Anthesis* |
|---|---|---|---|---|---|---|---|---|---|
| 1. 9 Sept. | 21 Sept. (12 d) | 17 Feb. (161 d) | (9 l) | 22 Apr. (225 d) | (11 l) | 30 May (263 d) | (14 l) | 20 June (284 d) | 29 June (293 d) |
| 2. 5 Oct. | 21 Oct. (16 d) | 14 Apr. (191 d) | (9 l) | 13 May (220 d) | (11 l) | 30 May (237 d) | (13 l) | 23 June (261 d) | 29 June (267 d) |
| 3. 21 Oct. | 8 Nov. (18 d) | 2 May (193 d) | (9 l) | 21 May (212 d) | (11 l) | 7 June (229 d) | (13 l) | 27 June (249 d) | 5 July (257 d) |
| 4. 8 Nov. | 9 Dec. (31 d) | 6 May (179 d) | (8 l) | 21 May (194 d) | (9 l) | 7 June (211 d) | (11 l) | 27 June (231 d) | 5 July (239 d) |
| 5. 10 Dec. | 14 Jan. (35 d) | 6 May (147 d) | (7 l) | 21 May (162 d) | (8 l) | 7 June (179 d) | (10 l) | 29 June (201 d) | 7 July (209 d) |
| 6. 14 Jan. | 3 Mar. (48 d) | 6 May (112 d) | (6 l) | 25 May (131 d) | (8 l) | 17 June (154 d) | (101) | 3 July (170 d) | 11 July (178 d) |
| 7. 7 Feb. | 17 Mar. (38 d) | 13 May (95 d) | (6 l) | 25 May (107 d) | (8 l) | 23 June (136 d) | (10 l) | 5 July (148 d) | 13 July (156 d) |
| 8. 9 Mar. | 7 Apr. (29 d) | 23 May (75 d) | (6 l) | 7 June (90 d) | (8 l) | 1 July (114 d) | (101) | 9 July (122 d) | 15 July (128 d) |
| 9. 29 Apr. | 12 May (13 d) | 5 July[†] (37 d) | (10 l) | 15 July[†] (47 d) | (12 l) | 21 July[†] | | 2 Aug.[†] | 16 Aug.[†] |

* Values in parentheses indicate the number of days from sowing or the number of emerged leaves on the mainstem.
[†] Very variable reproductive development (see text).
(From Hay 1986)

continued throughout the growth of the crops with the result that, for plants sown at dates 6 months apart, the emergence of the flag leaf took place within 1 month and, for all the sowings, anthesis was complete within a period of 16 days. This confirms the common observation that the date of maturity and harvest of autumn-sown cereals is little affected by sowing date. Finally, the ninth sowing (29 Apr.) took place during a period of rising temperatures, unfavourable for rapid vernalization. Only about half of the plants achieved double ridge during July (more slowly than the March sowing), the remainder remaining vegetative. However, in contrast to this interpretation of plant response to variation in sowing date, which relies heavily upon the need for vernalization, Jones and Allen (1986) have proposed a scheme by which the same responses can be explained in terms of daylength alone.

In the absence of a requirement for vernalization, the effect of delay in sowing on the phenology of spring-sown cereals is a simple telescoping of development resulting in the near synchrony of crop maturity. For example, for spring wheats sown at different dates between 1 March and 17 April, the dates of achievement of successive developmental stages converged such that ear emergence in each case fell within a period of 20 days, and harvest dates would have been closer (Stern and Kirby 1979).

## 6.3.1. SOWING DATE AND THE COMPONENTS OF GRAIN YIELD

### 6.3.1.1 NUMBER OF EARS PER UNIT AREA

The acceleration of the development of late-sown crops means that the duration of each stage of development is progressively reduced with increasing delay in sowing. For example, the number of leaves per stem declines (Table 6.4) because the length of the period of leaf initiation decreases, and this, in turn, tends to reduce the maximum number of tillers initiated, which is broadly related to the number of leaf nodes (Kirby and Ellis 1980). However, this reduction in the number of potential ear-bearing tillers per plant does not necessarily lead to lower ear population densities. As discussed elsewhere in this chapter (sect. 6.2 and 6.4), the number of ears per unit area is also strongly dependent upon the established plant population density and upon the environmental conditions around the time of terminal spikelet, which determine the proportion of the tillers which will survive to bear ears (Fig. 6.11).

In practice, most studies have shown a fall in ear population density with delay in the sowing of both winter and spring cereals. For example, in a three-season investigation of winter barley sown in September, October and November, Harris (1984) found a progressive decline in ear population density with later sowing, owing to reductions in plant establishment and in the number of fertile tillers per plant (Table 6.5). Green *et al.* (1985) give similar results for winter wheat. However, in many cases, the effects are less clear-cut as, for example, in a study of three sowings of dual-purpose (i.e. reasonably hardy, but normally spring sown) barley cultivars, in late October, mid-March and late April (Kirby 1969). As a result of unfavourable soil conditions and subsequent winter mortality, plant population densities for the autumn-sown crops were 20 to 45 per cent lower than in the spring crops, but final ear populations were virtually identical for the October and March crops, and significantly lower for the April sowing, because the number of fertile tillers per plant decreased progressively with later sowing. It is not known whether this effect was the consequence of less prolific tillering or of higher tiller mortality. In general, spring cereals in the UK tend to give highest ear numbers when sown in early March, although higher populations can be recorded in late March to early April, with a progressive decrease with further delay in sowing (e.g. Jessop and Ivins 1970).

**Table 6.5** Demography of crops of winter barley cv. Igri sown at three different dates in southern England 1980–2. Values in brackets for fungicide-treated crops

| | *September* | *October* | *November* |
|---|---|---|---|
| Plants $m^{-2}$ | 178 (181) | 162 (136) | 144 (126) |
| Ears/plant | 4.7 (5.2) | 4.1 (4.7) | 4.1 (4.8) |
| Ears $m^{-2}$ | 839 (941) | 655 (633) | 591 (601) |

(From Harris 1984)

Overall, it is clear that the results obtained from experiments investigating the influence of sowing date upon the first component of grain yield are determined, first, by the success of crop establishment and, secondly, by the number of fertile tillers per plant, which normally decreases with delay in sowing (i.e. as the period available for plant development decreases). In most studies of sowing date effects, tiller demography has not been studied in sufficient detail to indicate whether tiller production or mortality is the major process in determining the number of tillers which survive to bear ears.

#### 6.3.1.2 NUMBER OF GRAINS PER EAR

Although the acceleration in the rate of crop development associated with increased plant population density, or with delay in sowing, means that the duration of the phase of spikelet initiation is reduced, the overall effects of these two management factors upon ear size are different. In the case of population density, the rate of spikelet initiation is relatively unaffected, with the result that ear size declines progressively with increasing seed rate (Fig. 6.6). In contrast, variation in sowing date is commonly found to have a negligible influence upon the number of grains per ear (e.g. Kirby 1969; Harris 1984), although Jessop and Ivins (1970) demonstrated a pronounced tendency towards larger ears at the latest of their three sowings of spring cereals.

This effect is at least partly explained by the fact that in cereal cultivars originating from a wide geographic range (Africa, Australia, North America, northern Europe), the rate of spikelet initiation increases continuously with photoperiod between 8 and 24 hours (e.g. Allison and Daynard 1976; Rahmann and Wilson 1977). Thus, in a detailed study of spring wheat growing in the UK, Stern and Kirby (1979) were able to show that, within the normal range of sowing dates, there was a remarkably precise inverse relationship between the rate and duration of spikelet initiation; the resulting stability of ear size over a range of sowing dates presumably holds only if daylength continues to increase throughout the period during which the successively sown crops become reproductive. Crops adapted to seasons in which the change of photoperiod is small or negative may respond differently to sowing date. However, it should be stressed that both spikelet and floret mortality after terminal spikelet can also affect the number of grains per ear (Fig. 6.5), and this may account for the substantial differences in ear size found by Jessop and Ivins (1970).

#### 6.3.1.3 INDIVIDUAL GRAIN WEIGHT

As a consequence of the convergence of development described in the previous sections, the start of grain filling is remarkably synchronous for crops sown over a wide range of dates (e.g. Table 6.4, where winter wheat crops sown on 9 Sept. and 10 Dec. reached anthesis within a period of 9 days) and, therefore, the conditions during grain-filling will tend to be very similar. However, because of other aspects of the acceleration of development (in particular, lower crop dry weight at anthesis), there may also be a tendency for later-sown crops to give lighter grains. Thus in most experiments carried out in the British Isles to investigate the influence of time of sowing within the normal range of dates, grain weight is either unaffected (e.g in spring cereals, Jessop and Ivins 1970; winter barley, Harris 1984; winter wheat, Green *et al.* 1985) or reduced by later sowing by up to about 10 per cent (e.g. in spring barley, Taylor and Blackett 1982; spring wheat, Angus and Sage 1980).

These generally modest effects tend to lend support to the proposal that the individual grain weight for a given cultivar is a relatively stable character (Gallagher *et al.* 1975). However, when delay in the start of grain-filling by a few days coincides with a rapid deterioration in the environment, much larger effects can be anticipated. Thus, for example, a range of spring wheat cultivars sown in late February gave individual grain weights which were, on average, only 8 per cent higher than those from late April sowings in 1975, but the difference was 18 per cent (and up to 28 per cent for one cultivar) in 1976, when the duration of grain filling was cut short by severe drought (Pearman *et al.* 1978). Similar effects can result from a heavy incidence of fungal infection or serious lodging.

In regions where the growing season is more clearly delimited than in the UK, for example by high temperatures or by the availability of soil moisture (e.g. India, Australia), delay in sowing can have a much greater influence on grain weight and can, in certain cases, result in the total loss of the grain yield. Such risks are most pronounced in areas where the climate is both marginal for the crop and highly variable from year to year (Hay 1981a). On the other hand, the optimum sowing period for winter cereals grown in areas where the winter is severe can also be relatively short (e.g. the last two weeks in August in Saskatchewan, Canada), because earlier or later sowing gives significantly lighter grains (Fowler 1983).

In summary, as far as northern European crops are concerned, grain yield generally declines with delay in sowing, principally as a consequence of decreases in ear population density, but also in some cases because of small decreases in individual grain weight. Where less consistent effects are recorded, these are normally the result of differences in established plant population density.

### 6.3.2 LATITUDE EFFECTS

The fundamental difference between the individual crops of a sowing date experiment is that, during the early stages of growth at least, plant development proceeds under different photoperiods and temperatures. As we have seen in Chapter 2 and the preceding sections of this chapter, variation in each of these environmental factors can have a profound influence on development and, although considerable progress has been made (Ch. 9), it is not yet possible to present a detailed interpretation of their interactions with one another and with other environmental factors. An alternative to the sowing date experiment is to grow cereal crops at different latitudes, but otherwise under identical management; in this way, the crops will be subject to different daylengths from emergence, although differences in temperature and other variables, such as rainfall, disease incidence etc., will depend upon the size of the difference in latitude as well as the particular pattern of weather during the test season.

In an unusually detailed investigation of this type involving spring barley grown at sites 4° of latitude apart (Cambridge, England, 52°11′N, and Edinburgh, Scotland, 55°53′N), the overall effects on grain yield were analogous to those resulting from variation in sowing date (Kirby and Ellis 1980; Ellis and Kirby 1980). Throughout the two growing seasons, 1976–7, photoperiod was longer at the northern site by up to an hour, but air temperatures were generally lower by 1–3°C and the rainfall was significantly higher. However, most of the differences between the crops could be explained in terms of temperature alone and it was not possible to identify any specific influence of the rather modest differences in daylength. Thus, the rate of primordium initiation was higher under the warmer conditions at the southern site, leading to the development of more potential spikelets; however, spikelet mortality was lower at the Scottish site such that the final numbers of grains per ear at harvest were very similar at the two sites (Table 6.6). It is proposed that this lower mortality was a consequence of lower rates of development and less intense intra-plant competition for assimilate after awn initiation, under cooler conditions, with similar receipts of solar radiation spread over longer days. Higher numbers of ear-bearing tillers per plant (leading to higher ear population densities) and heavier grains at the northern site (Table 6.6) were also attributed to the lower rate of development under cool conditions. Overall, an increase in latitude of 4° was associated with an increase in grain yield of between 15 and 120 per cent (Table 6.6). Similar effects of latitude on cereal development have been reported by Cottrell *et al.* (1985).

From an ecological point of view, it is clear that at least some modern spring barley cultivars bred for use in the UK are better adapted to the northern part of their range. Comparing these results with those from sowing date experiments (sect. 6.3.1), it can be concluded that, within limits, the longer the period available for crop development (early sowing, higher latitudes, cool temperatures, adequate water supply) or, alternatively, for the interception of solar radiation (sect. 6.6), the higher will be the potential yield.

### 6.3.3 OPTIMUM SOWING DATE IN PRACTICE

The year-to-year variation in plant establishment, pest and disease incidence, and in winter-kill makes it very difficult to predict optimum sowing dates for cereals on purely physiological grounds. In practice, recommended dates are normally drawn up from the results of long-running series of agronomic experiments, which can give mean sowing dates for highest yield together with a realistic estimate of expected yield penalties for each week of delay in sowing (e.g. for spring barley in Scotland, Rodger 1982). However, in accepting such guidelines, several reservations must be appreciated, in addition to the fact that use of the recommended date is not a guarantee of highest yield for that season. First, there can

**Table 6.6** The components of grain yield of crops of two cultivars of spring barley grown at two sites in the UK in 1976 and 1977

| | *1976* | | *1977* | | |
|---|---|---|---|---|---|
| | *Golden Promise* | *Maris Mink* | *Golden Promise* | *Maris Mink* | *SEM* |
| *Grain yield tDM ha$^{-1}$* | | | | | |
| Cambridge | 4.7 | 5.7 | 4.8 | 4.5 | 0.37 |
| Edinburgh | 6.8 | 6.5 | 10.5 | 10.6 | |
| *Ears m$^{-2}$* | | | | | |
| Cambridge | 764 | 934 | 748 | 722 | 13.6 |
| Edinburgh | 1022 | 986 | 1182 | 1187 | |
| *Grains per ear* | | | | | |
| Cambridge | 23 | 20 | 20 | 18 | 0.2 |
| Edinburgh | 24 | 22 | 23 | 21 | |
| *Individual grain weight, mg* | | | | | |
| Cambridge | 25 | 29 | 32 | 35 | 0.2 |
| Edinburgh | 28 | 31 | 38 | 42 | |

(From Ellis and Kirby 1980)

be very large differences in the pattern of response to sowing date among cultivars; for example, two winter wheat varieties (Norman and Moulin) tested at the Plant Breeding Institute, Cambridge, have shown optima at very different dates in a single season (29 Sept. and 19 March 1981, respectively) as well as considerable variation from year to year (e.g. 9 Sept. in both cases in 1982) (Bingham *et al.* 1983).

Secondly, because of the speed with which agricultural practice can change, it can happen that there exists insufficient agronomic information on which to base such recommended sowing dates. For example, owing to improvements in weed control and soil management, and the abandoning of traditional rotations, farmers in the UK now have the opportunity to sow winter cereals much earlier in September and even in August in some areas. Until about 1980, very few experiments had been sown before the last week of September, and in view of the considerable weed, drought, pest and disease (including virus disease) risks associated with early sowing in warmer soils, several seasons elapsed before it became possible to give realistic recommended sowing dates. Thirdly, the interactions between plant diseases and sowing date are not fully understood. For example, late-sown spring crops usually, but not invariably, develop more mildew than those sown nearer the optimum (Last 1957), as do winter crops sown too early in the autumn (Jenkyn and Bainbridge 1978). Last (1957) has proposed that increased susceptibility is associated with the faster development of late crops but, on the other hand, crops emerging in late autumn or early spring are exposed to little inoculum, and later-formed leaves are normally more resistant (Jones and Hayes 1971). Finally, in practice, sowing date is commonly determined by other factors such as the availability of seed; this problem is becoming more acute as the interval between harvest and sowing narrows, and may result in the favouring of cultivars with a very short duration of seed dormancy.

## 6.4 NITROGEN FERTILIZER AND YIELD

The growth, development and yield of cereal crops can be adversely affected by a deficient, or excessive, supply of any one of a series of essential macronutrients, as well as other toxic substances (Russell 1973; Hay 1981b). However, in intensive agriculture, soil and tissue levels of phosphorus, potassium, calcium, magnesium, sulphur and trace elements can usually be adjusted to fall within the optimal range, leaving nitrogen as the major nutrient factor determining crop yield. Nitrogen plays a central role in plant biochemistry as an essential constituent of cell walls, cytoplasmic proteins, nucleic acids, chlorophyll and a vast array of other cell components; consequently, a deficiency in the supply of nitrogen has a profound influence upon crop growth and can lead to a total loss of grain yield in extreme cases. Thus, in many fertilizer experiments where the control plots receive no fertilizer, the yield responses associated with the first increment of nitrogen are often simply the result of relief of severe nitrogen deficiency.

As far as crop physiology is concerned, nitrogen fertilizer influences cereal crops in four, largely interrelated, ways. First, as discussed at length in section 2.2, increased nitrogen supply, through its effects upon leaf size and longevity and upon tiller formation and survival, results in increases in the size and duration of the crop canopy (leaf area index and leaf area duration). In turn, these increases result in higher rates of crop dry-matter production since, in the absence of nitrogen deficiency, the level of nitrogen fertilization does not normally affect the rate of photosynthesis (but see sect. 3.4.5.2), although the mutual shading of leaves at high levels can give depressed net assimilation rates (Pearman *et al.* 1979). An example of the very close relationship which exists between crop nitrogen uptake and dry-matter production, indicating that increased uptake in spring tends to precede the acceleration of crop growth around growth stage 30, is given in Fig. 6.8.

Secondly, the amount and timing of nitrogen fertilizer treatments can also influence the development of the individual plants of the stand, with important implications for the components of grain yield (sect. 6.4.2). For example, the timing of application can be crucial in determining the proportion of tillers which survives to bear ears, whereas the supply of nitrogen at the end of the period of spikelet initiation, when the demand by the plant is increasing sharply (Fig. 6.8a), can affect spikelet and floret survival. Thirdly, the

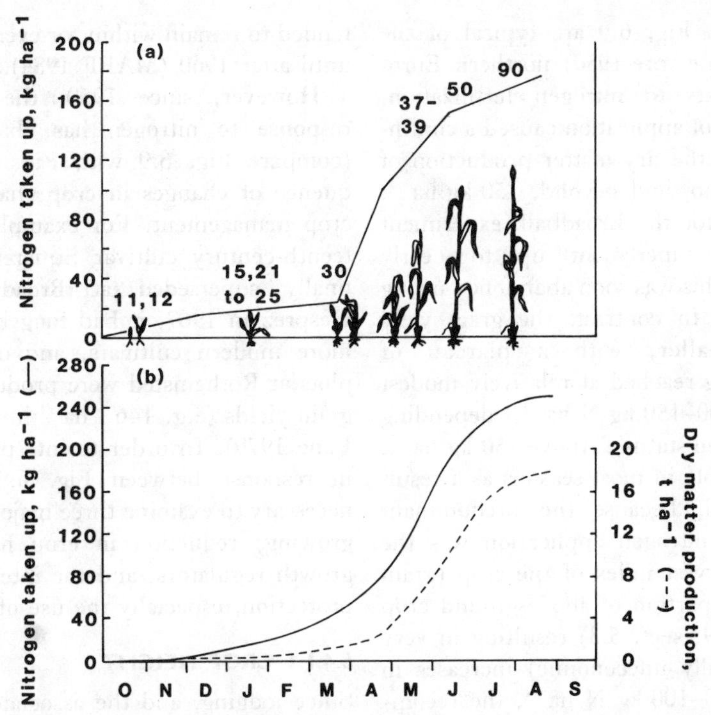


**Fig. 6.8** Relationships between the time course of nitrogen uptake by a wheat crop and (a) crop development, indicated by Decimal code (Table 6.1), and (b) crop dry-matter production (adapted from Widdowson 1979; Lidgate 1984). See also Fig. 5.5.

quality as well as the quantity of harvested grain is determined by fertilizer practice; for example, late and/or heavy nitrogen applications can result in barley grains which are unacceptable to the malting industry because of their high nitrogen/protein content. In contrast, West German farmers following the Schleswig-Holstein system of growing winter wheat apply nitrogen late for the express purpose of raising the nitrogen content of the grain, thereby ensuring a premium price (sect. 6.6). In general, there is an inverse relationship between grain yield and protein content, in spite of the heavier nitrogen applications required to achieve higher grain yield (Holmes 1982; Riggs 1984). Finally, the larger leaves and taller stems of heavily-fertilized crops are mechanically much weaker, leading to potential yield losses by lodging and various types of pathogenic attack.

## 6.4.1 DRY-MATTER PRODUCTION, GRAIN YIELD AND NITROGEN FERTILIZATION

The number of publications dealing with crop responses to nitrogen fertilizers is immense, but a large fraction of the available information is of little use in the physiological analysis of yield because it is derived from poorly-documented fertilizer trials in which grain yield alone was recorded. Furthermore, the formulation of generalizations from this mass of data is hampered by the fact that the pattern of crop response depends strongly upon the nitrogen status of the unfertilized soil, the type of cultivar used (e.g. tall traditional, semi-dwarf, profuse-tillering etc.), the management of the crop (timing of fertilizer, use of growth regulators, fungicides etc.), as well as the climate (e.g. determining losses of fertilizer nitrogen by leaching or denitrification).

The wheat yields harvested from the long-term Broadbalk experiment at Rothamsted provide a convenient baseline for a review of cereal crop responses to nitrogen, for a number of reasons; these include the infrequent, stepwise changes in crop management and cultivar, giving long periods of uniform data, and the very low nitrogen status of the unfertilized plots which have received no additional fertilizer or organic nitrogen for the last 150 years (Garner and Dyke 1969). The results presented in Fig. 6.9 are typical of the response of older (i.e. pre-1950) northern European wheat cultivars to nitrogen fertilization. Increasing the level of application caused a continuous stimulation of the dry matter production of the whole crop up to, and beyond, 150 kg ha$^{-1}$. (The original plan for the Broadbalk experiment included a further increment up to nearly 200 kg N ha$^{-1}$ but this was soon abandoned owing to severe lodging.) In contrast, the grain yield response was smaller, with a plateau of 2.8–3.0 t ha$^{-1}$ being reached at relatively modest fertilizer levels (100–150 kg N ha$^{-1}$, depending upon the soil nitrogen status). Above 150 kg ha$^{-1}$, grain yield fell sharply in most seasons as a result of lodging. Overall, because the predominant effect of increased nitrogen application was the reduction in the harvest index of the crop (grain dry weight as a proportion of above-ground crop dry weight; Fig. 6.9; sect. 5.3) resulting in very modest (and generally uneconomic) increases in grain yield above 50–100 kg N ha$^{-1}$, the recommended levels of fertilizer nitrogen application

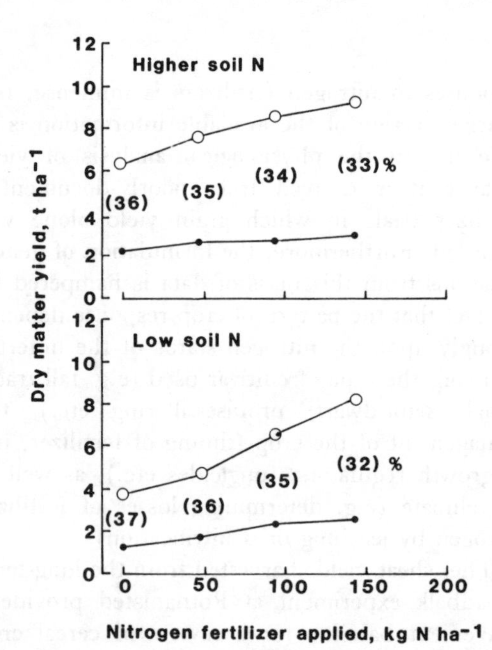


**Fig. 6.9** The influence of nitrogen fertilization on the total dry-matter (○) and grain (●) yields of crops of winter wheat cv. Squareheads Master grown in soils of differing nitrogen status at Rothamsted. The points represent mean values over thirty-one seasons (1935–64) and the values in parenthesis indicate mean harvest index (adapted from Garner and Dyke 1969).

tended to remain within, or even below, this range until after 1960 (MAFF 1984).

However, since 1950, the pattern of crop response to nitrogen has changed significantly (compare Fig. 6.9 with Fig. 6.10) as a consequence of changes in crop characteristics and in crop management. For example, when the nineteenth-century cultivar Squareheads Master was finally superseded at Broadbalk by Capelle Desprez in 1967, it had long been outclassed by more modern cultivars, and other experimental plots at Rothamsted were producing much higher grain yields (e.g. 4–6 t ha$^{-1}$ in 1962, Benzian and Lane 1979). In order to interpret the differences in response between Fig. 6.9 and 6.10, it is necessary to examine three major changes in cereal growing: reduction in crop height, the use of growth regulators, and the intensification of crop protection, especially the use of fungicides.

#### 6.4.1.1 CROP HEIGHT

Since lodging, and the associated grain losses by reduced leaf area duration, variation in grain maturity, increased disease incidence and predation by pests, set a limit to the quantity of nitrogen which could be applied to older cereal varieties, one aim of plant breeders over the last 30 years has been to generate cultivars with shorter, stronger stems. The success of this approach in the UK can be illustrated by Table 6.7 which shows that, by incorporating dwarfing genes (e.g. from the Japanese Norin 10 cultivar), the straw length of winter wheat crops has decreased by up to 50 per cent since Little Joss was first introduced in 1908. Similarly, of twenty-four spring barley cultivars included in the Scottish Colleges' cereal variety trials in 1983, none was classed as tall (> 1 m), twenty-three as semi-dwarf (0.7–0.9 m) and one, Tweed, as dwarf (0.5–0.7 m) (COSAC 1984), whereas most of the material bred before 1950 exceeded 1 m in height (Riggs *et al.* 1981).

As it has turned out, the improvements in harvest index associated with the programme of breeding for shorter stature (e.g. 0.3 to 0.45 for wheat, Table 6.7) (Riggs 1984) have probably been more important than the reduction of lodging, because they can act to give higher grain yields at all levels of fertility. Furthermore, the harvest index of a given cultivar appears to be a

**Table 6.7** Characteristics of a range of winter wheat cultivars

| *Cultivar* | *Date of introduction* | *Stem length to base of ear (cm)* | *Relative grain yield* | *Harvest index (%)* |
|---|---|---|---|---|
| Little Joss | 1908 | 130 | 100 | 30 |
| Holdfast | 1935 | 112 | 94 | 31 |
| Maris Widgeon | 1964 | 115 | 122 | 34 |
| Maris Ranger | 1966 | 88 | 130 | 39 |
| Maris Huntsman | 1972 | 95 | 148 | 40 |
| Maris Kinsman | 1975 | 82 | 145 | 38 |
| Maris Hobbit | 1977 | 67 | 166 | 45 |

(From Austin 1978; Austin *et al.* 1980b)

conservative characteristic under modern crop management, varying little even under stressful conditions (Biscoe and Willington 1984a). In recent studies of new wheat and barley cultivars, harvest indices of up to, and above, 0.5 have been recorded (Dyke *et al.* 1983), but it is important to emphasize that over the last 100 years of cereal breeding there has been little, if any, improvement in total potential dry-matter production (Austin *et al.* 1980b; Riggs *et al.* 1981). Indeed, Chevalier spring barley (nineteenth-century variety), grown with modern intensive methods and prevented from lodging, produced more dry matter than most of the more modern barleys studied by Riggs *et al.* (1981), including Plumage (1900), Spratt Archer (1933), Proctor (1953), Golden Promise (1966) and Triumph (1980), but its grain yield was about 30 per cent lower than that of the most modern cultivars. Because of the need to support the ear physically and to provide it with photosynthate from the leaf canopy, there must be an upper limit to cereal crop harvest indices, estimated by Austin (1980) to be between 0.6 and 0.65; since this is the extreme theoretical limit, it seems unlikely that the increasing yield of cereals can be maintained much longer by increases in harvest index, and that breeding programmes in the future will have to be much more directed towards improvement in total dry-matter production, while maintaining the high harvest indices already achieved (MAFF 1984).

The cause of these recent improvements in harvest index is not necessarily a simple relocation of assimilate to give, for example, more grains per ear or heavier grains; in the case of semi-dwarf wheats, incorporation of the Rht2 gene from Norin 10 has resulted in relatively consistent increases in the number of grains formed per spikelet (Gale 1979), but in barley the improvement is commonly due to increased numbers of fertile tillers per plant and, in both species, the increase in the grain yield and harvest index of any given cultivar can be caused by changes in any of the three components of grain yield, alone or in combination (Austin *et al.* 1980b; Riggs *et al.* 1981). Finally, in the context of this section, it is important to note that, in the absence of lodging, the optimum level of nitrogen fertilization tends to be similar for more modern (i.e. post-1950) short- and long-stemmed cultivars, even though the actual grain yields may differ considerably (e.g. Pearman *et al.* 1978).

#### 6.4.1.2 CROP GROWTH REGULATORS

Because the breeding of cultivars with improved characteristics such as shorter, stronger stems is very costly in terms of time and resources, alternative approaches to the control of lodging have been investigated, leading to the development of commercial plant growth regulators. Since stem extension is a plant response to increased levels of endogenous gibberellins, interest has concentrated upon substances which can interfere with their biosynthesis or action. Of the very many chemical species which have been screened for activity, only the following three have, so far, been developed for use in practical agriculture:

(a) Chlormequat (manufactured in different formulations with additives under a variety of trade-names). As a result of wide-ranging trials throughout Europe and elsewhere

(Humphries 1968; Woolley 1982), the use of chlormequat is now an established part of the husbandry of intensively-managed wheat crops in the UK, if taller cultivars are grown and there is a serious risk of lodging (i.e. high-fertility sites in the wetter arable areas).

(b) Ethephon, which influences stem extension by the release of ethylene, is not widely used alone, but more commonly in formulations which also contain:

(c) Mepiquat chloride, which has been developed much more recently, especially for use in barley crops, which do not, in general, respond well to chlormequat treatment.

The effects of these growth regulators on crop height and the incidence of lodging are very well documented, and especially well for chlormequat (Thomas 1982). Treatment of wheat at the appropriate growth stage (commonly before growth stage 31, one node detectable, Table 6.1) gives a consistent reduction in crop height and in the incidence of lodging, if it occurs. However, the height reduction achieved depends strongly upon the level of nitrogen applied and upon the cultivar used. For example, Lovett and Kirby (1971) reported very modest changes in the height of semi-dwarf wheat plants treated with chlormequat. From the very large number of trials carried out to investigate grain yield effects, the general conclusion appears to be that consistent increases in the yield of wheat crops occur mainly when lodging is prevented, thus permitting the crop to receive higher applications of nitrogen (e.g. the long series of experiments carried out by ADAS in England and Wales, reviewed by Woolley 1982) (Herbert 1983). The fact that beneficial results are generally obtained only when control plots suffer lodging indicates that plant growth regulators do not affect the total dry matter production or the harvest index of the crop (see previous section on crop height); however, there do exist dependable reports of modest grain yield increases in the absence of lodging (e.g. Humphries *et al.* 1965; Woolley 1982) which are commonly the result of increased tiller and spikelet fertility (Cartwright and Waddington 1982). Overall, since gibberellins take part in most plant development processes and the precise timing of the treatments will inevitably tend to vary between experiments, it is perhaps not surprising to find that results can be inconsistent, between sites, crops and years (Child *et al.* 1983). Furthermore, it is probably still too early to gain any useful perspective of the value of mepiquat chloride/ethephon treatment of barley crops, since the results appear to be less consistent and the chemical is much more expensive than chlormequat (Paterson *et al.* 1983; Harris 1984).

#### 6.4.1.3 FUNGICIDES

In recent years, a number of factors have combined to increase the importance of fungal diseases in determining grain yield; these include the expansion of continuous monoculture of cereals, the evolution of fungal strains which are resistant to specific fungicides, the genetic plasticity of important fungal pathogens which can rapidly overcome new resistance genes (Wolfe and Schwarzbach 1978), the commercial value of cultivars like Golden Promise spring barley (for malting) which are particularly susceptible to disease, and the general relief of other constraints. In addition, the high plant population densities and high rates of nitrogen fertilization of intensively-managed crops promote the development and spread of fungal diseases by providing a suitably humid microclimate and mechanically weaker, more susceptible leaf tissues (Jenkyn 1977). Although the economically important diseases of cereals caused by biotrophic fungi (especially mildews and rusts) have been shown to affect many aspects of crop physiology, development and growth, their most important effect is the reduction of leaf area index and duration (thereby effectively reversing the effects of nitrogen fertilization) (Walters 1985). Yield losses in excess of 10 per cent are not uncommon in unprotected crops.

Consequently, the use of fungicides, applied either to the seed (mainly systemic) or to the crop stand by spraying, has become standard practice among cereal growers, so that they can obtain the maximum benefit from the nitrogen fertilizer applied. For example, it has been estimated that, in 1982, nearly 80 per cent of all winter cereal crops grown in the UK received at least one foliar fungicide spray (Cook 1984). Originally, fungicides were applied whenever the disease achieved a threshold incidence within a given crop but, increasingly, prophylactic treatments at pre-arranged dates or stages of development (whether the disease is present or not) have become part of cereal crop-growing systems. This approach can give very substantial increases in grain yield in certain years (e.g. Harris 1984) but, in view of the cost of fungicides, the tendency in the future will probably be towards managed disease control or the development of more reliable disease-forecasting techniques (Cook 1984); the continuing development of resistant cultivars will presumably be a feature of the former approach, for use in mixtures of cultivars of the same species but of differing resistance to a given pathogen.

Together, these improvements in harvest index, lodging control and crop protection have led to increased grain yield at a given level of nitrogen fertilizer application and to increases in the quantity of fertilizer that can be applied to a crop to give a positive grain yield response. These changes can be illustrated by comparison of the response of the Squareheads Master crops (Fig. 6.9) with those of two more recent winter wheat cultivars (Capelle Desprez, Maris Fundin) grown at Rothamsted in 1975 using modern management systems (Fig. 6.10). Two major differences are apparent: first, changes in harvest index have caused at least a doubling of grain yield at each nitrogen fertilizer level up to 150 kg N $ha^{-1}$ and, secondly, because of improved control of lodging and crop disease, larger quantities of nitrogen were applied in the expectation of achieving a positive yield response, in the case of the semi-dwarf Maris Fundin cultivar in Fig. 6.10 up to 180 kg $ha^{-1}$. Naturally, crop responses to nitrogen, even under modern management, can vary widely among sites and years, according to the prevailing weather and soil fertility. This problem is considered in detail in section 6.4.3. However, the major features of Fig. 6.10 (a near-linear response of grain yield to increasing nitrogen treatment from 0 to 150 kg $ha^{-1}$ followed by a somewhat unstable yield plateau up to about 200 kg N $ha^{-1}$; George 1984) are common to many intensively-managed winter cereal crops (Boyd *et al.* 1976); spring cereals give similar patterns of response at lower nitrogen levels (sect. 6.4.3). Overall, in farming practice in the UK at least, modern crops are capable of benefiting from at least 50 kg $ha^{-1}$ more nitrogen fertilizer than crops growing 30 years ago.

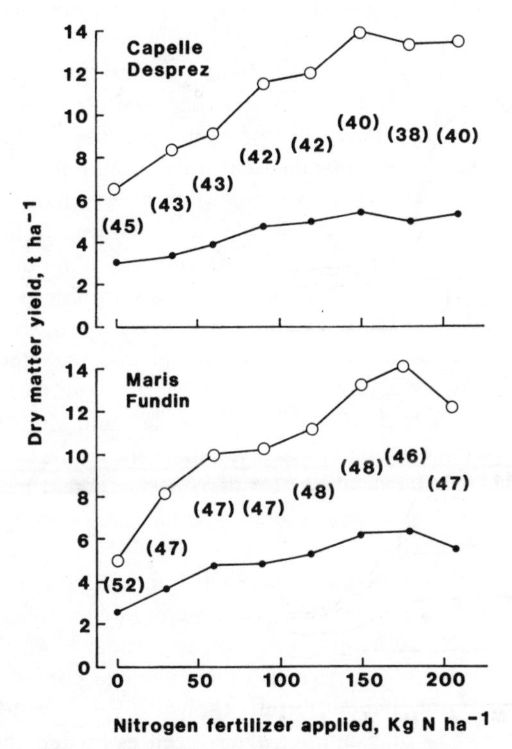


**Fig. 6.10** The influence of nitrogen fertilization on the total dry-matter (○) and grain (●) yields of crops of winter wheat cv. Capelle Desprez and Maris Fundin grown at Rothamsted in 1975. The values in parenthesis indicate mean harvest index (adapted from Pearman *et al.* 1978).

## 6.4.2 NITROGEN FERTILIZER AND THE COMPONENTS OF GRAIN YIELD

Because of these changes in cereal crop characteristics, their management and their responses to nitrogen fertilization over the last 30 years, the following discussion of grain yield components is, for the sake of clarity, largely confined to work carried out since 1975.

### 6.4.2.1 NUMBER OF EARS PER UNIT AREA

Nitrogen fertilization of cereal crops causes increases in tiller population density and/or tiller fertility, with the overall effect determined by the rate and timing of the application (Fig. 6.11; Simons 1982). Consequently, with few exceptions, increased nitrogen application gives increased ear population density at harvest (e.g. for spring barley, Table 6.8) usually across the full range of normal experimental rates (0–200 kg N $ha^{-1}$; Pearman *et al.* 1978). Indeed, when working with an irrigated wheat crop in West Australia, Whingwiri and Kemp (1980) observed a positive response between 100 and 300 kg $ha^{-1}$.

The sharp rise in the requirement of a cereal crop for nitrogen just before stem extension (Fig. 6.8) suggests that the considerable variation in the response of ear population density to nitrogen among seasons and sites (e.g. Table 6.8) might be the consequence of variation in the timing of nitrogen application and availability in relation to crop demand. This idea has been investigated by means of a variety of experiments in which both the timing and the seasonal distribution of fertilization was varied (e.g. comparison between winter cereal crops receiving the same quantity of fertilizer at different growth stages, with or without autumn fertilizer, e.g. Ellen and Spiertz 1980). In winter cereals, provided that the crop has a sufficient nutrient supply from soil reserves or from autumn fertilization for the

**Table 6.8** The effect of nitrogen application on the components of grain yield of crops of spring barley (5 cultivars) grown for three seasons in Northern Ireland. Results expressed as the per cent difference between 0 and 80 kg N $ha^{-1}$

| | *No. of ears* $m^{-2}$ | *No. of grains per ear* | *Individual grain weight* |
|---|---|---|---|
| 1972 | +21 | +5 | −19 |
| 1973 | +37 | +5 | −3 |
| 1974 | +24 | +8 | −2 |
| Three-season mean | +26 | +7 | −7 |

(From Lynch *et al.* 1979)

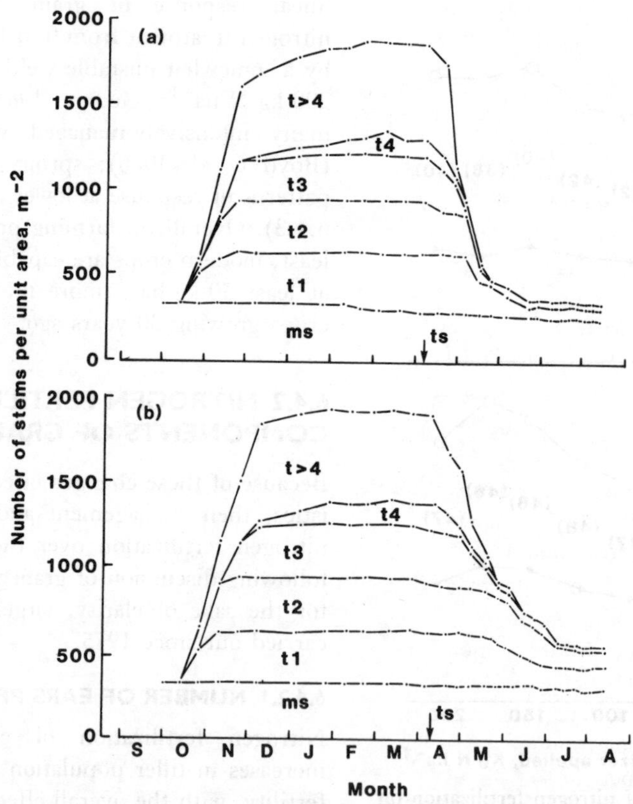


**Fig. 6.11** Time courses of stem population density of crops of winter wheat cv. Avalon grown in SE England at different levels of nitrogen fertilization: (a) 90 kg N $ha^{-1}$; (b) 330 kg N $ha^{-1}$. ms refers to the mainstem; t1 to t4, tillers at successive nodes; t >4, tillers at higher nodes; and ts indicates the timing of the mainstem terminal spikelet stage. The final numbers of fertile tillers were (a) 389 and (b) 526 $m^{-2}$ (adapted from Biscoe and Willington 1984b).

normal proliferation of tillers by early spring, the most important factor determining final ear population density appears to be tiller mortality during the period just before, and including, the achievement of crop growth stage 31 (GS31; Table 6.1) and mainstem apex terminal spikelet (Fig. 6.11; Baker and Gallagher 1983; Hay 1986).

In several cases, this mortality has been reduced by accurately-timed application of fertilizer, presumably because terminal spikelet marks the beginning of the most rapid phase of crop growth, when intra-plant competition for nitrogen is at its greatest (Kirby and Appleyard 1984b; Fig. 6.8). Thus, in a winter wheat experiment where half of the applied nitrogen was given at GS2, the highest number of ears per unit area was achieved by applying the remainder at GS2 or 30. There was a modest penalty to pay for delay up to GS32, which increased to 5–10 per cent if the additional nitrogen was applied at GS39 or 61 or withheld altogether (Darwinkel 1980a,b). Furthermore, the penalties for delayed fertilization were increased by mildew infection. In a subsequent experiment in which very high levels of nitrogen were used (210 kg N $ha^{-1}$), a pronounced optimum at or before GS30 was observed, although the response to the timing of additional fertilizer did vary considerably among the three cultivars studied (Darwinkel 1983). Similar results have been obtained by Biscoe and Willington (1984b) and by George and Skinner (1984). However, to place these findings in perspective, it should be emphasized that, in practice, few farmers in the UK apply nitrogen later than GS30–31, except for the purpose of increasing the nitrogen content of the grain.

In spring cereals, where development and growth are more rapid, the demand for nitrogen is high from crop emergence onwards, and the risk of nitrogen losses by leaching is much less than with winter cereals, there would appear to be much less scope for increasing ear population density by the timing of fertilization. This is confirmed by a range of agronomic trials which have shown no economic benefit to be gained from the splitting or delaying of nitrogen application, except, perhaps, in the case of very early-sown crops (sect. 6.4.3). Similarly, in a detailed physiological study of the effects of variation in the timing of nitrogen application to spring barley, the first component of grain yield was relatively unaffected by application date over a range of 70 days from sowing (Fig. 6.12).

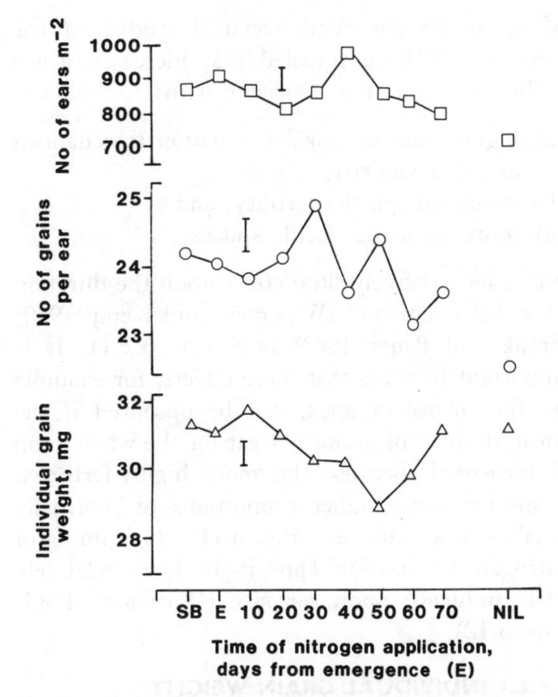


**Fig. 6.12** The influence of time of application of nitrogen fertilizer on the components of the grain yield of crops of spring barley cv. Midas grown for three seasons (1978–80) in Northern Ireland. In each case, the fertilizer applied was split between a seedbed treatment (SB) and a later application (total 60–75 kg N $ha^{-1}$). The results without nitrogen fertilization are entered under 'NIL', and the vertical bars indicate LSD at $p = 0.001$ (from Easson 1984).

#### 6.4.2.2 NUMBER OF GRAINS PER EAR

The influence of nitrogen fertilization on the number of grains per ear is also normally positive, but the effects tend to be smaller than for ear population density (e.g. for spring barley, Table 6.8; winter wheat, Pearman *et al.* 1978; Blacklow and Incoll 1981) and may be not statistically significant, or even negative in a few cases (Evans 1977). These differences are probably the result of interactions between cultivar and nitrogen effects; for example, Pearman *et al.* (1978) found that, in Capelle Desprez winter wheat, ear size was relatively unaffected by nitrogen application across the range 0 to 210 kg N $ha^{-1}$, whereas, in Maris Fundin, the same treatments gave increases of up to 75 per cent. Detailed studies of ear development have revealed that increased availability of nitrogen is associated with:

(a) higher rates of spikelet initiation (mechanism as yet unknown);
(b) improved spikelet fertility; and
(c) more grains per fertile spikelet;

but it has relatively little effect upon the duration of spikelet initiation (Whingwiri and Kemp 1980; Frank and Bauer 1982) (see sect. 6.2.1). It is important to stress that these effects, for example on the mainstem apex, can be obscured if the mean number of grains per ear for the whole crop is measured, because the more highly-fertilized crops can have higher proportions of (normally smaller) tiller ears (e.g. Fig. 6.11). The timing of nitrogen fertilization appears to have relatively little influence upon ear size (Darwinkel 1983; Fig. 6.12).

#### 6.4.2.3 INDIVIDUAL GRAIN WEIGHT

Individual grain weight at harvest is determined by the supply of assimilate, from current photosynthesis or from storage, during the period of grain-filling, from just after anthesis to maturity. Since nitrogen fertilization causes increases in dry-matter production and leaf area duration, it might be expected that cereal grains would tend to be heavier with increasing applications of nitrogen. However, by the time of anthesis, the level of fertilization has already determined the number of grains per unit area, by the degree of stimulation of tiller fertility, spikelet initiation and floret fertility; since grain population density can rise very steeply with nitrogen fertilization, the ultimate grain weight will depend upon source/sink relationships (Gallagher *et al.* 1975) and, in particular, upon the leaf (and green ear) area duration per grain (measured by very few investigators). Individual grain weight, therefore, appears to be determined primarily by the duration rather than the rate of grain-filling (sect. 5.2). As an extreme example, a heavily-fertilized crop, whose canopy senesces rapidly after anthesis because of drought, could give a reasonable overall grain yield but which was made up of a large number of very light grains.

The complex problems of source/sink relationships during grain-filling are treated in much greater detail in Chapter 5, but it is clear even from a superficial analysis that individual grain weight can be increased, reduced or unaffected by nitrogen fertilization, according to the grain population density achieved and the degree of environmental stress (drought, disease incidence, lodging etc.) determining the duration of grain-filling after anthesis. The picture is further complicated by the widespread use of grain weights, averaged over the entire crop, which can mask treatment effects because the more highly-fertilized crops will tend to carry a higher proportion of lighter grains on subsidiary tiller ears (Whingwiri and Kemp 1980; Fig. 6.11).

In practice in northern Europe, cereal crops grown using modern cultivars and management systems (absence of serious lodging, efficient crop protection) show either modest reductions in individual grain weight with increased nitrogen application (e.g. Pearman *et al.* 1978; Table 6.8) or, more commonly, relative stability of grain weight (Evans 1977; Spiertz 1980; Blacklow and Incoll 1981; Hay 1982), although there are well-documented cases of much greater reductions under conditions of stress (e.g. Table 6.8, 1972). Several workers have described important interactions between the effects of nitrogen and those of cultivar (Pearman *et al.* 1978) or disease incidence (Spiertz 1980), whereas the timing of fertilization appears to have a relatively small effect upon individual grain weight, but a very marked influence upon grain nitrogen content (Darwinkel 1983; Fig. 6.12). In contrast, there is evidence from Australia, where wheat is grown at lower population densities and with higher inputs of radiant energy, that nitrogen fertilization can cause significant increases in grain size (Whingwiri and Kemp 1980).

### 6.4.3 NITROGEN FERTILIZATION IN PRACTICE

In most arable areas of northern Europe, the application of nitrogen fertilizer is now the most important management factor in determining grain yield, giving, for example, yield responses of up to 150 per cent at Rothamsted (Fig. 6.10). It is, therefore, important to be able to predict, with some degree of accuracy, what the optimum level of fertilization for a given crop will be. This can be extremely difficult for a number of reasons. First, the farmer is interested in maximizing the economic, rather than the biological, yield of his crop; this requires a sound knowledge of grain yield response to nitrogen fertilizer which, for a given site, can either be similar to the ideal curve of Fig. 6.10 or, according to the fertility of the soil (previous crop), soil type and depth, and the recent weather (degree of leaching of nitrate by rain), quite different. For example, *loss* of potential yield even at the first increment of fertilizer is not uncommon (Batey 1976).

A wide range of possible crop response curves is presented in MAFF (1984). Since the cost of fertilizer treatment increases linearly with increased application and the grain yield response is usually curvilinear, the maximum economic grain yield is almost always lower than the maximum biological yield, as is the level of nitrogen application at which it is achieved (Fig. 6.13). Calculation of the economic maximum is further complicated by the fact that, at higher nitrogen levels, it becomes necessary to use fungicides and growth regulators so as to guarantee the maximum beneficial effect of the applied fertilizer. The most economic level of nitrogen application will therefore vary from season to season according to the price of the harvested grain and the cost of the fertilizer and other agrochemicals (Whitear 1976). For example, Murphy (1984) estimates that, for winter wheat grown intensively in Eastern England in 1982/3, the economic optimum yield was between 6.5 and 7.5 t $ha^{-1}$, well below the highest achievable yields of around 10 t $ha^{-1}$. Overall, there is an increasing tendency among farmers achieving high yields to estimate the yield potential of a given site and to set the fertilizer application at a level which will ensure that the crop receives at least the minimum quantity of nitrogen necessary for the estimated yield.

Secondly, having decided how much fertilizer to apply to the crop, it is then necessary to decide how the application should be distributed through the growing season to give maximum effect at the chosen level of application. This involves an understanding of the potential losses of nitrogen, especially by leaching and denitrification, during winter and spring (e.g. Selman 1983), as well as the seasonal variation in crop requirement for nitrogen (Fig. 6.8). There is some evidence that yields of winter cereals can benefit from the application of nitrogen at precise growth stages in

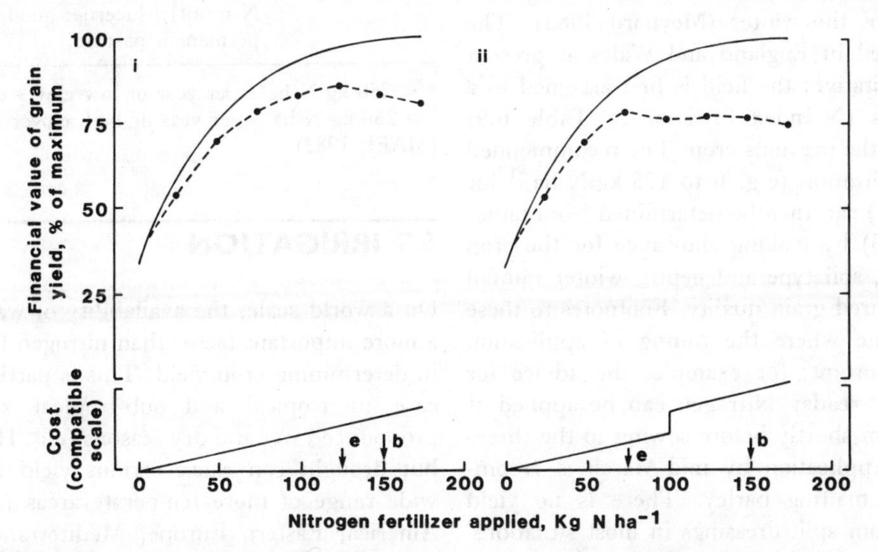


**Fig. 6.13** Model diagrams to illustrate the difference between maximum biological (——) and maximum economic (---) grain yield of a cereal crop, both expressed in financial terms, by taking into account the cost of applying fertilizer nitrogen. In (i), the cost of fertilizing the crop increases linearly with increased application, so that the application for maximum economic yield (e) falls about 25 kg N $ha^{-1}$ below that required to maximize biological yield (b). This difference increases considerably (ii) if fungicide and growth regulator treatments are necessary to secure the benefit of nitrogen application above 100 kg N $ha^{-1}$. See text for a fuller account.

spring (see above), and the timing of spring nitrogen application according to crop development is well established in France and other Northern European countries (e.g. Meynard 1983). However, these ideas have not yet been incorporated into recommended growing systems for the UK, and widespread adoption of such procedures can prove difficult because soil conditions are not always favourable for mechanized traffic at the recommended date every year, and farmers need to acquire a deeper understanding of crop development. Thirdly, bearing in mind the fact that the price received for malting barley or milling wheat will depend upon the protein content of the grain, it is important to plan nitrogen applications (quantity and timing) to maximize the quality of the grain; this can involve further sacrifice of biological yield in the pursuit of economic yield.

The primary difficulty in devising a programme to solve these problems is how to predict the potential of a given soil to supply nitrogen to the crop at different times during the growing season (MAFF 1984). In France, spring applications of nitrogen to winter cereals are based upon prediction equations which rely upon measurements of the mineral nitrogen present in the rooting zone at the end of the winter (Meynard 1983). The approach used in England and Wales at present is less quantitative; the field is first assigned to a fertility class (N-Index 0, 1 or 2, Table 6.9) according to the previous crop. The recommended level of fertilization (e.g. 0 to 175 kg N $ha^{-1}$ for winter wheat) can then be determined from tables (MAFF 1983) by making allowance for the crop to be grown, soil type and depth, winter rainfall and the required grain quality. Footnotes to these tables indicate where the timing of application may be important; for example, the advice for spring barley reads: 'Nitrogen can be applied at any time from shortly before sowing to the three-leaf stage. Application by mid-March is recommended for malting barley. There is no yield advantage from split dressings in most situations' (MAFF 1983). The Scottish Colleges of Agriculture use a similar, but simpler, system; for example, the advised level of fertilizer for spring barley in 1982 varied between 60 and 130 kg N $ha^{-1}$ according to the previous crop (Rodger 1982). In view of the substantial research investment in measuring and simulating the various components of the nitrogen cycle (rooting characteristics, leaching, denitrification, mineralization etc.), it is likely that a more quantitative approach to the fertilization of cereal crops in the UK will be developed in the near future.

**Table 6.9** The classification of soils using nitrogen indices (measures of the ability of the soil to supply nitrogen to a crop, on the basis of the previous crop)

| *Nitrogen index* | *Previous crop* |
|---|---|
| 0 | Cereals; harvested forage crops and short leys; grazed short-term leys with low N input;* maize; poor-quality permanent pasture; sugar beet completely harvested; vegetable crops receiving less than 200 kg N $ha^{-1}$. |
| 1 | Any crop receiving organic manures; beans; grazed forage crops and short leys (high N input);† long leys (low N input);* oilseed rape; peas; potatoes; sugar beet (tops ploughed in); vegetable crops receiving more than 200 kg N $ha^{-1}$. |
| 2 | Any crops receiving heavy dressings of organic manures; long leys (high N input);† lucerne; good-quality permanent pasture. |

* $< 250$ kg N $ha^{-1}$ per year or low clover content.
† $> 250$ kg N $ha^{-1}$ per year or high clover content.
(MAFF 1983)

## 6.5 IRRIGATION

On a world scale, the availability of water can be a more important factor than nitrogen fertilization in determining crop yield. This is particularly the case in tropical and sub-tropical zones with pronounced wet and dry seasons (e.g. Hay 1981a), but drought can cause serious yield losses in a wide range of more temperate areas (e.g. North America, Eastern Europe, Mediterranean zones, Australia etc.). In the more humid climate of northern Europe where precipitation tends to be distributed more evenly through the growing season, there was little need for routine irrigation of arable crops as long as potential yields remained low (sect. 6.4.1, Fig. 6.9). However, with the development of high-yielding cultivars, the removal of other constraints (lodging, disease, fertilization etc.) and the recent experience of three very dry years (1975, 1976, 1984), it is now clear that, in certain areas, water supply is limiting production, especially on light, sandy soils. For example, Day (1981) estimates that the yield of some arable crops growing in the South East of England can benefit from irrigation in 8 years out of 10. On the other hand, yield depressions under irrigation are not uncommon (sect. 6.6).

Shortage of water can affect most aspects of the growth and physiology of cereal crops. In particular, the rate of net assimilation can be depressed by stomatal closure and increased respiration (sect. 3.4.5.1) on the one hand, and by reduced leaf area expansion and accelerated tissue senescence (lower $L$ and leaf area duration) on the other (sect. 2.2; Day *et al.* 1981). In turn, the increased intra-plant competition for a limited supply of assimilate can affect tiller survival, spikelet development and grain-filling (sect. 5.2), and the duration of the last process can also be determined by the rate of canopy senescence and by other biochemical processes (Gallagher *et al.* 1976). Thus, according to the timing, duration and severity of the period of water stress, drought can reduce grain yield through its influence upon any one of the components of yield (or their combinations) although there is good evidence to support the idea that retranslocation of stored assimilate can compensate for reduced assimilate supply after anthesis, as long as the stem and ear remain green (sect. 5.1.2; Day *et al.* 1978; Lawlor *et al.* 1981; Austin *et al.* 1980a).

However, in practice in most areas of the UK, cereal yield response to irrigation is normally very modest (and uneconomic) compared with other crops (e.g. 18 kg DM $ha^{-1}$ per mm compared with 25 kg for grass, 80 kg for potatoes, 130 kg for sugar beet and 140 kg for cabbage; MAFF 1982b) although much higher responses would be expected in years such as 1976 or 1984. Consequently, it is rare for irrigation systems to be installed to supply water to cereals alone, although many farmers in eastern areas of the UK take the opportunity to use systems acquired for the benefit of more responsive crops to irrigate cereals when the opportunity arises (Bailey 1984). Overall, the practice is very limited in the UK, with only 14,760 ha of irrigated cereals out of a total irrigated area of 100,000 ha (all crops, Cooksley 1984) and a national cereal area of 4 million ha (MAFF 1984).

## 6.6 THE ACHIEVEMENT OF HIGH GRAIN YIELDS IN PRACTICE

Reviewing the effects of variation in the environment or in management on the components of grain yield (sect. 6.2 to 6.5), it becomes clear that, in general, ear population density is the most sensitive component (e.g. Tables 6.2, 6.6 and 6.8; Fig. 6.4) and although the number of grains per ear is much less variable (e.g. Tables 6.2, 6.6 and 6.8), there can be very pronounced effects with variation in plant population density (Figs 6.4 and 6.6). Consequently, since individual grain weight is a relatively stable characteristic of a given cultivar (sect. 6.2.1, 6.3.1 and 6.4.2; Gallagher *et al.* 1975) except under lodging (e.g. Fig. 6.4) or other conditions which curtail the duration of the canopy (Tables 6.6 and 6.8), variation in grain yield can commonly be attributed to variation in *grain population density*, the product of the first two components (Ellen and Spiertz 1980; McLaren 1981; Fig. 6.14). It would, therefore, be predicted, according to the ideas of source/sink relations (Ch. 5), that there is a maximum level of grain population density for a given crop, beyond which grain yield will not increase further. Working with wheat crops in the Netherlands, Ellen and Spiertz (1980) did find that yield reached a maximum (around 7 t $ha^{-1}$) at 18,000 grains $m^{-2}$ but, in similar investigations in England, yield was still rising linearly at grain population densities in excess of 20,000 (McLaren 1981; Fig. 6.14). (It is worth noting, in passing, that the existence of such high grain population densities is evidence against the idea of compensation among components, although, as shown in section 6.2.1, it can occur in certain cases.)

High grain yield is, therefore, achieved by managing the crop to give a target grain population density and to avoid the stresses (drought, lodging, leaf diseases) which can cut short the duration of the canopy. However, an alternative approach indicates that variation in cereal grain yield is caused primarily by differences in crop

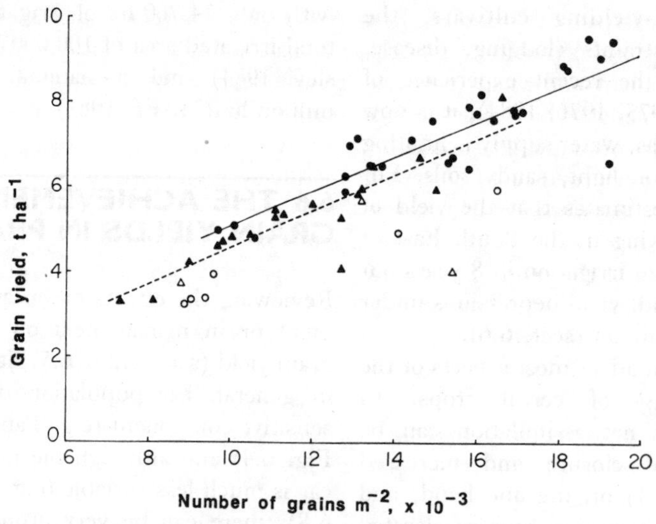


**Fig. 6.14** Relationships between grain yield and grain population density for thirty-one crops of winter wheat cv. Maris Huntsman (●) and twenty-two of cv. Capelle Desprez (▲) grown at a number of sites in the Midlands and south of England between 1962 and 1981. The continuous line is the linear regression line for M. Huntsman, the broken line that of Capelle Desprez, and the open symbols (○, △) represent data for the drought year of 1976 (from Gales 1983).

biomass which, in turn, are the consequence of differences in the quantity of PAR which is intercepted by the canopy (McLaren 1981; sect. 2.4). By this approach, high yield is achieved by maximizing the interception of PAR without necessarily paying attention to the components of grain yield. How can these two approaches be reconciled?

In reality, of course, the approaches are basically the same. Any change in management or in the environment which tends to stimulate canopy development or prolong its duration will also stimulate tiller production, tiller fertility, and spikelet and floret numbers. More specifically, since individual grain weight and harvest index are both relatively stable characteristics of a cultivar (sect. 6.4.1), then grain population density is linearly related to crop biomass *via* grain yield; furthermore, the linear relationship between crop biomass and the quantity of PAR intercepted by the canopy will fail in seasons such as 1976, when individual grain weight is depressed (Fig. 6.14). These ideas are also compatible with the finding that grain yield can be predicted from biomass at anthesis (Dyson 1977; Midmore *et al.* 1984).

In the UK, as in other northern European countries, cereal yields have risen sharply during the last 50 years, not only in variety trials and under experimental conditions but also in farmers' fields (Austin *et al.* 1980b; Riggs *et al.* 1981; Silvey 1981), as a result of improvements in cultivars and in crop management. As discussed in section 6.4.3, the economic optimum yield of a given cultivar at a given site will vary from season to season according to the costs of growing the crop and the price of the grain (Fig. 6.13). In practice, the yields of cereals on specialist arable farms are, on average, equal to or slightly lower than variety trial yields which, because of the high level of management, are an index of the maximum yield which can be achieved at the test site using existing cultivars and growing systems. These mean yields seldom exceed 50–60 per cent of the maximum potential yield, calculated from records of available PAR, but the averaging procedure conceals a long tail of very poor yield, whereas a few farmers can consistently achieve yields which equal or even exceed the maximum potential yield (i.e. 13–14 t ha$^{-1}$ for winter wheat; Austin 1982; Hay *et al.* 1986).

These findings suggest that improvement in mean farm yield (in particular, improvement in the performance of below-average crops) depends upon the identification of those factors which are responsible for farm-to-farm and field-to-field variation in yield (Holmes 1982; Church and

Austin 1983). One approach to this problem has been to construct generalized simulation models of crop growth, development and yield, such as the AFRC winter wheat model, which is described in detail in Chapter 9. In principle, it should be possible to predict the performance of a cereal crop at any site, from environmental records (temperature, available solar radiation, etc.) and details of the management of the crop. The input data could then be varied in a systematic way to give the maximum (or economic optimum) grain yield, thereby establishing the best system of management for the crop at that site. In practice, this procedure has not yet proved to be possible because of the complexity of the model which would be required to simulate all aspects of growth and development, especially below the soil surface. This is particularly important because, as the level of husbandry improves, many of the remaining factors contributing to yield depression appear to operate in the soil (Gales 1983; and see below). Furthermore, it is an inherent deficiency of such simulation models that they operate on historic rather than current environmental data, although it is possible to update the simulation continuously as the season proceeds, thus improving its ability to predict yield.

An alternative approach is to carry out multifactorial experiments on a number of representative soil types, in which the interactions of different aspects of crop management, with one another and with soil and site, in determining crop yield can be evaluated over a number of seasons. For example, the Rothamsted multidisciplinary winter wheat experiment (at two sites examining the effects of method and date of sowing, rate and timing of nitrogen fertilization, irrigation, application of growth regulators and crop protection chemicals, and the previous crop) has established the consistently positive influence of fungicide treatment and the strongly negative influence of barley as a preceding crop, although in more than one of the seasons since 1978/9 early sowing, protection against aphids and the *withholding* of irrigation have also been beneficial (e.g. Tinker and Widdowson 1982; Prew *et al.* 1983). Although agronomic experiments of this kind are ultimately only of local value and should be repeated on a range of soils in different environments, they do serve to identify complex and poorly understood sources of yield variation, mainly originating in the soil (e.g. the effect of the previous crop, see above), which can be investigated in greater detail.

Most of the results of the physiological and agronomic experiments reviewed in this chapter have by now influenced the way in which cereal crops are grown in northern Europe, and many of the well-known systems of cereal production are founded upon crop development and physiology. For example, the relatively low input system devised by Laloux to give consistently profitable, rather than maximum, yields of winter wheat (6–7 t $ha^{-1}$ at 15 per cent mc) from year to year in Belgium relies on the management of sowing and establishment to give plant populations in the region of 200 plants $m^{-2}$, and correctly timed (in relation to plant development) moderate applications of fertilizer nitrogen (three treatments totalling 140 kg N $ha^{-1}$), chormequat and fungicide, to give ear populations of at least 400 ears $m^{-2}$. Overall, the emphasis of the system is on careful crop husbandry, based on detailed evaluation of crop development by the farmer (Biscoe 1979; Laloux *et al.* 1980). In contrast, in the more stable continental climate of Schleswig-Holstein (North Germany), cultural practices can be timed according to calendar dates because the rate of crop development varies much less from year to year than in Belgium or the UK. In this area, the system adopted has been much more intensive and is directed towards high yields (≥8.5 t $ha^{-1}$ at 15 per cent mc) whose profitability depends upon a premium price for high grain nitrogen content (achieved by late application of nitrogen). The targets for plant and ear populations are higher (300–400 plants $m^{-2}$; 500–600 ears $m^{-2}$), the crops receive 200 kg N $ha^{-1}$ in three dressings and there is an intensive, and expensive, programme of sprays (growth regulators, herbicides and fungicides) (Biscoe 1978).

# CHAPTER 7 POTATOES

*. . . dry matter distribution within the potato plant, particularly early in its life, exercises the dominant role in determining both total dry matter production and yield. The key to growth and development in the potato crop, and to their relationship with yield, lies in the balance struck between the conflicting demands of foliage and tuber growth; the key to successful agronomic management of the crop lies in manipulating this balance to best advantage.*

(Ivins and Bremner 1965)

*Ivins and Bremner (1965) formulated a model of growth which suggested that leaf growth was determined by plant size at tuber initiation, and initiation before "the basis of a large leaf area is established" resulted in smaller leaf surfaces which senesced earlier. A conflict between leaf and tuber growth was regarded as inherent in the growth of the crop, with any factor promoting the growth of one at the expense of the other. . . . Thus, while the pattern of growth which Ivins and Bremner (1965) describe does occur, it is by no means universal. There is good evidence that in most temperate environments yield is directly related to total dry matter yield.*

(Allen and Scott 1980)

## 7.1 INTRODUCTION – THE LIFE HISTORY OF THE POTATO PLANT

The life-history of a potato tuber begins when, under a favourable combination of photoperiod, irradiance, temperature and nutrient supply, an underground, diageotropic stolon (i.e. with neutral response to gravity), branching from a normal stem of a potato plant, begins to accumulate starch in a localized swelling at about eight to twelve internodes from the stolon apex. In practice, it is difficult to determine the precise timing of the onset of tuberization and it is not yet clear whether initiation is induced by growth substances, by the availability of surplus photosynthate or by a combination of factors (Menzel 1985). Once initiation has taken place, the development of the tuber proceeds acropetally; starting at the internode furthest from the apex, successive internodes of the stolon bud grow longitudinally and radially by cell division and cell expansion to give the fully-grown tuber, each of whose eyes, arranged according to a spiral phyllotaxis, normally contains three axillary buds bounded by the scar of one of the twelve or so original leaves of the stolon apex (Fig. 7.1). Meanwhile, the initiation of further leaf primordia by the axillary buds continues so that, for example, the original stolon apex (now the apical bud of the tuber) will normally contain at least twelve leaf primordia and partly-developed leaves at tuber harvest.

The time course of dry-matter accumulation by the tubers of a typical potato crop is shown in

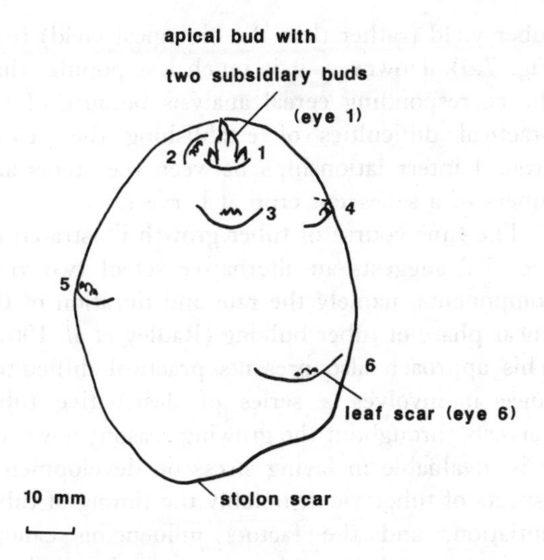


**Fig. 7.1** A sprouting seed tuber cv. Pentland Javelin, showing apical dominance.

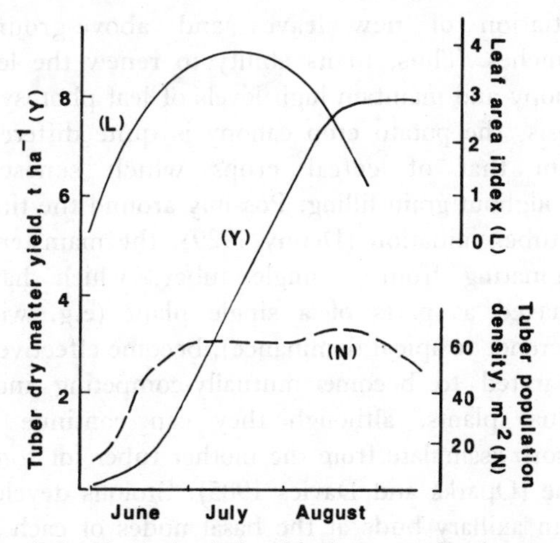


**Fig. 7.2** Typical relationship between the time courses of tuber numbers, tuber dry weight and leaf area index for an early-maturing crop of maincrop potatoes. In a cooler area, the canopy would persist into September (from Scott and Wilcockson 1978).

Fig. 7.2. Within about two weeks of the initiation of the first tubers, generally on the longer stolons branching from the lower nodes of the stem, all of the tubers which will ultimately contribute to yield have been formed, and dry-matter accumulation by the tubers (initially from both the mother tuber and from canopy photosynthesis) increases to a steady rate (known as the rate of tuber bulking), which is maintained up to the time when $L$ falls to 1 through natural senescence, disease or deliberate defoliation. However, this overall rate of bulking is the sum of the different individual growth rates of the daughter tubers, with the highest rates and largest tubers commonly associated with stolons growing from the lower nodes (Gray 1973). During the bulking phase, the import of assimilate by the tubers is relatively insensitive to variation in environmental factors (sect. 7.1.1 and 7.7), although water stress can cause severe disruption. After the cessation of bulking, the tubers undergo a series of biochemical and anatomical changes, including the development of a thicker, corky periderm, which improve the ability of the fully-mature tubers to survive damage during harvest and long-term storage.

The second phase of the life history of a tuber takes place during storage. At harvest, all of its axillary buds are innately dormant and cannot resume growth for several weeks, even under favourable conditions. The duration of this period of innate dormancy varies considerably between varieties, and is also influenced by the date of harvest and the state of maturity of the tubers when they enter storage. Once dormancy has been broken, the rate and pattern of bud (sprout) growth are determined by the duration and temperature of storage. For example, short, warm storage tends to favour apical dominance, giving seed tubers with a single long apical sprout, whereas long, cool storage favours the slower growth of sprouts from several eyes. The physiology of sprout development is discussed in greater detail in section 7.2.

After planting in the field, the rate of sprout growth is determined primarily by soil temperature (MacKerron and Waister 1985) but the number of sprouts which emerge to form mainstems can be less than the number of sprouts at planting, owing to damage or to the re-imposition of apical dominance. The final above-ground stem population density can be supplemented by the emergence of secondary stems, branching below ground from mainstems (which are growing directly from tuber buds), and also by stolons which cease to be diageotropic if they emerge from the soil. Leaf growth continues throughout much of the lifetime of the crop, initially from the leaf primordia carried by the tuber sprouts but subsequently as a result of the initiation of new leaves and above-ground branches. Thus, in its ability to renew the leaf canopy and maintain high levels of leaf photosynthesis, the potato crop canopy is quite different from that of cereal crops which senesces throughout grain-filling. Possibly around the time of tuber initiation (Denny 1929), the mainstems originating from a single tuber, which have behaved as parts of a single plant (e.g. with reference to apical dominance), become effectively separated to become mutually-competing individual plants, although they can continue to absorb assimilate from the mother tuber for some time (Oparka and Davies 1985). Stolons develop from axillary buds at the basal nodes of each of these three classes of aerial stem, although those attached to mainstems are normally the most numerous and important, leading to a new cycle of tuber initiation and development. (Note that, strictly speaking, these underground diageotropic stems are rhizomes, but the use of the incorrect term stolon has been accepted for several decades.) Although in some seasons potato crops can complete the full cycle of sexual reproduction, leading to the production of viable seeds, vegetative propagation by tuber production is by far the more important process in normal agricultural production. However, true seed does have an important role to play in the selection and breeding of new potato varieties.

### 7.1.1 THE COMPONENTS OF TUBER YIELD

By analogy with cereal crops (sect. 6.1), it is possible to split the tuber yield of a potato crop into three components:

| tuber dry matter (or fresh weight) yield | = plant population density | × number of tubers per plant | × mean tuber dry weight (or fresh weight) |
|---|---|---|---|

Alternatively, if population density is expressed in terms of the number of stems per unit area, the second component will be number of tubers per stem. As we shall see in subsequent sections, this approach has been used in a number of investigations because it stresses the importance of the number and size distribution of the tubers of a given crop or variety in relation to the saleable tuber yield (rather than the biological yield) (e.g. Fig. 7.9). However, it is much less popular than the corresponding cereal analysis because of the practical difficulties of establishing the below-ground interrelationships between the stems and tubers of a senescent crop at harvest.

The time course of tuber growth illustrated by Fig. 7.2 suggests an alternative set of two yield components, namely the rate and duration of the linear phase of tuber bulking (Radley *et al.* 1961). This approach also presents practical difficulties since it involves a series of destructive tuber harvests throughout the growing season; however, it is invaluable in laying stress on developmental aspects of tuber yield, notably the timing of tuber initiation, and the factors influencing canopy persistence and the prolongation of tuber bulking, rather than the stem and tuber population measurements of the preceding analysis. Care must be exercised in the use of this approach because some detailed studies have suggested that the time course of tuber bulking is described more precisely by an asymptotic curve, indicating a trend towards lower rates as the season progresses (Wurr 1974). We shall return to the linearity or otherwise of these curves in section 7.7, but in the intervening sections it is sufficiently accurate in most cases to consider the relationship to be linear. It must also be borne in mind that the rate of bulking of the individual tubers attached to a given stem will not be identical and that their relative rates of growth can also change (Ahmed and Sagar 1981). Finally, the term bulking rate can be used to describe the rate of increase of either fresh or dry weight of tubers; in some cases, this can cause confusion.

At this point, it is important to point out that, in contrast to cereals, where yield is normally expressed as the weight per unit area of grain dried to 15 per cent moisture content (or grain dry weight in physiological experiments), potato yields are usually expressed as the fresh weight of tubers per hectare. The use of fresh weight has several practical advantages, since tubers have a very high water content and tend to caramelize on heating; however, it must be borne in mind that variation in crop management, especially the level of fertilization (sect. 7.5), can have a large effect upon tuber water content. Furthermore, tuber yields are normally divided into size classes by means of a series of riddles. The ware yield, which represents the fraction of the crop which is sold for domestic consumption, is usually determined by means of a pair of riddles with holes of 45 and 80 mm in diameter. Tubers passing through the 45 mm riddle are discarded as chats, whereas those retained by the 80 mm riddle are oversize. Since the tubers of most commercial varieties are not spherical but ovoid, those passing through the 80 mm riddle will have a cross-section of diameter 80 mm or less but, of course, they can be longer than 80 mm. The grading criteria for early, seed and canning tubers are different (sect. 7.3.3).

This chapter is principally concerned with the influence of environmental and management factors on tuber yield and on these different sets of yield components.

## 7.1.2 THE CLASSIFICATION OF POTATO VARIETIES INTO MATURITY CLASSES

In many potato-growing areas, the varieties used are classed as early, second-early or maincrop, according to the normal time of commercial harvesting. For example, until the beginning of the large-scale import of potatoes from the Mediterranean region, substantial areas of early potatoes were grown on the relatively frost-free west coast of the British Isles (e.g. Cornwall, Pembrokeshire, Ayrshire) for lifting in May to July, when the tuber yield was much lower than that of the mature crop, but its high monetary value ensured an economic yield. These enterprises persist on a much-reduced scale. In contrast, the aim of maincrop potato growers (principally in the eastern, arable areas such as Cambridgeshire, Lincolnshire, Fife and Angus) is to maximize the ware yield of crops maturing from September onwards, when there is no price premium for earliness. Second-early growing enterprises tend to be intermediate both geographically and in tuber yield.

These practices have led to the selection and breeding of physiological types or maturity classes suited to each growing system. For example, desirable characteristics for an early variety include rapid canopy development and tuber initiation in cool, short days, and the formation of few tubers per plant, to maximize the proportion of large, saleable tubers at an early harvest. The desirable characteristics of a maincrop variety are different: slower tuber initiation to ensure a large canopy when bulking begins, a long-lived canopy, higher (but not too high) tuber numbers per plant, as well as resistance to the major diseases. As a result of these differences, the tuber yield of an early crop is generally higher than that of a maincrop variety during the summer months but, owing to a higher rate and longer duration of bulking, the maincrop variety should overtake the early variety progressively to give a much higher total yield at maturity in autumn (Fig. 7.3). Realization of this higher yield potential will, of course, depend upon protection from

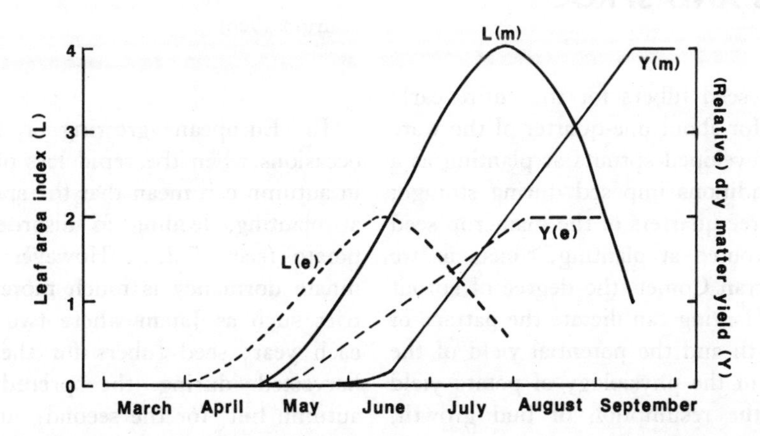


**Fig. 7.3** Idealized comparison of the seasonal pattern of leaf area index and dry-matter production of early (---) and main (——) crops of potatoes in the UK.

diseases which can cause premature defoliation of the crop.

Although this classification is usefully employed by growers, plant breeders and research workers, there are varieties which do not fit naturally into one or other of the three classes, and whose seasonal pattern of growth and yield depends strongly upon the treatment of the seed tubers before planting. For example, depending upon the degree of sprouting (physiological age, sect. 7.2.2) of the seed tubers, the seasonal pattern of tuber yield of the early variety Arran Comet can follow either the early or the maincrop trend shown in Fig. 7.3, and when permitted to grow on to maturity, crops of this variety can sometimes outyield maincrop varieties. In contrast, the variety Home Guard tends to behave consistently according to the 'early' pattern, indicating that, for this variety, high early yield is not compatible with high yield later in the season (Griffith *et al.* 1984). It is clear that more attention must be paid to the physiological age of the seed tubers in trials to determine the maturity class and yield potential of both new and existing varieties. Except where otherwise stated (especially sect. 7.2), the remainder of this chapter will concentrate upon the physiological aspects of growing main (ware) crops of potatoes in the UK.

## 7.2 THE PHYSIOLOGY OF SPROUTING AND SPROUT GROWTH

In the UK, the seed tubers for the entire early potato crop and for about one-quarter of the ware crop have well-developed sprouts at planting as a result of the conditions imposed during storage; the remaining three quarters of the maincrop seed tubers are unsprouted at planting. Since, as we have seen for Arran Comet, the degree of sprout development at planting can dictate the pattern of subsequent growth and the potential yield of the crop, any study of the physiology of potato yield must begin at the resumption of bud growth, whether this occurs in storage or after planting in the field.

### 7.2.1 TUBER BUD DORMANCY AND SPROUT GROWTH

In the past, study of the dormancy of potato tuber buds was complicated by the use of ill-defined terms such as the resting phase (reviewed by Burton 1963). In storage, seed tubers undergo only two forms of dormancy: immediately after harvest, there is a period of several weeks of innate dormancy during which the buds cannot be induced to grow by alteration of the environmental conditions, and this can be followed by a state of induced dormancy if growth of the otherwise non-dormant tuber buds is suppressed by low temperature. The duration of innate dormancy can vary considerably among batches of the same variety from different harvest dates, growing seasons or storage temperatures, and it is usually shorter for early than maincrop varieties (Krijthe 1962; Toosey 1964; Holmes and Gray 1972). However, even within a maturity class there are pronounced inter-varietal differences (Table 7.1).

**Table 7.1** The duration of innate dormancy (weeks) of four British potato varieties under different storage temperatures

| *Maturity class* | *Variety* | *Storage temperature (° C)* | | |
|---|---|---|---|---|
| | | *4* | *10* | *23* |
| Early | Home Guard | 12 | 5 | 3 |
| | Ulster Prince | 14 | 14 | 8 |
| Maincrop | King Edward | 16 | 6 | 5 |
| | Majestic | >28 | 12 | 8 |

(Burton 1966)

In European growing systems, there are occasions when the rapid loss of tuber dormancy in autumn can mean that the sprouts are very old at planting, leading to disorders such as little potato (sect. 7.2.2). However, the duration of innate dormancy is much more critical in countries such as Japan where two crops are grown each year; seed tubers for the spring crop are harvested during the preceding summer or autumn but, for the second, autumn, crop, they will be too old, whereas the newly-harvested tubers will generally still be dormant (Kawakami

1962). This has led to considerable interest in the physiology and biochemistry of tuber dormancy but, apart from establishing that a wide range of growth regulators and related compounds can relieve or prolong dormancy, these studies have yet to elucidate the detailed mechanism of dormancy control.

Once dormancy has been broken (sprouts >2 mm), tuber bud growth is controlled primarily by temperature. Several workers (e.g. Headford 1962; Short and Shotton 1970) have shown that the rate of sprout extension increases with temperature, with a maximum at about 25 °C, whereas the duration of extension in storage is longest at lower temperatures (Fig. 7.4). The net effect of these two responses is that, over the normal period of sprouting in the UK, the longest sprouts are induced by storage temperatures of about 15 °C. However, there is an important interaction between the temperature and the lighting of a potato store in the determination of sprout length. Tubers held in total darkness give very long, easily-damaged, etiolated sprouts, but this effect can be abolished by very low irradiances of the order of 40 KJ $m^{-2}$ $day^{-1}$; increasing the irradiance does not normally have any additional effect on sprout growth (Short and Shotton 1968). Sprout extension can also be controlled by a range of natural and synthetic growth inhibitors, for example as have been used in the storage of tubers for human consumption.

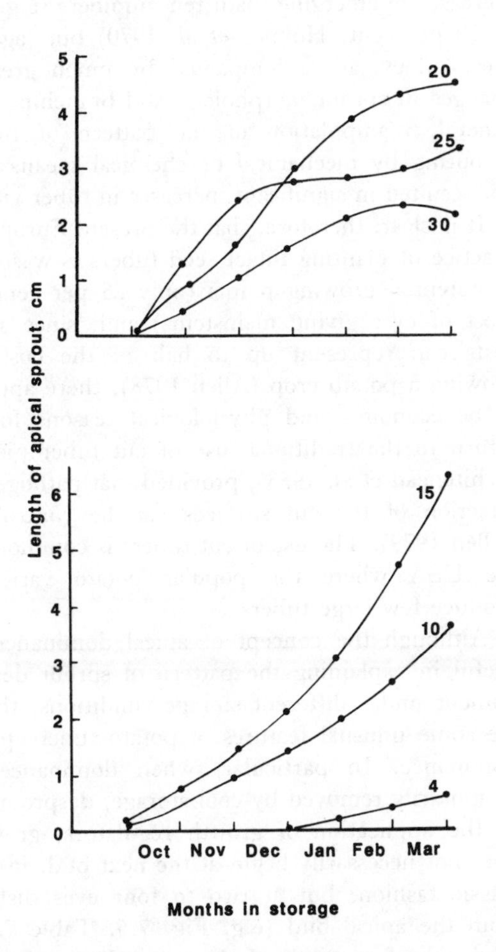


**Fig. 7.4** The influence of the temperature (4 to 30 °C) and duration of storage on the extension of the apical sprout of potato tubers cv. Arran Pilot. Note that above 15 °C, extension ceased during the period of storage (adapted from Headford 1962).

Sprouts developing at different temperatures differ in other characteristics in addition to length. First, there can be pronounced differences in morphology such that, above 15 °C, sprouts tend to have thicker bases with visible stolon buds. In some cases, the apex of the sprout may become necrotic, leading to branching, and the sprouts are generally more susceptible to mechanical damage during handling. These observations, which indicate that they are at a more advanced stage of development than those of the same chronological age stored at lower temperatures, have contributed to the concept of physiological age discussed in section 7.2.2.

Secondly, storage temperature influences the distribution of growth among the various buds of the tuber. A wide range of experiments (Toosey 1964; Moorby and Milthorpe 1975) has shown that the number of growing buds per tuber depends primarily upon the apical dominance exerted over the other eyes by the apical sprout. Apical dominance tends to be greatest when tuber dormancy is first broken, but diminishes during storage; as a result, tubers which sprout early under high temperatures usually carry a single apical sprout (e.g. Fig. 7.1), whereas those held at 5 °C throughout the storage period tend to produce more sprouts (commonly up to six per tuber) when returned to higher temperatures (Fig. 7.5). A range of intermediate sprout numbers can be induced by manipulating the temperature and duration of storage. Within this general pattern, sprout growth can vary among batches of seed tubers according to variety or tuber size. For example, Griffith *et al.* (1984) found a three-fold difference in the number of sprouts longer than 3 mm between batches of Home Guard (1.3 sprouts/tuber) and Juliver (3.6) stored under identical conditions. Variation in tuber size causes differences in sprout numbers because larger tubers tend to carry more potential sprouts (Allen 1978) and apical dominance diminishes with increase in tuber size (Goodwin 1967).

The fact that the tuber yield of a potato crop is dependent upon stem population density (sect. 7.3) has stimulated interest in alternative methods of breaking apical dominance to give more growing points at planting. For example, Holmes and Gray (1971) found that removal of all sprouts from tubers which had been induced to grow early in the storage period increased the number of sprouts which developed subsequently, but no more than did cool long-term storage. There were marked differences in the branching of sprouts between the treatments, but the most important conclusion of experiments of this kind is that the number of sprouts at planting does not provide a reliable guide to the number of mainstems which subsequently emerge. In their experiment, Holmes and Gray (1971) showed that the re-imposition of apical dominance meant that approximately three out of every four expanded sprouts (on both desprouted and late sprouted tubers) did not emerge to give aerial shoots. The same effect is shown in Fig. 7.5 where sprouts growing from eyes 5, 6 and 9 had begun to expand but ceased growth, presumably under the influence of the apical bud, leaving only three potential mainstems. The treatment of seed tubers with growth substances, especially gibberellic acid, to reduce apical dominance can result in modest increases in emerging mainstem numbers (e.g. up to 10 per cent, Holmes *et al.* 1970) but, again, these effects are accompanied by much greater changes in sprout morphology and branching. In general, manipulation of the pattern of tuber sprouting by mechanical or chemical means has not resulted in significant increases in tuber yield.

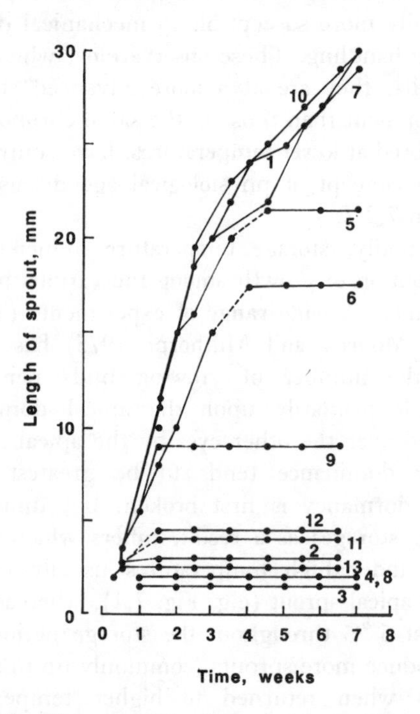


**Fig. 7.5** The extension of sprouts from different eyes (numbered 1 to 13 as in Fig.7.1, but different from the original) of potato tubers cv. Arran Pilot stored at 1 °C for 12 weeks after harvest and then transferred to 20 °C, each phase in the dark (adapted from Goodwin 1967).

It is clear, therefore, that the present European practice of planting intact seed tubers is wasteful of potential growing points (only 25 per cent at most of eyes giving mainstems) and, since seed costs can represent up to half of the cost of growing a potato crop (Allen 1978), there appear to be economic and physiological reasons for a return to the traditional use of cut tuber pieces (Whitehead *et al.* 1953), provided that pathogenic infection of the cut surfaces can be controlled (Allen 1979). The use of cut tubers is common in the USA where the popular potato varieties produce few large tubers.

Although the concept of apical dominance is useful in explaining the pattern of sprout development under different storage conditions, there are some unusual features of potato tuber apical dominance. In particular, when dominance is temporarily removed by cool storage, desprouting or the application of growth regulators, growth does not necessarily begin at the next bud, in the classic fashion, but at two to four eyes distant from the apical bud (e.g. Fig. 7.5, Table 7.2). The lack of response of the eyes adjacent to the apical sprout has not been explained, although Goodwin (1967) has proposed that there is a relationship between the degree of expansion of the tissues around a given eye and its potential for subsequent growth. Another feature which is

**Table 7.2** Demography of sprouts (20 March 1986) growing from tubers of Pentland Javelin which were stored from harvest at 5 °C (Young) or stored at 20 °C until four weeks after the break of dormancy, followed by storage at 5 °C (Old)

| *Physiological age* | *% of tubers with sprouts (>2 mm) at different eyes (see Fig. 7.1)* | | | | | | *% of tubers with different total numbers of sprouts* | | | | *Number of tubers measured* |
|---|---|---|---|---|---|---|---|---|---|---|---|
| | *1* | *2* | *3* | *4* | *5* | *6* | *1* | *2* | *3* | *>6* | |
| Old | 99 | 7 | 8 | 12 | 25 | 29 | 47 | 21 | 17 | 3 | 212 |
| Young | 90 | 39 | 51 | 60 | 78 | 48 | 7 | 11 | 21 | 20 | 213 |

(Hay, unpublished data)

poorly understood is the considerable variation in the pattern of sprout development among the tubers of a uniformly-treated batch; for example in Table 7.2, although, on average, apical dominance was a feature of the 'old' tubers, 32 per cent of the population had three or more sprouts and, on the other hand, 7 per cent of the 'young', apparently multisprouted tubers carried only one sprout. Since few workers explain how they have selected within a treated batch for tuber measurement and planting, there remains the possibility that a sizeable proportion of the more variable tubers has been excluded. This could well limit the practical relevance of such experimental work.

Finally, the results of a long series of experiments in which sprouting tubers have been exposed to a variety of environments (air, compost, nutrient solutions, e.g. Moorby 1967; Morris 1967; Davies 1984) suggest that apical dominance is not the only form of interaction among the sprouts of a single tuber, and that competition for a limited supply from the mother tuber of an, as yet unidentified, nutritional factor can limit the rate of sprout elongation. This competition, and the related cessation of extension of older sprouts (Fig. 7.4), is unlikely to be for carbohydrate or protein, under normal conditions; the most likely candidate at present appears to be the calcium ion (e.g. Davies 1984).

## 7.2.2 THE PHYSIOLOGICAL AGE OF POTATO TUBERS

The concept of physiological age, expressing the degree of development of the sprouts on a tuber and their potential for growth (Wurr 1980), was recognized by growers of early potato crops long before it could be precisely defined and measured. For example, Krijthe (1962) advocated the use of four stages of development: (a) one sprout, (b) multiple sprouts, (c) sprout branching and (d) the development of stolons with daughter tubers before planting ('little potato'), whereas Toosey (1964) considered that the physiological age of a tuber depended upon the date of tuber initiation, the environmental conditions during tuber bulking in the field, its state of maturity at harvest and the temperature of storage. These original ideas were inadequate for the precise description and prediction of sprout development; for example, there was no discrimination between sprouts of different lengths, and there appeared to be an overlap between the periods of innate dormancy and tuber ageing.

As a result of more recent work, it is now recognized that physiological ageing does not begin until innate dormancy has been broken, that the physiological age of a tuber increases thereafter with time (i.e. its chronological age) but that the process can be accelerated by increasing the temperature of storage. These findings indicate that, in common with other indices of plant development (e.g. sect. 2.2.2), physiological age can be expressed in terms of the accumulated temperature, in day degrees above 4 °C (a typical minimum temperature for sprout growth, Fig. 7.4) which the stored tuber has experienced since the end of dormancy. This has been confirmed in a series of investigations (e.g. O'Brien *et al.* 1983) which have established linear relationships between sprout extension (in particular, the length of the longest sprout per tuber) and accumulated temperature above 4 °C (Fig. 7.6). Use of this approach has both advantages and disadvantages.

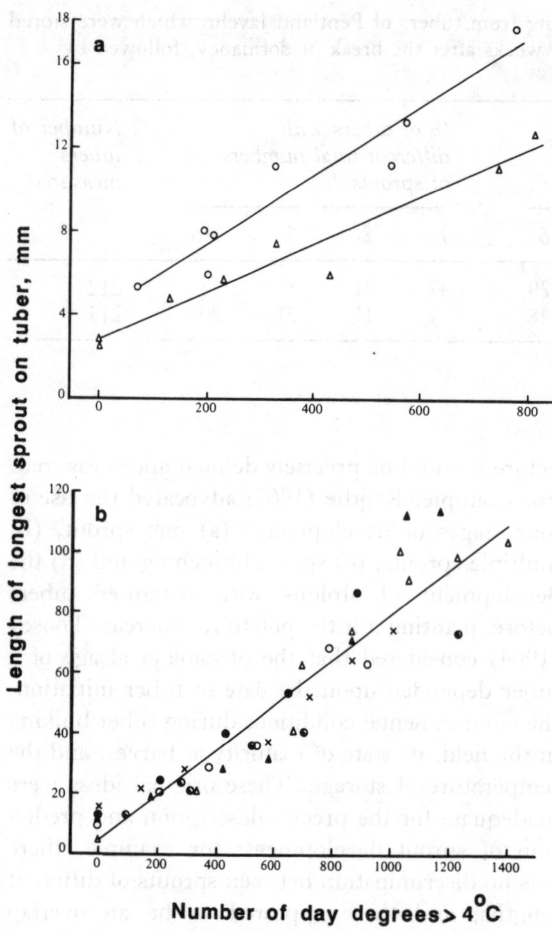


**Fig. 7.6** Relationships between the physiological age of stored potato tubers, expressed in terms of accumulated temperature, and the mean length of the longest sprout per tuber. (a) Comparison of varieties (Pentland Javelin, △; Vanessa, ○); (b) comparison of different populations of cv. Home Guard (from O'Brien *et al.* 1983).

The principal difficulty is that the ageing response to accumulated temperature can vary among varieties (e.g. Fig. 7.6a), although there is evidence that different batches of a single variety tend to age in a broadly similar way (Fig. 7.6b). On the other hand, the technology for the monitoring and control of potato store temperatures is highly developed, there is no need for laborious measurement of large numbers of tuber sprouts, and the system can easily be extended to include older tubers which continue to age although their sprouts have ceased to extend (Fig. 7.4).

### 7.2.3 PHYSIOLOGICAL AGE AND TUBER YIELD

Stems originating from pre-sprouted (old) tubers begin the annual life-cycle earlier than those from younger, non-sprouted, tubers. As a result, all stages of their development in the field (emergence, canopy development, tuber initiation, canopy senescence, tuber maturity) occur earlier in the season (Toosey 1964); indeed, with very old seed tubers planted into cold soils, tuber initiation can occur even before emergence ('little potato') with disastrous consequences for tuber yield. Early potato growers, therefore, plant older seed tubers to ensure early tuber initiation and maturity. However, such early initiation is associated with lower canopy leaf area indices (lower temperatures during leaf growth; Fig. 2.15; sect. 7.4) and incomplete interception of the available solar radiation; this, in turn, results in lower rates of tuber bulking. (Note that the earlier theory, which assumed that bulking rate is determined by leaf area index at tuber initiation *via* intraplant competition for assimilate, has now been generally abandoned; sections 7.5.1 and 7.7.) The practical importance of these ideas is demonstrated by the interaction between physiological age and harvest date in the determination of tuber yield; for example, in Fig. 7.7 the highest early (15 June) crop yield of Home Guard was obtained using very old seed tubers, but if harvest took place on 30 June or 18 July, yield was maximized from much younger tubers. Growers need, therefore, to plan storage regimes with a given harvest date, or series of dates, in mind.

The use of seed of a specified physiological age improves the planning and timing of early potato growing rather than increasing the biological yield of the crop. The strategy adopted by the 25 per cent of maincrop growers who use pre-sprouted seed tubers is subtly different; in general, they farm in areas where there is a considerable risk of premature canopy senescence owing to factors such as drought or blight, and they, therefore, require early-maturing crops of moderate to high yield (compared with the biological potential of crops grown on to maturity in the absence of stress) (Short 1983). This requirement influences varietal choice as well as tuber storage regimes.

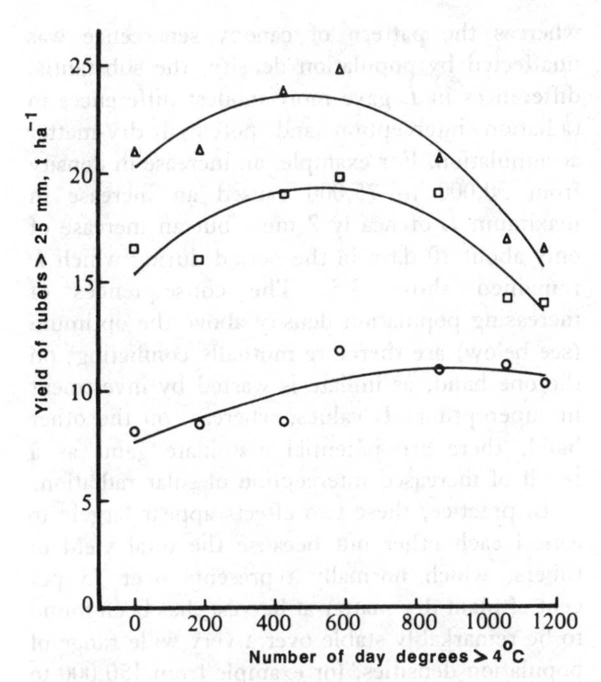


**Fig. 7.7** Relationships between the yield of tubers >25 mm (at three harvest dates: ○, 15 June; □, 30 June; △, 18 July) and the physiological age of the seed tubers at planting. Data from early crops of cv. Home Guard grown in Pembrokeshire, West Wales, in 1977 (from O'Brien *et al.* 1983).

As discussed in section 7.2.1, the pre-sprouting of seed tubers not only influences the rate of crop development but also determines the pattern of sprout and stem development according to the extent of apical dominance. The implications of these effects for tuber yield are explored further in section 7.3, but it is important to point out at this stage that discussion of bulking alone is not sufficient to explain the overall influence of tuber age on commercial yield because a large number of small tubers, however early, will not be saleable. In early potato production it is, therefore, important to employ: (a) seed tubers in which apical dominance has not been broken; (b) varieties which give few tubers per stem, as well as (c) larger seed tubers whose more extensive storage reserves lead to earlier crop emergence and more vigorous early growth.

## 7.3 PLANT POPULATION DENSITY AND YIELD

Within a few weeks of crop emergence, each of the mainstems originating from the eyes of the mother tuber becomes, in turn, the mainstem of an independent plant which may or may not carry secondary stems by branching below ground (sect. 7.1). There can, therefore, be competition for solar radiation, nutrients or water between the stems of a given 'hill' (i.e. those originating from the same mother tuber) as well as between stems of adjacent 'hills'. Consequently, the population density of potato crops is most correctly expressed in terms of the number of stems per unit area, although there may be some disagreement as to whether secondary stems should be included. This question is of little relevance to the many main crops in which mainstems predominate (Allen and Wurr 1973) but is of considerable importance in crops, grown from apically-dominant or physiologically old seed tubers (e.g. Holmes *et al.* 1970), in which secondary stems are commoner but difficult to distinguish from mainstems in the field without destructive harvesting. However, there is evidence that tubers initiate and develop on well-developed secondary stems in much the same way as they do on mainstems and it is, therefore, possible to avoid these practical difficulties by using the total number of above-ground stems as a measure of population density (e.g. Wurr 1974). This is analogous to the use of ear population density in cereals (sect. 6.1).

In most seed-bearing species, a plateau of crop biomass is reached at a modest plant population density, and further increase in density is associated with a progressive reduction in individual plant or stem size (e.g. in cereals, Fig. 6.3; Harper 1977). Responses to increase in population density are different in potato crops, whose crop biomass can continue to increase up to higher densities, although reductions in stem weight, by decreased branching, are observed (Bremner and El Saeed 1963). This effect is illustrated by Fig. 7.8 in which a stepwise increase in population density from 25,000 to 75,000 seed tubers ha$^{-1}$ resulted in progressively higher rates of canopy development and maximum leaf area indices.

However, because increases in density above 30,000 seed tubers $ha^{-1}$ brought forward the date of 90 per cent interception of incident solar radiation ($L$ = 3–3.5, Fig. 2.20) by only a few days, whereas the pattern of canopy senescence was unaffected by population density, the substantial differences in $L$ gave more modest differences in radiation interception and potential dry-matter accumulation. For example, an increase in density from 50,000 to 75,000 caused an increase in maximum $L$ of nearly 2 units but an increase of only about 10 days in the period during which $L$ remained above 3.5. The consequences of increasing population density above the optimum (see below) are therefore mutually conflicting; on the one hand, assimilate is wasted by investment in superoptimal $L$ values, whereas, on the other hand, there are potential assimilate gains as a result of increased interception of solar radiation.

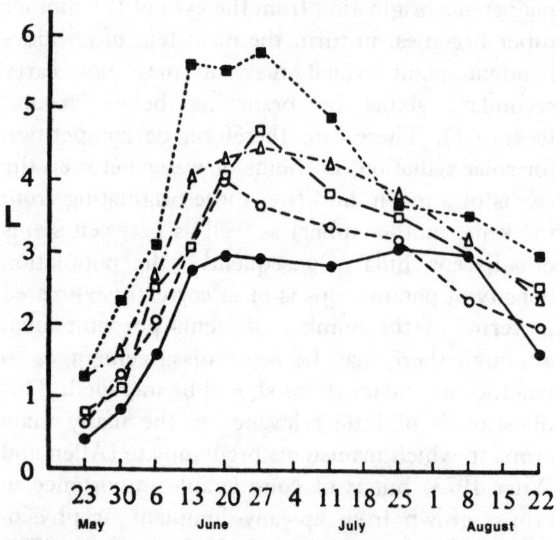


**Fig. 7.8** Seasonal pattern of leaf area index of maincrops of potatoes (cv. Désirée and Maris Piper) grown at different plant population densities in Wales in 1974 (where ●, ○, □, △ and ■ indicate 25, 30, 37.5, 50 and 75 × $10^3$ plants $ha^{-1}$, (from Allen and Scott 1980).

In practice, these two effects appear largely to cancel each other out because the total yield of tubers, which normally represents over 75 per cent of plant dry matter at harvest, has been found to be remarkably stable over a very wide range of population densities, for example from 150,000 to over 400,000 stems $ha^{-1}$ in main crops of Pentland Crown and Maris Piper (Fig. 7.9). Similar yield plateaux have been recorded in a long series of seed tuber spacing and size experiments over the last century which has been fully reviewed by Holliday (1960) and Allen (1978). For example,

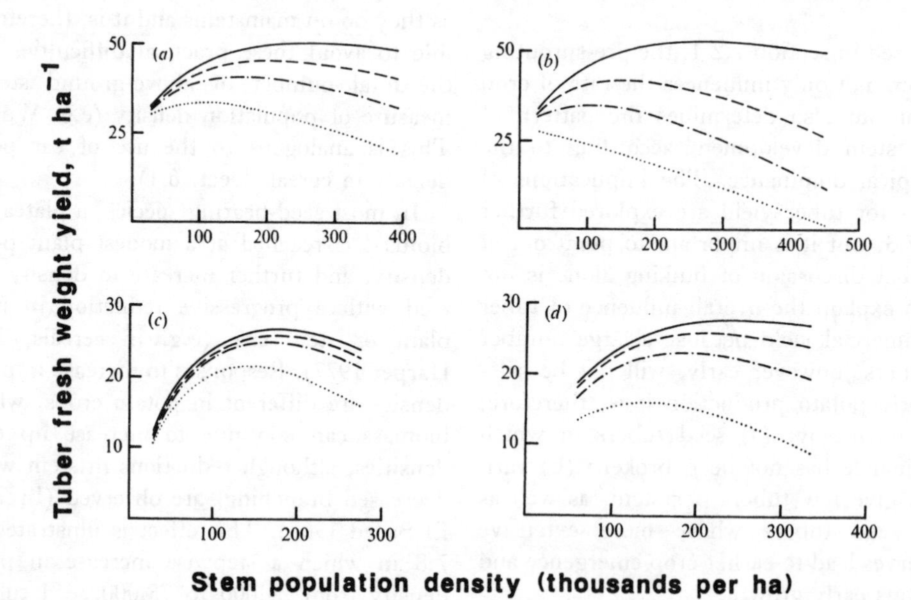


**Fig. 7.9** The influence of stem population density on total tuber fresh weight yield (——) and on the size distribution of the tubers harvested from two cv. of potatoes grown in two seasons in South-East England: (a) Pentland Crown, 1969; (b) Maris Piper, 1969; (c) Pentland Crown, 1970; (d) Maris Piper, 1970, where --- indicates yield > 38 mm -.-.- > 44 mm and . . . . . > 51 mm (from Wurr 1974).

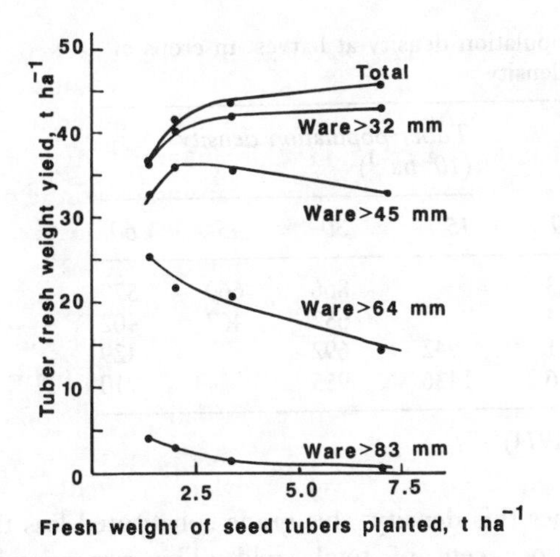


**Fig. 7.10** The influence of seed tuber population density on the total tuber fresh weight yield, and on the size distribution of the tubers harvested from main crops of potatoes in Southern England (adapted from Holliday 1960).

Fig. 7.10 shows a typical yield response to increase in population density, expressed as weight of seed tubers $ha^{-1}$; as we shall see in section 7.3.3, recommended population densities are still drawn up on the basis of such response curves, but use of the stem as the unit of population density is much more precise because the number of stems formed per unit of seed tuber weight will vary from season to season. However, total yield of tubers is only one aspect of crop yield; variation in population density also has a profound influence upon the size distribution of the harvested tubers, as discussed in the following section.

The assumption, implicit in the preceding paragraph, that crop response to variation in stem population density is the same, irrespective of the cause of variation (seed tuber spacing or size), is largely supported by experiment (sect. 7.3.2), although there are reports of interactions between spacing and size, especially at more extreme densities (Allen 1978). However, although the appropriate experiments have not been done, it is unlikely that the yields of a series of crops, whose stem population densities have been manipulated by means of the physiological age of the seed tubers, would be simply related to stem population density (as in Fig. 7.9) because there will be a pronounced interaction between stem population density and canopy duration (earlier senescence of crops grown from older seed tubers).

## 7.3.1 POPULATION DENSITY AND THE COMPONENTS OF TUBER YIELD

### 7.3.1.1 TUBER BULKING

There have been few detailed studies of the influence of population density upon the rate and duration of tuber bulking. From the growth curves which are available, it appears that the time of tuber initiation is relatively unaffected; for example, the use of larger seed can give only a few days advantage over small seed (Bremner and El Saeed 1963; Wurr 1974). However, Wurr (1974) showed that, although increases in stem population density could be associated with modest increases in total bulking rate, the rate of growth of individual tubers was determined by a complex interaction among population density, variety and the size distribution of the growing tubers. This serves to underline the usefulness of the alternative set of yield components (sect. 7.1.1) in the analysis of population density effects.

### 7.3.1.2 NUMBER OF TUBERS PER STEM

The number of tubers per stem is primarily under genetic control, with the extremes among normal British maincrop varieties represented by Pentland Crown, which can produce a small number of outsize tubers under the highest yielding conditions, whereas King Edward can give a high proportion of unsaleable chats in less favourable years. For example, Allen (1979) recorded 5–8 tubers per stem for King Edward and 2.5–4 for Pentland Crown grown under identical management. The expression of this genetic potential is, in turn, controlled by a number of crop and environmental factors such as the level of fertilization (sect. 7.5), the incidence and severity of water stress (sect. 7.6) and the physiological age of the seed tubers (mainly through its influence on stem numbers).

In a few investigations, the number of tubers per stem has proved to be insensitive to changes in population density; for example, working with whole and cut tubers of King Edward and Majestic, Moorby (1967) showed that there was no change in this component over a four-fold variation in stem population density and a three-

**Table 7.3** Number of tubers/stem and tuber population density at harvest in crops of different potato cultivars at varying population density

| *Spacing (cm)* | *No. of tubers/stem* | | | | *Tuber population density ($10^3$ ha$^{-1}$)* | | | |
|---|---|---|---|---|---|---|---|---|
| | *15* | *30* | *45* | *60* | *15* | *30* | *45* | *60* |
| King Edward | | 3.6 | 4.0 | 4.3 | | 866 | 660 | 579 |
| Majestic | | 4.4 | 5.0 | 5.4 | | 652 | 467 | 402 |
| P. Crown | 4.2 | 5.5 | | 7.1 | 942 | 692 | | 429 |
| M. Piper | 5.1 | 6.4 | | 8.6 | 1436 | 955 | | 710 |

(Data from Bremner and El Saeed 1963, Wurr 1974)

fold variation in stem weight. By far the more common response is an inverse relationship between stem population density and the number of tubers per stem (Table 7.3), leading to the suggestion by Wurr (1974) that the results of Moorby (1967) and others may not conflict with the consensus but may be the result of different interpretations of the terms 'stem' and 'tuber'. However, whichever response holds for a particular investigation, the variation in number of tubers per stem is relatively small over a wide range of stem population densities; consequently, increase in population density is associated with a steep rise in tuber population density (Allen 1978) (Table 7.3).

#### 7.3.1.3 INDIVIDUAL TUBER WEIGHT

As stem population density increases, the total yield of tubers rises rapidly to a sustained plateau (Figs 7.9 and 7.10), whereas the number of tubers per unit area continues to rise (Table 7.3) even at densities well above commercial optimum ranges (sect. 7.3.3). As a result, mean tuber weight falls steadily with increasing population density (e.g. Thompson and Taylor 1974). As shown in Figs 7.9 and 7.10, the yield of a given crop is not a set of uniform tubers but a range of sizes whose distribution varies with population density (as well as other factors, as discussed in subsequent sections). Thus, the observed reductions in mean tuber weight are largely the consequence of the proliferation of poorly filled tubers, to the detriment of the larger size classes. For example, approximately half of the yield of a Maris Piper crop grown at 100,000 stems ha$^{-1}$ in 1969 took the form of large tubers which would not pass through a 5.1 cm riddle, whereas at four times this density, this grade constituted less than 10 per cent of total yield. The corresponding proportions of (normally unsaleable) chats ($\leqslant$3.8 cm) were 11 and 40 per cent respectively (Fig. 7.9). This study also indicates that there are distinct differences in response between varieties (the size distribution of tubers produced by crops of Pentland Crown being less affected than that of the more prolific Maris Piper) and seasons (the effects being more pronounced in higher-yielding years) (Fig. 7.9). Similar results have been reported from a wide range of seed tuber size and spacing experiments (e.g. Fig. 7.10; Allen 1978).

### 7.3.2 THE INFLUENCE OF SPATIAL ARRANGEMENT

It has been pointed out by several authors that the use of larger seed tubers will give a more clumped arrangement of stems than would the closer spacing of smaller tubers. Bleasdale and Thompson (1966) and Svensson (1972) have examined the influence of spatial arrangement on yield by means of seed tubers or tuber pieces carrying a single sprout, arranged in different configurations within the ridge, with conflicting results. Svensson's results, from a relatively early harvest in August, indicated that tuber yield per stem was significantly affected by its position within the 'hill', with central stems being suppressed by surrounding stems, but the earlier work of Bleasdale and Thompson over a longer growing season gave no support for the idea that stem yield was affected by its position apart from the fact that greening of tubers was more common in clumped arrangements. It is clear that more investigation of this topic is necessary (Allen 1978) but the fact that, in more recent experiments, it has been found possible to derive yield response curves (e.g. Fig. 7.9) which are common to both types of stem population variation indicates that spatial arrangement of stems is relatively unimportant. This, and the fact that tuber yield is unaffected by wide variation in row width at the same population density, is not surprising in view of the broad yield plateau response to population density (Figs 7.9 and 7.10).

### 7.3.3 OPTIMUM POPULATION DENSITY IN PRACTICE

Because, in practice, it has so far been found difficult to establish target stem population densities, recommended population densities continue to be based on empirical relationships between the weight of seed tubers planted, the total yield of tubers and the proportion of the yield which falls within the saleable ware grade. The shapes of such response curves (e.g. Fig 7.10) have been established for many decades and, although some allowance must be made for seed tuber size and the variety grown (e.g. MAFF 1982a), the 'seed rate' for commercial maincrops in the UK has changed very little over the last 30 years (generally 2.5–2.9 t $ha^{-1}$; Harris 1983). Clearly, in view of the high financial value of the seed tubers and the declining proportion of ware-grade tubers with increasing population density (Figs 7.9 and 7.10), the 'seed rate' for maximum economic return (in contrast to the maximum yield of saleable tubers) will lie below the rate corresponding to the achievement of the yield plateau.

If, through the manipulation of the physiological age, weight and spacing of seed tubers, it becomes feasible to predict stem population density in the field, then relationships such as Fig. 7.9 offer the possibility of more precise control, not only of ware yield, but of the yield of seed tubers (4–5 cm) or tubers for canning (2–3 cm) by the use of progressively higher stem population densities. In the case of the yield of tubers for canning, of course, it is also necessary to use a prolific tuber-forming variety like the second-early Maris Peer (Gray 1973; Thompson and Taylor 1974). Finally, there is evidence to suggest that tuber quality (i.e. per cent dry matter content) may be enhanced at higher population densities; these observations require further examination (Allen 1978).

## 7.4 PLANTING DATE AND YIELD

Because the yield of a crop is determined ultimately by the quantity of solar radiation intercepted by its canopy, any reduction in the actual growing season (emergence to complete senescence) will lead to a loss of potential yield (MacKerron and Waister 1985). In the case of the potato maincrop, it has commonly been found that delay in planting beyond a critical period of 2–3 weeks in spring causes a progressive reduction in final tuber yield (e.g. Dyke 1956). The existence of this critical period and its timing can be explained primarily by the response of the seed tubers to soil temperature; in mild maritime areas like Pembrokeshire, soil temperatures in February are sufficiently high for the rapid emergence of early crops but, in the cooler maincrop areas, the sprout growth of crops planted before the beginning of March will be very slow, with the result that they will be overtaken by later plantings. In addition, sprouted tubers planted in cold soils can be susceptible to disorders such as little potato and coiled sprout.

Within the overall constraint of soil temperature, account must also be taken of the incidence of freezing temperatures and of soil/water relations. Depending upon its severity, frost can cause the death or, more commonly, the temporary defoliation of the crop, resulting in the loss of any potential advantage from early planting (Radley 1963). On the other hand, since tuber planting must be preceded by a period of intensive soil cultivation and traffic, it is not feasible to begin work until a soil moisture deficit has developed, irrespective of the soil temperature. Although each of these factors can vary considerably from season to season (e.g. the last killing frost can occur at any time over several weeks), the requirement for a combination of growing temperatures, freedom from frost and soil moisture deficit means that the timing of critical period is broadly similar from season to season. Taking an alternative approach, Allen (1977) argues that planting within the critical period ensures higher yield in most seasons because the development of the canopy is in phase with the period of maximum solar radiation input (Fig. 2.19; sect. 7.7).

For early crops in mild areas where the aim is to maximize saleable tuber yield at a harvest well before canopy senescence, the advantages of early planting are clear, as long as the crop is not defoliated by frost (Jones and Allen 1983). However, as shown in Fig. 2.15, the exploitation of any potential yield benefit from the timely planting of maincrops depends more critically upon the size and longevity of the canopy, since early planting normally leads to early senescence. In a dry, but not extreme season (1975), development of the canopy of later-planted crops (9 April, 2 May) was progressively delayed relative to the first planting (19 March), but the peak value of $L$ was progressively higher, possibly owing to higher temperatures during leaf growth and more extensive branching; in the case of the early variety, Home Guard, and, to a lesser extent, the maincrop variety, Désirée, the overall leaf area duration of the third-sown crop was considerably greater than that of earlier plantings. Consequently, the third sowing tuber yields overtook those of the earlier crops during August (Jones and Allen 1983). The relative yields of the three sowings at final harvest, therefore, depended upon the date of defoliation and the cessation of bulking which, in turn, were determined by the nitrogen and water status of the crops and the incidence of late blight. In the case of the other maincrop variety (Maris Piper), which developed generally more extensive canopies, the overall trends in $L$ were similar, but the higher $L$ values of the third-sown crop in the second half of the growing season did no more than compensate for the late development of the canopy, with the result that tuber yield in September was not significantly affected by planting date (Jones and Allen 1983).

By contrast, in the drought season of 1976, when all of the crops senesced prematurely (Fig. 2.15), the higher $L$ of late-sown crops was not fully expressed. In general, there was no effect of planting date on yield but there was a small, if statistically insignificant, yield benefit from early planting of Maris Piper.

The work of Jones and Allen (1983) (Fig. 2.15) is invaluable in explaining why the results of planting date experiments vary with season and management, and it underlines the need for several years' experience in drawing up recommended planting dates for a given site. It also emphasizes the importance of intervarietal differences in response to variation in planting date; overall differences between early and maincrops have already been noted (e.g. the economic yield benefit of planting early varieties as soon as possible, compared with the critical planting period for maincrops, each in its appropriate environment), but there are also pronounced differences within each maturity class (Bremner and Radley 1966). For example, Jones and Allen (1983) have shown that the relative tuber yields of early varieties in Pembrokeshire are dependent upon their ability to emerge, develop and maintain their leaf canopies in a hostile and variable physical environment; these, in turn, are dependent upon the length of the seed tuber sprouts (e.g. Home Guard tends to carry longer sprouts than other varieties of the same physiological age), soil temperatures after planting, and the incidence of frost.

Finally, although early planting and the sprouting of seed tubers are both methods of advancing the growth of early varieties, it is not yet possible to say to what extent an increase in physiological age can compensate for delay in planting. Allen (1977) found that crops grown from sprouted and unsprouted seed tubers tended to respond quite differently to delay in planting, but his conclusions must be treated with caution since they are derived from comparisons between seasons. As far as maincrops are concerned, there is evidence that differences in yield potential between crops planted as sprouted or unsprouted seed tubers are expressed only after the critical planting period; for example, in an experiment at Rothamsted, the loss of yield associated with delay in planting up to mid-May was avoided by using sprouted seed tubers (Dyke 1956).

### 7.4.1 PLANTING DATE AND THE COMPONENTS OF TUBER YIELD

According to the ideas of Ivins and Bremner (1965), delay in planting delays the date of tuber initiation, but the more rapid growth of the canopy of later-planted crops ensures that $L$ at initiation and, therefore, the tuber bulking rate, are higher than for earlier plantings. This does occur in some cases (e.g. in Ulster Chieftain, Bremner and Radley 1966; and in some of the crops studied by Jones and Allen 1983), but in many other cases, bulking rate has been found to be relatively unaffected by planting date; indeed, Allen (1977) gives an example of a Maris Piper crop whose bulking rate was depressed by delay in planting. Furthermore, it appears that, in many of the earlier experiments upon which the Ivins and Bremner theory was founded, the date of tuber initiation was not actually recorded, but estimated by extrapolation. Consequently, as discussed earlier, the leaf area index at tuber initiation is not now thought to be a limiting factor, and the relative yields of the crops in a planting date experiment are dependent upon the date of harvest and the duration of the bulking period (incidence of drought, nitrogen deficiency, late blight) as well as the maturity class and variety employed (sect. 7.7).

There have been few attempts to investigate the influence of planting date on stem and tuber numbers, although Allen (1977) has shown that for five varieties in two seasons the effects on each were negligible. Where yield was depressed by late planting, the difference between plantings was expressed in terms of mean tuber weight and there was no indication of substantial differences in the size distribution of tubers. Elsewhere, pronounced changes in the proportion of ware tubers have been recorded only for very late crops (Dyke 1956; Radley 1963).

### 7.4.2 OPTIMUM PLANTING DATES IN PRACTICE

Because of the variability of the environment and the intervarietal differences in response to variation in planting date discussed in the foregoing sections, it is clearly not practicable to calculate optimum planting dates from physiological principles. Recommended planting dates for maincrops, normally within the period from mid-March to mid-April, are drawn up on the basis of long-term agronomic experiments, and the probable yield penalties for delay can be computed (e.g. 0.75 t $ha^{-1}$ for every week's delay from mid-April to mid-May, increasing steeply thereafter, Radley 1963). However, in spite of the fact that these optima have been established for several decades, it is estimated that a substantial fraction of the UK maincrop is still planted after mid-April (Allen and Scott 1980), primarily as a result of problems with soil/water relations. Determination of optimum dates for early varieties is much more hazardous in view of the irregular incidence of frost and wet weather.

## 7.5 NITROGEN FERTILIZER AND TUBER YIELD

Potato crops need the full range of plant nutrients for normal growth and development but, in intensively-managed soils where there is no risk of trace element deficiency, tuber yield is normally dependent upon the supply of nitrogen, potassium and phosphorus. The requirements of the crop for the latter two elements are considerably higher than that for nitrogen (MAFF 1983), and there is evidence that the potassium or phosphorus status of the crop can influence several aspects of its physiology (e.g. canopy development and senescence; Harris 1978a; Perrenoud 1983). However, owing to the accumulation of substantial reserves of available potassium and phosphorus in most arable soils in the UK and northern Europe, potato crops receiving the recommended levels of these nutrients do not normally respond to increased P and K application. The major limiting nutrient element is, therefore, nitrogen.

Apart from relieving deficiency, the primary effect of the application of nitrogen fertilizer to a potato crop is to increase the size and number of its leaves, the latter effect principally through increased above-ground branching (sect. 2.2 *et seq.*; Millard and MacKerron 1986), with the result that canopy development is more rapid, and progressively higher peak $L$ values are attained (Fig. 7.11). By contrast, much of the available evidence from early-maturing maincrops suggests that the longevity of the canopy is relatively unaffected by the level of nitrogen fertilization (Dyson and Watson 1971); for example, Fig. 7.11 shows that the final date of defoliation in mid-August was the same for crops receiving 0 to 200 kg N $ha^{-1}$ and only a few days later at 300 kg N $ha^{-1}$. This pattern does not necessarily hold for crops whose canopies persist well into

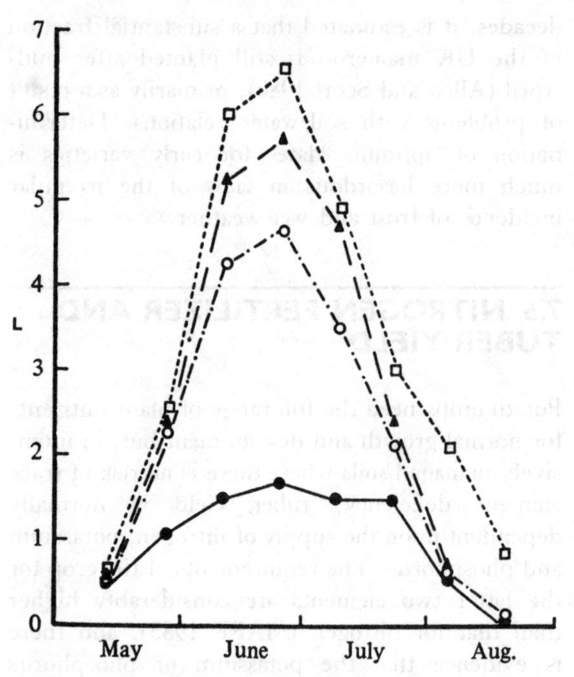


**Fig. 7.11** Typical seasonal patterns of leaf area index of early-maturing main crops of potatoes grown at different levels of nitrogen fertilization (●, ○, ▲, □ indicating 0, 100, 200, 300 kg N $ha^{-1}$ respectively) in Southern England (from Allen and Scott 1980).

September; for example, the generalized curves presented by Ivins and Bremner (1965), based on several seasons' results, indicate that leaf senescence was substantially delayed by nitrogen application, leading to enhanced leaf area duration and increased tuber yield in September. Millard and MacKerron (1986) conclude from more recent experiments that the application of nitrogen changes the pattern of canopy senescence from a synchronous decline of all the leaves to a progressive death of leaves of different ages.

Whichever pattern of leaf senescence applies, the overall effect of increasing the nitrogen supply is the progressive increase in gross leaf area duration owing to earlier canopy development, higher peak $L$ and/or delayed leaf senescence. However, since an $L$ of 3 to 3.5 is sufficient to maximize the interception of solar radiation, any increase in $L$ above about 4 will not increase the rate of dry-matter production (see sect. 2.4). Consequently, if leaf area duration is estimated by assigning values over 3 as 3, then Fig. 7.11 indicates that, in the crop under study, there was little to be gained in terms of intercepted solar radiation by applying more than 100 kg N $ha^{-1}$.

Application of nitrogen has an important secondary effect: delaying tuber bulking. Detailed field studies (Dyson and Watson 1971) have shown that this is caused not so much by a delay in tuber initiation but by a prolongation of the time between initiation and the establishment of the linear bulking phase. These results have been used to support the concept of intraplant competition between shoots and tubers; for example, increased nitrogen availability was thought to lead to the investment of surplus carbohydrate in the production of additional leaf area rather than in the early growth of tubers (Ivins and Bremner 1965). Although some experimental results can be interpreted in this way, there are reports of delayed tuber filling without any accompanying increase in stem or leaf weight (Harris 1978a). Furthermore, solution culture experiments have shown that tuber initiation can be suppressed by nitrate ions, irrespective of the nitrogen status of the plant (Sattelmacher and Marschner 1979).

This delay, and the fact that nitrogen promotes leaf growth, means that the early growth of tubers occurs at higher $L$ values than in crops receiving less nitrogen fertilizer; therefore, following the ideas of Ivins and Bremner (1965), it would be predicted that increased application of nitrogen fertilizer would lead to higher rates of tuber bulking. This has been confirmed for a range of potato crops (Gunasena and Harris 1969; Dyson and Watson 1971) but, reviewing the available evidence, there are clear indications that rates of bulking rise only up to fertilizer levels sufficient to ensure maximum interception of solar radiation ($L > 3$) at tuber initiation (i.e. 50–100 kg N $ha^{-1}$, depending upon season, location, soil etc.). For example, in an experiment carried out in the south of England, identical rates of bulking were observed at $L$ values varying from 3.5 (100 kg N $ha^{-1}$) to 5 (300 kg N $ha^{-1}$) and the rate was depressed below this value only at $L = 1.5$ in the unfertilized crop (Harris 1978a). Consequently, increase in fertilization above this level can increase tuber yield only if there is a substantial increase in leaf area duration at the end of the growing season (bearing in mind the delay in onset of bulking). These ideas are discussed in greater detail in section 7.7.

Finally, increases in the level of applied nitrogen fertilizer can have an effect upon the susceptibility of the crop to pests and diseases, although the responses are much more complex than in cereals (sect. 6.4), and on the quality of the harvested tubers, especially dry-matter content (per cent of fresh weight) as discussed in section 7.5.1.

In spite of the considerable investment in time and resources in fertilizer trials over the last century, very little of the information obtained is of use in understanding how these different effects combine to determine tuber yield. As in similar experiments on other crops, these trials tend to involve few crop or environmental measurements other than final yield, the range of levels of fertilization is limited, and the recorded responses vary considerably according to season, variety and location. However, the interpretation of potato investigations involves further, unique, difficulties. In particular, it is now clear from maximum yield studies that, in many fertilizer trials, yield is not limited simply by nitrogen supply but by the date and cause of haulm senescence (which may or may not be determined by nitrogen supply), the incidence of pests and diseases and by the water relations of the crop. Furthermore, the expression of yield in terms of fresh weight of tubers may be very important to the farmer but it can be seriously misleading to the crop physiologist; as discussed in section 7.5.1, the increase in tuber yield associated with an increment of nitrogen fertilizer can be the result of a higher yield of tuber water with no actual increase in dry matter production. Finally, it is difficult to apply the concept of harvest index, which is so useful in cereals, to the potato crop, because it is unusual for both haulm and tuber dry weights to be recorded and, in any case, under natural senescence, much of the haulm dry matter will have been lost by decomposition by the time of crop maturity.

Concentrating, therefore, on total fresh weight of tubers, it is clear that, because of the many factors influencing yield, the pattern of response to increasing nitrogen application varies considerably within and between experiments (e.g. Fig. 7.12). There have been numerous attempts to describe this family of yield response curves using quadratic expressions (e.g. Birch *et al.* 1967), but it has now been generally accepted that most of these curves can, in practice, be interpreted either as asymptotic relationships or as a pair of straight lines (Figs 7.12 and 7.13). The yield plateau may show a small but persistent response to further nitrogen application, and there can be an ultimate decline in yield at very high levels of fertilization. The rate of increase in yield below the yield plateau is generally determined by soil fertility (Fig. 7.12) but, for intensively-managed soils in northern Europe, the point of inflection tends to fall between 100 and 150 kg N applied per hectare (e.g. Perrenoud 1983).

Studies combining measurement of tuber dry-

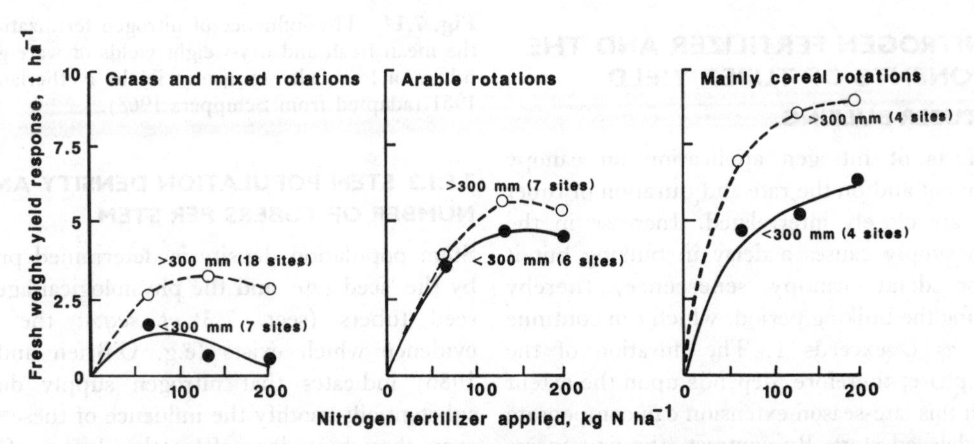


**Fig. 7.12** The influence of previous cropping and preceding winter rainfall (mm, Oct.–Mar.) on the response to nitrogen application of main crops of potatoes cv. Majestic grown at forty-six sites throughout the UK during the years 1956–62. Each curve shows the yield increment over that from unfertilized soil (adapted from Birch *et al.* 1967).

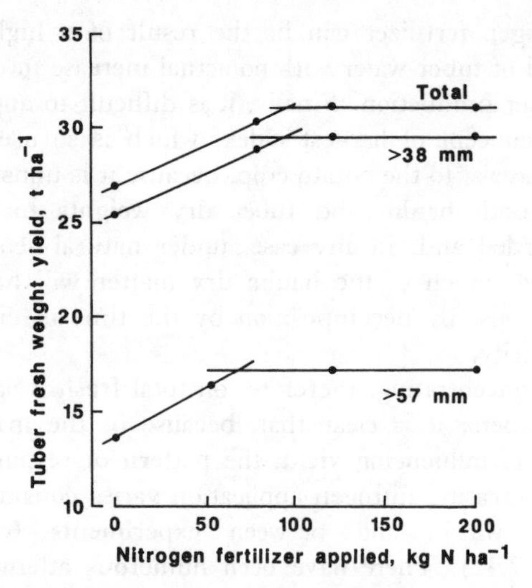


**Fig. 7.13** The influence of nitrogen fertilization on the mean tuber fresh weight yield of main crops of potatoes cv. Majestic grown at forty-two sites throughout the UK in the years 1956–62 (from data of Birch *et al.* 1967).

matter yield with a sufficient range of fertilizer levels to give a full response curve are relatively rare. However, the available data show similar asymptotic patterns of response but with the point of inflection at lower levels of nitrogen application (≤ 100 kg N ha$^{-1}$, Fig. 7.14; Inkson and Reith 1966). The observed differences between the curves of fresh and dry-weight yield are, of course, a measure of the variation in tuber water content (Schippers 1968).

## 7.5.1 NITROGEN FERTILIZER AND THE COMPONENTS OF TUBER YIELD

### 7.5.1.1 TUBER BULKING

The effects of nitrogen application on canopy development and on the rate and duration of tuber bulking are closely interrelated. Increase in the nitrogen supply causes a delay in bulking but it can also delay canopy senescence, thereby prolonging the bulking period, which can continue as long as $L$ exceeds 1. The duration of the bulking phase, therefore, depends upon the extent to which this late-season extension can compensate for the delayed start. By contrast, the first increment of nitrogen (50–100 kg ha$^{-1}$) is usually sufficient to give maximum interception of solar radiation around the time of tuber initiation, with the result that the rate of bulking does not increase, in relation to that of the unfertilized control crop, with further nitrogen application. These effects are sufficient to account for most dry matter yield response curves (e.g. Fig. 7.14). The increase in tuber dry weight associated with the first increment of nitrogen (80 kg ha$^{-1}$) is presumably the consequence of an increased rate of bulking, with a possible contribution from increased duration of bulking. Above 80 kg ha$^{-1}$, the bulking rate remains constant but, owing to the progressive delay in tuber initiation, the duration of bulking decreases, such that yield begins to decline very slowly at the highest level of application (240 kg ha$^{-1}$). The pattern of tuber fresh-weight yield (Figs 7.12 and 7.13) can be interpreted in the same way, bearing in mind the decrease in tuber dry-matter content with increasing nitrogen application.

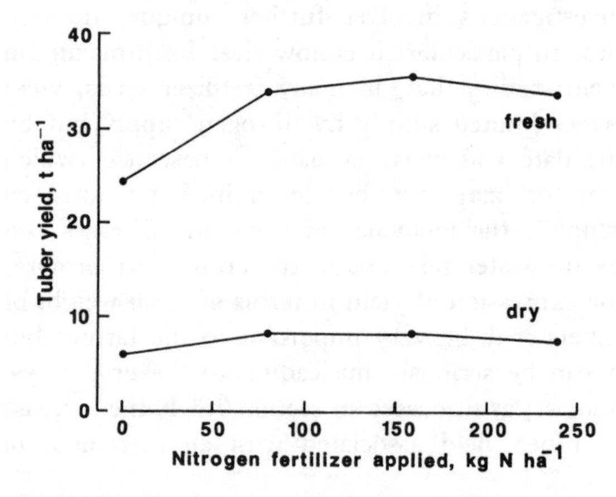


**Fig. 7.14** The influence of nitrogen fertilization on the mean fresh and dry-weight yields of ware-grade tubers of four cultivars grown in the Netherlands in 1961 (adapted from Schippers 1968).

### 7.5.1.2 STEM POPULATION DENSITY AND NUMBER OF TUBERS PER STEM

Stem population density is determined primarily by the 'seed rate' and the physiological age of the seed tubers (sect. 7.3 *et seq.*); the limited evidence which exists (e.g. O'Brien and Allen 1986) indicates that nitrogen supply does not substantially modify the influence of these factors, even though it does affect the degree of above-ground branching. In the experiments of Gunasena and Harris (1969, 1971), the application of nitrogen fertilizer before tuber initiation increased the number of tubers per plant (values per stem not available) by as much as 60 per cent over unfertilized control values, but the difference had decreased considerably by harvest and there was no effect if fertilization was delayed beyond the start of initiation. However, using a much wider range of nitrogen treatments, Clutterbuck and Simpson (1978) found that tuber number per plant responded differently in three crops over two growing seasons (little effect or reduction in 1970, increase in 1971). This lack of consistency within and between experiments may be a varietal effect (i.e. due to differing responses between Craig's Royal and Pentland Crown), or the consequence of differences in the definition of 'a tuber'. For example, Gunasena and Harris (1969, 1971) made no attempt at definition whereas Clutterbuck and Simpson's (1978) findings hold equally for all tubers, and for number of tubers excluding chats (≤ 3 cm). More recent work with the classic early variety Home Guard has shown a pronounced depression in tuber numbers per plant and per stem, but only at very high levels of fertilization (O'Brien and Allen 1986).

#### 7.5.1.3 INDIVIDUAL TUBER WEIGHT AND TUBER WATER CONTENT

There is more agreement about the role of nitrogen fertilizer in increasing tuber weight. For example, in a series of experiments with Pentland Crown, a variety which produces relatively few large tubers, Clutterbuck and Simpson (1978) found that a 32 per cent increase in tuber fresh-weight yield associated with the application of the first 100 kg N $ha^{-1}$ was caused by a 23 per cent increase in mean tuber fresh weight, together with a smaller increase (7 per cent) in tuber numbers. Tuber yield, mean tuber weight and tuber numbers were relatively stable between 100 and 300 kg N $ha^{-1}$. Grading the yields indicated that the proportion of ware tubers (≥ 4.5 cm) was unaffected by the level of fertilization and that the increase in mean tuber weight was the result of an increase in the proportion of large to outsize tubers (> 7.6 cm) (e.g. from 27 to 57 per cent by weight of ware-grade tubers with the application of the first increment of 100 kg N $ha^{-1}$) (Table 7.4). Very similar results have been obtained by other workers; for example, in Fig. 7.13, larger ware tubers (> 5.7 cm) of Majestic accounted for 51 per cent by weight of the total yield of tubers harvested from unfertilized crops, rising to 55 per cent with 134 kg N $ha^{-1}$, although the proportion of ware grade tubers (> 3.8 cm) remained constant at 95 per cent. In general, it appears that increasing the nitrogen supply up to 100–150 kg $ha^{-1}$ increases the proportion of large tubers without substantially altering the combined proportions of ware and outsize tubers harvested.

**Table 7.4** Components of the tuber yield of irrigated (I) and control (C) crops of Pentland Dell potatoes grown in south-east Scotland in 1970

| | | *Nitrogen fertilizer applied (kg N $ha^{-1}$)* | | | |
|---|---|---|---|---|---|
| | | *0* | *100* | *200* | *300* |
| Total tuber fresh-weight yield (t $ha^{-1}$) | C | 36.3 | 47.8 | 42.7 | 43.2 |
| | I | 38.1 | 47.0 | 50.2 | 48.5 |
| Yield of ware-grade (>45 mm) tubers (t $ha^{-1}$) (% of total) | C | 35.5 (98) | 47.2 (99) | 41.1 (96) | 41.9 (97) |
| | I | 36.3 (95) | 46.4 (99) | 49.7 (99) | 47.4 (98) |
| Number of ware-grade tubers per plant | C | 4.2 | 4.5 | 3.9 | 4.1 |
| | I | 5.5 | 4.9 | 4.9 | 5.2 |
| Fresh-weight yield of large tubers (>76 mm) as a % of ware yield | C | 27 | 57 | 58 | 55 |
| | I | 8 | 35 | 46 | 38 |

(Recalculated from Clutterbuck and Simpson 1978)

This can lead to the production of increasing yields of unsaleable, outsize tubers in varieties such as Pentland Crown.

Several investigations have shown that the dry matter content of tubers (per cent of fresh weight, commonly expressed in terms of specific gravity, Cole 1975) falls with increasing nitrogen application. For example, Fig. 7.14 illustrates the gain in fresh weight yield which can be obtained by supplying nitrogen at levels higher than that required to maximize tuber dry matter yield (here 80 kg N $ha^{-1}$). Although a great deal is known about the effects of variety and tuber size (at a given fertilizer level) on tuber specific gravity, there are few published data on tubers of different grades grown at different nitrogen levels (Cole, 1975; Ifenkwe and Allen 1978; O'Brien and Allen 1986). It is, therefore, not possible to say whether the observed effects are caused by the proportional increase in large tubers of low dry matter content or by a general effect on all grades.

### 7.5.2 NITROGEN FERTILIZATION IN PRACTICE

The results shown in Figs 7.13 and 7.14 are typical of fertilizer experiments carried out on maincrop potatoes in intensively managed arable soils in the UK. Dry-matter yield generally reaches a maximum with the application of 75–100 kg N $ha^{-1}$, but fresh-weight yield and mean tuber size continue to increase with the addition of a further 50 kg N $ha^{-1}$. Since the trade in potatoes is based on fresh weight of tubers, and quality aspects such as specific gravity are normally important only in tubers for processing (a minor, but increasing fraction of the national yield, Jones 1983), it appears that the commercial yield plateau will be reached with applications in the range 100–150 kg N $ha^{-1}$. These values are well below the mean level of fertilization in practice (approximately 200 kg N $ha^{-1}$). Although the economic return per unit of nitrogen fertilizer would be expected to be higher below the fresh-weight yield plateau, the situation is more complex than that for cereals (sect. 6.4.3) because tuber size (another index of tuber quality and, therefore, of the financial value of the crop) can continue to increase with nitrogen applications up to that sufficient to maximize fresh-weight yield (Fig. 7.13). Consequently, the fertilizer treatment for maximum financial yield depends not only upon the costs of growing the crop and its value but also upon the variety used; for example, using a variety such as Pentland Crown, it may prove necessary to adjust the population density (seed rate) at higher levels of nitrogen fertilization to reduce the risk of producing outsize, unsaleable tubers (sect. 7.3.1).

However, maincrop potatoes are grown in a variety of soils and crop rotations and although the shapes of the response curves tend to be similar for most crops (Fig. 7.13), the quantity of nitrogen fertilizer required to achieve the yield plateau can vary considerably (e.g. from 70 to 200 kg $ha^{-1}$, according to rotation and degree of nitrogen loss by leaching, Fig. 7.12). Because these variations are largely the result of differences in the nitrogen status of the soil before fertilization, recommended levels of application are, as with cereal crops, based on the soil type and the nitrogen index of the relevant field (sect. 6.4.3; Table 6.9). Thus, for maincrops grown on mineral soils, the recommended application can vary between 220 (index 0) and 100 kg N $ha^{-1}$ (index 2), whereas for fen peats the corresponding values are 130 and 50 kg $ha^{-1}$ (MAFF 1983). A similar, but simpler, system of prediction is used in Scotland (ESCA 1983).

Ivins (1963) suggested that since the application of nitrogen delays tuber initiation, then the withholding of most of the fertilizer nitrogen until just after initiation might advance the start of tuber bulking without seriously affecting canopy development. This, and the suggestion that late application of fertilizer might delay senescence and prolong bulking, led to a series of experiments investigating the influence of the timing of application (e.g. split applications, late foliar sprays etc., Harris 1978a). Apart from cases where spring rainfall has caused serious leaching, these treatments have not given consistent yield increases in unirrigated crops (sect. 7.6), and it is normally recommended that all fertilization should take place at planting (MAFF 1983). However, experiments on the value of slow-release nitrogen fertilizer compounds to ensure adequate supplies of nitrogen during the second half of the growing season continue, with mixed results (e.g. Penny *et al.* 1984). Finally, it should be stressed that all the levels of application discussed refer to fertilizer broadcast over the soil, ridged or flat, at planting; these may be reduced somewhat if it is banded near to the seed tubers, although this can involve chemical damage to the tuber and its sprouts under some conditions (Harris 1978a; MAFF 1983).

## 7.6 IRRIGATION

Exposure of potato crops to water stress can lead to loss of potential yield by reducing both the interception of solar radiation and the efficiency of canopy photosynthesis. As a result of shortage of water during canopy development, the area of individual leaves is reduced but, in the potato crop at least, the number of leaves can also be reduced because of depressed rates of leaf unfolding (Munns and Pearson 1974). Furthermore, drought can lead to premature leaf senescence. Thus, depending upon the severity and timing of the period of water stress, the crop will suffer a loss of leaf area duration caused by delayed canopy growth, lowered peak leaf area index, premature defoliation, or a combination of these effects (Fig. 7.15). However, it should be stressed that the alleviation of these effects by irrigation may be a consequence not only of improved water supply but also of increased availability of nitrogen and other nutrients (Perrenoud 1983).

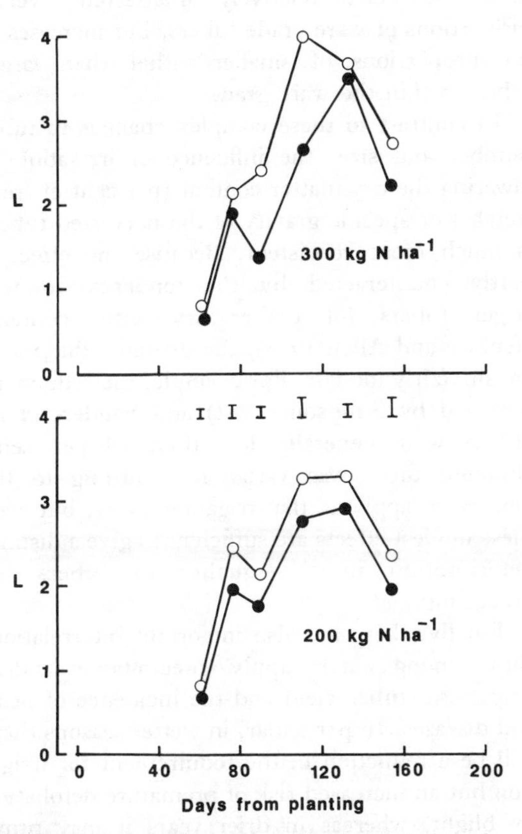


**Fig. 7.15** Seasonal patterns of leaf area index of main crops of potatoes cv. Pentland Crown grown with (○) or without (●) irrigation at two levels of nitrogen fertilization in south-east Scotland in 1970. The vertical bars indicate SEM for all four curves (from data of Clutterbuck and Simpson 1978).

In addition to its influence upon canopy size, relatively mild water stress can have a marked influence on the physiology of fully-expanded potato leaves. Thus, lowering of the water potential of the soil by as little as 0.025 MPa (Epstein and Grant 1973) or of the leaf by 0.5 MPa (Munns and Pearson 1974) was sufficient to give significantly lower rates of $CO_2$ exchange owing to partial closure of stomata. This indicates that the potato is among the more sensitive of mesophytic plant species (Fitter and Hay 1987). There will also be a premature decline in the photosynthetic potential of leaves senescing under water stress (sect. 3.4.5.1).

### 7.6.1 IRRIGATION AND TUBER YIELD

Because the potato crop is so sensitive to water stress, many irrigation experiments in the major maincrop growing areas of the UK have shown positive yield responses to applied water. However, as a result of variation in the timing and severity of drought, in the threshold soil matric potential at which irrigation is started and in the methods of applying irrigation water, the observed increase in tuber yield per unit of applied water can vary considerably, and there are cases of yield depression under poorly-designed or poorly-managed irrigation schemes. Harris (1978b) tabulates a wide range of published irrigation efficiencies from −0.1 to 1.4 t $ha^{-1}$ $cm^{-1}$ of applied water and estimates that responses of up to 2–2.5 t $ha^{-1}$ $cm^{-1}$ are possible. However,. it should be emphasized that these efficiencies are expressed in terms of applied water and that the water-use efficiency of the crop (response per unit of water transpired) will lie at the upper limits of this range. Some workers have observed interactions between irrigation and nitrogen fertilizer level mainly because of the increased demand for nitrogen by higher-yielding irrigated crops. For example, in a season of average length in south-east Scotland, irrigation of a crop of Pentland Crown increased the level of fertilizer required for maximum tuber yield from 100 to 200 kg N $ha^{-1}$ (Table 7.4) whereas, in an unusually long season, both irrigated and control crops gave the highest yield at 300 kg N $ha^{-1}$ (Clutterbuck and Simpson 1978). However, as will be clear from the following discussion of the influence of irrigation on the components of yield, there are few data to support these observations, mainly because controlled irrigation experiments are so difficult to carry out. Furthermore, it should be noted that the canopy leaf areas achieved by the crops in Clutterbuck and Simpson's experiment were low considering the high levels of fertilization (compare Fig. 7.15 with Fig. 7.11), suggesting that factors other than water and nitrogen supply were controlling plant growth and yield.

In the same series of experiments, the water stress experienced by the unirrigated crops was not extreme, tuber bulking ceased at a similar date in the different treatments and the increases in yield shown by the irrigated crops appeared to be the consequence of delayed tuber initiation (although this was not measured) but higher bulking rates. There are few data on the effect of water supply on the time of tuber initiation but it is clear from a wide range of investigations that bulking rate is very sensitive to variation in water supply (compared with the effects of variation in irradiance or temperature) (Moorby *et al.* 1975; Asfary *et al.* 1983). Thus, under moderate to severe stress, bulking can cease altogether, resuming when the stress is lifted. However, renewed tuber expansion does not occur uniformly, resulting in the various tuber deformities known collectively as secondary growth (dolls, physiological cracks, etc.). Increase in yield under irrigation can, therefore, be the result of higher and/or maintained bulking rate, extended duration of bulking, as well as the prevention of tuber deformities.

Reviewing a large body of evidence from Europe and North America, Salter and Goode (1967) concluded that irrigation early in the life of the crop (probably before tuber initiation in most cases) generally reduced the yield of ware-grade tubers owing to the proliferation of poorly-filled tubers, but that treatment after initiation resulted in increased tuber size, with little effect on tuber numbers. This is a useful generalization which is used widely in practical advice to farmers (e.g. Hart 1983), but it should be emphasized that the overall result of an irrigation treatment depends upon season, the incidence of drought in relation to crop development, the timing and method of irrigation, level of nitrogen fertilization and the variety grown, as well as other factors such as disease incidence. Thus, for example, it has already been noted that irrigation can actually depress yield and, as shown in Table 7.4, after tuber initiation it can raise tuber yield as a result of unexpected combinations of effects (increased tuber numbers, relatively unaffected overall proportions of ware-grade tubers, but increases in the proportions of smaller rather than larger tubers within the ware grade).

In contrast to these complex changes in tuber number and size, the influence of irrigation in lowering the dry-matter content (per cent of fresh weight) or specific gravity of the harvested tubers is much more consistent. Because the effect is partly counteracted by the tendency towards larger tubers (of higher dry-matter content, Ifenkwe and Allen 1978), the overall reduction is usually fairly modest. For example, the reductions recorded by Simpson (1962) and Sanders *et al.* (1972) were generally less than 10 per cent, although there was variation according to the method of applying the irrigation water, but even these modest effects are sufficient to give a distinct deterioration in the quality of tubers for processing.

Finally, there are also important interrelationships among water supply (precipitation and/or irrigation), tuber yield and the incidence of pests and diseases. In particular, in wetter seasons there will be a reduction in the requirement for irrigation but an increased risk of premature defoliation by blight, whereas in drier years it may prove necessary to irrigate certain varieties to prevent the incidence of scab. Depending upon the season, these and other pests and diseases can reduce both the yield and quality of tubers. Overall, it can be concluded that, although irrigation can have a substantial role in ensuring high yield in dry areas (MacKerron *et al.* 1982 estimate that between 50 and 60 per cent of the potential yield of maincrops in England is lost as a consequence of water stress), very little is known about the effect of water relations on the physiology of the potato crop.

### 7.6.2 IRRIGATION IN PRACTICE

Recommended procedures for the irrigation of maincrops of potatoes in the UK have been drawn up using the results of long-term experiments, some lasting up to 30 years (MAFF 1982b; Hart 1983). Once the tubers have reached the marble stage (10–20 mm), water should be applied to raise the soil to field capacity whenever the soil moisture deficit reaches 25 mm (for sandy soils only; the limiting deficiency is higher, but variable, for soils with heavier textures). This should be continued throughout the bulking phase, although the treatment should cease before defoliation to promote tuber maturity. If the treatment is designed to prevent scab, it may prove necessary to maintain wetter conditions (irrigation at a smd of 15 mm before and during initiation using small applications of 5–10 mm) and care should be exercised with varieties such as Maris Peer, Maris Piper and Record which are particularly susceptible to drought. In the UK, yield increases (variously quoted at 2–4 t ware tubers $ha^{-1}$ per 25 mm or 80 kg $ha^{-1}$ $mm^{-1}$) can be anticipated if the crop receives sufficient fertilizer; in some cases, this may require the splitting of nitrogen application between planting and tuber initiation (MAFF 1983; Penny *et al.* 1984). Crop surveys indicate that the irrigated area in the UK is expanding rapidly (e.g. from 3 to 15 per cent between 1973 and 1978, Harris 1983).

## 7.7 THE ACHIEVEMENT OF HIGH YIELDS OF TUBERS

Reviewing the conclusions of sections 7.2 to 7.6, it is clear that the factors contributing to high total tuber yields of potato crops (physiologically young, multisprouted seed tubers, Fig. 7.7; optimum stem population density, Figs 7.9 and 7.10; planting not later than the critical period, sect. 7.4; optimum nitrogen fertilization, Figs 7.12 to 7.14; irrigation at a threshold soil moisture deficit; together with protection from defoliating pests and diseases) act primarily to extend the life of the crop canopy and the duration of tuber bulking, rather than by increasing the rate of tuber bulking. The same conclusion can be reached by plotting the total dry matter yield of a series of potato crops against the total amount of solar radiation intercepted by the crop canopy over the growing season (e.g. Fig. 7.16). Since the harvest index of the potato crop (tuber dry weight as a fraction of the dry weight of above-ground parts and tubers combined) appears to be *relatively* stable (Allen and Scott 1980; MacKerron

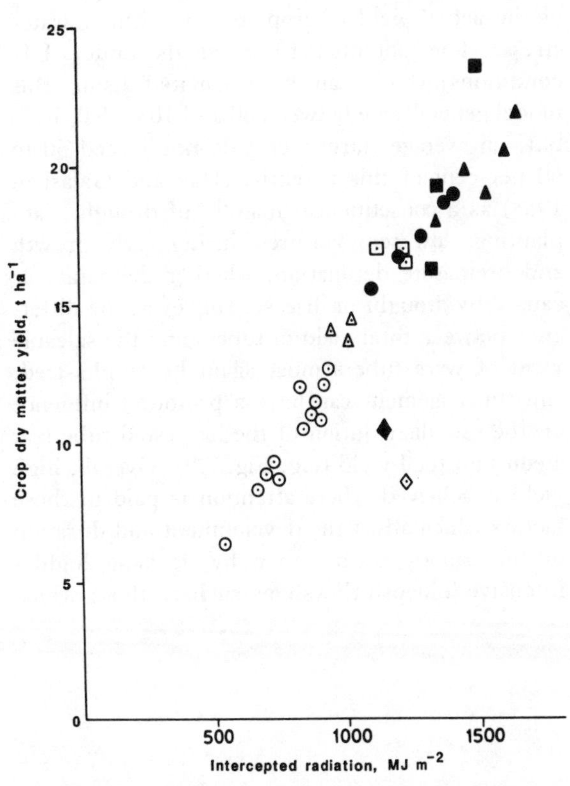


**Fig. 7.16** Relationship between total dry-matter production of maincrops of potatoes (open symbols) and sugar beet (closed symbols) and the total quantity of solar radiation intercepted by the crop canopies in Southern England during several seasons (○, 1964; △, 1972; □, 1973; ▲, ●, early and late planted, 1971; ■, 1973–5). Note that the results for the drought season of 1976 (◇, ◆) do not conform to the linear relationship which holds for each of the other seasons (adapted from Scott and Wilcockson 1978).

and Heilbronn 1985), then the linear relationship obtained will also hold for tuber dry weight yield, confirming that the duration of the crop canopy determines tuber yield, except in extreme years such as 1976, when the efficiency of dry matter production (and rate of tuber bulking) is reduced by water stress (Fig. 7.16).

Further support for this conclusion comes from the simulation of the growth, development and yield of the potato crop carried out by MacKerron and Waister (1985) and MacKerron (1985). Their model, which uses simple agrometeorological data (daily records of temperature and incident solar radiation) from planting to canopy senescence to estimate radiation interception by the canopy, gave yield estimates which were within 6 per cent of the actual yield of crops grown without water stress. The potential tuber yields under UK conditions which can be calculated using this model generally lie between 60 and 105 t FW ha$^{-1}$ but, on average, farm yields do not exceed 50 to 60 per cent of this potential (Hay and Galashan 1988) as a consequence, mainly, of drought, late planting, low temperatures during early growth and premature defoliation, whether deliberate or caused by drought or disease. However, the difference between total yield of tubers and the saleable yield of ware tubers must again be emphasized, since management can have a profound influence on the size distribution of the harvested tubers at a constant total yield (e.g. Fig. 7.9). Overall, high yield is achieved where attention is paid to those factors which affect the development and duration of the canopy, rather than by devising highly-intensive 'blueprint' systems such as those which were fashionable in the 1970s but which proved not to be widely applicable to different soil types, climates or varieties (Evans *et al.* 1978).

At least one outstanding physiological problem remains – the linearity of tuber bulking rate in the face of variation in the supply of incident solar radiation. This is at least partly explained by the fact that bulking rates are normally calculated from destructive harvests at intervals of one week or longer. There is, therefore, a considerable degree of statistical smoothing of the data since, during the midsummer months of May to July, the total incident radiation per week will vary much less than from day to day. Thus, if the canopy is intercepting >90 per cent of the incident radiation during this period, an approximately linear relationship would be anticipated. However, following the same argument, there should be a progressive decline in the rate of bulking in August and September in phase with the seasonal decline in incident solar radiation. In their review of seventeen crops and treatments over several seasons, Allen and Scott (1980) did find considerable variation in the rate of bulking, with a distinct tendency towards lower rates from August onwards, and Wurr (1974) measured bulking curves which were distinctly curvilinear. Although it is not possible at this stage to rule out effects of the early stages of leaf senescence on the efficiency of dry matter production, these findings go some way to reconciling the generally accepted (but increasingly questioned) linearity of the bulking phase with results such as those presented in Fig. 7.16. It is clear that a great deal of detailed work needs to be done to clarify this subject.

# CHAPTER 8 GRASSLAND

*Grazing . . . is always an inefficient way of harvesting, leading to a loss of 40–50 per cent of the shoot growth produced. Systems involving grazing, therefore, introduce interrelationships with the animal which often dominate the direct responses of the plant to defoliation.*

(Milthorpe and Davidson 1966)

*. . . the normal method of measuring production – as dry matter yields of herbage harvested from cut plots – has both resulted in questionable advice on grassland management and played no small part in the separation of plant and animal, which has long characterized much of our research thinking and effort. . . . Grazing and conservation are key factors in the economic realization of the potential of grass for animal production; they thus need to be understood fully. We know less than we should about the sward/animal interface in the pasture, and the techniques which are presently available for measuring sward growth and herbage intake leave much to be desired.*

(Lazenby 1981)

Grassland farming is by far the most extensive agricultural enterprise in the world, extending over 2500 to 3000 million ha and supporting some 2100 million head of livestock (compare with 700 million ha of all cereals) (Langer 1979; Lazenby and Down 1982). Although there are few areas of true natural grassland in the UK, grass now accounts for up to 75 per cent of the more intensive agricultural area of the country (Holmes 1980), wherever water relations and topography make arable cropping impractical. This area of anthropogenic grassland can be subdivided in several ways: into temporary (rotational) grass, permanent grass and rough grazing or, according to the species present, the method of harvesting (grazing, cutting for conservation as hay or silage or a combination of both), or on the basis of yield potential. For example, annual dry-matter production can range from 1 t $ha^{-1}$ for *Molinia* hill grass heath (HFRO 1979) to 15–20 t $ha^{-1}$ for intensively managed lowland swards of *Lolium perenne* (Lazenby and Down 1982) but with a national average, excluding rough grazings, of only 6 t $ha^{-1}$ (Robson 1981). For simplicity, the present chapter concentrates mainly upon the most thoroughly studied and characterized lowland grasslands in which *L. perenne* tends to be the dominant species. However, mixtures of species and less-favoured environments are also considered where appropriate (e.g. sect. 8.4).

The grass sward presents the crop physiologist with a number of unique problems, of which the principal difficulty is how to measure yield in a meaningful way. Since grass is grown expressly for consumption by ruminant livestock, then the true yield is the intake of grass, or of digestible dry matter, per unit area of sward, irrespective of whether it is grazed or cut. Alternatively, yield may be expressed in terms of the live weight gain of the test animal. However, not only are such methods very costly and difficult to carry out, but they also involve aspects of animal physiology, nutrition and behaviour which may not be of strict relevance to the physiology of the crop. Consequently, the productivity of grass swards is commonly assessed by means of a series of harvests, by cutting, at regular time intervals, although it must always be borne in mind that the recorded yields will vary with cutting regime, and the seasonal pattern of growth may be different for cut and grazed swards (sect. 8.2).

Other difficulties are associated with the perennial nature of the crop. For example, the canopy structure, yield and species composition of a sward can differ markedly between the establishment year and subsequent seasons. Furthermore, in contrast to the crops studied in Chapters 6 and 7, where yield is accumulated in specialized grains or tubers, the leaves are effectively both source and sink for photosynthate, in the absence of flowering. There is, therefore, a conflict between crop harvesting and the maintenance of photosynthetic tissue, although this is somewhat modified by the relatively short life of individual leaves (sect. 8.1 and 8.2.1). Finally, since the harvest index of grass crops is higher than that of cereals ($\geq 0.6$), the grass breeder is forced to pursue higher rates of biomass production as a primary aim; this is intrinsically more difficult than increasing the harvest index of cereals (sect. 3.5.2 and 6.4.1; Robson 1981).

## 8.1 GRASS SWARDS AND THE SEASONAL PATTERN OF DRY-MATTER PRODUCTION

The unit of population in grassland studies is the tiller. Because the species used for intensive production in the lowlands of northern Europe (*Dactylis glomerata*, *Lolium* spp., *Phleum pratense*) are generally tufted, with new tillers developing close to the parent stem, it is essential to maintain high plant population densities and prevent damage to the sward by poaching, pests (e.g. *Tipula* spp., moles) or diseases. In contrast, several of the so-called 'secondary' grass species (e.g. *Agrostis* spp., *Poa pratensis*), which are of greater importance in less intensively managed permanent pastures, upland grassland and in sports turfs, are capable of colonizing bare areas by means of stolons or rhizomes.

As inter-plant competition develops following the establishment of a grass sward, plant population density falls (Langer *et al.* 1964) but, because of the proliferation of tillers, tiller population density increases to a relatively stable level (e.g. Fig. 8.1) determined by management, which tends to decline with the age of the sward. For example, hay meadows tend to carry relatively few large tillers, whereas a continuously-grazed sward is composed of a very high population of small tillers (Fig. 8.4, Table 8.1). The relative stability of tiller numbers from season to season under the same management appears to suggest that each tiller, which can normally live for up to a year, is replaced, on average, by a single daughter tiller, and that the majority of tiller buds in a sward do not develop. However, this overall view obscures the fact that new tillers are generated throughout the year and that there is a continuing turnover of short-lived vegetative tillers (Fig. 8.1); this can result in important short-term fluctuations in tiller numbers, for example as associated with the midsummer depression in yield discussed below. The effect of management on the tiller dynamics

**Table 8.1** Tiller population densities of perennial ryegrass swards under contrasting management

| *Method of defoliation* | *Nitrogen fertilizer applied* (*kg N ha$^{-1}$ yr$^{-1}$*) | *Tiller population density* (*10$^3$ m$^{-2}$*) | *Source* |
|---|---|---|---|
| Continuous grazing: | | | |
| $L = 1$ | 360 | 22–65 | Parsons *et al.* 1983a |
| 3 | 360 | 20–40 | |
| Cutting at intervals of: | | | |
| 3 weeks | 0 | 4.5 | Wilman and Wright 1983 |
| | 263 | 5.3 | |
| | 525 | 6.3 | |
| 10 weeks | 0 | 3.8 | |
| | 263 | 3.2 | |
| | 525 | 2.8 | |

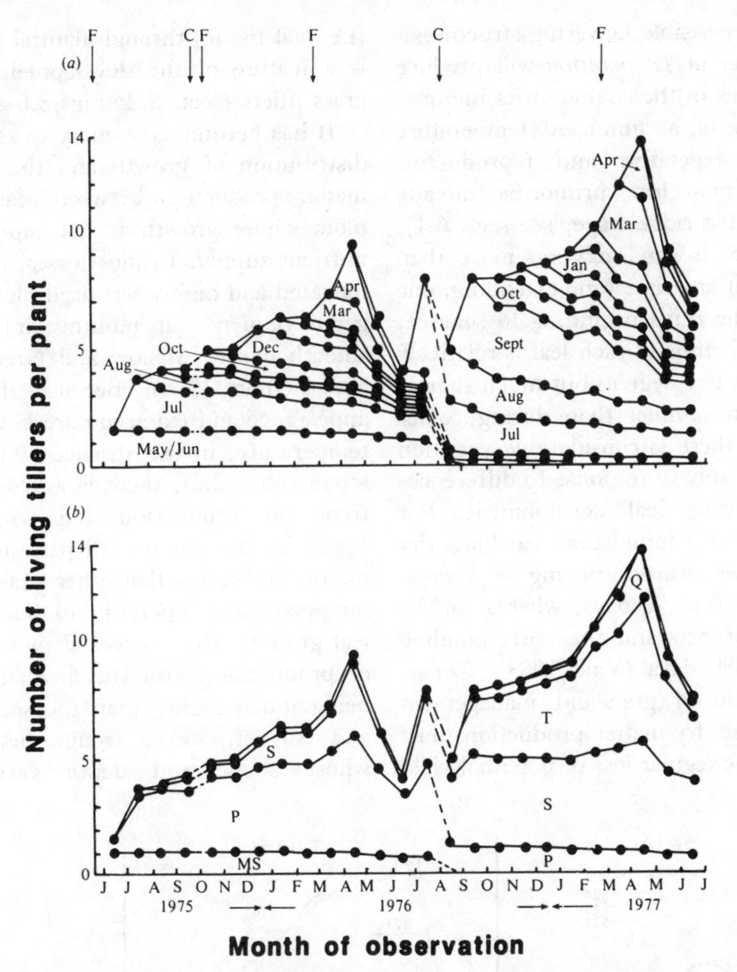


**Fig. 8.1** Changes with time in the tiller demography of a sward of perennial ryegrass cv. S23, established in North Wales in May 1975, expressed in terms of: (a) the month of tiller initiation, and (b) tiller order group (MS, mainstem; P, primary, or first-order, branching from the mainstem; S, secondary, or second-order, branching from a primary tiller; T, tertiary or third-order; Q, quaternary or fourth-order). The arrows marked C and F indicate the dates of cutting (once per season for hay) and fertilizer application (from Colvill and Marshall 1984).

of grass swards is considered in greater detail in subsequent sections (8.2 and 8.3)

Long-lived tillers formed in late spring, summer or autumn remain vegetative (with the stem apex at or below the soil surface generating leaf primordia only) until vernalized by low temperature, short daylengths or a combination of these factors (sect. 6.3). Thereafter, the tiller, but not the whole plant, becomes reproductive; the initiation of new leaves ceases, the stem apex produces only spikelet primordia and the tiller will eventually undergo internode extension to produce a spike or panicle in the late spring or summer of the second year (Jewiss 1981). These developments are very similar to those described for cereals in section 6.1, with the important difference that new vegetative tillers can continue to develop from the lower nodes of the flowering stem. Of course, this pattern refers only to crops grown for conservation (hay or silage) and does not strictly apply to grazed swards; the continuous removal of extending stem and spike or panicle tissues means that the development of reproductive tillers is not fully expressed in grazed swards. However, it is important to emphasize that these tillers are still reproductive even though they do not produce visible flowering structures.

Typically, a tiller of *L. perenne* will produce more than ten leaves in the course of its lifetime, at regular intervals of accumulated temperature during both the vegetative and reproductive phases (latterly from leaf primordia already existing at the double ridge stage, see sect. 6.1). Consequently, since it can carry no more than three green leaves at any time (one extending, one fully mature and the third beginning to senesce; Davies 1977), the lifetime of each leaf is relatively constant in terms of day degrees but much shorter in calendar time in summer than during colder periods. However, there is considerable variation between areas, probably in response to differences in temperature during leaf development. For example, Davies (1977) found that leaf longevity in perennial ryegrass swards growing at Aberystwyth varied from 30 to 90 days, whereas in the cooler south-east of Scotland the corresponding range was 50 to 180 days (Vine 1983). Consequently, any system of grassland management designed to optimize dry-matter production must take into account the regular loss of potential yield (i.e. leaf tissue) through natural senescence which is a feature of the development and growth of grass tillers (sect. 8.2; Figs 8.6 and 8.7).

It has become customary to assess the seasonal distribution of growth and the variation in dry matter production between seasons using grass plots whose growth is not limited by water or nutrient supply. In most cases, up to four sets of irrigated and highly-fertilized plots are cut in staggered sequence at monthly intervals to give a smooth curve (Anslow and Green 1967; Corrall and Fenlon 1978). Because the rates of leaf appearance and extension are both determined by temperature, in the absence of water or nutrient stress (sect. 2.2), there is a pronounced seasonal trend in production (e.g. for *L. perenne*, Fig. 8.2), but the overall pattern is highly asymmetric, indicating that other factors in addition to temperature are operating to determine the rate of leaf growth. Most curves show two distinct peaks of production, with the first, in early summer, being much higher than the second during July and August, and a trough between the peaks whose shape and depth varies considerably

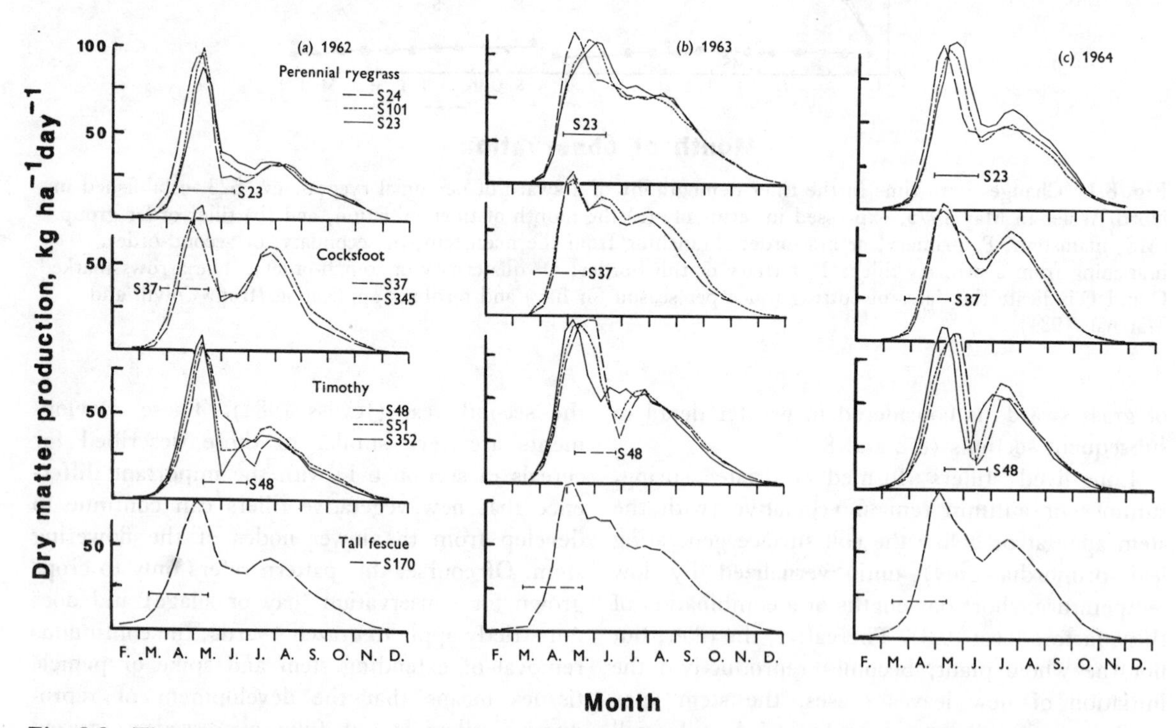


**Fig. 8.2** The seasonal distribution of dry-matter production of pure swards of several cultivars (S24 etc.) of four grass species grown in Southern England during three seasons. The horizontal bars indicate the period from initiation to emergence of inflorescences of one cv. per species (adapted from Anslow and Green 1967).

between seasons. Thus, the response of the sward to temperature alters between spring and late summer, and there is a distinct midsummer depression of growth. Other grass species show similar patterns of response, differing only in the relative heights of the two peaks and in the timing and severity of the midsummer depression (Fig. 8.2).

This seasonal asymmetry has been investigated intensively over the last 15 years. Peacock (1975b) showed that the rate of leaf extension was significantly higher in spring than in autumn, with the enhancement rising from 0 per cent at 0 °C to 100 per cent at 13 °C, and he proposed that the difference in response to temperature was associated with the change from a sward dominated by reproductive tillers (spring) to one containing vegetative tillers only (midsummer to winter). This was confirmed by Parsons and Robson's (1980) study of leaf extension during the transition from vegetative to reproductive development early in the year. Plants of an early variety of perennial ryegrass (S24) changed abruptly from the 'autumn' rate of leaf extension to the faster 'spring' rate in mid-February at a time when the apical dome had begun to elongate before the development of double ridges. Although the mechanism of this change has not been studied in detail, it appears to be related to an increased mobilization of fructosan reserves (sect. 3.4.4).

However, Parsons and Robson (1981a) found that the photosynthetic potential of the ryegrass canopy also began to rise rapidly at the same transition because the newly-extended leaves, which were not shaded by the prostrate rosette of older leaves and which developed in a period of steadily increasing solar radiation, had a higher photosynthetic potential than those formed in late autumn (Woledge and Leafe 1976; Woledge 1977). This effect was maintained throughout the subsequent life of these reproductive tillers because progressive sheath and stem extension ensured that each new leaf developed at the top of the canopy in full sunlight. Using a crop simulation model, they were able to show that 63 per cent of the enhancement of dry-matter production (over that of the vegetative sward) was the consequence of the increased photosynthetic potential of the individual leaves, the remainder being caused by the combination of increased leaf area index and improved radiation interception by the canopy (i.e. by changes in the extinction coefficient, sect. 2.4.3). Analysis of these and subsequent experiments (Parsons and Robson 1981b) indicated that the faster growth of reproductive swards is not caused by differences in the partitioning of dry matter between root and shoot, nor by the increased sink capacity of the extending stem (Deinum 1976). The asymmetry is, therefore, the consequence of the more rapid rates of photosynthesis and growth of reproductive tillers than vegetative tillers, when exposed to similar levels of temperature and irradiance.

It was originally thought that the midsummer trough between the two peaks of production was caused primarily by drought but, although water stress can be an important contributory factor in many years, the trough also appears in wet seasons. Detailed studies of tiller dynamics have revealed that the depression in growth coincides with a period of low tiller population density caused by the shading and death of young vegetative tillers developing at the base of a canopy dominated by reproductive tillers undergoing stem extension and heading (compare Figs 8.1 and 8.2). Furthermore, because the leaves of the surviving vegetative tillers have developed under shaded conditions, they have a generally low photosynthetic potential (Woledge 1977). Midsummer depression of dry-matter production is, therefore, the result of a combination of lower tiller numbers, lower leaf area indices and the lower photosynthetic potential of individual leaves, and it can be intensified by water (or other) stress. Production can recover only by the proliferation of new vegetative tillers and the production of new (unshaded) leaves of higher potential.

The Corrall and Fenlon (1978) method of measuring the seasonal distribution of growth has been used widely for comparative purposes (e.g. between seasons and countries, Corrall 1984), but it has only a limited relevance in the interpretation and prediction of grassland production. In particular, the method of harvesting (staggered cuts at intervals of 4 weeks) bears no relation to frequently defoliated grazed swards or to grassland cut for hay after heading, and it is a rather poor approximation to the two or three cuts per year for conservation as silage. The frequency of defoliation is similar to that of some systems of rotational grazing but, as discussed below, there are important differences between the effects of cutting and grazing on the sward. Long-term studies of dry-matter production under grazing are much less common, for practical reasons; however, the few available accounts of the seasonal distribution of production under continuous defoliation (e.g. Fig. 8.3) tend to show a smoother curve without a pronounced midsummer depression, although growth is still significantly faster in spring and early summer. These observations are generally compatible with the findings from cut swards. In particular, the generally more gradual change in the rate of growth under grazing during the spring and summer is the consequence of a more gradual decline in the proportion of reproductive tillers (removed whenever the apex rises into the zone of grazing) and of the development of new vegetative tillers in a less competitive and less shaded environment.

Each of the curves shown in Figs 8.1 to 8.3 was constructed using data from swards consisting of a single variety. Because there are distinct differences in response between varieties and between species (Fig. 8.2), the seasonal pattern of production can be altered by the use of grass mixtures. For example, the high-yielding reproductive phase can be extended to a certain extent, without necessarily increasing total yield, by a judicious combination of early and late varieties, whereas the midsummer trough, which is a pronounced feature of timothy varieties, can be reduced by the inclusion of a proportion of other species. Such mixtures are a normal feature of both grazed swards and hay meadows, but are less important in grassland cut for silage, where synchrony of development and quality is necessary.

In less intensive systems of production, it may be necessary to encourage the growth of clover. This requires a high degree of grassland management because the seasonal pattern of clover growth is quite different from that of its companion grasses; even in lowland areas, clover growth is slower than grass growth under low spring temperatures, but there is a single, symmetric peak of production during the five months May to September (Haycock 1984) which can compensate, to a certain extent, for the midsummer depression in grass production. This and other features of grass/clover mixtures are considered in more detail in sections 8.3 and 8.4.

Finally, superimposed upon these seasonal patterns of dry-matter production, there are also seasonal trends in the chemical composition of the crop which affect its quality in relation to the dietary requirements of ruminant livestock. Relevant indices of quality include the metabolizable energy and the digestible crude protein contents (Holmes 1980), but most workers use a simpler measure, the *D* value (digestible organic matter as a percentage of total dry matter). In general, the *D* value of a perennial ryegrass sward falls from around 75 per cent in spring to 60 per cent at heading in June, in parallel with the increasing proportions of indigestible components such as lignin in the stem tissues. The *D* value of

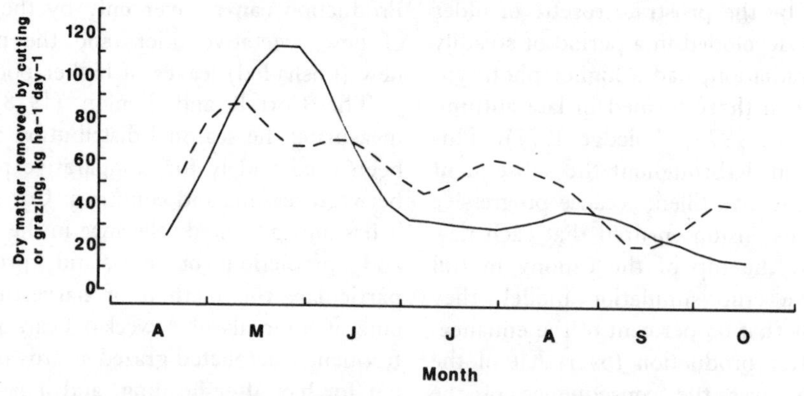


**Fig. 8.3** The seasonal distribution of dry-matter production of similar swards of perennial ryegrass cut according to the method of Corrall and Fenlon (1978) (see text) (——), or continuously grazed by sheep to mean sward heights of 3, 5 or 7 cm (---, representing the mean of the three treatments) (A J Parsons: in press; data from Hurley, Southern England).

regrowth tends to be intermediate between these two values. Thus, overall, the inclusion of the digestibility factor serves to increase the seasonal asymmetry of production. Less is known about grazed swards, for practical reasons (Raymond 1969), but the continual removal of extending stems, which contain higher proportions of indigestible supporting tissues, can maintain the digestibility at a relatively high level (> 69 per cent; Lazenby 1981).

The fact that this highly-asymmetric seasonal pattern of grassland production is out of phase with the seasonal pattern of demand by livestock is the primary problem of animal production from grass (Spedding 1971), and this has led to the development of various systems for the conservation of excess production during the first half of the season.

## 8.2 THE INFLUENCE OF CUTTING AND GRAZING ON PHYSIOLOGY AND YIELD

Although it is much easier to study the growth and physiology of cut swards, a substantial proportion of the UK grass production is harvested directly by the grazing animal, even in intensively managed areas. It is, therefore, important to understand the physiological factors determining yield under each system and to be able to make meaningful comparisons of their productivity. Cut and grazed grasslands vary in a number of ways. First, grazing involves the cycling and conservation of inorganic nutrients, whereas the cutting and removal of grass for hay or silage is associated with substantial net exports of all nutrients from sward and soil. On the other hand, the fouling and contamination of grass by the grazing animal, especially cattle, can lead to reduced utilization of the standing crop and actual sward damage, in addition to unevenness in soil fertility. Secondly, the sward can also be damaged by treading, especially under wet conditions. This, again, tends to be more serious with cattle and can result in substantial areas of unproductive bare soil where livestock congregate to drink, shelter or to receive supplementary feeding. Thirdly, the physical details of the harvesting process differ between the two systems. In contrast to the uniform stubble left after cutting, a grazed sward is harvested at varying heights, there can be selection for and against certain species and stages of development, and the process of grazing involves pulling rather than cutting, with the risk of plant and sward damage.

These factors combine to make grazed swards much less uniform in appearance and productivity, and can cause substantial losses of potential yield. For example, fouling by dairy herds can lead to the rejection of up to 20 per cent of available herbage by the grazing animals (Marsh and Campling 1970), and Brown and Evans (1973) give yield depressions of around 30 per cent associated with relatively mild treading damage in New Zealand. However, from the physiologist's point of view, the fifth, and most important, difference between the two systems of management is in the frequency and timing of defoliation which, in turn, influence tiller demography, tissue senescence and radiation interception. The remainder of this section will, therefore, concentrate upon this last difference, although the cycling of nutrients will be considered again in section 8.3.

### 8.2.1 THE RESPONSE OF THE SWARD TO DEFOLIATION

Until about 1975, detailed physiological studies concentrated upon cut swards and it was widely assumed, from the results of experiments in which the frequency and height of cutting were varied, that continuously stocked grassland was less productive than that managed for cutting, although the difference might be reduced considerably if digestibility were taken into account. However, because cut swards are inefficient at intercepting solar radiation (suboptimal $L$ for some weeks after harvest and, commonly, superoptimal values at later stages, Fig. 2.18), the highest productivity of a given sward should, in theory, be achieved under grazing, if it can be managed to give a canopy which consistently intercepts 90 per cent or more of incident PAR (i.e. sward $L$ of 3–4). This hypothesis has led to a series of field investigations of the influence of the defoliation regime on sward structure and physiology.

For example, Jones *et al.* (1982) studied the

effects of two harvesting methods, cutting four times per year and continuous stocking with sheep, on the structure of heavily-fertilized and irrigated swards of perennial ryegrass. It was intended that the grazing be controlled to maintain 90 per cent interception throughout the season but, in practice, the levels measured fell between 50 and 80 per cent. These values are probably underestimates because of the difficulty of locating sensors under such short swards. The time courses of $L$ (Fig. 2.18) show that differences in the overall appearance of the two stands developed almost immediately, but the radiation interception results indicate that there were also early differences in extinction coefficient, presumably because of differences in leaf angle and sheath length (sect. 2.4.3) (e.g. $L$ for 90 per cent interception = 8 for the tall, erect canopy before cutting, but 4 for the shorter, grazed treatment). These differences in $L$ were associated with differences of similar magnitude in herbage mass.

However, the most important changes were in stem characteristics; there was evidence of increased tiller numbers under grazing within one month of the imposition of the treatments and, by the end of the first season, the continuously stocked sward carried 30–35,000 compared with 5–10,000 tillers $m^{-2}$ under cutting management. In parallel experiments at the same site, it was shown that the degree of stimulation of tiller numbers was dependent upon the intensity of defoliation (Fig. 8.4). This work has been extended by Grant *et al.* (1983) who showed that tiller numbers, which reached their highest levels in swards maintained at 2–3 cm in height, declined if grazing pressure was increased or relaxed. As in other crops (sect. 6.2 and 7.3), there is an inverse relationship between stem population density and individual stem weight, commonly following the 3/2 power law (Harper 1977; Hodgson *et al.* 1981). For example, tillers from the cut sward examined by Jones *et al.* (1982) were three times heavier than the corresponding grazed tillers before harvest in June.

Although the rate of removal of leaf laminae depends upon the grazing intensity or cutting frequency, it tends to be higher for grazed (removed by the grazing animal) than for cut swards (removed by senescence). It would, therefore, be expected that the herbage mass under grazing would contain less dead material, but Jones *et al.* (1982) found the converse to be true. However, they did not identify the senescent tissues and it seems certain that they were the accumulating sheaths of older leaves whose laminae had already been lost to grazing (Parsons *et al.* 1983a).

Overall, any increase in the frequency of defol-

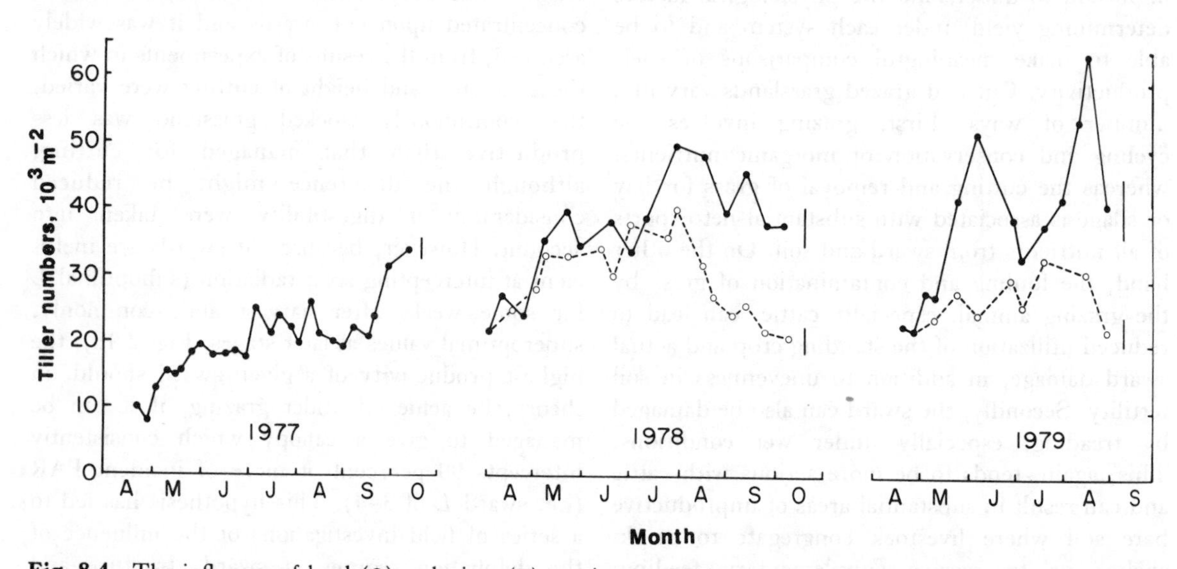


**Fig. 8.4** The influence of hard (●) or lenient (○) continuous grazing by sheep on the tiller population density of swards of perennial ryegrass grown in Southern England during three seasons, 1977–9 (see text for details). The vertical bars indicate SEM (from Parsons *et al.* 1983a).

iation, whether by cutting or grazing, results in a wide range of structural responses by the sward, including reduction in height and $L$, increase in tiller population density, reduction in the weight of individual tillers, alteration in the pattern of senescence (less senescent leaf but more senescent sheath material present in the herbage mass), and reduction in pseudostem height at a given leaf area (Hodgson *et al.* 1981).

In the UK, two major groups have investigated the influence of these differences in canopy structure on physiology and productivity. For example, Hodgson *et al.* (1981) examined the relationship between $L$ and canopy net photosynthesis in stands of perennial ryegrass under different harvesting regimes. At low leaf area index ($\leqslant 2$), continuously grazed swards (whether by sheep or cattle) performed distinctly better than cut swards, presumably because of the higher photosynthetic potential of their (unshaded) leaves and, possibly, because the more prostrate grazed canopy was more effective at radiation interception (Fig. 8.5). A different result might be anticipated under high irradiance, when the more horizontally disposed leaves would run the risk of light saturation (sect. 2.4.3). However, above a value of 3, the rate of canopy net photosynthesis at a given $L$ and irradiance was largely independent of the method by which the level of $L$ had been achieved (grazing by sheep or cattle, cutting at different frequencies and heights).

In an extension of this work, Bircham and Hodgson (1983a, b) developed stands of a mixture of *L. perenne*, *Poa annua* and *T. repens* of herbage mass varying from 500 to 1700 kg organic matter $ha^{-1}$, corresponding to $L$ values of 1 to 4.5, by manipulating the stocking density (sheep). Measurements of individual tillers during short periods when stock were excluded (i.e. brief periods of release from grazing) showed that the net production of the sward was virtually unchanged over a wide range of stocking densities, herbage masses or $L$ values (e.g. 2–5, Fig. 8.6). This broad optimum range of $L$ values was the consequence of the balance struck between increasing dry-matter production and increasing loss of dry matter by senescence.

Parsons *et al.* (1983a, b) approached the same problem in a slightly different way. The photosynthesis and respiration of perennial ryegrass swards maintained at $L$ values of 1 (hard grazing) and 3 (lenient) were measured in small enclosure chambers by $CO_2$ exchange techniques, and the actual intake of grass by the grazing sheep was monitored using chromic oxide as an indigestible marker. The measurements did not involve the

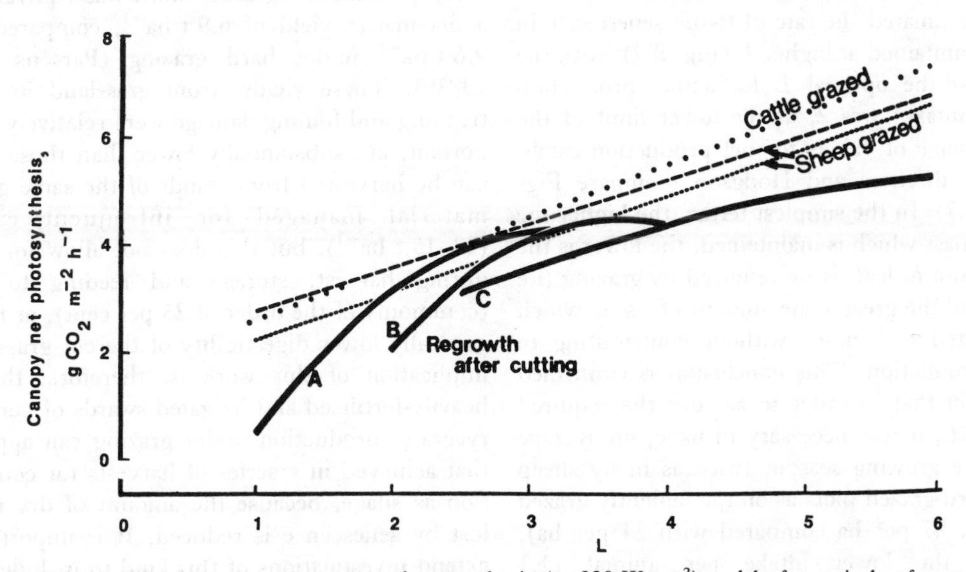


**Fig. 8.5** Relationships between canopy net photosynthesis (at 320 W $m^{-2}$) and leaf area index for swards, dominated by perennial ryegrass, which were defoliated in different ways: continuous grazing by sheep or cattle; A – cut weekly to 2 cm; B – cut every 3 weeks to 4 cm; C – cut weekly to 4 cm). Data from Southern Scotland (from Hodgson *et al.* 1981).

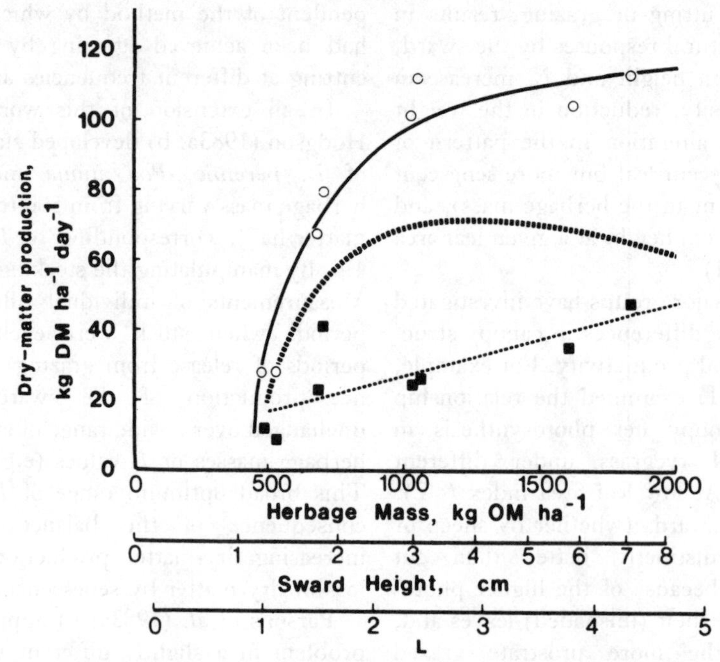


**Fig. 8.6** Relationships between leaf area index (or herbage mass/sward height) of a perennial ryegrass sward, maintained by continuous grazing by sheep, and the rates of dry-matter production (○), tissue senescence (■) and net production (middle curve, representing the dry matter available for grazing intake, after subtracting losses by senescence). Data from Southern Scotland (from Bircham and Hodgson 1983b).

disruption of the grazing treatments. Their most important conclusion was that the whole-plant measurements of Bircham and Hodgson (1983a, b) underestimated the rate of tissue senescence in swards maintained at higher *L* (Fig. 8.7), with the result that the optimal *L* for actual production (grazing intake) was 2, at the lower limit of the optimal range of values for net production established by Bircham and Hodgson (compare Figs 8.6 and 8.7). In the simplest terms, the higher the herbage mass which is maintained, the lower is the total amount of leaf tissue removed by grazing (i.e intake) and the greater the amount of tissue which is permitted to senesce without contributing to animal production. This conclusion is confirmed by the fact that in order to achieve the required values of *L*, it was necessary to have, on average during the growing season, twice as many sheep on the hard-grazed plots as on the leniently-grazed grass (i.e. 47 per ha compared with 24 per ha), although the lower intake per animal (1.1 compared with 1.6 kg DM ha$^{-1}$ day$^{-1}$) meant that the individual sheep on the leniently-grazed plots achieved the target live weight more quickly.

Over the growing season, the intake of grass from the leniently-grazed sward was equivalent to a dry-matter yield of 6.9 t ha$^{-1}$, compared with 9.6 t ha$^{-1}$ under hard grazing (Parsons *et al.* 1983b). These yields, from grassland in which treading and fouling damage were relatively unimportant, are substantially lower than those which can be harvested from stands of the same genetic material managed for infrequent cutting (12–15 t ha$^{-1}$), but this does not allow for losses during harvest, storage and feeding to stock (commonly of the order of 25 per cent), or for the generally lower digestibility of the cut grass. The implication of this work is, therefore, that for heavily-fertilized and irrigated swards of perennial ryegrass, production under grazing can approach that achieved in a series of harvests for conservation as silage, because the amount of dry matter lost by senescence is reduced. It is important to extend investigations of this kind to include other species, different levels of fertility and grazing systems other than continuous stocking.

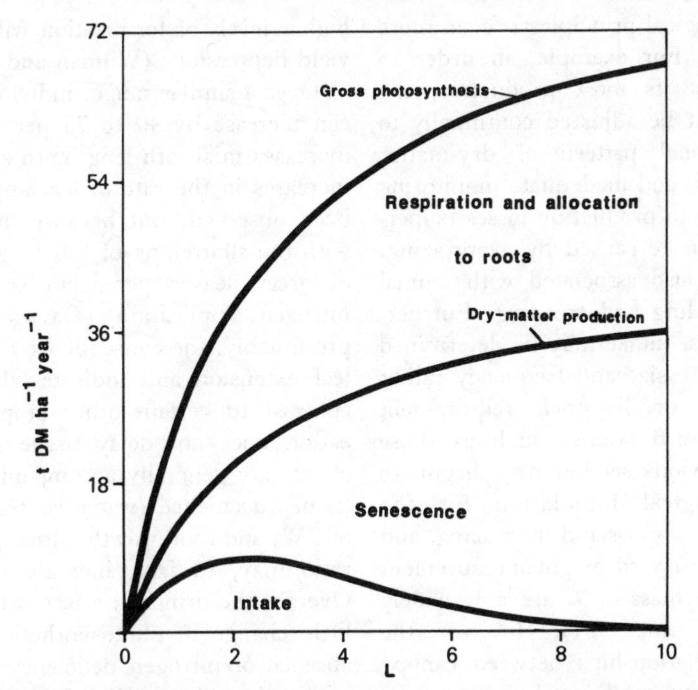


**Fig. 8.7** Generalized relationships between the rates of gross photosynthesis, dry-matter production and intake of herbage for swards of perennial ryegrass maintained at different (constant) leaf area indices by continuous grazing by sheep at different stocking densities. The loss by senescence is assessed by subtracting intake (measured in this experiment) from dry-matter production. Data from Southern England (adapted from Parsons *et al.* 1983b; Parsons 1985). Compare with Fig. 8.6.

## 8.2.2 DEFOLIATION IN PRACTICE

As shown in section 8.1, decisions about the timing and frequency of harvests for conservation involve compromises between the quantity of harvested material and its nutritional quality. This is particularly true of the first half of the growing season (the reproductive phase) during which the rapid accumulation of dry matter is associated with a progressive deterioration in its digestibility which accelerates sharply at heading. In the UK, this problem has generally been resolved, as far as the first cut of silage is concerned, by establishing a target minimum digestibility of 67 per cent (67*D*), which for perennial ryegrass is reached, on average, at half-ear emergence, although there can be a variation of ± 10 days among varieties. Hay is normally cut at a much later stage of flowering to give a heavier yield of lower-quality forage, but with a higher content of fibre which may be required for winter feeding. However, the heavier hay crop shades younger tillers very severely, with the result that the hay harvest is followed by a period of at least 4 weeks of recovery before the aftermath can be grazed intensively. In contrast, because the depression of both tiller numbers and the photosynthetic potential of surviving leaf tissues is much less severe under the silage regime (lower intensity and shorter duration of shading), one or more further cuts for conservation can be made, depending upon environmental conditions and the grass variety grown. However, such estimates of optimum cutting date are of little value if soil conditions are unsuitable for the traffic of harvesting machinery, or the immediate weather forecast indicates unfavourable conditions for hay drying. For example, irrespective of the estimated compromises between quantity and quality of the material to be harvested, many fields of grass grown for conservation in the exceptionally wet year of 1985 still awaited harvest in October.

The management of continuous grazing systems

according to physiological principles is even more difficult in practice. For example, in order to maintain a sward at its most productive, the stocking density must be adjusted continually to allow for the seasonal pattern of dry-matter production (Fig. 8.3), and inadequate monitoring of the sward can lead to production losses (super-optimal *L*) or to damage caused by overgrazing. Intensive grazing is also associated with animal health problems, fouling and treading. Furthermore, in certain cases, intake may be determined by factors such as bite size and frequency rather than herbage mass or livestock requirement (Hodgson 1981). Sward studies such as those described in the previous section have begun to provide a physiological foundation for the optimum exploitation of grassland by grazing, and techniques for regular sward height measurement to determine herbage mass or *L* are now widely used (reviewed by Frame 1981). However, the nature of the interrelationships between canopy photosynthesis, herbage intake and tissue senescence, which is explored in these studies (Figs 8.6 and 8.7), suggests that there is limited scope for increased production by manipulation of continuous grazing. As a result, the work has progressed to consider swards which have been released from continuous grazing for different periods of time (e.g. Bircham and Hodgson 1984). This should give important information for the management of rotational and aftermath grazing.

## 8.3 NITROGEN FERTILIZER AND YIELD

Apart from relieving the symptoms of deficiency, the application of nitrogen fertilizer affects grass crops in a number of ways. As with cereals (sect. 6.4), the predominant effect is the progressive increase in leaf area per plant and in *L* with increasing levels of application, as a consequence of increased leaf dimensions (Fig. 2.10) and tiller population density. The size of the response will depend upon a series of other factors including plant population density, P and K status and the system of defoliation but, for example, the tiller population density in a stand of perennial ryegrass can increase at rates of from 3 to 7 tillers $m^{-2}$ $kg^{-1}$ N up to 500 kg N $ha^{-1}$, although higher levels of fertilization will normally lead to yield depressions (Wilman and Wright 1983).

Over a similar range, individual leaf blade area can increase by 40 to 75 per cent, with parallel increases in sheath length. In some cases, modest increases in the rate of leaf appearance have also been observed, but because these are associated with the shortening of leaf longevity, the number of green leaves per stem is little affected by nitrogen application (Davies 1977). This is, presumably, the consequence of increased rates of leaf extension and indicates that for grasses, in contrast to certain other crops, nitrogen application does not delay tissue senescence. These effects are generally accompanied by increases in tissue succulence (water content as a percentage of FW) and root growth, although the shoot : root ratio may, in fact, increase (Whitehead 1970). Overall, the principal effect is to increase *L*, with little change in photosynthetic efficiency, in the absence of nitrogen deficiency or severe shading within the canopy (Woledge and Pearse 1985; but see sect. 3.4.5.2). In the remainder of this section, interest will be focussed upon the responses to nitrogen of pure stands of perennial ryegrass; the more complex responses of grass/clover grassland will be considered in section 8.4.

The extent to which such increases in *L* are exploited by the sward to give increased dry-matter yield depends upon the system of harvesting and the crop/water relations (Whitehead 1970; sect. 8.5). There is ample information in the literature from grassland fertilizer trials (Holmes 1968, 1980) but the majority of this information is of little use in physiological analysis of dry-matter yield because the experiments were poorly documented and restricted to a single location and system of management. The need for a more comprehensive approach to the subject in the UK was recognized in the early 1960s, leading to a series of collaborative nitrogen fertilizer experiments over several seasons at twenty-one sites throughout lowland England and Wales, which included several systems of management (cutting at a 'grazing frequency', GM20; cutting four times for conservation, GM21; comparison of grazing with frequent cutting, GM22; management to maintain clover, GM23; the last series of experiments being considered in sect. 8.4). The results published to date (Morrison *et al.* 1980) come mainly from the plots of perennial ryegrass which were cut at a 'grazing frequency' (six cuts per year at 4-week intervals, beginning at dates varying from 6 to 16 May, according to latitude) over five seasons, 1970–4; nitrogen fertilizer was applied at six different rates from 0 to 750 kg N $ha^{-1}$ $yr^{-1}$, in four different patterns of timing. As far as was practical, all other conditions were optimized except that the swards were not irrigated.

The precise shape of the (total annual dry matter) yield response curve varied considerably from site to site but, in general, it could be described most satisfactorily by an inverse polynomial of the form shown in Fig. 8.8 rather than, for example, a quadratic expression or a pair of straight lines. However, because the shape (but not the actual yields) did not vary markedly from season to season at a given site, the following five characteristics could be derived for each site, from response curves constructed using 4-year means:

$Y_{max}$ – the highest dry-matter yield achieved (overall mean 11.9 t $ha^{-1}$)

$N_{max}$ – the quantity of fertilizer nitrogen required to give $Y_{max}$ (overall mean 624 kg N $ha^{-1}$ $yr^{-1}$)

$Y_{10}$ – an index of 'optimum yield', here defined as the yield at the point on the response curve where the rate of increase in dry matter per unit of nitrogen fell below a threshold (set at 10 kg DM $kg^{-1}$ N in this investigation) (overall mean 10.9 t $ha^{-1}$)

$N_{10}$ – the quantity of fertilizer nitrogen required to give $Y_{10}$ (overall mean 388 kg N $ha^{-1}$ $yr^{-1}$)

Res $N_{300}$ – the gradient of the response curve in its linear section, normally from 0 to 300 kg N $ha^{-1}$ $yr^{-1}$ (overall mean 23 kg DM $kg^{-1}$ N)

The results from the two most extreme sites are presented in full in Table 8.2. Similar patterns of yield response have been recorded in several single-site experiments (Holmes 1968; Richards 1977); for example, Reid (1970), working in the south-west of Scotland, observed a linear response up to 336 kg N $ha^{-1}$, an $N_{max}$ of 560 kg N $ha^{-1}$ and an $N_{opt}$ of 450–500 kg N $ha^{-1}$ (based on a threshold gradient of 5.67 kg DM $kg^{-1}$ N calculated from current fertilizer prices, rather than 10 kg DM $kg^{-1}$ N as used above).

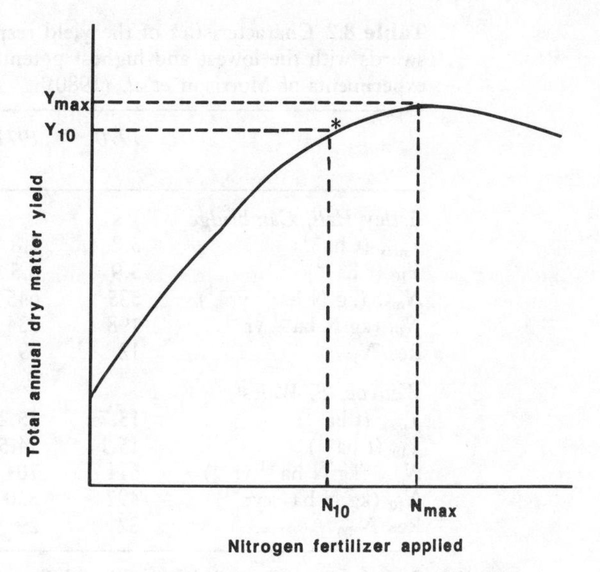


**Fig. 8.8** Model curve to illustrate the influence of the amount of nitrogen fertilizer applied on the total annual dry-matter yield of a sward of perennial ryegrass (see text for details). * marks the point on the curve where the yield response to nitrogen has fallen to 10 kg DM per kg N applied (adapted from Morrison *et al.* 1980).

Several important conclusions can be drawn from these data. First, each of the five indices of sward response to nitrogen varied considerably, especially among sites; regression analysis identified the total amount of available water from June onwards as the most important factor determining variation between sites ($r = 0.76$), but there were no significant correlations with soil moisture deficit or environmental temperature during particular periods, and the factors responsible for variation among seasons at a given site could not be identified. There can be no doubt that some of the variation encountered was the consequence of differences in the capacity of the unfertilized soil to supply nitrogen, but investigations of this kind are hampered by the lack of an appropriate measurement or index of soil nitrogen status throughout the long season of grass growth, and there are important interactions between water and nitrogen supplies. For example, it is generally held that the application of irrigation water during periods of soil moisture deficit increases dry-matter yield as much by increasing nitrogen

**Table 8.2** Characteristics of the yield response curves (see Fig. 8.8) of the grass swards with the lowest and highest potential* for dry-matter production in the experiments of Morrison *et al.* (1980)

| | *1970* | *1971* | *1972* | *1973* | *Four-season mean* |
|---|---|---|---|---|---|
| *Astley Hall, Cambridge* | | | | | |
| $Y_{max}$ (t ha$^{-1}$) | 5.2 | 7.8 | 5.0 | 8.0 | 6.51 |
| $Y_{10}$ (t ha$^{-1}$) | 3.9 | 6.5 | 3.3 | 7.3 | 5.24 |
| $N_{max}$ (kg N ha$^{-1}$ yr$^{-1}$) | 535 | 645 | 514 | 466 | 540 |
| $N_{10}$ (kg N ha$^{-1}$ yr$^{-1}$) | 198 | 334 | 183 | 324 | 260 |
| Res $N_{300}$ | 12 | 19 | 10 | 15 | 14.0 |
| *Wenvoe, S. Wales* | | | | | |
| $Y_{max}$ (t ha$^{-1}$) | 15.7 | 15.2 | 13.8 | 15.4 | 15.04 |
| $Y_{10}$ (t ha$^{-1}$) | 15.1 | 14.5 | 13.0 | 14.7 | 14.35 |
| $N_{max}$ (kg N ha$^{-1}$ yr$^{-1}$) | 611 | 704 | 705 | 690 | 678 |
| $N_{10}$ (kg N ha$^{-1}$ yr$^{-1}$) | 497 | 550 | 536 | 536 | 530 |
| Res $N_{300}$ | 32 | 29 | 26 | 28 | 28.8 |

* Highest mean production. The highest annual yield recorded was at Seale Hayne (17.9 t ha$^{-1}$).

supply as by the relief of water stress, whereas heavy rainfall or waterlogging can lead to losses of soil nitrogen by leaching and denitrification. Overall, as shown by Table 8.2, the potential for dry-matter production by unirrigated swards tends to be very low in the drier south-east of England, even at the highest levels of nitrogen fertilization, but very much higher in the wetter west of the British Isles.

Secondly, grass crops can benefit from nitrogen application up to rates far beyond the maxima for other crops (e.g. 150–175 kg N ha$^{-1}$ for cereals, sect. 6.4.1; 100–150 kg N ha$^{-1}$ for potatoes, sect. 7.5.1), although yield depressions are observed at very high levels of application above $N_{max}$ (normally 600 kg N ha$^{-1}$), as a result of severe mutual shading within the canopy, diseases and pests. Part of this difference can, of course, be explained by the longer growing season and perennial nature of grassland. The optimum rates of nitrogen application ($N_{10}$) established during these investigations were, on average, nearly 250 kg N ha$^{-1}$ yr$^{-1}$ lower than the corresponding values of $N_{max}$, but the difference was much less at the more productive sites (e.g. 114 kg N ha$^{-1}$ yr$^{-1}$ at Wenvoe, Table 8.2) because the linear response to nitrogen was maintained up to much higher rates of fertilization.

Thirdly, it was found that by alteration of the proportion of the total amount of nitrogen applied at the start of growth and after each cut, it was possible to alter the seasonal pattern of production slightly, without significantly affecting total annual yield. In particular, a more even distribution of production could be achieved over the first half of the growing season (Fig. 8.2) by restricting nitrogen supply during the early peak of growth. In a subsequent, more rigorous analysis, Reid (1984) has demonstrated how, and to what extent, the seasonal pattern of dry-matter production can be manipulated by the sequence of fertilizer applications.

Unfortunately, the results from the plots cut four times per season for conservation have not yet been published in full, in spite of the fact that this system of defoliation relates more to agricultural practice. Preliminary comparison of the two parallel experiments (GM20 and GM21, Morrison 1973) at a single site suggests that the shapes of the yield response curves are essentially the same for the two systems (Fig. 8.8), but that $Y_{max}$ and $N_{10}$ may be slightly higher with less frequent cutting in the first season. This effect tends to be less important in subsequent seasons, apparently because of sward damage at higher rates of nitrogen application. Overall, for practical purposes, it seems reasonable to use the response curves for frequently cut swards as guides to crop response to nitrogen under intensive management for conservation, at least until the full results of these experiments are published. As a result of the general trend towards conservation as silage, there have been no recent comprehensive studies of nitrogen fertilization for conservation as hay; the limited information available indicates that yield is maximized at relatively low rates of application (e.g. 100 kg N $ha^{-1}$; Wilcox 1962).

Jackson and Williams (1979) present the results of the parallel, but smaller-scale, comparison of rotational grazing and frequent cutting (GM22) at six sites and three rates of nitrogen fertilization. As in other experiments (sect. 8.2), it was found to be difficult to impose a realistic grazing treatment using cattle; for example, the animals spent only 6 days per annum on the test plots (1 day per defoliation), there were behavioural problems associated with intake, and it was difficult to assess the importance of senescence of stubbles between grazing harvests. However, bearing these limitations in mind, the results (Table 8.3) do illustrate several important points. In particular, the grazed swards were less responsive to nitrogen application, with $N_{max}$ values at about 200 kg N $ha^{-1}$ in each season after the first, whereas the cut swards gave significant yield responses up to at least 400 kg N $ha^{-1}$. This effect was, presumably, the consequence of nutrient cycling under grazing. However, although the dry-matter production from cut swards was generally higher, grazed swards did outyield the corresponding cut grassland at two sites at the lowest rate of application (200 kg N $ha^{-1}$). In other experiments in south-west England, Richards (1977) has shown very similar responses of grazed and cut swards (grazed slightly superior) up to 300 kg N $ha^{-1}$ ($N_{max}$ for grazing) although the yield under cutting did continue to rise slowly up to 600 kg N $ha^{-1}$. Elsewhere, less satisfactory experiments have shown yield responses (usually in terms of grazing days) to higher rates of application (Holmes 1968, 1980). No allowance was made in any of these investigations for differences in digestibility or harvest losses, but the results provide further support for the suggestion that grassland can be as productive under grazing as under cutting management (Parsons *et al.* 1983b; sect. 8.2), particularly at rates of application below 300 kg N $ha^{-1}$.

In addition to these yield responses, there are also well-established effects of nitrogen application on the nutritional quality of the herbage (decrease in crude fibre content, increases in nitrogen content and digestibility; Waite 1970), although at the highest levels of fertilization, a significant proportion of the total nitrogen content (normally expressed as 'crude protein') will be in the form of nitrate ions rather than amino acids or proteins. Other aspects of crop quality can also cause concern; for example, because of the high moisture content of heavily-fertilized grass for conservation, the crop must be carefully wilted to reduce the quantity of silage effluent.

**Table 8.3** Total annual dry-matter yield (t $ha^{-1}$) from comparable swards of perennial ryegrass at six sites in England and Wales, harvested by cutting (C) or grazing (G) (see text for details). Means over four seasons at three levels of nitrogen fertilization

| *Site* | | *Nitrogen fertilizer applied (kg $ha^{-1}$ $yr^{-1}$)* | | | *Standard error* |
|---|---|---|---|---|---|
| | | *200* | *400* | *600* | |
| Bridget's, S. England | C | 6.4 | 9.3 | 9.6 | 0.41 |
| | G | 3.1 | 4.2 | 4.2 | 0.27 |
| Gleadthorpe, Midlands, England | C | 5.6 | 9.6 | 10.3 | 0.35 |
| | G | 6.7 | 8.0 | 7.6 | 0.43 |
| High Mowthorpe, N. England | C | 6.8 | 8.3 | 7.9 | 0.34 |
| | G | 5.4 | 5.7 | 5.3 | 0.38 |
| Great House, NW England | C | 9.0 | 11.5 | 12.2 | 0.42 |
| | G | 5.5 | 6.6 | 6.5 | 0.32 |
| Rosemaund, E. Wales | C | 6.3 | 9.8 | 11.2 | 0.47 |
| | G | 5.6 | 6.6 | 6.4 | 0.26 |
| Trawsgoed, W. Wales | C | 7.6 | 10.6 | 11.0 | 0.36 |
| | G | 8.7 | 8.8 | 9.1 | 0.56 |

(Jackson and Williams 1979)

### 8.3.1 NITROGEN FERTILIZATION IN PRACTICE

In the absence of drought or nutrient stress (P, K, Mg, trace elements), dry-matter production by grass swards is limited by the supply of nitrogen, and the experiments carried out since 1950 (reviewed in sect. 8.3) indicate clearly how to achieve high yields of dry matter in the more productive, wetter areas of the UK. Annual dry-matter yield increases linearly up to at least

300 kg N $ha^{-1}$ $yr^{-1}$ and, depending upon the unit cost of fertilizer, economic yield responses can be obtained up to 400 to 500 kg N $ha^{-1}$ $yr^{-1}$. However, even though fertilizer application to grassland has increased dramatically since 1950, the majority of intensively managed swards (whether for grazing or cutting) in the UK still receive less than 200 kg N $ha^{-1}$ $yr^{-1}$ (Church 1985). Many of the reasons for the reluctance of grassland farmers to apply more nitrogen are not relevant to a study of crop physiology (management of stocking density, use of concentrate feeds, capital costs of housing livestock and grass conservation etc.) but it should be stressed that these fertilizer statistics are a poor guide to the nitrogen nutrition of grassland because they do not include the very substantial (but rarely measured) returns of nutrients to the sward directly by the grazing animal or by the spreading of slurry or manure (or the contribution from clover, sect. 8.4). The economic and potential environmental costs of applying high rates of nitrogen (in either form) to grassland are indicated by the fact that the recovery of fertilizer nitrogen in the harvested herbage can vary widely (5 to 70 per cent; Morrison *et al.* 1980; Van der Meer 1983), whereas the proportion of the nitrogen intake which is contained in the final product (milk, meat etc.) rarely exceeds a few per cent. Clearly, there is a need for improved understanding of the complex nitrogen relations of grassland systems (sward, soil, available water and livestock) under contrasting management (e.g. Whitehead and Dawson 1984).

Because of the variability of grassland management systems, advice on rates of fertilizer application are generally given on the basis of an overall nitrogen policy (e.g. 'high' – 250–400 kg N $ha^{-1}$ $yr^{-1}$, 'moderate' – 125–250, or 'low' – 50–120; SAC 1983) and the harvesting system, with adjustments being made for the nitrogen status of the soil, mean annual precipitation and the quantities of nutrients recycled to the sward. Thus the rates recommended by the Scottish Agricultural Colleges (SAC 1983) range from 360 kg N $ha^{-1}$ $yr^{-1}$ for three cuts for silage under a high nitrogen policy down to less than 100 kg N $ha^{-1}$ $yr^{-1}$ for grazing or hay under a low nitrogen policy. The recommendations also include suggested patterns of application to exploit the full potential of the sward after each defoliation.

One aspect of the timing of nitrogen fertilization which has received considerable attention over the last 20 years has been the date of the first application in spring, especially for grazing but also for conservation. Application too early in the year, before the crop can make use of the fertilizer, results in the loss of nitrogen by leaching or denitrification, whereas late application can lead to substantial yield losses, especially since the potential for dry-matter production is at its highest during reproductive growth in spring (sect. 8.1). In the course of a long series of experiments in the Netherlands, Van Burg (1968) found that the recovery of applied nitrogen in the harvested herbage rose sharply after the middle of February and that this change was associated with the start of visible sward growth and the achievement of an accumulated mean air temperature from 1 January of 250 °C days, excluding negative values (Jagtenberg 1970). This value, which was subsequently revised to 200 °C days (termed the Dutch T-sum 200 in the UK), has been used with great success in the Netherlands as a tool for predicting the optimum first date of fertilization (dates varying from the first week of February to the middle of April) (Fig. 8.9).

However, it must be stressed that this is an empirical relationship without a strict theoretical foundation and, therefore, it cannot be expected to apply without modification elsewhere in Europe. Thus, although it has proved to be useful in parts of England (Daly and MacKenzie 1983), it can predict unrealistically early dates in western and northern parts of the UK because maritime winters are generally milder and more variable than in the more continental climate of the Netherlands. However, since the concept is simple and useful for advisory purposes, a modified calculation has been proposed in which temperature accumulation (up to 80 °C days) begins on 1 February (to avoid unrealistic dates in January/early February) and which uses soil (10 cm) rather than air temperatures (to damp climatic variation) to give a 'T-value 80'. (An alternative calculation which makes use of the more widely measured soil temperature at 30 cm has not been generally adopted.) However, recent tests at thirty-five sites in Scotland (Swift *et al.* 1985) have shown that

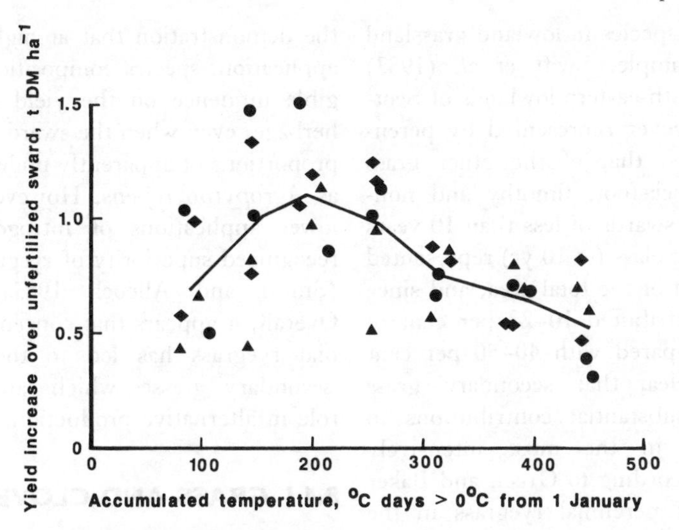


**Fig. 8.9** The influence of the timing of the first spring nitrogen application on the yield response of swards of perennial ryegrass grown on different soil types (●, peat; ▲, sand; ■, clay) in the Netherlands in 1979. The line is drawn through the mean response (adapted from van Burg *et al.* 1980).

use of the T-value 80 (ensuring 92 per cent of maximum yield over three seasons) did not give greater precision than the Dutch T-sum 200 (94 per cent).

## 8.4 SPECIES MIXTURES

Within the dozen or so genera of temperate grasses which have potential for cultivation (Spedding and Diekmahns 1972), there is considerable variation, among species and varieties, in those characteristics which determine productivity (height and growth habit, seasonal distribution of production, canopy structure, palatability and digestibility, response to defoliation and fertilization etc.) (e.g sect. 2.4.4). The performance of these grasses in pure stands and in mixtures has been the subject of intense investigation in small plots, under controlled conditions and in natural grasslands (Donald 1963; Rhodes 1970; Harper 1978) and there exists a great deal of information on their competitiveness, mutual compatibility and persistence. Some of this information has been of direct use in grassland production; for example, it has already been noted that the seasonal distribution of production under grazing can be improved, without necessarily altering total annual yield, by the use of mixtures of species or varieties of different maturity types (SAC 1982). However, much of this work has been of interest only to ecologists because over the last 50 years in the UK, grassland agronomists have tended to concentrate on the development of intensively managed lowland swards dominated by *Lolium perenne*; the main exceptions to this trend are short-term leys (1–3 years) for which the faster-growing, but non-persistent, *L. multiflorum* (Italian ryegrass) has been favoured. For example, perennial ryegrass varieties are the principal grass components of the seed mixtures recommended by the Scottish Agricultural Colleges (SAC 1982) for all grassland of duration longer than 2 years (whether for grazing, silage or hay) apart from a single specialist grazing mixture for the west of Scotland based on meadow fescue (*Festuca pratensis*).

The reasons for this trend are clear: perennial ryegrass combines a wide range of favourable and useful characteristics including vigorous tillering, responsiveness to nitrogen fertilization, persistence, palatability and high digestibility, and the existence of different morphological types within the species has permitted the selection of varieties suited to different systems of defoliation. Furthermore, perennial ryegrass is generally compatible with white clover (see below). However, in spite of this concentration upon ryegrass over half a century and the resulting improvement in its status (Lazenby 1981), it is still not necessarily the most important grass species in lowland grassland in the UK. For example, Swift *et al.* (1983) showed that in the south-eastern lowlands of Scotland the percentage cover represented by perennial ryegrass exceeded that of the other grass species combined (cocksfoot, timothy and non-sown grasses) only in swards of less than 10 years of age. Since this older class (≥ 10 yr) represented more than 30 per cent of the total area, and since non-sown grasses contributed 10–22 per cent of younger swards (compared with 40–50 per cent of ryegrass), it is clear that 'secondary' grass species make very substantial contributions to animal intake even in the most intensively managed systems. According to Green and Baker (1981), the status of perennial ryegrass in the lowland grasslands of England and Wales declines with age in a similar way and, even where the species does persist, there is evidence that the surviving genotypes can be distinctly different from the sown varieties within as little as one season (e.g. Charles 1972). In general, the tendency to heterogeneity is more rapid under grazing (sect. 8.2), and the contribution of other species is even greater in upland and hill grassland.

This concentration upon perennial ryegrass led to the development of grassland management systems which have only recently begun to be reassessed in the light of the species diversity described above. For example, because the widely-recognized decline in dry-matter production during the first few years of a newly-sown sward is normally associated with a progressive decline in the proportion of perennial ryegrass, it was concluded that the most productive systems should involve temporary grass resown at intervals of around 5 years. However, although it is clear that short-term leys can produce more dry matter and animal products per unit area than permanent grass in many cases, this advantage may not be sufficient to compensate for the cost of resowing and the losses in production during the establishment years. For example, in a farm-scale comparison carried out over five seasons under a high nitrogen policy (300 kg N $ha^{-1}$ $yr^{-1}$) in the north of England, the production of milk from permanent grass was found to be significantly more profitable than from temporary grass (Bastiman and Mudd 1971). The superiority of temporary grass has been further questioned by the demonstration that at high levels of nitrogen application, species composition can have a negligible influence on the yield and quality of the herbage, even when the sward includes substantial proportions of apparently undesirable species such as *Agropyron repens*. However, the results with lower applications of nitrogen show the well-recognized superiority of ryegrass/clover mixtures (Smith and Allcock 1985a,b) (sect. 8.4.1). Overall, it appears that concentration upon perennial ryegrass has led to the neglect of other 'secondary' grasses which can play an important role in alternative production systems.

### 8.4.1 GRASS AND CLOVER MIXTURES

Mixed swards of grasses and legumes are of considerable importance for two reasons. First, legumes have access to an additional source of nitrogen from the atmosphere, as a result of the symbiotic relationship between their roots and *Rhizobium* bacteria. This reduces the requirement for fertilizer nitrogen not only for the legume itself but also for its companion grasses *via* the soil organic matter, and possibly by the leakage from the nodules into the rhizosphere of amino acids and other chemical species. Secondly, because of the growth habit of most successful legumes in temperate grassland, the material harvested by cutting or grazing is mainly leaf blades with varying amounts of petiole according to the height of the canopy. This material is of very high digestibility (*D* values commonly greater than 80 per cent, sect 8.1; Spedding and Diekmahns 1972) because of the low proportion of fibrous stem tissues, and it can improve the overall digestibility of mixed herbage to a considerable extent. Legumes can also contribute to herbage quality as a consequence of their higher mineral content, but there is a risk of digestive problems in some cases ('bloat') if they make up too high a proportion of the diet of ruminants.

White clover (*Trifolium repens*) is by far the most important forage legume in the UK although, in some areas, red clover and lucerne are grown for conservation. The primary reason for its pre-eminence is probably its compatibility with the most important grass species, *L. perenne* (sect. 8.4); indeed, in a detailed demographic study of a semi-natural grassland in North Wales,

Turkington and Harper (1979) showed a positive association of the two species. This mutual compatibility appears to be the consequence of the complementarity of their growth habits (tufted grass; shorter, spreading clover) and seasonal patterns of production (the peak of clover production coincides with the midsummer depression in grass production, sect. 8.1) and the high nutrient requirements of both species, but it should be emphasized that the relationship is dynamic, based upon a regeneration cycle (Turkington and Harper 1979; Burdon 1983). Other useful features include its resistance to, and commonly its proliferation under, grazing, since the stolons are generally close to the soil surface, and the high degree of variability within and between natural populations which has facilitated the selection of several commercial varieties, differing mainly in leaf size, plant height and time of flowering (Burdon 1983). Unfortunately, in some cases these improved characteristics may have been gained at the expense of persistency (see below). Other problems associated with white clover include its sensitivity to drought, low temperatures and freezing stress, its susceptibility to clover rot (*Sclerotinia trifoliorum*) and virus diseases, and the need to ensure infection of the roots with an effective strain of *Rhizobium trifolii* (Spedding and Diekmahns 1972).

Annual dry-matter yields of white clover in the UK can be as high as 12 t $ha^{-1}$ from pure stands and 5–6 t $ha^{-1}$ in grass/clover mixtures (Spedding and Diekmahns 1972), but the total yield of herbage and the proportion of each component are strongly dependent upon the level of nitrogen fertilization. These effects can be illustrated by the results of a three-season comparison of a mixed S23 perennial ryegrass/S100 white clover sward with a pure grass stand (each cut five times per season) carried out over a very wide range of fertilizer applications, 0–900 kg N $ha^{-1}$ $yr^{-1}$, by Reid (1970) (Fig. 8.10). Without nitrogen fertilizer, clover represented at least 30 per cent of the material harvested from the mixed sward, which outyielded the pure grass by 160–180 per cent. This difference can be attributed to the nitrogen-fixing activities of the legume. However, as the level of nitrogen application increased, giving progressively taller grass tillers with larger leaf laminae, the proportion of clover in the mixture declined progressively (see below) with the result that, in the second and third seasons, clover made no contribution to the harvested herbage in swards receiving more than 300 kg N $ha^{-1}$. Over the range 0–300 kg N $ha^{-1}$, the yield responses of each sward type were essentially linear, but convergent, and although the mixed sward did continue to outyield the pure grass sward up to the highest nitrogen application, the differences were negligible above 400 kg N $ha^{-1}$. As indicated in the previous section, broadly similar results were obtained from comparisons of ryegrass/clover swards with a series of grass mixtures at 50 and 300 kg N $ha^{-1}$ $yr^{-1}$ (Smith and Allcock 1985b); in general, species composition appears to be particularly important at lower nitrogen levels

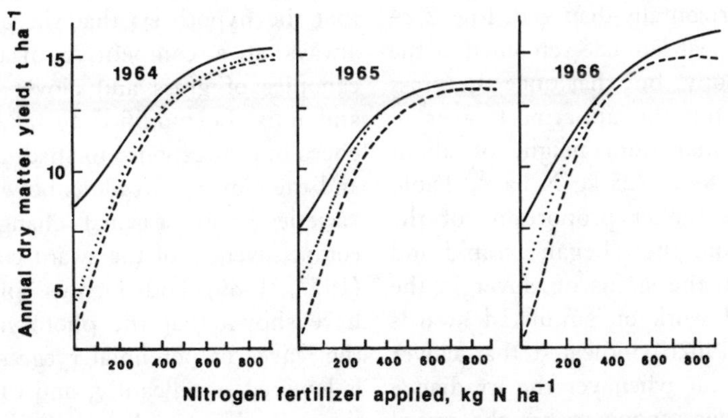


**Fig. 8.10** The influence of nitrogen fertilization on the yields of grass swards grown in south-west Scotland during three seasons, 1964–6, where (——) indicates the total yield of a mixed perennial ryegrass/white clover sward, (..........) indicates the contribution to the yield of the mixed sward made by the ryegrass, and (-------) indicates the comparable yield of a pure ryegrass sward (adapted from Reid 1970).

whereas, at higher levels, similar yields can be achieved from a variety of sward types.

It can be concluded from these and a number of other field experiments that where production is maintained under a high nitrogen policy (sect. 8.3.1), there can be no yield advantage from including white clover in the seed mixture as it will disappear rapidly from the sward. However, it is clear that at lower levels of application ($\leqslant$ 220 kg N ha$^{-1}$ yr$^{-1}$ according to Reid 1970), grass/clover mixtures will yield more dry matter than pure grass swards at a given rate of fertilization. Since grassland in the UK receives, on average, less than 200 kg N ha$^{-1}$ yr$^{-1}$, white clover should play an important role in production but, on the contrary, it is generally in decline, as shown by the sharp fall in the demand for clover seed.

The principal reason for this decline appears to be simply the difficulty in maintaining white clover in the sward, even at moderate to low rates of nitrogen application, and attempts to develop management systems to improve the persistency and to maintain the required proportions of white clover in swards have revealed how little is known about the physiology of the clover plant under competition in mixed stands (Davidson *et al.* 1986; Davidson and Robson 1986).

Many of the results of earlier studies (reviewed by Burdon 1983) were quantified and extended in a series of classic experiments carried out in Australia by Stern and Donald (1962a,b); they were able to show that clover remained a major component of a mixed sward as long as a substantial fraction of its (horizontally-disposed, Figs 2.24 and 2.25; Table 2.2) leaf laminae remained at the top of the sward canopy, but that once the grass leaves began to overtop the uppermost layer of clover leaves, at a maximum height of about 40 cm (e.g. after day 84 at 225 kg N ha$^{-1}$, Table 8.4), and to capture higher proportions of the incident solar radiation, there began a rapid and progressive decline in the status of clover in the stand. More detailed work on simulated swards (Stern and Donald 1962b) indicated that clover would tend to disappear whenever the irradiance at the top of the clover canopy within the sward fell to approximately 30 W m$^{-2}$. This work, therefore, suggested that the maintenance of clover was simply a matter of adjusting the level of fertilization and pattern of defoliation to ensure that the clover component was not affected by severe shading stress. However, as discussed above, this is clearly not the whole story since it is not uncommon for clover to disappear even at very moderate levels of nitrogen application. Furthermore, the grass/legume system used in Stern and Donald's experiments was not the same as that used in the UK; the species were different (subterranean clover, *T. subterraneum*, and Wimmera ryegrass, *L. rigidum*), both are annuals, and the clover tends to be more prostrate than most white clover varieties, whereas the grass is taller than perennial ryegrass.

**Table 8.4** Contributions to canopy leaf area index made by the grass and clover components of a *Lolium rigidum/Trifolium subterraneum* sward grown at the Waite Institute, South Australia in 1957

| *Days from sowing* | | *Nitrogen fertilizer applied (kg N ha$^{-1}$)* | | | |
|---|---|---|---|---|---|
| | | *0* | *25* | *75* | *225* |
| 84 | Grass | 0.9 | 1.6 | 3.4 | 4.7 |
| | Clover | 5.0 | 5.0 | 4.1 | 3.7 |
| | Total | 5.9 | 6.6 | 7.5 | 8.4 |
| 133 | Grass | 0.4 | 0.8 | 0.8 | 6.0 |
| | Clover | 14.3 | 10.8 | 10.4 | 0.7 |
| | Total | 14.7 | 11.6 | 11.2 | 6.7 |

(Stern and Donald 1962a)

More recent work in the UK using the perennial ryegrass/white clover system has indicated that the hypothesis that the clover component is always at a competitive disadvantage in dense canopies of grass and clover is a simplification, and must be modified to take account of differences in the response to shading between the grass and the clover, as well as between different clover varieties, and seasonal changes in the relative competitiveness of the sward components. Woledge (1977, 1986) and Dennis and Woledge (1983) have shown that the photosynthetic potential of the leaves of perennial ryegrass and white clover is lowered significantly, and to a similar extent, if they develop in shade. This effect has already been used to help explain the midsummer depression in grass production (sect. 8.1). However, although clover leaves grow from stolons near the base of the canopy, and therefore begin their development in shade, measurements of the photosynthetic potential of fully-expanded clover laminae in the field show little difference between those growing under normal conditions and those artificially exposed to full solar radiation (Dennis and Woledge 1982). This apparent paradox arises because the critical period for shading seems to occur when the clover laminae have achieved about half the mature leaf area, by which time petiole extension has carried the developing laminae into higher levels of irradiance nearer to the top of the canopy. This effect, which has been confirmed in parallel experiments by Boller and Nösberger (1985), means that when clover growth is not limited by low temperature, the photosynthetic potential of the clover leaves in a mixed canopy tends to be at least as high as grass leaves of the same age and higher than those developing from shaded tillers, with the result that the clover component can contribute disproportionately to the photosynthetic activity of the sward (Davidson and Robson 1984). During the summer, this competitive advantage (not a disadvantage, as proposed by Stern and Donald 1962a) can lead to increases in the proportion of the legume in the sward, even in dense swards such as that shown in Fig. 8.11. Thus, as long as the clover petioles can extend sufficiently to place the developing laminae near the top of the canopy, and there is some evidence to suggest that more recent varieties such as Blanca show improvement in this characteristic, there is no reason for clover to decline under appropriate grazing or cutting management during the period of rapid growth.

However, since the temperature threshold for growth (including petiole extension) is higher for clover than for ryegrass, the relative competitiveness of clover declines sharply in autumn, winter and spring (Davidson and Robson 1986). Thus, as shown by Dennis and Woledge (1985), the normal practice of applying nitrogen in early spring as soon as the grass is able to respond (sect. 8.3.1) can lead to a rapid reduction in the clover contribution to the canopy, presumably because of the shading of mature leaves and the inability of clover petioles to place developing laminae in a favourable radiation environment (Davies and Evans 1982). Later in the season, the proportion of clover can increase as a consequence of its competitive superiority at higher temperatures, but it is now clear why clover can decline very quickly after several seasons during which it maintained a stable status (e.g. Stewart and Haycock 1983); for example, a severe decline in the proportion of clover during a prolonged cold spring followed by a summer drought (clover is more susceptible than ryegrass) could result in the disappearance of the legume from the sward. The fact that a sudden decline could also be the result

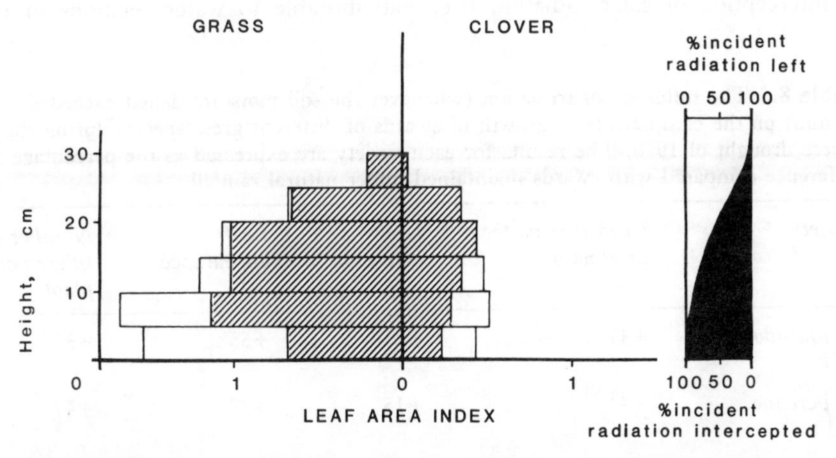


**Fig. 8.11** Stratified contributions to canopy leaf area index made by the components of a *Lolium perenne*/*Trifolium repens* sward grown in Southern England (Hurley) in 1984 (total $L = 6.7$, grass = 5.1, clover = 1.6). The sward was fertilized (80 kg N $ha^{-1}$) at the start of growth in March and the measurements made 6 weeks later. The sward structures were measured as described in section 2.4.3. The hatched areas represent laminar areas, the unshaded areas sheaths or petioles (J. Woledge, personal communication).

of a high incidence of disease or pests emphasizes the need for a multifactorial approach to the development of systems for the maintenance of clover, and it seems clear that research on clover decline can be meaningful only if it is conducted over at least five successive seasons.

## 8.5 IRRIGATION

In contrast to arable crops, grass swards maintain a relatively high $L$ throughout the year (e.g. Fig. 2.18) and can, therefore, be affected by drought over much longer periods. For example, during a cycle of soil drying in May, the demand made on soil water by a grass canopy ($L>3$) will be much greater than those made by spring-planted cereal or potato crops ($L \doteq 1$) (sect. 2.3.1). This effect may be intensified by the relative concentration of roots in the top 30 cm of soil, although the importance of the shallow rooting of grasses was probably overstressed in earlier work (Garwood and Sinclair 1979). Furthermore, the ability of grassland plants to survive unusually dry years such as 1975, 1976 or 1984 is an essential requirement for production in subsequent years (Garwood *et al.* 1979). Overall, water supply is a crucial factor in determining the productivity of grassland even under the relatively humid conditions of the British Isles.

Water stress can affect crop growth rate by reducing the interception of solar radiation (i.e. by reducing $L$) and/or by reducing the efficiency of the photosynthetic apparatus. The relatively few detailed field experiments which have been carried out in the UK indicate that, for grasses, the former is the more important effect (D'Aoust and Tayler 1969). Thus, although cell and leaf expansion in *L. perenne* were shown by Leafe *et al.* (1980) to be not particularly sensitive to stress (serious reductions beginning only at $-1$ MPa, with decreasing rates down to $-1.5$ MPa, presumably owing to osmotic adjustment, sect. 2.2.3), the overall effect of a prolonged summer drought was a pronounced reduction in $L$ (fewer, smaller leaves and tillers) associated with the development of new, drought-resistant leaves, whose rates of photosynthesis per unit area were not distinguishable from non-stressed leaves. Other workers have shown similar trends in canopy leaf area, laying particular emphasis on changes in the number of tillers per plant and specific leaf weight, and the differences in response between species and varieties (Norris 1982, Table 8.5).

The overall effects of drought on annual dry-matter yield will depend upon the duration and severity of the period of water stress, and upon the interactions between nitrogen and water supply discussed in section 8.3. For example, as shown in Table 8.2, all aspects of sward response to nitrogen, of identically managed swards of perennial ryegrass, were superior at the wetter site, leading to an improvement in potential yield attributable to water relations of up to 131 per

**Table 8.5** The influence of irrigation (whenever the soil moisture deficit exceeded 43 mm) on the components of growth of swards of different grass species during the severe drought of 1976. The results for each variety are expressed as the percentage difference compared with swards maintained under natural rainfall

| *Variety* | *Crop dry-matter production* | *Leaf extension* | *Leaf appearance* | *Number of tillers per plant* |
|---|---|---|---|---|
| *L. multiflorum* RVP | +47 | +25 | +55 | +53 |
| *L. perenne* S23 | +41 | +15 | +27 | +47 |
| *D. glomerata* | | | | |
| Bc 6392 | −7 | +7 | +7 | −3 |
| Bc 6394 | +15 | +9 | −10 | −1 |

(Data from Aberystwyth, Wales, from Norris 1982)

cent. In general, the quantity of available water over the growing season accounted for 60 per cent of the variation in annual yield among the twenty-one sites, but it was not possible to explain year-to-year variation in terms of water supply. There are several other reports of substantial yield responses to irrigation from the drier areas of the UK; for example, Garwood *et al.* (1979) found that the increase in dry-matter yield of swards grown in southern England under a high nitrogen policy which could be attributed to irrigation varied from 12–14 to 78–93 per cent according to the pattern of rainfall. However, the mean yield response to irrigation in the UK (0.025 t DM $ha^{-1}$ $mm^{-1}$ of applied water; MAFF 1982b) is low, and only slightly higher than that of cereals (sect. 6.5).

When animal intake is taken into account, the increase in annual production under irrigation is seen to be less important than the improvement in the seasonal pattern of production, although, as shown in section 8.1, irrigation can only ameliorate and not abolish the midsummer depression in growth. This smoothing of the seasonal pattern and gain in possible grazing days (Lazenby and Down 1982) permits a higher stocking rate during the summer months which, in turn, can lead to a more efficient use of the early-season peak of production, and less need for grass to be held in reserve against the risk of drought. Overall, the paramount importance of water relations for grassland production is demonstrated by the close statistical correlation between mean effective transpiration and grassland as a proportion of farmland or stocking density, established using data from all agricultural holdings in England and Wales (Smith 1983).

In summary, it is clear that irrigation of grassland should be considered only in areas where the soil moisture deficit exceeds 25 mm for considerable periods, and for intensive livestock enterprises (especially dairying, using high rates of nitrogen fertilizer and high stocking rates). Using a mathematical model, Doyle (1981) has shown that irrigation will be profitable only where the mean summer rainfall is less than 350 mm (higher if the water-holding capacity is very low) and the rate of fertilization exceeds 300 kg N $ha^{-1}$ $yr^{-1}$. Since only a small percentage of grassland enterprises in the UK fall into this category, it can be assumed that irrigation of grass will be carried out almost exclusively by farmers who have acquired the equipment for other, more drought-sensitive crops such as vegetables or potatoes (sect. 7.6).

## 8.6 HIGH-YIELDING SWARDS

If a grass sward intercepts all the available solar radiation, and the mean efficiency of conversion of solar to chemical potential energy is 3–4 per cent (sect. 3.2), then its maximum potential yield under UK conditions will lie within the range 25–30 t DM $ha^{-1}$ $yr^{-1}$ (Cooper 1970; Wilson 1982). Such levels of production are by no means unrealistic; several small-scale experiments using irrigated and highly-fertilized perennial ryegrass plots have given yields in excess of 25 t $ha^{-1}$, and an annual production of 23 t DM $ha^{-1}$ has been recorded at the field scale, without serious deterioration in herbage quality (Adams *et al.* 1983). However, in very few field experiments, far less farmers' fields, does production achieve half of the calculated potential (Tables 8.2 and 8.3; Fig. 8.10). For example, Robson (1981) estimates that the national (England and Wales) average yield over all levels of fertilization, but excluding rough grazings, was only 6 t DM $ha^{-1}$ $yr^{-1}$, and these low levels of plant production are matched by low yields of animal products (Johnson and Bastiman 1979). British grassland farmers are, therefore, much less successful than their arable counterparts who can achieve average cereal yields above 50 per cent of the potential, and some of whom have been able to equal or exceed the calculated maximum yield (Hay *et al.* 1986).

Many of the reasons for these low yields have been explored in the foregoing sections (especially level of nitrogen fertilization, water stress and method of defoliation, but also pests and diseases), and there are other constraints (Johnson and Bastiman 1979) which are not strictly relevant to a study of crop physiology. Although the relatively poor performance of British grassland farmers is a serious problem, it could be ameliorated considerably by the simple expedient of applying more fertilizer nitrogen. However, the variability of dry-matter yield from season to season at a given site is much more serious, in practice, because of the difficulty of matching stock numbers to the quantity of grass available. In turn, this can cause further reductions in grass intake and in the production of milk and meat. For example, even in the maximum yield experiments of Adams *et al.* (1983) where the growing conditions were optimized, the seasonal variation represented 30 per cent of the 5-year mean yield. This probably remains the most intractable problem for grassland physiologists and agronomists, although the various ways of smoothing the seasonal pattern of production (sect. 8.3.1, 8.4.1 and 8.5) have made some contribution.

The genetic potential of available herbage varieties is, therefore, poorly exploited at present, and there are considerable opportunities for improvement through changes in crop husbandry (i.e. from 6 to 20 t DM $ha^{-1}$ $yr^{-1}$). However, as with all agricultural crops, there has been continued interest in the possibility of raising their genetic potential for dry-matter production. This problem is quite different from that of arable crops such as cereals; in particular, there is less scope for improvement in: (a) the harvest index (already over 60 per cent); (b) interception of solar radiation (evergreen crop with substantial $L$ throughout the year); and (c) partition of dry matter between root and shoot. Clearly, any significant improvement must come from increased biomass production and, in particular, from increased rates of canopy net assimilation (Wilson 1982). Although this approach has met with little success with many crop species, there are indications that there exists sufficient genetic variation among grasses in the components of net assimilation rate (gross photosynthesis, photorespiration, dark respiration) to make possible the development of higher-yielding varieties; for example, as discussed fully in section 4.4.2, more productive ryegrass genotypes with lower rates of maintenance respiration have been identified.

# PART 3 CROP SIMULATION

# CHAPTER 9 MATHEMATICAL MODELS AND CROP PHYSIOLOGY

*Those who seek the direct road to truth should not bother with any object of which they cannot have a certainty equal to the demonstrations of arithmetic and geometry.*

(R Descartes, 1596–1650)

*Equations prove nothing: they simply facilitate the consideration of a number of attributes conjointly.*

(F L Engledow, 1925)

## 9.1 INTRODUCTION

Models of one sort or another are a familiar part of everyday life. They are devices which help us to interpret and understand the world in which we live. To most people, the word probably suggests a small-scale, three-dimensional representation of something, such as a railway engine, a group of buildings or an area of land. The model engine makes it easier to understand the workings of the real thing because it can be handled and the parts manipulated; the model buildings may give an appreciation of the design and scale of a proposed new building in relation to existing ones; and features of the landscape, such as the drainage pattern or the siting of settlements, may more readily be interpreted from a model of the topography than from observations on the ground.

We are in no doubt about the identity and meaning of such models: from an early age we are able to comprehend the idea of scale. However, these examples differ considerably in the amount of abstraction involved in their translation from the originals. The model engine, for example, may be a miniature replica of the real thing, made of the same materials and working in the same way, while the model landscape is more abstract, constructed from a variety of materials merely to represent the topography, vegetation, rivers, buildings, etc. This process of abstraction can be continued much further, however, to create a different type of model: a two-dimensional representation of the landscape, which we call a map. The interpretation of a map relies on the acceptance of certain conventions: for example, that contour lines join points of equal height and are drawn at intervals of, say, 10 metres, while other information is conveyed by words and symbols. The information contained in the map is, because of its abstraction, less readily accessible to the uninitiated, but once the conventions of presentation are understood there is no difficulty in interpreting it in terms of that part of the real world which it represents. The map is then a more convenient way of conveying information than the less abstract relief model.

Scientists and engineers use various sorts of models as tools to help them to understand the phenomena which they are studying. At the lowest level of abstraction there are the physical models: of aircraft, for example, which in wind tunnels mimic certain aspects of the behaviour of real aircraft in flight; or of molecules, particularly large ones, showing the spatial distribution of atoms and the bonding between them. The utility of such models is well established, both as experimental tools and as aids to understanding concepts and explaining observations. Perhaps the most celebrated endorsement of the value of thinking about molecular structure in terms of three-dimensional models is Watson's (1968) account of the elucidation of the structure of DNA.

To understand the structure of atoms themselves, however, we resort to what we may call a conceptual model. In its simplest form, this represents the atom as a dense, central nucleus with electrons orbiting about it, and is derived from observations of the behaviour of atoms. However, as our abilities to observe have become more refined, the model has become more complex, and now includes various sub-atomic particles. The existence of neutrons, protons and electrons as discrete, physical entities is questionable: they are merely devices which allow us to relate the properties of atoms to a physical structure which our imaginations are able to comprehend. Familiarity with the simple model, and the fact that for many purposes it provides a structure consistent with observations, make us forget that it is merely a model which approximates reality. Discoveries of the more fundamental particles of matter have now highlighted the inadequacies and crudeness of this basic model.

This digression about atomic structure is a reminder that conceptual models are a familiar part of the fabric of science: they help us to come to terms with otherwise intractable phenomena and, inevitably, this involves an element of simplification. The wave-particle duality of light provides another familiar example, offering two fundamentally different models, both of which are necessary to help us to account for the observed properties of light (sect. 3.2). However, no matter how profound the concept, the phrase 'conceptual model' means nothing more than 'idea', and ideas or hypotheses are the stock-in-trade of every experimental scientist: few experiments are done other than to test a preconceived idea. Conceptual models attain an added significance, however, when they provide a quantitative, rather than merely qualitative, description; in other words, when they are expressed as mathematical equations. In such a form, the relationships between the relevant components are formally and rigorously defined, providing not only a quantitative description under one set of conditions, but also the means of predicting the effect of changing one or more of those conditions.

As an example, consider the Monsi-Saeki equation:

$$I = I_o \, e^{-kL} \qquad [9.1]$$

which describes the irradiance at a given horizontal level within a crop canopy, $I$, as a function of the irradiance above the canopy, $I_o$, the interposed leaf area index, $L$, and an extinction coefficient, $k$, which largely reflects canopy structure and the sun's position in the sky (sect. 2.4.3). This apparently arose as a conceptual model which drew an analogy between crop canopies and coloured solutions as absorbers of monochromatic light, to be refined into the formal mathematical statement as a modification of Beer's Law. The equation, or mathematical model, defines precisely the way in which photosynthetically active radiation (PAR) is attenuated with depth in the canopy, in relation to the density of leaves and their properties as interceptors and absorbers of radiation. It therefore provided, for the first time, the means of predicting the irradiance received by successive layers of leaves in the canopy, and thus made possible the quantitative description of canopy photosynthesis, an essential step in the development of mathematical models of crop growth.

At this point we should define what is meant by a mathematical model and explain the role of such models in crop physiology. A mathematical model is an equation, or set of equations, which quantitatively describes the behaviour of a system. The 'system' may be comparatively simple, such as a crop canopy as an absorber of solar radiation, so that a single equation represents it adequately. However, as the complexity of the system increases, so does the number of equations, although this does not necessarily imply a change in the physical identity of the system. Thus, the crop canopy as a producer of dry matter is a more complex system than the canopy merely as an absorber of radiation and, in addition to equations describing the distribution of light within the canopy, equations defining the photosynthetic light response curve and the respiratory losses are required. The partitioning of dry matter to the various sinks adds another layer of complexity, and more equations. Furthermore, if the model, or simulated, crop is to develop and grow realistically, it must be represented as a dynamic system; in other words, elements of the system must change with time. Most obviously, the size of the canopy will change, and the model must be able to simulate the rate of leaf appearance, of leaf growth and of senescence. Developmental switches must also be incorporated, controlling, for example, the transition from the vegetative to the reproductive state, or the onset and cessation of tillering. The system is thus widened to take into account the controlling role of environmental factors, such as temperature and daylength.

And so the process of defining the system can be continued, bringing more and more components, or sub-systems, into play. The essential characteristic of systems is that they comprise components which may be treated separately but which are interlinked: any system is itself a subsystem in a higher level of organization, and can itself be analysed in terms of components at lower levels of organization. The rate of net photosynthesis per unit leaf area, for example, is part of the system of economic yield production by a crop, and is itself the result of $CO_2$ diffusion into the leaf, carboxylase activity, photorespiration and respiration. Clearly, although any single equation can be considered as a mathematical model in its own right, it may equally well be a sub-model, representing just one part of a larger, more complex system.

Indeed, it is the complexity of many biological systems which has prompted, or even necessitated, their study by mathematical models. Biology, or more specifically crop physiology, can be studied at several levels within a hierarchy of organization: at the crop, plant, organ, tissue, cell, organelle or molecular level. Biologists, as Passioura (1979) and Thornley (1980) have observed, tend to be reductionists, trying to explain observations made at one level of organization in terms of the behaviour of the system at lower levels. This approach is inevitable and desirable because it offers explanation, but it may also be self-perpetuating. The individual crop physiologist may be restricted, by inclination, resources and time, in the extent to which he follows the reductionist path, but the discipline as a whole may well worship the 'cult of omnipotent reductionism', as Thornley has it. If this is the case, the research effort becomes unbalanced, with a disproportionate attention to the molecular (biochemical) level and, at a time of greatly reduced funding for science, a resultant neglect of problems at higher levels.

Ultimately, crop physiology must deal with the behaviour of communities, and although an explanation of that behaviour depends on an understanding of what happens at lower levels of organization, simple extrapolation, from organ to whole plant or from laboratory to field, for example, rarely takes into account the interactions which influence the contribution of each subsystem to the whole. Unfortunately, field trials, in which only a few of the many controlling factors are measured, and in which little attempt is made to explain the influence of those factors on yield, have little general applicability. Engledow and Wadham (1924) recognized this over 60 years ago: 'The results of a yield trial may be usefully indicative; they can never be decisive. Geographical differences of soil and climate and seasonal fluctuations limit the applicability of results to the locality and year of the experiment.' Thus, although an understanding of the performance of the crop in the field, particularly in relation to the weather, can come only from an understanding of the physiology of the crop's sub-systems, this approach on its own is inadequate: it does not give an integrated explanation of the performance of the whole system and, therefore, any measure of the significance of sub-system processes.

It is the need for such integration which has fostered the development of mathematical models. Model building is an adjunct to crop physiological research, a way of making the most effective and efficient use of the large body of information derived from field and laboratory experiments. It may also be the only way of assessing quantitatively the effect of changes at sub-system level on the performance of the crop. Models therefore play the same general role in crop physiology as they do in other disciplines which also require the analysis of complex systems. Our expectations about the 'greenhouse effect' or the 'nuclear winter', for example, are derived from mathematical models. The complexity of such systems, which operate, or threaten to operate, on a world scale, is not questioned, but we must not underestimate the complexity of the crop/environment system. Here we are dealing with events which span an enormous organizational range, from the molecular to the community level. Consideration only of those factors which are defined on a field scale, such as population density, the water and nutrient status of the soil, growth regulators and crop protection chemicals, and all aspects of the weather, suggests a multiplicity of interactions whose effects could never be satisfactorily unravelled by field experiments. Mathematical modelling is therefore a way of putting back together the parts of a system dismantled by reductionist reseach.

There is a certain irony in this. Reductionism is philosophically akin to Cartesianism, named after René Descartes, the seventeenth-century philosopher and mathematician, who advocated that physical phenomena could be accounted for by mathematical theory. In this sense, the reductionist, or Cartesian, approach, by ultimately reducing the essence of a process to a mathematical equation, thereby provides the means to reverse the trend: to build up an explanation of the behaviour of a system at high levels of organization by assembling a series of sub-models which accurately represent processes at lower levels. A model such as this is described as mechanistic, because it provides a quantitative description through an understanding of the mechanisms of the component processes and their interactions. This contrasts with an empirical model, which makes no assumptions about the components of a system or their workings, giving only a description, necessarily at a single level of organization.

Having recourse to an empirical model, which essentially implies making an informed guess at an appropriate equation, may be unavoidable if the workings of the system are not fully understood. An empirical model may well be simpler than its mechanistic counterpart because it is not constrained by the assumptions upon which the mechanistic model is built. However, it has the disadvantage that it contains parameters without biological meaning or whose meaning is unclear. For example, the Monsi-Saeki equation is an empirical model in which it may be difficult to specify precisely the features of canopy architecture or radiation environment responsible for a particular extinction coefficient (sect. 2.4.3).

This distinction between empirical and mechanistic models prompts us now to consider in more detail the roles of mathematical models, both in agriculture generally and in crop physiology in particular. Simple empirical models, used directly by farmers or their advisers, may be perfectly satisfactory as aids to crop management. For example, various measures of accumulated temperature allow the first application of nitrogen fertilizer to grassland in the spring to be made at a stage of growth when the plants are best able to use that nitrogen (sect. 8.3.1). Similarly, reliable estimates of potato yield have been calculated using a simple linear polynomial equation combining temperature and insolation data (Hartz and Moore 1978); and temperature and rainfall data are routinely used to predict the appearance of blight in potato crops, so that control measures can be taken at an appropriate time. At the other extreme, complex mechanistic models may be used as a means of integrating and evaluating knowledge about processes operating at different levels of organisation within the crop/environment system. Such models, as we have noted already, are devised predominantly as research tools to further the pursuit of knowledge: indeed, they may be deliberately highly speculative as a means of challenging hypotheses or of innovating ideas rather than representing established ones. Between these extremes there are, of course, models of varying complexity, designed for a range of purposes and comprising different combinations of mechanistic and empirical components. It should be noted that practically all mechanistic models contain some elements of empiricism where sound knowledge is lacking.

Leaving aside the simplest empirical models and focussing here on those which have at least some mechanistic, or 'process-level', content, models may contribute in one or more of the following general areas:

(a) As we have suggested already, modelling is the most effective way of integrating sub-systems and thereby of evaluating their significance in the functioning of the whole system. The hierarchical nature of crop physiology makes this a most valuable facility in the quest for an understanding of the physiological bases of development, growth and yield. This, and indeed its value in the analysis of the functioning of individual sub-systems, is itself sufficient justification for a modelling approach. Furthermore, physiology's potential as an aid to effective plant breeding will be furthered by the rigorous appraisal which modelling provides. Thus, the effects on yield of changes in such attributes as the maintenance respiration coefficient, the pattern of assimilate partitioning, or the size and rigidity of leaves can readily be assessed in a range of circumstances.

(b) The power of models to predict the effects of specific changes at sub-system level may also have more immediate application in crop management or in helping agronomists to specify improved husbandry practices. Management may be aided simply through the provision of advance knowledge of likely harvest dates or final yield; or, more actively, by knowing with certainty the timing of phenological events so that fertilizer, growth regulators or herbicides are applied at the most appropriate times. It must be said, however, that the financial, or even yield, benefits of such 'fine-tuning' are by no means certain. More important from a practical point of view are the models which aid management decision-making. In crop protection, for example, there may be high financial and environmental costs of unnecessary application of chemicals, whereas inactivity or tardiness may lead to catastrophic crop losses. Thus, a disease management model has been used for wheat crops in the Netherlands, helping farmers to make tactical decisions about spraying according to the incidence of various diseases in their own fields (Zadoks 1983). Pest management may also benefit from a modelling approach. For example, a simple mechanistic model of potato growth has been developed, which simulates the effect of defoliation at various times during the season on tuber yield (Johnson *et al.* 1986).

On a larger scale, strategic decisions may be made with the help of models in the absence of experimental information, or where such information would be difficult, and therefore expensive, to obtain. One such application was the need to evaluate the economic benefits of cloud-seeding to increase rainfall in north-western Victoria, Australia. For this, a model of the development, growth and yield of the wheat crop, with particular emphasis on a water-balance sub-model, has been developed (O'Leary *et al.* 1985).

(c) The reliance of model building on a set of justifiable assumptions necessarily means that a thorough review of relevant knowledge is an early step. This, and the need to quantify relationships, may reveal gaps in current knowledge, and experiments can then be designed specifically to make good these deficiencies. In this way, model building can help to determine the direction of research. A related, but perhaps less tangible, advantage is that team-work may be fostered, and the individual's experimental results may be put in context. Model building can thus be a way of squeezing the most information from the experimental data, particularly when this involves the work of several individuals.

In assessing the value of model building, we must not forget that it is a comparatively young technique in crop physiology, much of the foundations having been laid by de Wit in the 1960s.

Early extravagant claims for the ability of mathematical models to solve all manner of problems gave the technique an unrealistic reputation and ample scope for its detractors' counter-claims. As a result, many physiologists remain sceptical about the usefulness of models, while others seem to believe that work which is not explicitly model-orientated betrays a limited intellectual vision. Needless to say, the latter view is as damaging to the status of model building as uninformed scepticism. We have tried to indicate some of the contributions which model building can make, but its more general acceptance depends on the proven success of the technique. This in turn depends on the correct balance being struck between realism and simplification, which will vary according to the purpose of the particular model.

The relatively short history of model building in crop physiology shows, as Moorby (1985) has pointed out, that the trend towards increased complexity which accompanied advances in knowledge (and indeed improvements in computer technology) has to some extent been reversed. Models of the distribution of radiation within the crop canopy, for example, became particularly complex, and disproportionately so compared with other sub-models. As far as the predictive value of a model is concerned, it is a waste of time, both human and computer, if complexity and precision are not consistent across all sub-models. Simplification of an unbalanced model may well make it more widely accessible without any loss of meaning. However, an over-simplified model may be successful as a predictive device without necessarily being realistic at the sub-model level if, for example, successive errors fortuitously cancel each other out. Here, simplicity compromises meaning, and such specious success is likely to be limited: if a model is to be used over several seasons or in a range of environments, its more general, and genuine, success will inevitably depend on realism at all levels.

Model builders are, of course, as aware of the pitfalls of their discipline as they are convinced of its usefulness. They are refreshingly candid about the limitations of their method of working, acknowledging that the simplifying assumptions upon which the models are built are 'based on a mix of data, knowledge and conjecture' (France and Thornley 1984). As Dent and Blackie (1979) have put it, modelling is 'an art which requires ingenuity, foresight, resourcefulness and, above all, integrity on the part of the model builder'; integrity is particularly important because the modeller's readership is often much less fluent in mathematics than the modeller himself. Space would not permit more than just a superficial outline here of the general principles and techniques of model building; the reader is referred instead to the books and review articles which give a detailed treatment (e.g. Thornley 1976; Loomis *et al.* 1979; France and Thornley 1984). The rest of this chapter will deal in detail with a whole-crop model, which will illustrate at least some of the techniques in practice. In total it is a large model, but its sub-models are straightforward and the mathematics not complex.

## 9.2 THE AFRC WINTER WHEAT MODEL

### 9.2.1 INTRODUCTION

The construction of a whole-crop model, having several sub-models, relies on team-work and is a valuable stimulus for collaboration between individuals. The model which we describe here well illustrates this multi-disciplinary approach, because it is the result of a joint project between (originally) four Institutes supported by the UK's Agricultural and Food Research Council (AFRC), and has involved several other establishments in its testing and evaluation. AFRC WHEAT is available, at no cost, for research and educational purposes, from Dr J R Porter, Long Ashton Research Station, Bristol.

The main aim of the model is to complement field experiments by providing a means of explaining variations in yield between sites and seasons. The reader should now be in no doubt about the complexity of the crop/environment system and the resultant difficulties in the interpretation of field experiments: extrapolation from specific treatments superimposed upon the pattern of weather experienced by the particular site can rarely be justified. Furthermore, multifactorial experiments cannot be comprehensive enough to furnish explanations of quite large variations in yield despite comparatively small differences in weather and management (sect. 6.6).

The influences of weather and management on grain yield must be interpreted through their effects on the physiological processes which determine yield, and this interpretation is made possible by simulating the crop's performance by means of mathematical models.

We have already noted other general benefits of model building; particularly important in the present context is the facility of the simulated crop to allow single factors, or groups of factors, to be altered at will. The simulated crop thus becomes a cheaper and more rewarding source of information than the real crop. It should be borne in mind that model building is itself a dynamic process, and the model described here will therefore inevitably be modified as new information becomes available and ideas evolve.

## 9.2.2 THE STRUCTURE OF THE MODEL

The model is represented schematically in Fig. 9.1, and is described in detail by Weir *et al.* (1984), Porter (1984) and Weir *et al.* (1985). It comprises five sub-models: (1) phenological development; (2) tiller and leaf production (canopy development); (3) root production; (4) light interception and photosynthesis (dry-matter production); and (5) dry-matter partitioning. The weather data required to run the model are daily maximum and minimum air temperature, daylength, daily wet and dry bulb temperatures, and daily net short-wave solar radiation: in model building terms, these are the driving variables.

Within the framework set by the phenological development sub-model, the simulated crop produces tillers, some of which survive to bear ears, and thereby generates a canopy of photosynthesizing leaves and a population of ear-bearing stems. Photosynthesis is calculated according to the amount of radiation intercepted by the canopy and a photosynthetic light response curve, and respiration subtracted to give the daily increase in carbohydrate. This is then partitioned between roots, leaves, stems and, later, ears, in proportions which vary with the stage of development.

The root growth sub-model will not be described here because, in its present form, the overall model applies to crops which are healthy and free from water and nutrient stress. In these circumstances, the roots affect the growth of the crop only through their role as a sink for assimilate. The assumption of a crop as free from environmental stress as possible is a reasonable starting point from which to develop a whole-crop model; further development of the model will introduce modifications to allow for water or nutrient limitations. However, the assumption was largely obligatory because of a lack of knowledge

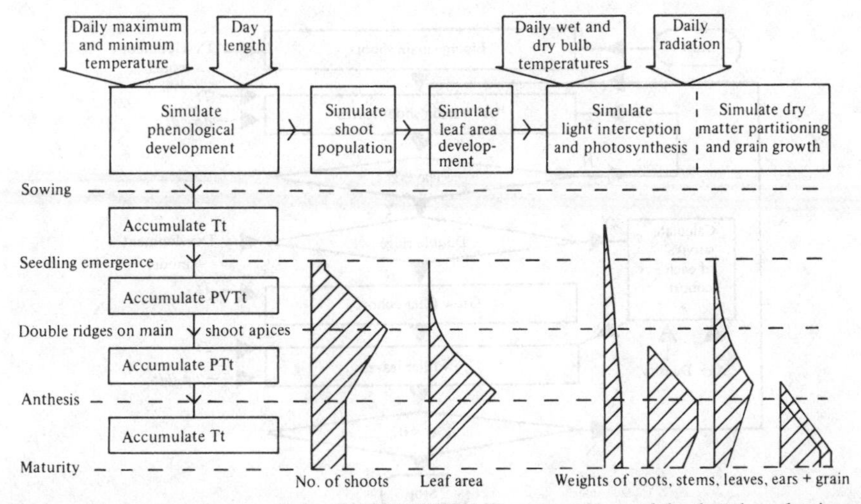


**Fig. 9.1** Diagrammatic representation of the AFRC WHEAT computer model, showing the interactions between the sub-models: phenological development, tiller and leaf growth, root growth, light interception and photosynthesis, dry-matter partitioning and grain growth. The arrowed boxes show the driving variables.

Abbreviations: Tt, thermal time; PTt, thermal time modified by photoperiod; PVTt, thermal time modified by photoperiod and vernalization (after Weir *et al.* 1984).

about the growth and development of roots, both in time and space. Subsequent work has begun to describe in detail the development of the root system in field-grown crops, and has become the basis for a model which simulates a realistic synchrony between root growth and development and shoot development (Porter *et al.* 1986). This model generates profiles of root length density (km of root per $m^2$ of soil in successive 10 cm layers), comprising roots of specified age, type (main axis, first-order lateral, second-order lateral) and origin (main shoot, tiller number), at specified times after sowing. Its further development will eventually allow sub-models of water and nutrient uptake to be combined with those of above-ground processes, so that growth, development and yield in non-ideal circumstances can be modelled.

#### 9.2.2.1 PHENOLOGICAL DEVELOPMENT SUB-MODEL

The timing of the major developmental events and, therefore, the duration of the intervening phases is determined primarily by accumulated temperature (thermal time). This is essentially the cumulative mean daily temperature above a base temperature of 1 °C (up to anthesis) or 9 °C (from anthesis to maturity), and has units of degree-days (°C days). Thus, the progression of the crop from one developmental phase to the next depends on the accumulation of pre-set totals of thermal time. For the period between emergence and the appearance of double ridges, thermal time is modified by photoperiod and vernalization factors (sect. 6.3), and from double ridges to anthesis by photoperiod. In addition to the major phenological stages shown in Fig. 9.1, the model also registers the timing of the following events, after pre-set intervals of thermal time, modified as appropriate by photoperiod and vernalization factors: first spikelet initiation, terminal spikelet, the beginnings of ear growth and of grain growth, and the end of grain growth (sect. 6.1).

#### 9.2.2.2 TILLER AND LEAF GROWTH SUB-MODEL

Figure 9.2 is a flow diagram which summarizes the decisions, calculations and driving variables involved in the simulation of crop canopy development. This sub-model is 'switched on', so that the main shoots emerge, by information from the phenological development section. Leaves then appear after pre-set intervals of thermal time, which depends on the rate of change of daylength at seedling emergence (Baker *et al.* 1980; sect. 2.2.2). New leaf production stops with the appearance of the flag leaf, which is the last one to reach maximum size by the time of anthesis.

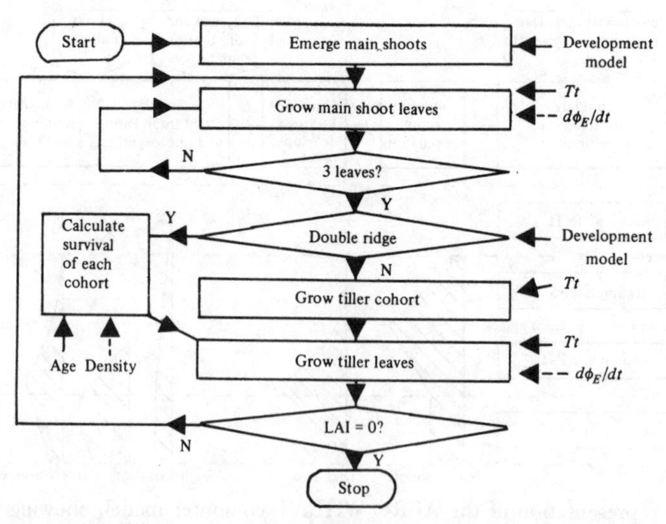


**Fig. 9.2** Flow diagram of the tiller and leaf growth sub-model. Solid arrowed lines indicate principal driving variable in a calculation; broken lines indicate modifiers.

Abbreviations: $d\phi_E/dt$, rate of change of daylength at crop emergence; LAI, green leaf area index; Y, yes; N, no (from Porter 1984).

After the appearance of the third leaf, groups (or cohorts) of tillers are produced each week, the number depending on an empirically derived rate of tiller production and the number of day degrees in the preceding week. Thus:

$$N = \sum_{j=k}^{j=k+6} T_j \cdot TPr \qquad [9.2]$$

where $N$ is the number of new tillers per $m^2$ initiated in a week (i.e. a cohort), $k$ and $k + 6$ are the days at the start and end of the week, $T_j$ is the daily mean temperature throughout the week (°C), and $TPr$ is a tiller production rate taken, from observations of crops grown at optimal supplies of nitrogen and largely free of disease, to be 2.63 tillers (per $m^2$)/°C day.

Tillering is set to stop at the double ridge stage, and some tillers die between then and anthesis (Fig. 6.11). Those which survive are considered to develop ears. The model calculates tiller survival empirically as:

$$p = \frac{1}{1 + \left(\dfrac{Tt/400}{(A/N_n)^{\alpha}}\right)^{\beta}}, \qquad [9.3]$$

where $p$ is the proportion of each cohort which survives, $Tt$ is the thermal time since double ridge

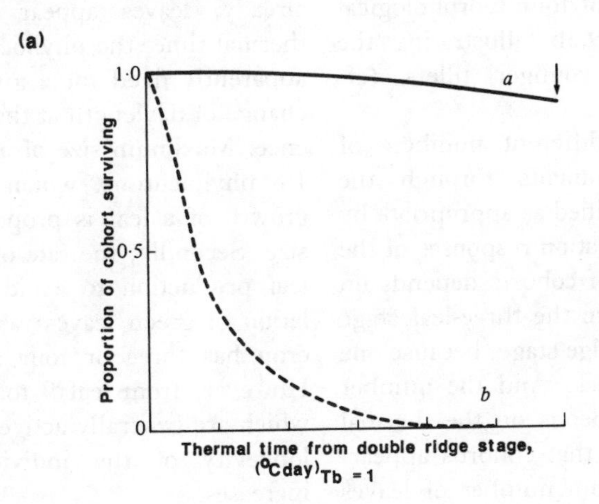


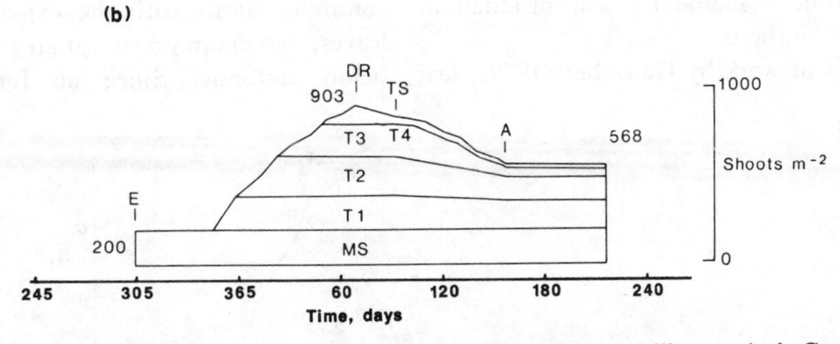


**Fig. 9.3** (a) The simulated effect of time of production of a tiller cohort on tiller survival. Curve *a*, $N_n$ = 200; curve *b*, $N_n$ = 1600. For details, see text and eqn [9.3], in which $A$ = 825, $\alpha$ = 1.46 and $\beta$ = 2.24, values derived empirically from data obtained from crops (cv. Hustler) grown with optimal supplies of nitrogen and free of disease. The arrow indicates the time of anthesis (from Porter 1984).

(b) Changes in shoot numbers throughout the life of a simulated crop (cv. Avalon) of 200 plants $m^{-2}$, sown on 16 October 1980 with daily weather for Littlehampton, Sussex, England. Time is measured in days, counting 1 January as day 1.

Abbreviations: MS, main stems; T1, T2, T3, T4, tiller classes 1, 2, 3, 4; E, emergence; DR, double ridges; TS, terminal spikelet; A, anthesis.

Numbers specify shoot numbers at emergence, double ridges and maturity (from Weir *et al.* 1985).

formation, and $N_n$ is the number of shoots (main shoots + tillers) produced before cohort *n*. The parameter $N_n$ is a measure of the age of the cohort as well as the overall density of shoots: the greater its value the younger the cohort and the lower the value of *p*. Thus, the proportion of each cohort which survives is inversely related to shoot density and positively related to cohort age, behaviour which resembles that of real crops. Curves produced by eqn [9.3] are shown in Fig. 9.3a. Clearly, a much higher proportion of the earlier produced cohort (curve *a*, $N_n = 200$) survives to anthesis compared with the younger cohort (curve *b*, $N_n = 1600$). An example of the integrated effect of this type of response for a simulated crop with tillers of four morphological classes is shown in Fig. 9.3b, illustrating the greater mortality of the younger tillers (cf. Fig. 6.11).

The model allows for different numbers of tillers in different environments through the effects of thermal time, modified as appropriate by the photoperiod and vernalization responses of the plants. Thus, the number of cohorts depends on the length of the period from the three-leaf stage (main stem) to the double ridge stage, because one cohort is produced each week. And the number of tillers in each cohort depends on the thermal time of the week preceding that cohort's appearance. Similarly, the maximum number of leaves produced by a shoot is linked to the environment through the time available for leaf production, which ceases at anthesis.

On the basis of work by Gallagher (1979), leaf growth, on main stems and tillers, is modelled by assuming a linear increase in lamina size with respect to thermal time, until a pre-set maximum size is reached. This is maintained for a period, after which there is a senescent decline to zero green area. The maximum size varies with leaf position on the stem: it is identical for the leaves which appear before the double ridge stage, and thereafter progressively increases up to leaf 12. After the double ridge stage, leaf sheaths also contribute to leaf area, but the photosynthetic area of ears is not included. Simple rules govern lamina growth and senescence in the model. First, the time required for the attainment of maximum size depends on the rate of leaf appearance. As noted already, leaves appear at regular intervals of thermal time, the phyllochron (interval), which is apparently fixed for a given plant by the rate of change of daylength at the time of seedling emergence. Maximum size of all leaves is reached after 1.8 phyllochrons, which means that the rate of growth of a leaf is proportional to its maximum size. Secondly, the rate of senescence is geared to leaf production to avoid an unrealistic accumulation of green leaves: as a result, the simulated crop has three or four active leaves per shoot. However, from leaf 9 to the flag leaf, the leaves which are generally active during grain-filling, the longevity of the individual leaf progressively increases.

The cycle of leaf production, growth and death continues until, with the expansion of the flag leaves, maximum green leaf area is reached shortly before anthesis. Since no further leaves are

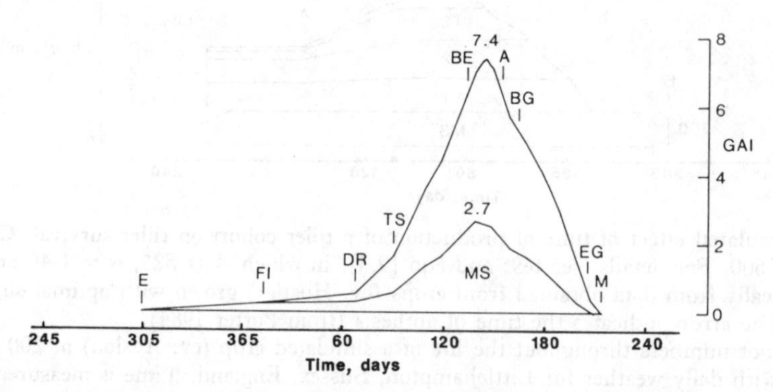


**Fig. 9.4** Changes in green leaf area index (GAI) throughout the life of a simulated crop of winter wheat cv. Avalon.

Details and abbreviations: as Fig. 9.3, and FI, floral initiation; BE, beginning of ear growth; BG, beginning of grain growth; EG, end of grain growth (from Weir *et al.* 1985).

produced, there follows a steady decline in green area throughout the period of grain growth, reaching zero at crop maturity (Fig. 9.4).

#### 9.2.2.3 PAR INTERCEPTION AND PHOTOSYNTHESIS SUB-MODEL

The sub-models of phenological and canopy development together allow the interception of solar radiation by the crop, throughout the season, to be simulated. The diurnal variation in incident PAR, on an hourly basis, is calculated from the daily values of net short-wave radiation, using an equation which describes a sine-wave distribution of radiation throughout the day (Charles-Edwards 1978). The mean PAR incident on the leaf surfaces within a horizontal layer of the canopy is then calculated for each layer and for each hour, using the equation:

$$I(z) = \frac{k}{(1 - m)} I_o e^{-kL(z)} \quad [9.4]$$

where $I(z)$ is the PAR incident on the leaf surfaces at level $z$, the mid-point of a canopy layer; $I_o$ is the PAR available at the top of the canopy; $L(z)$ is the cumulative green leaf area index down to level $z$; $k$ is an extinction coefficient (0.44); and $m$ a leaf transmission coefficient for PAR (0.1). The number of layers into which the canopy is divided is simply defined by the number of horizons at which $I(z)$ is calculated.

This is clearly the Monsi-Saeki equation, modified by the factor $k/(1 - m)$, which allows for the fact that the irradiance available for photosynthesis by a layer of leaves is not the same as that incident on a horizontal surface at the same level. This is because the leaves themselves may not be horizontal, and some of the radiation which is absorbed is transmitted through the leaves. The small correction made by the divisor $(1 - m)$ takes into account the fact that the amount of radiation incident on a leaf is greater than the amount absorbed, and allows eqn [9.4] to be used in conjunction with a model of the photosynthetic light response based on absorbed, rather than incident, PAR.

Equation [9.4] can be derived from the simple Monsi-Saeki form by considering the passage of light down through an element of the canopy having a leaf area of $\Delta L$. If we assume that intercepted light which is not absorbed is transmitted through the leaf (thus ignoring the complications of reflected light), then $(1 - m)$ is the fraction of intercepted light which is absorbed, and the reduction in irradiance due to absorption by the leaves, $\Delta I$, can be written as:

$$-\Delta I = \Delta L\,(1 - m)\,I_L \quad [9.5]$$

where $I_L$ is the irradiance incident on the leaf surfaces, and the negative sign shows that there is a decrease in downward irradiance. If this equation is rearranged so that:

$$I_L = \frac{-(\Delta I/\Delta L)}{(1 - m)} \quad [9.6]$$

and $\Delta I/\Delta L$ is replaced by the derivative $dI/dL$, we have:

$$I_L = \frac{-(dI/dL)}{(1 - m)} \quad [9.7]$$

which is the irradiance incident on the leaves and therefore available for photosynthesis. $dI/dL$ is obtained from eqn [9.1] by differentiating $I$ with respect to $L$:

$$dI/dL = -kI_o e^{-kL} \quad [9.8]$$

Substituting this into eqn [9.7] gives:

$$I_L = \frac{kI_o e^{-kL}}{(1 - m)} \quad [9.9]$$

which is identical to eqn [9.4].

The response of photosynthetic rate, $P$ (net or gross), to irradiance, $I(z)$, has often been modelled by an equation of the form:

$$P = \frac{\alpha I(z) P_{max}}{\alpha I(z) + P_{max}} \quad [9.10]$$

where $\alpha$ is a constant, the photosynthetic efficiency, having units of, for example, mg $CO_2$ $J^{-1}$, and $P_{max}$ is the value of $P$ at saturating irradiance. This produces the photosynthetic light response curve shown in Fig. 9.5, a curve known as a rectangular hyperbola. Equation [9.10] is sometimes encountered in the form:

$$P = \frac{1}{a + [b/I(z)]} \quad [9.11]$$

where $a = 1/P_{max}$ and $b = 1/\alpha$.

The rectangular hyperbola provides a good fit to data obtained under some circumstances, such

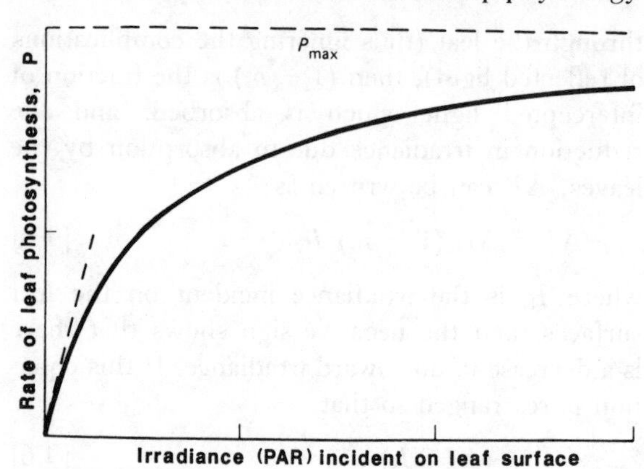


**Fig. 9.5** Photosynthetic light response curve of a leaf according to eqn [9.10]. The broken line at $P_{max}$ denotes the light-saturated rate of photosynthesis. The initial slope of the response curve is $\alpha$ (after France and Thornley 1984).

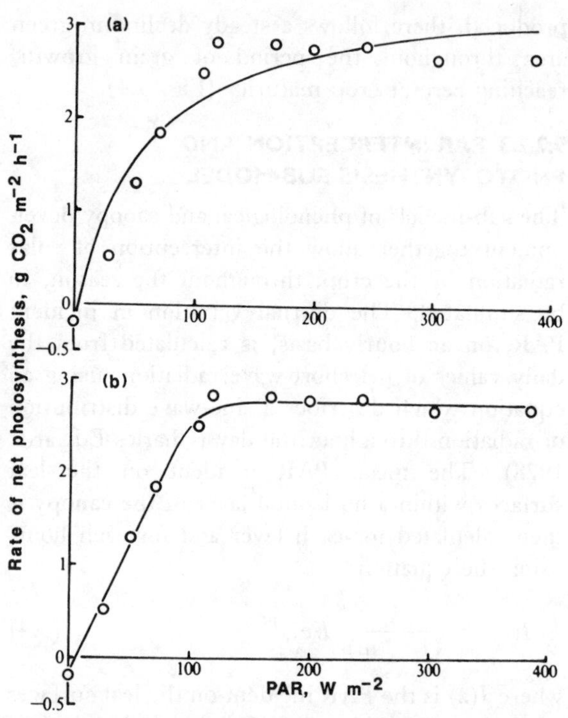


**Fig. 9.6** (a) The measured photosynthetic light response of a fully expanded flag leaf of winter wheat in the field (○), with the fitted rectangular hyperbola (full line).

(b) The same measurements fitted by a curve described by eqn [9.12], modified to include terms which account for respiration (from Marshall and Biscoe 1980).

as in controlled environments. There is, however, evidence that leaves of plants growing in the field may have a somewhat different photosynthetic light response. This is illustrated by some measurements of the rate of net photosynthesis of a fully expanded flag leaf in a winter wheat crop (Fig. 9.6). Here, a linear relationship between photosynthesis and irradiance up to about 100 W $m^{-2}$ gives way quite abruptly to light saturation above 150 W $m^{-2}$ (sect. 2.4.3). The fitted rectangular hyperbola misrepresents this response: it overestimates both the slope of the linear part of the relationship ($\alpha$) and the asymptote ($P_{max}$), and underestimates the rate of photosynthesis at the transition from light limitation to light saturation. Such discrepancies are not surprising when we consider that the rectangular hyperbola model of photosynthesis is derived from the Michaelis-Menten relation between the rate of an enzyme-catalysed reaction and the concentration of its substrate. A model based only on the biochemistry of photosynthesis, taking no account of $CO_2$ transfer from the atmosphere to the chloroplasts, cannot be expected, in many cases, to be more than an approximation.

The quadratic equation:

$$\theta P_g^2 - P_g [P_{max} + \alpha I(z)] + \alpha I(z) P_{max} = 0 \qquad [9.12]$$

where $P_g$ is the rate of gross photosynthesis, net of photorespiration, and $\theta$ is the ratio of physical to total resistance to $CO_2$ transfer $[(\Sigma_r - r_x)/\Sigma_r]$, is more flexible. When $\theta$ is zero, it reduces to eqn [9.10], describing a rectangular hyperbola: this is to be expected because a low value of $\theta$ implies that $r_x$, the carboxylation or biochemical resistance, is much greater than the resistances to physical $CO_2$ transfer. On the other hand, when the physical resistances are dominant, so that the value of $\theta$ approaches unity, eqn [9.12] describes what is referred to as a Blackman-type response, in which an extensive, linear, light-limited region is abruptly succeeded by the asymptotic, light-saturated rate. Marshall and Biscoe (1980) derived a more complex version of eqn [9.12], including terms allowing for respiration so that net photosynthesis was modelled, and found that it gave a good fit to their field measurements of photosynthesis (Fig. 9.6b). Daily mean values of $\theta$ were found to be within the range 0.85–1.0, and showed no trend with leaf age.

Equation [9.12] was therefore adopted to model the photosynthetic light response in the AFRC model. $\alpha$ was assigned a value of 0.009 mg $J^{-1}$, and $\theta$ 0.995 (dimensionless). $P_{max}$ is given by:

$$P_{max} = 0.995\ C_a/r_a + r_s + r_m \quad (\text{mg m}^{-2}\ \text{s}^{-1}) \qquad [9.13]$$

where $C_a$ is the ambient $CO_2$ concentration, taken to be 600 mg $m^{-3}$, and $r_a$, the boundary layer resistance, and $r_m$, the mesophyll resistance, have fixed values of 30 and 400 s $m^{-1}$, respectively; $r_s$, the stomatal resistance, varies with both irradiance and the atmospheric vapour pressure deficit, $D$ (kPa), according to the form:

$$r_s = 1.56 \times 75[1 + 100/I(z)]\ (1 - 0.3D) \quad (\text{s m}^{-1}) \qquad [9.14]$$

The crop is assumed to be free from water stress, which avoids the need to take into account leaf water status in the estimation of $r_s$. Figure 9.7 provides an example of the response of the photosynthetic rate to irradiance generated by eqn [9.12], using the values of the parameters given above, and letting $D$ be 0.7 kPa; for comparison, the calculations have also been made with $\theta = 0.5$ and 0.1.

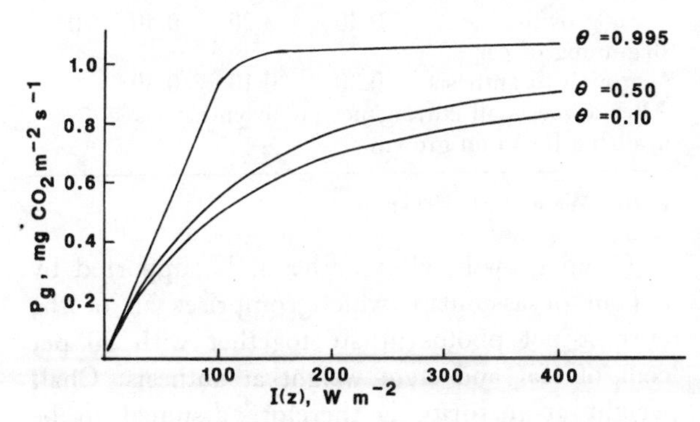


**Fig. 9.7** The response of the rate of gross photosynthesis to irradiance, calculated according to the equation

$$\theta P_g^2 - [P_{max} + \alpha I(z)]P_g + \alpha I(z)P_{max} = 0$$

and using the parameter values given in the text.

The equation is of the form $ax^2 + bx + c = 0$, where $a = \theta$, $b = [P_{max} + \alpha I(z)]$ and $c = [\alpha I(z)P_{max}]$, and was solved for $P_g$ by applying the standard formula for a quadratic equation, so that:

$$P_g = \frac{[P_{max} + \alpha I(z)] - \sqrt{[P_{max} + \alpha I(z)]^2 - 4\theta\alpha I(z)P_{max}]}}{2\theta}$$

Any model, or sub-model, is always a compromise between, on the one hand, simplicity and manageability and, on the other, mechanistic realism and accuracy. The photosynthesis sub-model used here is slightly more complex than the simple rectangular hyperbola, which many modellers find satisfactory, but it has the flexibility, through the value of $\theta$, to give a better representation of photosynthesis in the field. One obvious simplification is that no allowance is made for the well-documented, gradual senescent decline in $P_{max}$ (see, for example, Fig. 3.20). However, it would be difficult to accommodate this feature when the unit of photosynthetic activity is a layer of the canopy, having a particular value of $L$ and composed of leaves of different ages. Presumably, at any one time, the age distribution of leaves is such that a single $P_{max}$ is a satisfactory average. Furthermore, $P_{max}$ is, of course, only one of the factors determining $P_g$, and the leaf layers expected to have a reduced $P_{max}$ would be the lower ones in which $P_g$ would, in any case, be reduced by lower irradiance. Environmental factors, which affect all leaves more or less equally, are easier to accommodate. Irradiance is, of course, the major driving variable, but eqn [9.12] is modified to take into account also the limitation of $P_{max}$ by low temperatures. The derivation of the modified equation is complex, however, and need not concern us here.

The rate of gross photosynthesis is therefore calculated for each layer of the canopy for each daylight hour by solving eqn [9.12] for $P_g$ (see Fig. 9.7), using the appropriate value of $I(z)$ (eqn [9.4]). This gives $P_g$ in units of mg $CO_2$ $m^{-2}$ (leaf) $s^{-1}$: from this is obtained the hourly value, which is multiplied by the leaf area index of the layer (= total $L$/number of layers) to give that layer's contribution to gross photosynthesis per $m^2$ of ground. The values for each layer are then summed to give the hourly canopy gross photosynthesis, $P_g(h)$, which is multiplied by 0.65 to convert it from $CO_2$ to carbohydrate ($CH_2O$) equivalent.

Daily respiration, $R$, is calculated as the sum of growth and maintenance respiration (see sect. 4.3). Growth respiration is a function of the daily canopy gross photosynthate production, expressed as $CH_2O$ equivalent. Maintenance respiration, $R_M$, is a function of total crop weight and is temperature dependent, so that for a $Q_{10}$ of 2:

$$R_M = bW2^{\bar{T}/10} = bW2^{T_{max} + T_{min})/2 \times 10} \quad [9.15]$$

where $b$ (day$^{-1}$) is the maintenance respiration coefficient, $W$ (g m$^{-2}$) the crop weight, $T$ the mean daily temperature ($= (T_{max} + T_{min})/2$, where $T_{max}$ and $T_{min}$ are the daily maximum and minimum temperatures, respectively). Respiration is therefore calculated as:

$$R = 0.65\ a \sum_{h=0}^{h=H} P_g(h) + bW\,2^{0.05(T_{max} + T_{min})} \quad (\text{g } CH_2O \text{ m}^{-2} \text{ d}^{-1}) \quad [9.16]$$

where $a$, the growth respiration coefficient, has a fixed value of 0.34, $H$ is the number of daylight hours, and $b$ has a value of 0.002 before anthesis and 0.001 after anthesis. The change in the value of $b$ after anthesis largely reflects the lower cost of maintenance when an increasing proportion of the crop's weight comprises relatively inert starch in the grains.

#### 9.2.2.4 DRY-MATTER PARTITIONING SUB-MODEL

Daily net photosynthate production, $P_n$ ($CH_2O$), is calculated as the difference between daily gross production and respiration:

$$P_n\ (CH_2O) = P_g\ (CH_2O) - R \quad (\text{g m}^{-2} \text{ d}^{-1}) \quad [9.17]$$

This dry matter is partitioned between leaves, roots, stems and ears in proportions which change with the stage of development (Table 9.1). In this early version of the model, leaf growth and tiller production are not constrained by assimilate supply: as we have seen, leaf and tiller appearance and growth proceed according to the phenological development sub-model and thermal time. The start of ear growth is signalled by the phenological development sub-model, and thereafter 30 per cent of new net photosynthate is allocated to the ears. The dry weight accumulated by the ears at anthesis determines the number of grains set: 10 mg of ear weight is assumed to be equivalent to one grain. This simple relationship means that grain number is directly related to the daily total of intercepted radiation. It is negatively correlated with mean daily temperature because of accelerated development (earlier anthesis), and reduced net assimilation due to higher maintenance respiration. Fischer (1985) has confirmed the dependence of grain number on ear weight at anthesis, which he found to be directly related to the ratio of mean daily intercepted radiation to mean temperaure (above 4.5 °C) during the 30 days preceding anthesis.

**Table 9.1** The partitioning of dry matter between the plant parts

| *Stage of development* | *Proportions of current photosynthate* | | | |
|---|---|---|---|---|
| | *Leaves* | *Roots* | *Stems* | *Ears* |
| Emergence to double ridge (DR) | 0.55 | 0.35 | 0.10 | 0 |
| DR to beginning of ear growth | 0.40 | 0.20 | 0.40 | 0 |
| Beginning of ear growth to anthesis | 0.30 | 0.10 | 0.30 | 0.30 |
| After anthesis all current net photosynthate is available for grain growth | | | | |

(After Weir *et al.* 1984)

Grain growth, after anthesis, is supported by a pool of assimilate which comprises all of the current net photosynthate together with 30 per cent of leaf and stem weight at anthesis. Chaff weight at maturity is therefore assumed to be equal to ear weight at anthesis. The period from anthesis to maturity, which lasts for 345 °C days (base temperature 9 °C), is divided (55 : 240 : 55) into three phases: initiation, linear growth, and mature. There is no grain growth during the initiation, or lag, phase, so net photosynthate accumulates in the pool. During the linear phase, each grain has the potential to grow at a temperature-dependent maximum rate, $G_{max}$:

$$G_{max} = 0.045\ (T_{max} + T_{min})/2 + 0.4 \quad (\text{mg grain}^{-1} \text{ d}^{-1}) \quad [9.18]$$

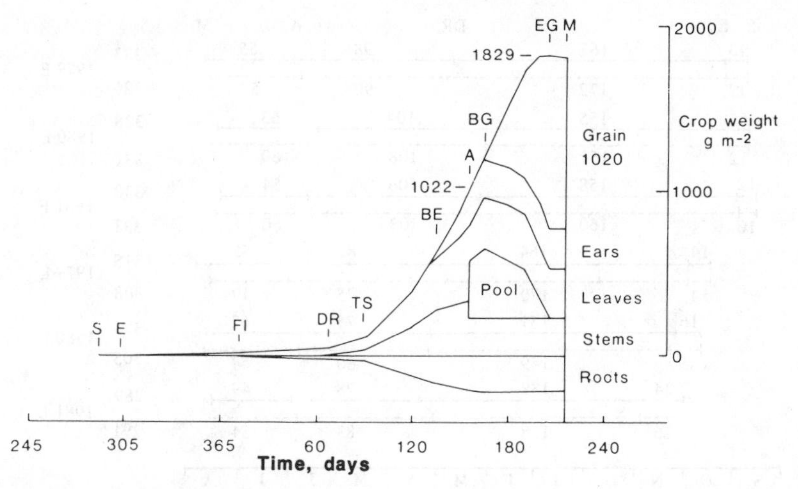


**Fig. 9.8** Partitioning of dry weight among the various plant parts of a simulated crop of winter wheat cv. Avalon.
Abbreviations: as Figs 9.3b and 9.4 (from Weir *et al.* 1985).

However, if there is not enough assimilate to meet the total demand, which is the product of $G_{max}$ and the number of grains, growth will be restricted. Final grain weight depends, of course, on the actual rate of growth and its duration (the time taken to reach 240 °C days).

An example of the pattern of dry-matter partitioning among the various plant parts in a simulated crop is given as Fig. 9.8.

## 9.2.3 THE BEHAVIOUR OF THE MODEL

In this section we are concerned primarily with comparisons of observed and simulated aspects of the crop's performance. This is a way of testing whether the biological assumptions of the model are justified, and whether the equations used are adequate representations of those assumptions. Such comparisons may identify failings requiring more detailed investigation, and perhaps necessitating refinements of the model. The evaluation of the model, in terms of its ability to fulfil its wider objectives, for a range of cultivars and environments, is an accompanying process, but one which presupposes methodological correctness. Here, we shall concentrate on tests of the basic soundness of the model, using the cultivar Hustler grown at Rothamsted. More rigorous tests of the assumptions upon which the phenological development sub-model is based have since been made, with the cultivar Avalon, by monitoring the development of crops sown simultaneously at ten sites throughout the UK (Porter *et al.* 1987; Kirby *et al.* 1987).

Figure 9.9 shows the timing of the major stages of phenological development for observed and simulated crops of Hustler in three successive seasons: the correspondence is good. The model can be adapted to deal with different cultivars by changing the thermal times between stages. For example, the interval between terminal spikelet and anthesis is about 100 °C days longer for Hustler than it is for Avalon. However, more field work is required in order to define and model such differences. Some work with Avalon at Rothamsted has, in fact, revealed discrepancies of up to 20 days between the simulated and observed timings of double ridge formation. This is a large error, but may be associated simply with low temperatures around the time of double ridge formation, which allow the passage of calendar time with little accumulation of thermal time (Porter 1985). On the other hand, the nationwide survey of Avalon development revealed that, at sites with relatively high mean temperatures during the emergence to double ridge period, floral initiation occurred before the model's estimated date of complete vernalization. This may have been a response to the initiation of a critical number of leaves, or to short days substituting for low temperatures.

The simulated changes in the number of shoots

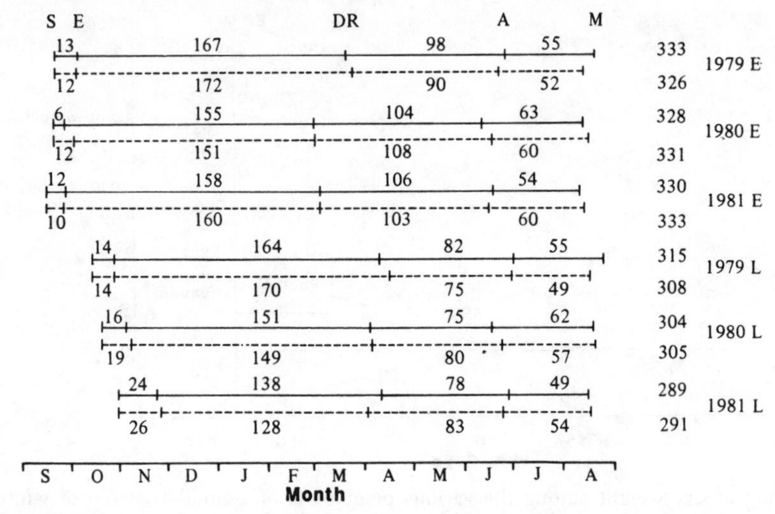


**Fig. 9.9** Diagram showing the timing of phenological development stages for observed (——) and simulated (---) wheat crops. In each season there was an early and late sown crop.

Abbreviations: S, sowing; E, emergence; DR, double ridge; A, anthesis; M, maturity. The numbers indicate durations between stages (days) (from Weir *et al.* 1984).

per $m^2$ and in the green leaf area index were also generally in good agreement with observations (Figs 9.10 and 9.11). However, the late-sown 1978–9 crop had a much higher maximum shoot number than was simulated, because tillering continued after the double ridge stage, at which time the model assumes it to have ceased. The circumstances in which this assumption is unrealistic are not clear. The death of tillers was adequately modelled, so that shoot numbers fell

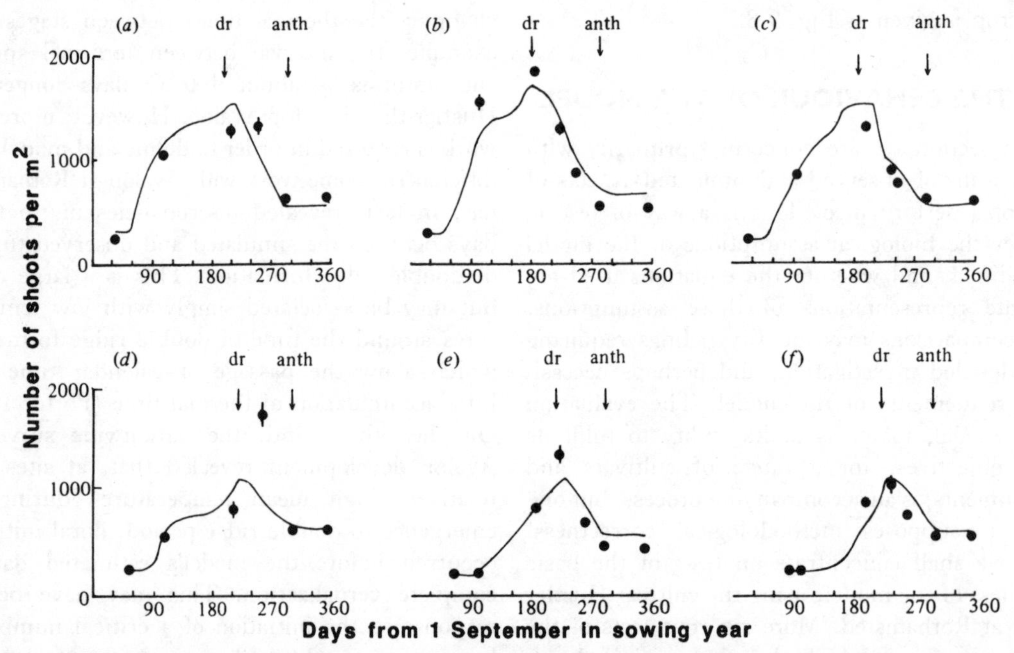


**Fig. 9.10** Observed (●) and simulated (——) changes in the number of shoots $m^{-2}$ for the crops referred to in Fig. 9.9 (a) 1978–9, early; (b) 1979–80, early; (c) 1980–1, early; (d) 1978–9, late; (e) 1979–80, late; (f) 1980–1, late. Arrows indicate the timing of the double ridge (dr) and anthesis (anth) stages (from Porter 1984).

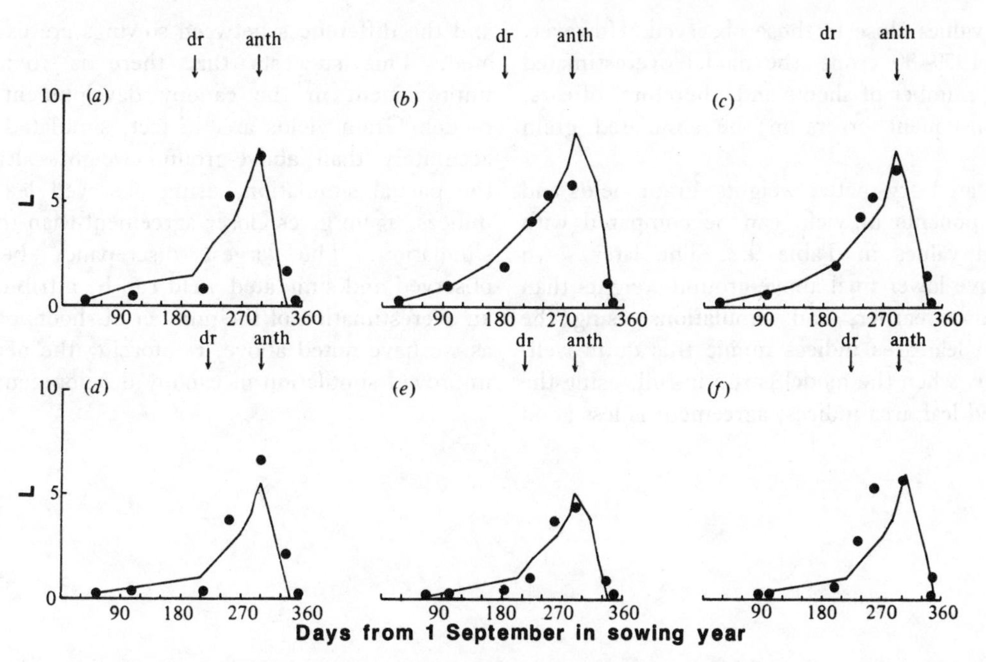


**Fig. 9.11** Observed (●) and simulated (——) changes in the green leaf area index ($L$) for the crops described in Fig. 9.10 (from Porter 1984).

**Table 9.2** Observed and simulated dry-matter weights and yield components for winter wheat cv. Hustler grown at Rothamsted, described also in Figs 9.9–9.11

| | | *1978–9* | | *1979–80* | | *1980–1* | |
|---|---|---|---|---|---|---|---|
| | | *E* | *L* | *E* | *L* | *E* | *L* |
| Top weight, maturity ($g\ m^{-2}$) | Observed | 1706 | 1663 | 1800 | 1540 | 2010 | 1780 |
| | Simulated 1 | 1960 | 1809 | 1965 | 1512 | 1910 | 1791 |
| | Simulated 2 | 2058 | 1575 | 2232 | 1523 | 2015 | 1473 |
| Grain yield ($g\ m^{-2}$) | Observed | 846 | 837 | 837 | 771 | 823 | 770 |
| | Simulated 1 | 862 | 838 | 934 | 750 | 870 | 869 |
| | Simulated 2 | 843 | 746 | 1127 | 833 | 895 | 773 |
| No. of ears per $m^2$ | Observed | 620 | 622 | 543 | 476 | 542 | 556 |
| | Simulated 2 | 655 | 616 | 635 | 576 | 601 | 627 |
| No. of grains per ear | Observed | 34.4 | 33.6 | 37.5 | 40.5 | 39.3 | 37.1 |
| | Simulated 2 | 34.8 | 32.5 | 42.4 | 35.6 | 38.1 | 31.6 |
| Grain weight ($mg\ grain^{-1}$) | Observed | 40.2 | 40.3 | 41.7 | 40.3 | 39.2 | 37.5 |
| | Simulated 2 | 37.0 | 37.3 | 41.9 | 40.6 | 39.1 | 39.0 |

*E* and *L*, early-sown and late-sown crops.
Simulated 1, using observed values of LAI; Simulated 2, using simulated values of LAI.
(After Weir *et al.* 1984)

to final values close to those observed. However, for the 1979–80 crops, the model overestimated the final number of shoots and, therefore, of ears, with consequent errors in the simulated grain yields.

Simulated dry-matter weights, grain yields and the components of yield can be compared with observed values in Table 9.2. The later sown crops have lower final above-ground weights than those sown earlier, and simulations using the observed leaf area indices mimic this quite well. However, when the model is run in full, using the simulated leaf area indices, agreement is less good and the differences between sowings are exaggerated. This suggests that there is room for improvement in the canopy development sub-model. Grain yields are, in fact, simulated more accurately than above-ground weight, although the partial simulation, using observed leaf area indices, again gives closer agreement than the full simulation. The largest discrepancy between observed and simulated yield can be attributed to an overestimation of the number of shoots per m$^2$, as we have noted above, reinforcing the need for improved simulation of canopy development.

# REFERENCES


Acock B, Thornley J H M, Warren Wilson J 1970 Spatial variation of light in the canopy. In Setlik I (ed) *Prediction and Measurement of Photosynthetic Productivity*. Centre for Agricultural Publishing, Wageningen, pp 91–102

Adams S N, Easson D L, Gracey H I, Haycock R E, O'Neil D G 1983 An attempt to maximise yields of cut grass in the field in Northern Ireland. *Record of Agricultural Research* **31**: 11–16

Ahmad I, Farrar J F, Whitbread R 1983 Photosynthesis and chloroplast functioning in leaves of barley infected with brown rust. *Physiological Plant Pathology* **23**: 411–19

Ahmed C M S, Sagar G R 1981 Volume increase of individual tubers of potatoes grown under field conditions. *Potato Research* **24**: 279–88

Allen E J 1977 Effects of date of planting on growth and yield of contrasting potato varieties in Pembrokeshire. *Journal of Agricultural Science, Cambridge* **89**: 711–35

Allen E J 1978 Plant density. In Harris P M (ed) *The Potato Crop*. Chapman and Hall, pp 278–326

Allen E J 1979 Effects of cutting seed tubers on number of stems and tubers and tuber yields of several potato varieties. *Journal of Agricultural Science, Cambridge* **93**: 121–8

Allen E J, Morgan D G 1972 A quantitative analysis of the effects of nitrogen on the growth, development and yield of oilseed rape. *Journal of Agricultural Science, Cambridge* **78**: 315–24

Allen E J, Scott R K 1980 An analysis of growth of the potato crop. *Journal of Agricultural Science, Cambridge* **94**: 583–606

Allen E J, Wurr D C E 1973 A comparison of two methods of recording stem densities in the potato crop. *Potato Research* **16**: 10–20

Allen P J 1942 Changes in the metabolism of wheat leaves induced by infection with powdery mildew. *American Journal of Botany* **29**: 425–35

Allen P J, Goddard D R 1938 A respiratory study of powdery mildew of wheat. *American Journal of Botany* **25**: 613–21

Allison J C S, Daynard T B 1976 Effect of photoperiod on development and number of spikelets of a temperate and some low-latitude wheats. *Annals of Applied Biology* **83**: 93–102

Andrews T J, Lorimer G H 1978 Photorespiration – still unavoidable? *FEBS Letters* **90**: 1–9

Angus W J, Sage G C M 1980 Patterns of yield production in two semi-dwarf and two conventional height European spring wheats. *Journal of Agricultural Science, Cambridge* **95**: 387–93

Anslow R C, Green J O 1967 The seasonal growth of pasture grasses. *Journal of Agricultural Science, Cambridge* **68**: 109–27

Araus J L, Tapia L, Azcón-Bieto J, Caballero A 1986 Photosynthesis, nitrogen levels, and dry matter accumulation in flag wheat leaves during grain filling. In Marcelle R, Clijsters H, Van Poucke M (eds) *Biological Control of Photosynthesis*. Martinus Nijhoff Publishers, Dordrecht

Asani S, Har-Nishimura I, Nishimura M, Akazawa T 1985 Translocation of photosynthates into vacuoles in spinach leaf protoplasts. *Plant Physiology* **77**: 963–8

Asfary A F, Wild A, Harris P M 1983 Growth, mineral nutrition and water use by potato crops. *Journal of Agricultural Science, Cambridge* **100**: 87–101

Auld B A, Dennett M D, Elston J 1978 The effect of temperature changes on the expansion of individual leaves of *Vicia faba* L. *Annals of Botany* **42**: 877–88

Auld B A, Kemp D R, Medd R W 1983 The influence of spatial arrangement on grain yield of wheat. *Australian Journal of Agricultural Research* **34**: 99–108

Austin R B 1978 Actual and potential yields of wheat and barley in the United Kingdom. *Agricultural*

*Development and Advisory Service (England and Wales) Quarterly Review* **29**: 76–87

Austin R B 1980 Physiological limitations of cereal yields and ways of reducing them by breeding. In Hurd R G, Biscoe P V, Dennis C (eds) *Opportunities for Increasing Crop Yields*. Pitman, pp 3–19

Austin R B, Bingham J, Blackwell R D, Evans L T, Ford M A, Morgan C L, Taylor M 1980b Genetic improvements in winter wheat yields since 1900 and associated physiological changes. *Journal of Agricultural Science, Cambridge* **94**: 675–89

Austin R B, Edrich J A, Ford M A, Blackwell R D 1977a The fate of the dry matter, carbohydrates and $^{14}C$ lost from the leaves and stems of wheat during grain filling. *Annals of Botany* **41**: 1309–21

Austin R B, Flavell R B, Henson I E, Lowe H J B 1986 *Molecular Biology and Crop Improvement*. Cambridge University Press

Austin R B, Ford M A, Edrich J A, Blackwell, R D 1977b The nitrogen economy of winter wheat. *Journal of Agricultural Science, Cambridge* **88**: 159–67

Austin R B, Ford M A, Edrich J A, Hooper B E 1976 Some effects of leaf posture on photosynthesis and yield in wheat. *Annals of Applied Biology* **83**: 425–46

Austin R B, Morgan C L, Ford M A, Blackwell R D 1980a Contributions to grain yields from pre-anthesis assimilation in tall and dwarf barley phenotypes in two contrasting seasons. *Annals of Botany* **45**: 309–19

Ayres P G 1979 $CO_2$ exchanges in plants infected by obligately biotrophic pathogens. In Marcelle R, Clijsters H, Van Poucke M (eds) *Photosynthesis and Plant Development*. W Junk, The Hague

Azcón-Bieto J 1983 Inhibition of photosynthesis by carbohydrates in wheat leaves. *Plant Physiology* **73**: 681–6

Azcón-Bieto J, Day D A, Lambers H 1983 The regulation of respiration in the dark in wheat leaf slices. *Plant Science Letters* **32**: 313–20

Azcón-Bieto J, Lambers H, Day, D A 1983 Effect of photosynthesis and carbohydrate status on respiratory rates and the involvement of the alternative pathway in leaf respiration. *Plant Physiology* **72**: 598–603

Azcón-Bieto J, Osmond C B 1983 Relationship between photosynthesis and respiration. The effect of carbohydrate status on the rate of $CO_2$ production by respiration in darkened and illuminated wheat leaves. *Plant Physiology* **71**: 574–81

Bailey R J 1984 When to water cereals. *Arable Farming* **April 1984**: 48–9

Baker C K, Gallagher J N 1983 The development of winter wheat in the field. 1. Relation between apical development and plant morphology within and between seasons. *Journal of Agricultural Science, Cambridge* **101**: 327–35

Baker C K, Gallagher J N, Monteith J L 1980 Daylength change and leaf appearance in winter wheat. *Plant, Cell and Environment* **3**: 285–7

Baker D A 1985 Regulation of phloem loading. In Jeffcoat B, Hawkins A F, Stead A D (eds) *Regulation of Sources and Sinks in Crop Plants*. Monograph 12, British Plant Growth Regulator Group, Bristol

Baker D N, Hesketh J D, Duncan W G 1972 Simulation of growth and yield in cotton: I. Gross photosynthesis, respiration and growth. *Crop Science* **12**: 431–5

Bange G G J 1953 On the quantitative explanation of stomatal transpiration. *Acta Botanica Neerlandica* **2**: 255–97

Bangerth F, Aufhammer W, Baum O 1985 IAA level and dry matter accumulation at different positions within a wheat ear. *Physiologia Plantarum* **63**: 121–5

Bangerth F, Ho L C 1984 Fruit position and fruit set sequence in a truss as factors determining final size of tomato fruits. *Annals of Botany* **53**: 315–19

Barling D 1982 Winter wheat production and crop development on Cotswold land. In *Yield of Cereals: Course Papers 1982*. Cereal Unit, National Agricultural Centre, Coventry, pp 23–36

Barnes A, Hole C C 1978 A theoretical basis of growth and maintenance respiration. *Annals of Botany* **42**: 1217–21

Bassham J A, Larson A L, Cornwell K L 1981 Relationships between nitrogen metabolism and photosynthesis. In Bewley J D (ed) *Nitrogen and Carbon Metabolism*. Martinus Nijhof/Dr W Junk. The Hague

Bastiman B, Mudd C H 1971 A farm scale comparison of permanent and temporary grass. *Experimental Husbandry* **20**: 73–83

Baszynski T, Brand J, Barr R, Krogmann D W, Crane F L 1972 Some biochemical characteristics of chloroplasts from mineral-deficient maize. *Plant Physiology* **50**: 410–11

Batey T 1976 Some effects of nitrogen fertiliser on winter wheat. *Journal of the Science of Food and Agriculture* **27**: 287–97

Begg J E, Turner N C 1976 Crop water deficits. *Advances in Agronomy* **28**:161–217

Belford R K, Cannell R Q, Thomson R J 1985 Effects of single and multiple waterloggings on the growth and yield of winter wheat on a clay soil. *Journal of the Science of Food and Agriculture* **36**: 142–56

Bennett J, Scott K J 1971 Inorganic polyphosphates in

the wheat stem rust fungus and in rust-infected wheat leaves. *Physiological Plant Pathology* **1**: 185–98

Benzian B, Lane P 1979 Some relationships between grain yield and grain protein of wheat experiments in South-east England and comparisons with such relationships elsewhere. *Journal of the Science of Food and Agriculture* **30**: 59–70

Berry J, Björkman O 1980 Photosynthetic response and adaptation to temperature in higher plants. *Annual Review of Plant Physiology* **31**: 491–543

Berüter J 1983 Effect of abscisic acid on sorbitol uptake in growing apple fruits. *Journal of Experimental Botany* **34**: 737–43

Bhagsari A S, Brown R H 1986 Leaf photosynthesis and its correlation with leaf area. *Crop Science* **26**: 127–32

Bingham J, Blackman J A, Angus W J, Newman R A 1983 *Wheat – a Guide to Varieties from the Plant Breeding Institute*. National Seed Development Organization, Cambridge

Birch J A, Devine J R, Holmes M R J, Whitear J D 1967 Field experiments on the fertilizer requirements of maincrop potatoes. *Journal of Agricultural Science, Cambridge* **69**: 13–24

Bircham J S, Hodgson J 1983a The influence of sward condition on rates of herbage growth and senescence in mixed swards under continuous stocking management. *Grass and Forage Science* **38**: 323–31

Bircham J S, Hodgson J 1983b Dynamics of herbage growth and senescence in a mixed-species temperate sward continuously grazed by sheep. *Proceedings of the 14th International Grassland Conference*, pp 601–3

Bircham J S, Hodgson J 1984 The effects of change in herbage mass on rates of herbage growth and senescence in mixed swards. *Grass and Forage Science* **39**: 111–15

Bird I F, Cornelius M J, Keys A J 1977 Effects of temperature on photosynthesis by maize and wheat. *Journal of Experimental Botany* **28**: 519–24

Bird I F, Cornelius M J, Keys A J 1980 Effect of carbonic anhydrase on the activity of ribulose bisphosphate carboxylase. *Journal of Experimental Botany* **31**: 365–9

Biscoe P V 1978 Schleswig-Holstein and its system. *Big Farm Management* **December 1978**: 6–10

Biscoe P V 1979 Gembloux's consistent wheat yields. *Big Farm Management* **November 1979**: 8–15

Biscoe P V, Clark J A, Gregson K, McGowan M, Monteith J L, Scott R K 1975a Barley and its environment I. Theory and practice. *Journal of Applied Ecology* **12**: 227–47

Biscoe P V, Gallagher J N 1977 Weather, dry matter production and yield. In Landsberg J J, Cutting C V (eds) *Environmental Effects on Crop Physiology*. Academic Press, pp 75–100

Biscoe P V, Gallagher J N 1978 Producing high yields of wheat. *Big Farm Management* **November 1978**: 21–8

Biscoe P V, Gallagher J N, Littleton E J, Monteith J L, Scott R K 1975b Barley and its environment IV. Sources of assimilate for the grain. *Journal of Applied Ecology* **12**: 295–318

Biscoe P V, Scott R K, Monteith J L 1975c Barley and its environment. III. Carbon budget of the stand. *Journal of Applied Ecology* **12**: 269–93

Biscoe P V, Willington V B A 1984a Environmental effects on dry matter production. In *The Nitrogen Requirements of Cereals*. Reference Book 385, Ministry of Agriculture, Fisheries and Food, HMSO, pp 53–65

Biscoe P V, Willington V B A 1984b Cereal crop physiology – a key to accurate nitrogen timing. In *Marketable Yield of Cereals, Course Papers December 1984*. Arable Unit, National Agricultural Centre, Coventry, pp 67–74

Black J N 1964 An analysis of the potential production of swards of subterranean clover (*Trifolium subterraneum* L.) at Adelaide, South Australia. *Journal of Applied Ecology* **1**: 3–18

Blacklow W M, Incoll L D 1981 Nitrogen stress of winter wheat changed the determinants of yield and the distribution of nitrogen and total dry matter during grain filling. *Australian Journal of Plant Physiology* **8**: 191–200

Blackman G E 1962 The limit of plant productivity. *Annual Report of East Malling Research Station for 1961*, pp 39–50

Bleasdale J K A, Thompson R 1966 Competition studies – potatoes. *Annual Report of the National Vegetable Research Station for 1965*, pp 39–40

Blenkinsop P G, Dale J E 1974 The effects of nitrate supply and grain reserves on fraction I protein level in the first leaf of barley. *Journal of Experimental Botany* **25**: 913–26

Boller B C, Nösberger J 1985 Photosynthesis of white clover leaves as influenced by canopy position, leaf age and temperature. *Annals of Botany* **56**: 19–27

Boyd D A, Yuen L T K, Needham P 1976 Nitrogen requirements of cereals. 1. Response curves. *Journal of Agricultural Science, Cambridge* **87**: 149–62

Boyer J S 1970 Leaf enlargement and metabolic rates in corn, soybean and sunflower at various leaf water potentials. *Plant Physiology* **46**: 233–5

Boyer J S, Bowen B L 1970 Inhibition of oxygen evolution in chloroplasts isolated from leaves with low water potentials. *Plant Physiology* **45**: 612–15

Breeze V, Elston J 1978 Some effects of temperature and substrate content upon respiration and the carbon balance of field beans (*Vicia faba* L.). *Annals of Botany* **42**: 863–76

Bremner P M, El Saeed A K 1963 The significance of seed size and spacing. In Ivins J D, Milthorpe F L (eds) *The Growth of the Potato*. Butterworths, pp 267–80

Bremner P M, Radley R W 1966 Studies in potato agronomy. 2. The effects of variety and time of planting on growth, development and yield. *Journal of Agricultural Science, Cambridge* **66**: 253–62

Bremner P M, Rawson H M 1978 The weights of individual grains of the wheat ear in relation to their growth potential, the supply of assimilate and interactions between grains. *Australian Journal of Plant Physiology* **5**: 61–72

Brix H 1962 The effects of water stress on the rates of photosynthesis and respiration in tomato plants and loblolly pine seedlings. *Physiologia Plantarum* **15**: 10–20

Brocklehurst P A 1977 Factors controlling grain weight in wheat. *Nature* **266**: 348–9

Brooking I R, Kirby E J M 1981 Interrelationships between stem and ear development in winter wheat: the effects of a Norin 10 dwarfing gene, Gai/Rht 2. *Journal of Agricultural Science, Cambridge* **97**: 373–81

Brooks A, Farquhar G D 1985 Effect of temperature on the $CO_2/O_2$ specificity of ribulose-1,5-bisphosphate carboxylase/oxygenase and the rate of respiration in the light. Estimates from gas change measurements on spinach. *Planta* **165**: 397–406

Brougham R W 1960 The relationship between the critical leaf area, total chlorophyll content and maximum growth rate of some pasture and crop plants. *Annals of Botany* **24**: 463–74

Brown K R, Evans P S 1973 Animal treading. *New Zealand Journal of Experimental Agriculture* **1**: 217–26

Brown R H, Blaser R E 1968 Leaf area index in pasture growth. *Herbage Abstracts* **38**: 1–9

Brun W A, Brenner M L, Schussler J 1985 Hormonal communication between sources and sinks in soybeans. In Jeffcoat B, Hawkins A F, Stead A D (eds) *Regulation of Sources and Sinks in Crop Plants*. Monograph 12, British Plant Growth Regulator Group, Bristol

Buchanan B B, Hutcheson S W, Magyarosy A C, Montalbini P 1981 Photosynthesis in healthy and diseased plants. In Ayres P G (ed) *Effects of Disease on the Physiology of the Growing Plant*. Cambridge University Press

Bunce J A 1977 Leaf elongation in relation to leaf water potential in soybean. *Journal of Experimental Botany* **28**: 156–61

Bunting A H, Drennan D S H 1966 Some aspects of the morphology and physiology of cereals in the vegetative phase. In Milthorpe F L, Ivins J D (eds) *The Growth of Cereals and Grasses*. Butterworths, pp 20–38

Burdon J J 1983 Biological flora of the British Isles – *Trifolium repens*. *Journal of Ecology* **71**: 307–30

Burton W G 1963 Concepts and mechanisms of dormancy. In Ivins J D, Milthorpe F L (eds) *The Growth of the Potato*. Butterworths, pp 17–40

Burton W G 1966 *The Potato* 2nd edn. Chapman and Hall

Camp P J, Huber S C, Burke J J, Moreland D E 1982 Biochemical changes that occur during senescence of wheat leaves. I. Basis for the reduction of photosynthesis. *Plant Physiology* **70**: 1641–6

Carmi A, Koller D 1977 Endogenous regulation of photosynthetic rate in primary leaves of bean (*Phaseolus vulgaris* L.). *Annals of Botany* **41**: 59–67

Carmi A, Koller D 1978 Effects of the roots on the rate of photosynthesis in primary leaves of bean (*Phaseolus vulgaris* L.) *Photosynthetica* **12**: 178–84

Carmi A, Koller D 1979 Regulation of photosynthetic activity in the primary leaves of bean (*Phaseolus vulgaris* L.) by materials moving in the water-conducting system. *Plant Physiology* **64**: 285–8

Cartwright P M, Waddington S R 1982 Growth regulators and grain yield in spring cereals. In Hawkins A F, Jeffcoat B (eds) *Opportunities for Manipulation of Cereal Productivity*. Monograph 7, British Plant Growth Regulator Group, pp 61–70

Chakravorty A K, Scott K J 1982 Biochemistry of host rust interactions. Part A: Primary metabolism: changes in the gene expression of host plants during the early stages of rust infection. In Scott K J, Chakravorty A K (eds) *The Rust Fungi*. Academic Press, London

Chapman E A, Graham D 1974 The effect of light on the tricarboxylic cycle in green leaves. I. Relative rates of the cycle in the dark and the light. *Plant Physiology* **53**: 879–85

Charles A H 1972 Ryegrass populations for intensively-managed leys. *Journal of Agricultural Science, Cambridge* **79**: 205–15

Charles-Edwards D A 1978 An analysis of the photosynthesis and productivity of vegetative crops in the United Kingdom. *Annals of Botany* **42**: 717–31

Chatterton N J, Silvius J E 1979 Photosynthate partitioning into starch in soybean leaves. I. Effects of photoperiod versus photosynthetic period duration. *Plant Physiology* **64**: 749–53

Chen-She S-H, Lewis D H, Walker D A 1975 Stimulation of photosynthetic starch formation by sequestration of cytoplasmic orthophosphate. *New Phytologist* **74**: 383–92

Child R D, Treharne K J, Hoad G V 1983 Growth regulator potential for improvement of cereal yields. In *Factors Affecting the Accumulation of Exploitable Reserves in the Cereal Plant*. Reference Book 222, Ministry of Agriculture, Fisheries and Food, HMSO, pp 14–26

Christy A L, Swanson C A 1976 Control of translocation by photosynthesis and carbohydrate concentrations of the source leaf. In Wardlaw I F, Passioura J B (eds) *Transport and Transfer Processes in Plants*. Academic Press

Church B M 1985 Use of fertilisers in England and Wales, 1984. *Annual Report of Rothamsted Experimental Station for 1984*, 277–84

Church B M, Austin R B 1983 Variability of wheat yields in England and Wales. *Journal of Agricultural Science, Cambridge* **100**: 201–4

Clifford P E, Offler C E, Patrick J W 1986 Growth regulators have rapid effects on photosynthate unloading from seed coats of *Phaseolus vulgaris* L. *Plant Physiology* **80**: 635–7

Clutterbuck B J, Simpson K 1978 The interactions of water and fertiliser nitrogen in effects on growth pattern and yield of potatoes. *Journal of Agricultural Science, Cambridge* **91**: 161–72

Cochrane M P 1983 Morphology of the crease region in relation to assimilate uptake and water loss during caryopsis development in barley and wheat. *Australian Journal of Plant Physiology* **10**: 473–91

Cochrane M P, Duffus C M 1981 Endosperm cell number in barley. *Nature* **289**: 399–401

Coffey M D, Marshall C, Whitbread R 1970 The translocation of $^{14}C$ labelled assimilates in tomato plants infected with *Alternaria solani* (Ell & Mant). Jones & Grout. *Annals of Botany* **34**: 605–15

Cole C S 1975 Variation in dry matter between and within potato tubers. *Potato Research* **18**: 28–37

Colman B, Espie G S 1985 $CO_2$ uptake and transport in leaf mesophyll cells. *Plant, Cell and Environment* **8**: 449–57

Colvill K E, Marshall C 1984 Tiller dynamics and assimilate partitioning in *Lolium perenne* with particular reference to flowering. *Annals of Applied Biology* **104**: 543–57

Constable G A, Rawson H M 1980 Effect of leaf position, expansion and age on photosynthesis, transpiration and water use efficiency of cotton. *Australian Journal of Plant Physiology* **7**: 89–100

Cook R J 1984 Managed disease control. *Agricultural Progress* **59**: 1–9

Cooke R C, Whipps J M 1980 The evolution of modes of nutrition in terrestrial plant pathogenic fungi. *Biological Reviews* **55**: 341–62

Cooksley J 1984 Think about irrigation a year ahead. *Arable Farming* **September 1984**: 32–4

Coombs J, Hall D O, Long S P, Scurlock J M O 1986 *Techniques in Bioproductivity and Photosynthesis* 2nd edn. Pergamon Press

Cooper J P 1970 Potential production and energy conversion in temperate and tropical grasses. *Herbage Abstracts* **40**: 1–15

Cornic G, Miginiac E 1983 Non-stomatal inhibition of net $CO_2$ uptake by +/− abscisic acid in *Pharbitis nil*. *Plant Physiology* **73**: 529–33

Corrall A J 1984 Grass growth and seasonal pattern of production under varying climatic conditions. In Riley H and Skjelvåg A O (eds) *The Impact of Climate on Grass Production and Quality*. The Norwegian State Agricultural Research Stations, Norway, pp 36–45

Corrall A J, Fenlon J S 1978 A comparative method for describing the seasonal distribution of production from grasses. *Journal of Agricultural Science, Cambridge* **91**: 61–7

COSAC 1984 Cereal variety trials, 1983. *Technical Note* 66, Council of the Scottish Agricultural Colleges

Cottrell J E, Dale J E 1984 Variation in size and development of spikelets within the ear of barley. *New Phytologist* **97**: 565–73

Cottrell J E, Easton R H, Dale J E, Wadsworth A C, Adam J S, Child R D, Hoad G V 1985 A comparison of spike and spikelet survival in mainstem and tillers of barley. *Annals of Applied Biology* **106**: 365–77

Cowton J M 1982 Winter wheat – should you drill early? *Arable Farming* **October 1982**: 60

Crookston R K, O'Toole J, Lee R, Ozbun J L, Wallace D H 1974 Photosynthetic depression in beans after exposure to cold for one night. *Crop Science* **14**: 457–64

Crosthwaite L M, Sheen S J 1979 Inhibition of ribulose-1,5-bisphosphate carboxylase by a toxin isolated from *Pseudomonas tabaci*. *Phytopathology* **69**: 376–9

Cutter E 1978 Structure and development of the potato plant. In Harris P M (ed) *The Potato Crop*. Chapman and Hall, pp 70–152

Dale J E 1976 Cell division in leaves. In Yeoman M M (ed) *Cell Division in Higher Plants*. Academic Press, pp 315–45

Dale J E 1982 *The Growth of Leaves*. Edward Arnold

Dale J E and Milthorpe F L 1983 General features of the production and growth of leaves. In Dale J E, Milthorpe F L (eds) *The Growth and Functioning of Leaves*. Cambridge University Press, pp 151–78

Daly J M 1976 The carbon balance of diseased plants: changes in respiration, photosynthesis and translocation. In Heitefuss R, Williams P H(eds) *Encyclopedia of Plant Physiology* Vol 4. Springer-Verlag, Berlin

Daly M, MacKenzie G H 1983 The timing of spring nitrogen application to grass, *Grass and Forage Science* **38**: 149

Dann P R, Axelsen A, Dear B S, Williams E R, Edwards C B H 1983 Herbage, grain and animal production from winter-grazed cereal crops. *Australian Journal of Experimental Agriculture and Animal Husbandry* **23**: 154–61

D'Aoust M J, Tayler R S 1969 The interaction between nitrogen and water in the growth of grass swards. 2. Leaf area index and net assimilation rate. *Journal of Agricultural Science, Cambridge* **72**: 437–43

Darwinkel A 1978 Patterns of tillering and grain production of winter wheat at a wide range of plant densities. *Netherlands Journal of Agricultural Science* **26**: 383–98

Darwinkel A 1980a Ear development and formation of grain yield in winter wheat. *Netherlands Journal of Agricultural Science* **28**: 156–63

Darwinkel A 1980b Grain production of winter wheat in relation to nitrogen and diseases. 1. Relationship between nitrogen dressing and yellow rust infection. *Zeitschrift für Acker- und Pflanzenbau* **149**: 299–308

Darwinkel A 1980c Grain production of winter wheat in relation to nitrogen and diseases. 2. Relationship between nitrogen dressing and mildew infection. *Zeitschrift für Acker- und Pflanzenbau* **149**: 309–17

Darwinkel A 1983 Ear formation and grain yield of winter wheat as affected by time of nitrogen supply. *Netherlands Journal of Agricultural Science* **31**: 211–25

Davidson I A, Robson M J 1984 The effect of temperature and nitrogen supply on the physiology of grass/clover swards. *Occasional Symposia of the British Grassland Society* **16**: 56–60

Davidson I A, Robson M J 1986 Effect of temperature and nitrogen supply on the growth of perennial ryegrass and white clover. 2. A comparison of monocultures and mixed swards. *Annals of Botany* **57**: 709–19

Davidson I A, Robson M J, Drennan D S H 1986 Effect of temperature and nitrogen supply on the growth of perennial ryegrass and white clover. 1. Carbon and nitrogen economies of mixed swards at low temperature. *Annals of Botany* **57**: 697–708

Davies A 1977 Structure of the grass sward. In Gilsenan B (ed) *Proceedings of an International Meeting on Animal Production from Temperate Grassland*. An Foras Taluntais, Ireland, pp 36–44

Davies A, Evans M E 1982 The pattern of growth in swards of two contrasting varieties of white clover in winter and spring. *Grass and Forage Science* **37**: 199–207

Davies H V 1984 Mother tuber reserves as factors limiting potato sprout growth. *Potato Research* **27**: 209–18

Davies I 1979 Developmental characteristics of grass varieties in relation to herbage production. 4. Effects of nitrogen on the length and longevity of leaf blades in primary growth of *Lolium perenne*, *Dactylis glomerata* and *Phleum pratense*. *Journal of Agricultural Science, Cambridge* **92**: 277–87

Day D A, Lambers H 1983 The regulation of glycolysis and electron transport in roots. *Physiologia Plantarum* **58**: 155–60

Day D A, de Vos O C, Wilson D, Lambers H 1985 Regulation of respiration in the leaves and roots of two *Lolium perenne* populations with contrasting mature tissue respiration rates and crop yields. *Plant Physiology* **78**: 678–83

Day W 1981 Water stress and crop growth. In Johnson C B (ed) *Physiological Processes Limiting Plant Productivity*. Butterworths, pp 199–215

Day W, Lawlor D W, Legg B J 1981 The effects of drought on barley: soil and plant water relations. *Journal of Agricultural Science, Cambridge* **96**: 61–77

Day W, Legg B J, French B K, Johnston A E, Lawlor D W, Jeffers W de C 1978 A drought experiment using mobile shelters: the effect of drought on barley yield, water use and nutrient uptake. *Journal of Agricultural Science, Cambridge* **91**: 599–603

Deinum B 1976 Photosynthesis and sink size: an explanation for the low productivity of grass swards in autumn. *Netherlands Journal of Agricultural Science* **24**: 238–46

Delécolle R, Gurnade J C 1980 Liaisons entre le développement et la morphologie du blé tendre d'hiver. 1. Stades de développement de l'apex, apparition des feuilles et croissance de la tige. *Annales d'Amélioration des Plantes* **30**: 479–98

Delrot S, Bonnemain J-L 1985 Mechanism and control of phloem transport. *Physiologie Végetale* **23**: 199–220

Dennett M D, Elston J, Milford J R 1979 The effect of temperature on the growth of individual leaves of *Vicia faba* L. in the field. *Annals of Botany* **43**: 197–208

Dennis W D, Woledge J 1982 Photosynthesis by white clover leaves in mixed clover/ryegrass swards. *Annals of Botany* **49**: 627–35

Dennis W D, Woledge J 1983 The effect of shade

during leaf expansion on photosynthesis by white clover leaves. *Annals of Botany* **51**: 111–18

Dennis W D, Woledge J 1985 The effect of nitrogenous fertiliser on the photosynthesis and growth of white clover/perennial ryegrass swards. *Annals of Botany* **55**: 171–8

Denny F E 1929 Role of mother tuber in growth of potato plant. *Botanical Gazette* **87**: 157–94

Dent J B, Blackie M J 1979 *Systems Simulation in Agriculture*. Applied Science Publishers

de Visser R, Brouwer K S, Posthumus F 1986 Alternative path mediated ATP synthesis in roots of *Pisum sativum* upon nitrogen supply. *Plant Physiology* **80**: 295–300

de Visser R, Lambers H 1983 Growth and the efficiency of root respiration of *Pisum sativum* as dependent on the source of nitrogen. *Physiologia Plantarum* **58**: 533–43

de Wit C T 1959 Potential photosynthesis of crop surfaces. *Netherlands Journal of Agricultural Science* **7**: 141–9

de Wit C T, Brouwer R, Penning de Vries F W T 1972 A dynamic model of plant and crop growth. In Wareing P F, Cooper J P (eds) *Potential Crop Production*. Heinemann

Dinar M, Rudich J 1985 Effect of heat stress on assimilate metabolism in tomato flower buds. *Annals of Botany* **56**: 249–57

Dinar M, Stevens M A 1982 The effect of temperature and carbon metabolism on sucrose uptake by detached tomato fruits. *Annals of Botany* **49**: 477–83

Dixon H H, Ball N G 1922 Transport of organic substances in plants. *Nature* **109**: 236–7

Doman D C, Geiger D R 1979 Effect of exogenously supplied foliar potassium on phloem loading in *Beta vulgaris* L. *Plant Physiology* **64**: 528–33

Donald C M 1961 Competition for light in crops and pastures. *Symposium of the Society of Experimental Biology* **15**: 282–313

Donald C M 1963 Competition among crop and pasture plants. *Advances in Agronomy* **15**: 1–118

Donald C M, Hamblin J 1976 The biological yield and harvest index of cereals as agronomic and plant breeding criteria. *Advances in Agronomy* **28**: 361–405

Donovan G R 1983 Effect of nitrogen concentration on endosperm DNA content and cell number in wheat ears grown in liquid culture. *Australian Journal of Plant Physiology* **10**: 569–72

Donovan G R, Jenner C F, Lee J W, Martin P 1983 Longitudinal transport of sucrose and amino acids in the wheat grain. *Australian Journal of Plant Physiology* **10**: 31–42

Doodson J K, Manners J G, Myers A 1964 Some effects of yellow rust (*Puccinia striiformis*) on the growth and yield of a spring wheat. *Annals of Botany* **28**: 459–72

Doodson J K, Manners J G, Myers A 1965 Some effects of yellow rust (*Puccinia striiformis*) on $^{14}$carbon assimilation and translocation in wheat. *Journal of Experimental Botany* **16**: 304–17

Downton J, Slatyer R O 1972 Temperature dependence of photosynthesis in cotton. *Plant Physiology* **50**: 518–22

Doyle C J 1981 Economics of irrigating grassland in the United Kingdom. *Grass and Forage Science* **36**: 297–306

Duncan W G 1971 Leaf angles, leaf area and canopy photosynthesis. *Crop Science*. **11**: 482–5

Duncan W G, Loomis R S, Williams W A, Hanau R 1967 A model for simulating photosynthesis in plant communities. *Hilgardia* **38**: 181–205

Duniway J M, Slatyer R O 1971 Gas exchange studies on the transpiration and photosynthesis of tomato leaves affected by *Fusarium oxysporum* f. sp. *lycopersici*. *Phytopathology* **61**: 1377–81

Dyer T A, Scott K J 1972 Decrease in chloroplast polysome content of barley leaves infected with powdery mildew. *Nature* **236**: 237–8

Dyke G V 1956 The effect of date of planting on the yield of potatoes. *Journal of Agricultural Science, Cambridge* **47**: 122–8

Dyke G V, George B J, Johnston A E, Poulton P R, Todd A D 1983 The Broadbalk wheat experiment 1968–78: yield and plant nutrients in crops growing continuously and in rotation. *Report of Rothamsted Experimental Station for 1982* **Part 2**: 5–40

Dyson P W 1977 An investigation into the relations between some growth parameters and yield of barley. *Annals of Applied Biology* **87**: 471–83

Dyson P W, Watson D J 1971 An analysis of the effects of nutrient supply on the growth of potato crops. *Annals of Applied Biology* **69**: 47–63

Eagles C F, Wilson D 1982 Photosynthetic efficiency and plant productivity. In Rechcigl M (ed) *CRC Handbook of Agricultural Productivity* Vol I *Plant Productivity*. CRC Press Inc, Florida

Easson D L 1984 The timing of nitrogen application for spring barley. *Journal of Agricultural Science, Cambridge* **102**: 673–8

Edwards G, Walker D A 1983 *$C_3$, $C_4$: Mechanisms, and Cellular and Environmental Regulation of Photosynthesis*. Blackwell Scientific Publications

Ehleringer J 1978 Implications of quantum yield differences on the distributions of $C_3$ and $C_4$ grasses. *Oecologia* **31**: 255–67

Ehleringer J, Björkman O 1977 Quantum yields for $CO_2$

uptake in $C_3$ and $C_4$ plants. Dependence on temperature, $CO_2$ and $O_2$ concentration. *Plant Physiology* **59**: 86–90

Ehleringer J, Pearcy R W 1983 Variations in quantum yield for $CO_2$ uptake among $C_3$ and $C_4$ plants. *Plant Physiology* **73**: 555–9

Ellen J, Spiertz J H J 1980 Effects of rate and timing of nitrogen dressings on grain yield formation of winter wheat (*T. aestivum* L.). *Fertilizer Research* **1**: 177–90

Ellis R J, Gatenby A A 1984 Ribulose bisphosphate carboxylase: properties and synthesis. In Lea P J, Stewart G R (eds) *The Genetic Manipulation of Plants and its Application to Agriculture*. Oxford University Press

Ellis R P, Kirby E J M 1980 A comparison of spring barley grown in England and Scotland. 2. Yield and its components. *Journal of Agricultural Science, Cambridge* **95**: 111–15

Engledow F L 1925 Investigations on yield in the cereals. II. A spacing experiment with wheat. *Journal of Agricultural Science, Cambridge* **15**: 125–46

Engledow F L, Wadham S M 1924 Investigations on yield in the cereals. I. *Journal of Agricultural Science, Cambridge* **14**: 287–345

Epstein E, Grant W J 1973 Water stress relations of the potato plant under field conditions. *Agronomy Journal* **65**: 400–4

ESCA 1983 Fertiliser Recommendations. *Bulletin* 28, East of Scotland College of Agriculture, Edinburgh

Evans J R 1983 Nitrogen and photosynthesis in the flag leaf of wheat (*Triticum aestivum* L.). *Plant Physiology* **72**: 297–302

Evans J R, Seeman J R 1984 Differences between wheat genotypes in specific activity of ribulose-1,5-bisphosphate carboxylase and the relationship to photosynthesis. *Plant Physiology* **74**: 759–65

Evans L T 1975 The physiologial basis of crop yield. In Evans L T (ed) *Crop Physiology*. Cambridge University Press

Evans L T, Dunstone R L 1970 Some physiological aspects of evolution in wheat. *Australian Journal of Biological Sciences* **23**: 725–41

Evans L T, Wardlaw I F 1976 Aspects of the comparative physiology of grain yield in cereals. *Advances in Agronomy* **28**: 301–59

Evans L T, Wardlaw I F, Fischer R A 1975 Wheat. In Evans L T (ed) *Crop Physiology*. Cambridge University Press, pp 101–49

Evans S A 1977 The influence of plant density and distribution and applied nitrogen on the growth and yield of winter wheat and spring barley. *Experimental Husbandry* **33**: 120–6

Evans S A, Nield J R A, Gunn J S 1978 Maximum yield of the potato crop – studies by ADAS 1971–1977. The 'Blueprint' specification and evaluation of the components. In *Maximising Yields of Crops*. Ministry of Agriculture, Fisheries and Food, HMSO, pp 174–84

Farquhar G D, von Caemmerer S, Berry J A 1980 A biochemical model of photosynthetic $CO_2$ assimilation in leaves of $C_3$ species. *Planta* **149**: 78–90

Farquhar G D, Sharkey T D 1982 Stomatal conductance and photosynthesis. *Annual Review of Plant Physiology* **33**: 317–45

Farrar J F 1980 The pattern of respiration in the vegetative barley plant. *Annals of Botany* **46**: 71–6

Farrar J F 1984 Effects of pathogens on plant transport systems. In Wood R K S, Jellis G J (eds) *Plant Diseases: Infection, Damage and Loss*. Blackwell Scientific Publications

Farrar S C, Farrar J F 1985a Carbon fluxes in leaf blades of barley. *New Phytologist* **100**: 271–83

Farrar S C, Farrar J F 1985b Fluxes of carbon compounds in leaves and roots of barley plants. In Jeffcoat B, Hawkins A F, Stead A D (eds) *Regulation of Sources and Sinks in Crop Plants*. Monograph 12, British Plant Growth Regulator Group

Fellows R J, Geiger D R 1974 Structural and physiological changes in sugar beet leaves during sink to source conversion. *Plant Physiology* **54**: 877–85

Fischer R A 1985 Number of kernels in wheat crops and the influence of solar radiation and temperature. *Journal of Agricultural Science, Cambridge* **105**: 447–61

Fischer R A, HilleRisLambers D 1978 Effect of environment and cultivar on source limitation to grain weight in wheat. *Australian Journal of Agricultural Research* **29**: 443–58

Fischer R A, Stockman Y M 1980 Kernel number per spike in wheat (*Triticum aestivum* L.): responses to pre-anthesis shading. *Australian Journal of Plant Physiology* **7**: 169–80

Fisher D B, Housley T L, Christy A L 1978 Source pool kinetics for $^{14}C$-photosynthate translocation in morning glory and soybean. *Plant Physiology* **61**: 291–5

Fisher D B, Outlaw W H 1979 Sucrose compartmentation in the palisade parenchyma of *Vicia faba*. L. *Plant Physiology* **64**: 481–3

Fitter A H, Hay R K M 1987 *Environmental Physiology of Plants* 2nd edn. Academic Press

Fletcher G M, Dale J E 1977 A comparison of main-stem and tiller growth in barley: apical development and leaf-unfolding rates. *Annals of Botany* **41**: 109–16

Fowler D B 1983 Influence of date of seeding on yield and other agronomic characters of winter wheat and rye growing in Saskatchewan. *Canadian Journal of Plant Science* **63**: 109–13

Fox T C, Geiger D R 1984 Effects of decreased net carbon exchange on carbohydrate metabolism in sugar beet source leaves. *Plant Physiology* **76**: 763–8

Frame J 1981 Herbage mass. In Hodgson J *et al.* (eds) *Sward Measurement Handbook*. British Grassland Society, pp 39–69

France J, Thornley J H M 1984 *Mathematical Models in Agriculture*. Butterworths

Frank A B, Bauer A 1982 Effect of temperature and fertiliser N on apex development in spring wheat. *Agronomy Journal* **74**: 504–9

Fung A K 1975 Carbohydrate metabolism in rust-infected plants. PhD Thesis, University of Sheffield

Gaastra P 1958 Light energy conversion in field crops in comparison with the photosynthetic efficiency under laboratory conditions. *Mededelingen van de Landbouwhogeschool, Wageningen* **58**(4)

Gaastra P 1959 Photosynthesis of crop plants as influenced by light, carbon dioxide, temperature and stomatal diffusion resistance. *Mededelingen van de Landbouwhogeschool, Wageningen* **59**(13): 1–68

Gale M D 1979 The effects of Norin 10 dwarfing genes on yield. *Proceedings of the Fifth International Wheat Genetics Symposium, Delhi*, 1978, 978–87

Gales K 1983 Yield variation of wheat and barley in Britain in relation to crop growth and soil conditions – a review. *Journal of the Science of Food and Agriculture* **34**: 1085–104

Gallagher J N 1979 Field studies of cereal leaf growth. 1. Initiation and expansion in relation to temperature and ontogeny. *Journal of Experimental Botany* **30**: 625–36

Gallagher J N, Biscoe P V 1978 A physiological analysis of cereal yield. 2. Partitioning of dry matter. *Agricultural Progress* **53**: 51–70

Gallagher J N, Biscoe P V 1979 Field studies of cereal leaf growth. 3. Barley leaf extension in relation to temperature, irradiance and water potential. *Journal of Experimental Botany* **30**: 645–55

Gallagher J N, Biscoe P V, Hunter B 1976 Effects of drought on grain growth. *Nature* **264**: 541–2

Gallagher J N, Biscoe P V, Scott R K 1975 Barley and its environment. 5. Stability of grain weight. *Journal of Applied Ecology* **12**: 319–36

Gallagher J N, Biscoe P V, Wallace J S 1979 Field studies of cereal leaf growth. 4. Winter wheat leaf extension in relation to temperature and leaf water status. *Journal of Experimental Botany* **30**: 657–68

Garner H V, Dyke G V 1969 The Broadbalk yields. *Report of Rothamsted Experimental Station for 1968* **Part 2**: 26–49

Garwood E A, Sinclair J 1979 Use of water by six grass species. 2. Root distribution and use of water. *Journal of Agricultural Science, Cambridge* **93**: 25–35

Garwood E A, Tyson K C, Sinclair J 1979 Use of water by six grass species. 1. Dry-matter yields and response to irrigation. *Journal of Agricultural Science, Cambridge* **93**: 13–24

Geiger D R 1975 Phloem loading. In Zimmermann M H, Milburn J A (eds) *Encyclopedia of Plant Physiology*, Vol 1. Transport in Plants. 1. Phloem Transport. Springer Verlag, Berlin

Geiger D R, Fondy B R 1980 Response of phloem loading and export to rapid changes in sink demand. *Berichte Deutschen Botanischen Gesellschaft* **93**: 177–86

Geiger D R, Giaquinta R T 1982 Translocation of photosynthate. In Govindjee (ed) *Photosynthesis* Vol II. *Development, Carbon Metabolism and Plant Productivity*. Academic Press

Geiger D R, Ploeger B J, Fox T C, Fondy B R 1983 Sources of sucrose translocated from illuminated sugar beet source leaves. *Plant Physiology* **72**: 964–70

George B J 1984 Design and interpretation of nitrogen response experiments. In *The Nitrogen Requirements of Cereals*, Reference Book 385, Ministry of Agriculture, Fisheries and Food, HMSO, pp 133–48

George B J, Skinner R J 1984 Determining the nitrogen requirements for winter wheat using an experiment of Simplex design. In *The Nitrogen Requirements of Cereals*, Reference Book 385, Ministry of Agriculture, Fisheries and Food, HMSO, pp 183–90

Gerhardt R, Heldt H W 1984 Measurement of subcellular metabolite levels in leaves by fractionation of freeze-stopped material in non-aqueous media. *Plant Physiology* **75**: 542–7

Giaquinta R T 1980 Translocation of sucrose and oligosaccharides. In Preiss J (ed) *The Biochemistry of Plants*, Vol 3. Academic Press

Giaquinta R T 1983 Phloem loading of sucrose. *Annual Review of Plant Physiology* **34**: 347–87

Gifford R M 1974 A comparison of potential photosynthesis, productivity and yield of plant species with differing photosynthetic metabolism. *Australian Journal of Plant Physiology* **1**: 107–17

Gifford R M, Bremner P M, Jones D B 1973 Assessing photosynthetic limitations to grain yield in a field crop. *Australian Journal of Agricultural Research* **24**: 297–307

Gifford R M, Jenkins C L D 1982 Prospects of applying knowledge of photosynthesis towards improving crop production. In Govindjee (ed) *Photosynthesis* Vol II. *Development, Carbon Metabolism and Plant Productivity*. Academic Press

Gifford R M, Thorne J H, Hitz W D, Giaquinta R T 1984 Crop productivity and photoassimilate partitioning. *Science* **225**: 801–8

Gleadow R M, Dalling M J, Halloran G M 1982 Variation in endosperm characteristics and nitrogen content in six wheat lines. *Australian Journal of Plant Physiology* **9**: 539–51

Good N E, Bell D H 1980 Photosynthesis, plant productivity, and crop yield. In Carlson P S (ed) *The Biology of Crop Productivity*. Academic Press

Goodwin P B 1967 The control of branch growth on potato tubers. *Journal of Experimental Botany* **18**: 78–86, 87–99

Gordon A J, Ryle G J A, Mitchell D F, Powell C E 1982 The dynamics of carbon supply from leaves of barley plants grown in long or short days. *Journal of Experimental Botany* **33**: 241–50

Gordon A J, Ryle G J A, Webb G 1980 The relationship between sucrose and starch during 'dark' export from leaves of uniculm barley. *Journal of Experimental Botany* **31**: 845–50

Grace J 1977 *Plant Response to Wind*. Academic Press

Grace J 1983 *Plant–Atmosphere Relationships*. Chapman and Hall

Graham D 1980 Effects of light on 'dark' respiration. In Davies D D (ed) *The Biochemistry of Plants*, Vol 2. Academic Press

Graham D, Reed M L 1971 Carbonic anhydrase and the regulation of photosynthesis. *Nature New Biology* **231**: 81–3

Graham J P, Ellis F B 1980 The merits of precision drilling and broadcasting for the establishment of cereal crops in Britain. *Agricultural Development and Advisory Service (England and Wales) Quarterly Review* **38**: 160–9

Grange R I 1984 The extent of starch turnover in mature pepper leaves in the light. *Annals of Botany* **54**: 289–91

Grange R I 1985 Carbon partitioning and export in mature leaves of pepper (*Capsicum annuum*). *Journal of Experimental Botany* **36**: 734–44

Grange R I, Peel A J 1975 A method for estimating the proportion of sieve tubes in the phloem of higher plants. *Planta* **124**: 191–7

Grant S A, Barthram G T, Torvell L, King J, Smith H K 1983 Sward management, lamina turnover and tiller population density in continuously-stocked *Lolium perenne*-dominated swards. *Grass and Forage Science* **38**: 333–44

Gray D 1973 The growth of individual tubers. *Potato Research* **16**: 80–4

Green C F, Paulson G A, Ivins J D 1985 Time of sowing and the development of winter wheat. *Journal of Agricultural Science, Cambridge* **105**: 217–21

Green J O, Baker R D 1981 Classification, distribution and productivity of UK grasslands. In Jollans J L (ed) *Grassland in the British Economy*. Centre for Agricultural Strategy Paper 10, pp 237–47

Greenland A J 1979 Invertases of healthy and rusted leaves of oat. PhD Thesis, University of Sheffield

Gregory P J, Marshall B, Biscoe P V 1981 Nutrient relations of winter wheat. 3. Nitrogen uptake, photosynthesis of flag leaves and translocation of nitrogen to the grain. *Journal of Agricultural Science, Cambridge* **96**: 539–47

Griffith R L, Allen E J, O'Brien S A, O'Brien P J 1984 Comparisons of growth and early yields of potato varieties of contrasting maturity classification at three sites. *Journal of Agricultural Science, Cambridge* **103**: 443–58

Gunasena H P M, Harris P M 1969 The effect of CCC and nitrogen on the growth and yield of the second early potato variety Craig's Royal. *Journal of Agricultural Science, Cambridge* **73**: 245–59

Gunasena H P M, Harris P M 1971 The effect of CCC, nitrogen and potassium on the growth and yield of two varieties of potatoes. *Journal of Agricultural Science, Cambridge* **76**: 33–52

Gunning B E S, Pate J S, Minchin F R, Marks I 1974 Quantitative aspects of transfer cell structure in relation to vein loading in leaves and solute transport in legume nodules. In Sleigh M A, Jennings D H (eds) *Transport at the Cellular Level. Symposia of the Society for Experimental Biology* **28**: 87–126. Cambridge University Press

Guy M, Reinhold L, Michaeli D 1979 Direct evidence for a sugar transport mechanism in isolated vacuoles. *Plant Physiology* **64**: 61–4

Hall J D, Barr R, Al-Abbas A H, Crane F L 1972 The ultrastructure of chloroplasts in mineral-deficient maize leaves. *Plant Physiology* **50**: 404–9

Hall J L, Flowers T J, Roberts R M 1974 *Plant Cell Structure and Metabolism*. Longman

Halliwell B 1981 *Chloroplast Metabolism*. Oxford University Press

Hammond J B W, Burton K S 1983 Leaf starch metabolism during the growth of pepper (*Capsicum annuum*) plants. *Plant Physiology* **73**: 61–5

Hammond J B W, Burton K S 1984 The control of leaf starch breakdown. *Annual Report of the Glasshouse Crops Research Institute for 1983*: 51–4

Hänsel H 1953 Vernalisation of winter rye by negative temperatures and the influence of vernalisation upon the lamina length of the first and second leaf in winter rye, spring barley and winter barley. *Annals of Botany* **17**: 417–32

Hansen G K 1978 Utilization of photosynthates for growth, respiration, and storage in tops and roots of *Lolium multiflorum*. *Physiologia Plantarum* **42**: 5–13

Hardham A R 1976 Structural aspects of the pathways of nutrient flow to the developing embryo and cotyledons of *Pisum sativum* L. *Australian Journal of Botany* **24**: 711–21

Harper J L 1977 *Population Biology of Plants*. Academic Press

Harper J L 1978 Plant relations in pastures. In Wilson J R (ed) *Plant Relations in Pastures*. CSIRO. Melbourne, Australia, pp 3–16

Harris P B 1984 The effects of sowing date, disease control, seed rate and the application of plant growth regulator and of autumn nitrogen on the growth and yield of Igri winter barley. *Research and Development in Agriculture* **1**: 21–7

Harris P M 1978a Mineral nutrition. In Harris P M (ed) *The Potato Crop*. Chapman and Hall, pp 196–243

Harris P M 1978b Water. *Ibid*. pp 244–77

Harris P M 1980 Agronomic research and potato production practice. In Hurd R G, Biscoe P V, Dennis C (eds) *Opportunities for Increasing Crop Yields*. Pitman, pp 205–17

Harris P M 1983 Physiological basis of production. In *Yield of Potatoes. Course Papers 1983*. Cereal Unit, National Agricultural Centre, Coventry, pp 33–53

Harris W, Rhodes I, Mee S S 1983 Observations on the environmental and genotypic influences on the over-wintering of white clover. *Journal of Applied Ecology* **20**: 609–24

Harrison J A C, Isaac I 1968 Leaf area development in King Edward potato plants infected with *Verticillium alboatrum* and *V. dahliae*. *Annals of Applied Biology* **61**: 217–30

Hart R 1983 Potato irrigation. In *Yield of Potatoes. Course Papers 1983*. Cereal Unit, National Agricultural Centre, Coventry, pp 75–84

Hartz T K, Moore F D 1978 Prediction of potato yields using temperature and insolation data. *American Potato Journal* **55**: 431–6

Havelka U D, Wittenbach V A, Boyle M G 1984 $CO_2$-enrichment effects on wheat yield and physiology. *Crop Science* **24**: 1163–8

Hawkins A F 1982 Light interception, photosynthesis and crop productivity. *Outlook on Agriculture* **11**: 104–13

Hay R K M 1978 Seasonal changes in the position of the shoot apex of winter wheat and spring barley in relation to the soil surface. *Journal of Agricultural Science, Cambridge* **91**: 245–8

Hay R K M 1981a Timely planting of maize – a case history from the Lilongwe Plain. *Tropical Agriculture, Trinidad* **58**: 147–55

Hay R K M 1981b *Chemistry for Agriculture and Ecology*. Blackwell Scientific Publications

Hay R K M 1982 The physiology of growth and yield in cereals. In Davies W J, Ayres P G (eds) *Biology in the '80s – Plant Physiology*. University of Lancaster Press, pp 85–96

Hay R K M 1985 The microclimate of an upland grassland. *Grass and Forage Science* **40**: 201–12

Hay R K M 1986 Sowing date and the relationships between plant and apex development in winter cereals. *Field Crops Research* **14**: 321–37

Hay R K M, Abbas Al-Ani M K 1983 The physiology of forage rye. *Journal of Agricultural Science, Cambridge* **101**: 63–70

Hay R K M, Galashan S 1988 The yields of arable crops in Scotland 1978–82. 2. Actual and potential yields of potatoes. *Research and Development in Agriculture* **5**:

Hay R K M, Galashan S, Russell G 1986 The yields of arable crops in Scotland 1978–82: actual and potential yields of cereals. *Research and Development in Agriculture* **3**: 159–64

Hay R K M, Heide O M 1983 Specific photoperiodic stimulation of dry matter production in a high-latitude cultivar of *Poa pratensis*. *Physiologia Plantarum* **57**: 135–42

Hay R K M, Tunnicliffe Wilson G 1982 Leaf appearance and extension in field-grown winter wheat plants: the importance of soil temperature during vegetative growth. *Journal of Agricultural Science, Cambridge* **99**: 403–10

Haycock R 1984 Dry-matter distribution and seasonal yield changes in five contrasting genotypes of white clover. *Journal of Agricultural Science, Cambridge* **102**: 333–40

Headford D W R 1962 Sprout development and subsequent plant growth. *European Potato Journal* **5**: 14–27

Hegarty T W 1973 Temperature relations of germination in the field. In Heydecker W (ed) *Seed Ecology*. Butterworths, pp 411–31

Heichel G H 1970 Prior illumination and the respiration of maize leaves in the dark. *Plant Physiology* **46**: 359–62

Herbert C D 1983 Interactions between nitrogen fertilisers and growth retardants in practical cereal production. In *Interactions between Nitrogen and Growth Regulators in the Control of Plant Development*, Monograph 9, British Plant Growth Regulator Group, pp 87–95

Herold A 1980 Regulation of photosynthesis by sink activity – the missing link. *New Phytologist* **86**: 131–44

Herzog H 1982 Relation of source and sink during grain filling period in wheat and some aspects of its regulation. *Physiologia Plantarum* **56**: 155–60

Hesketh J D, Baker D W, Duncan W G 1971 Simulation of growth and yield in cotton: respiration and the carbon balance. *Crop Science* **11**: 394–8

Hewitt H G, Ayres P G 1975 Changes in $CO_2$ and water vapour exchange rates in leaves of *Quercus robur* infected by *Microsphaera alphitoides* (powdery mildew). *Physiological Plant Pathology* **7**: 127–37

Hewitt J D, Dinar M, Stevens M A 1982 Sink strength of fruits of two tomato genotypes differing in total fruits solids content. *Journal of the American Society of Horticultural Science* **107**: 896–900

HFRO 1979 *Science and Hill Farming. HFRO 1954–1979.* Hill Farming Research Organisation, Penicuik, Scotland

Higgins C M, Manners J M, Scott K J 1985 Decrease in three messenger RNA species coding for chloroplast proteins in leaves of barley infected with *Erysiphe graminis* f. sp. *hordei*. *Plant Physiology* **78**: 891–4

Hipps L E, Asrar G, Kanemasu E T 1983 Assessing the interception of photosynthetically-active radiation in winter wheat. *Agricultural Meteorology* **28**: 253–9

Ho L C 1976a The relationship between the rates of carbon transport and of photosynthesis in tomato leaves. *Journal of Experimental Botany* **27**: 87–97

Ho L C 1976b The effect of current photosynthesis on the origin of translocates in old tomato leaves. *Annals of Botany* **40**: 1153–62

Ho L C 1979 Regulation of assimilate translocation between leaves and fruits in the tomato. *Annals of Botany* **43**: 437–48

Ho L C, Baker D A 1982 Regulation of loading and unloading in long distance transport systems. *Physiologia Plantarum* **56**: 225–30

Ho L C, Gifford R M 1984 Accumulation and conversion of sugars by developing wheat grains. V. The endosperm apoplast and apoplastic transport. *Journal of Experimental Botany* **35**: 58–73

Ho L C, Shaw A F 1977 Carbon economy and translocation of $^{14}C$ in leaflets of the seventh leaf of tomato during leaf expansion. *Annals of Botany* **41**: 833–48

Ho L C, Sjut V, Hoad G V 1983 The effect of assimilate supply on fruit growth and hormone levels in tomato plants. *Plant Growth Regulation* **1**: 155–71

Hodanova D 1970 Contribution to dynamics of development of photosynthetic systems. In Setlik I (ed) *Prediction and Measurement of Photosynthetic Productivity*. Centre for Agricultural Publishing, Wageningen, pp 134–5

Hodgson J 1981 Varitions in the surface characteristics of the sward and the short term rate of herbage intake by calves and lambs. *Grass and Forage Science* **36**: 49–57

Hodgson J, Bircham J S, Grant S A, King J 1981 The influence of cutting and grazing management on herbage growth and utilisation. In Wright C E (ed) *Plant Physiology and Herbage Production. Occasional Symposia of the British Grassland Society* **13**: 51–62

Holliday R 1960 Plant population and crop yield. *Field Crop Abstracts* **13**: 159–67, 247–54

Holligan P M, Chen C, Lewis D H 1973 Changes in the carbohydrate composition of leaves of *Tussilago farfara* during infection by *Puccinia poarum*. *New Phytologist* **72**: 947–55

Holligan P M, Chen C, McGee E M M, Lewis D H 1974 Carbohydrate metabolism in healthy and rusted leaves of coltsfoot. *New Phytologist* **73**: 881–8

Holmes J C 1982 Optimising yield and quality in wheat and barley. *Journal of the National Institute for Agricultural Botany* **16**: 1–6

Holmes J C, Gray D 1971 A comparison of desprouted and late sprouted seed potatoes in relation to apical dominance and tuber production. *Potato Research* **14**: 111–18

Holmes J C, Gray D 1972 Carry-over effects of sprouting and haulm destruction in the potato seed crop. *Potato Research* **15**: 220–35

Holmes J C, Lang R W, Singh A K 1970 The effect of five growth regulators on apical dominance in potato seed tubers and on subsequent tuber production. *Potato Research* **13**: 342–52

Holmes W 1968 The use of nitrogen in the management of pasture for cattle. *Herbage Abstracts* **38**: 265–77

Holmes W 1980 *Grass, its Production and Utilization.* Blackwell Scientific Publications

Housley T L, Pollock C J 1985 Photosynthesis and carbohydrate metabolism in detached leaves of *Lolium temulentum* L. *New Phytologist* **99**: 499–507

Hsiao T C 1973 Plant response to water stress. *Annual Review of Plant Physiology* **24**: 519–70

Hsiao T C, Acevedo E, Fereres E, Henderson D W 1976 Stress metabolism. *Philosophical Transactions of the Royal Society of London* **B273**: 479–500

Hubbard K R 1976 Effects of husbandry on yield of cereals. *Journal of the National Institute for Agricultural Botany* **14**: 190–6

Huber S C 1983 Role of sucrose-phosphate synthase in partitioning of carbon in leaves. *Plant Physiology* **71**: 818–21

Huber S C, Israel D W 1982 Biochemical basis for partitioning of photosynthetically fixed carbon between starch and sucrose in soybean (*Glycine max* Merr.) leaves. *Plant Physiology* **69**: 691–6

Huffaker R C, Peterson L W 1974 Protein transport in plants and possible means of its regulation. *Annual Review of Plant Physiology* **25**: 363–92

Humphries E C 1968 CCC and cereals. *Field Crop Abstracts* **21**: 91–9

Humphries E C, French S A W 1963 The effects of

nitrogen, phosphorus, potassium and gibberellic acid on leaf area and cell division in Majestic potatoes. *Annals of Applied Biology* **52**: 149–62

Humphries E C, Welbank P J, Witts K J 1965 Effect of CCC (chlorocholine chloride) on growth and yield of spring wheat in the field. *Annals of Applied Biology* **56**: 351–61

Hutmacher R B, Krieg D R 1983 Photosynthetic rate control in cotton. Stomatal and nonstomatal factors. *Plant Physiology* **73**: 658–61

Ifenkwe O P, Allen E J 1978 Effects of tuber size on dry matter contents of tubers during growth of two maincrop potato varieties. *Potato Research* **21**: 105–12

Inkson R H E, Reith J W S 1966 Estimating optimal nutrient rates for potatoes. *Transactions Comm. II and IV, International Society of Soil Science* pp 377–84

Innes P, Blackwell R D 1983 Some effects of leaf posture on the yield and water economy of winter wheat. *Journal of Agricultural Science, Cambridge* **101**: 367–76

Ivins J D 1963 Agronomic management of the potato. In Ivins J D, Milthorpe F L (eds) *The Growth of the Potato*. Butterworths, pp 303–10

Ivins J D, Bremner P M 1965 Growth, development and yield in the potato. *Outlook on Agriculture* **4**: 211–17

Jackson M A, Webster R K 1981 Effects of infection with *Rhynchosporium secalis* on some components of growth and yield in two barley cultivars. *Hilgardia* **49**: 1–14

Jackson M V, Williams T E 1979 Response of grass swards to fertiliser nitrogen under cutting and grazing. *Journal of Agricultural Science, Cambridge* **92**: 549–62

Jacobson B S, Fong F, Heath R L 1975 Carbonic anhydrase of spinach. Studies on its location, inhibition, and physiological function. *Plant Physiology* **55**: 468–74

Jagtenberg W D 1970 Predicting the best time to apply nitrogen to grassland in spring. *Journal of the British Grassland Society* **25**: 266–71

Jenkins G I, Woolhouse H W 1981 Photosynthetic electron transport during senescence of the primary leaves of *Phaseolus vulgaris* L. *Journal of Experimental Botany* **32**: 467–78

Jenkyn J F 1976 Effects of mildew (*Erisiphe graminis*) on green leaf area of Zephyr spring barley, 1973. *Annals of Applied Biology* **82**: 485–8

Jenkyn J F 1977 Nitrogen and leaf diseases of spring barley. In Mengel K (ed) *Fertiliser Use and Plant Health*. International Potash Institute, pp 119–28

Jenkyn J F, Bainbridge A 1978 Biology and pathology of cereal powdery mildews. In Spencer D M (ed) *The Powdery Mildews*. Academic Press, pp 283–321

Jenner C F 1970 Relationships between levels of soluble carbohydrates and starch synthesis in detached ears of wheat. *Australian Journal of Biological Sciences* **23**: 991–1003

Jenner C F 1974 Factors in the grain regulating the accumulation of starch. In Bieleski R L, Ferguson A R, Cresswell M M (eds) *Mechanisms of Regulation of Plant Growth. Royal Society of New Zealand Bulletin* **12**: 901–8

Jenner C F, Rathjen A J 1975 Factors regulating the accumulation of starch in ripening wheat grain. *Australian Journal of Plant Physiology* **2**: 311–22

Jenner C F, Rathjen A J 1978 Physiological basis of genetic differences in the growth of six varieties of wheat. *Australian Journal of Plant Physiology* **5**: 249–62

Jessop R S, Ivins J D 1970 The effect of date of sowing on the growth and yield of spring cereals. *Journal of Agricultural Science, Cambridge* **75**: 553–7

Jewiss O R 1981 Shoot development and number. In Hodgson J *et al.* (eds) *Sward Measurement Handbook*. British Grassland Society, pp 93–114

Johnson J, Bastiman B 1979 Some factors to explain why the potential of grazed grassland is not achieved in practice. *ADAS Quarterly Review* **35**: 211–22

Johnson K B, Johnson S B, Teng P S 1986 Development of a simple potato growth model for use in crop pest management. *Agricultural Systems* **19**: 189–209

Johnson R, Frey N M, Moss D N 1974 Effect of water stress on photosynthesis and transpiration of flag leaves and spikes of barley and wheat. *Crop Science* **14**: 728–31

Jones C P 1983 Quality requirements: the supermarket. In *Yield of Potatoes. Course Papers, 1983*. Cereal Unit, National Agricultural Centre, Coventry, pp 163–7

Jones H G 1973 Moderate-term water stress and associated changes in some photosynthetic parameters in cotton. *New Phytologist* **72**: 1095–105

Jones H G 1983 *Plants and Microclimate*. Cambridge University Press

Jones H G, Osmond C B 1973 Photosynthesis by thin leaf slices in solution I. Properties of leaf slices and comparison with whole leaves. *Australian Journal of Biological Sciences* **26**: 15–24

Jones I T, Hayes J D 1971 The effect of sowing date on adult plant resistance to *Erisiphe graminis* f.sp. *avenale* in oats. *Annals of Applied Biology* **68**: 31–9

Jones J L, Allen E J 1983 Effects of date of planting on plant emergence, leaf growth and yield in contrasting potato varieties. *Journal of Agricultural Science, Cambridge* **101**: 81–95

Jones J L, Allen E J 1986 Development in barley

(*Hordeum sativum*). *Journal of Agricultural Science, Cambridge* **107**: 187–213

Jones M B 1981 A comparison of sward development under cutting and continuous grazing management. In Wright C E (ed) *Plant Physiology and Herbage Production*. British Grassland Society Occasional Symposium **13**: 63–7

Jones M B, Collett B, Brown S 1982 Sward growth under cutting and continuous stocking managements: sward canopy structure, tiller density and leaf turnover. *Grass and Forage Science* **37**: 67–73

Jones M B, Leafe E L, Stiles W, Collett B 1978 Pattern of respiration of a perennial ryegrass crop in the field. *Annals of Botany* **42**: 693–703

Jordan W R, Ritchie J T 1971 Influence of soil water stress on evaporation, root absorption and internal water status of cotton. *Plant Physiology* **48**: 783–8

Kaiser G, Heber U 1984 Sucrose transport into vacuoles isolated from barley mesophyll protoplasts. *Planta* **161**: 562–8

Kakie T 1969 Effect of phosphorus deficiency on the photosynthetic carbon dioxide fixation products in tobacco plants. *Soil Science and Plant Nutrition* **15**: 245–51

Karamanos A J, Elston J, Wadsworth R M 1982 Water stress and leaf growth of field beans (*Vicia faba* L.) in the field: water potentials and laminar expansion. *Annals of Botany* **49**: 815–26

Kawakami K 1962 The physiological degeneration of potato seed tubers and its control. *European Potato Journal* **5**: 40–9

Keating B A, Evenson J P, Fukai S 1982 Environmental effects on growth and development of cassava (*Manihot esculenta* Crantz.). 1. Crop development. *Field Crops Research* **5**: 271–81

Kemp D R 1980 The location and size of the extension zone of emerging wheat leaves. *New Phytologist* **84**: 729–37

Kemp D R, Auld B A, Medd W 1983 Does optimising plant arrangements reduce interference or improve the utilization of space? *Agricultural Systems* **12**: 31–6

Kemp D R, Blacklow W M 1980 Diurnal extension rates of wheat leaves in relation to temperatures and carbohydrate concentrations of the extension zone. *Journal of Experimental Botany* **31**: 821–8

Keys A J, Whittingham C P 1981 Photorespiratory carbon dioxide loss. In Johnson C B (ed) *Physiological Processes Limiting Plant Productivity*. Butterworths

Khurana S C, McLaren J S 1982 The influence of leaf area, light interception and season on potato growth and yield. *Potato Research* **25**: 329–42

King R W, Evans L T 1967 Photosynthesis in artificial communities of wheat, lucerne and subterranean clover. *Australian Journal of Biological Sciences* **20**: 623–35

King R W, Wardlaw I F, Evans L T 1967 Effect of assimilate utilization on photosynthetic rate in wheat. *Planta* **77**: 261–76

Kirby E J M 1967 The effect of plant density upon the growth and yield of barley. *Journal of Agricultural Science, Cambridge* **68**: 317–24

Kirby E J M 1969 The effect of sowing date and plant density on barley. *Annals of Applied Biology* **63**: 513–21

Kirby E J M 1973 The control of leaf and ear size in barley. *Journal of Experimental Botany* **24**: 567–78

Kirby E J M, Appleyard M 1982 Development of the cereal plant. In *Yield of Cereals, Course Papers 1982*. Arable Unit, National Agricultural Centre, Coventry, pp 1–11

Kirby E J M, Appleyard M 1984a *Cereal Development Guide* 2nd edn. Arable Unit, National Agricultural Centre, Coventry

Kirby E J M, Appleyard M 1984b Cereal plant development – assessment and use. In *The Nitrogen Requirements of Cereals*. Reference Book 385, Ministry of Agriculture, Fisheries and Food, HMSO pp 21–38

Kirby E J M, Appleyard M, Fellowes G 1982 Effect of sowing date on the temperature response of leaf emergence and leaf size in barley. *Plant, Cell and Environment* **5**: 477–84

Kirby E J M, Appleyard M, Fellowes G 1983 Rate of change of daylength and leaf emergence. *Annual Report of the Plant Breeding Station, Cambridge, for 1982*, p 115

Kirby E J M, Ellis R P 1980 A comparison of spring barley grown in England and Scotland. 1. Shoot apex development. *Journal of Agricultural Science, Cambridge* **95**: 101–10

Kirby E J M, Faris D G 1970 Plant population induced growth correlations in the barley plant main shoot and possible hormonal mechanisms. *Journal of Experimental Botany* **21**: 787–98

Kirby E J M, Faris D G 1972 The effect of plant density on tiller growth and morphology in barley. *Journal of Agricultural Science, Cambridge* **78**: 281–8

Kirby E J M, Porter J R, Day W, Adam J S, Appleyard M, Ayling S, Baker C K, Belford R K, Biscoe P V, Chapman A, Fuller M P, Hampson J, Hay R K M, Matthews S, Thompson W J, Weir A H, Willington V B A, Wood D W 1987 An analysis of primordium initiation in Avalon winter wheat crops with different sowing dates and at ten sites in England and Scotland. *Journal of Agricultural Science, Cambridge* **109**: 123–34

Knight R 1983 Some factors causing variation in the yield of individual plants of wheat. *Australian Journal of Agricultural Research* **34**: 219–28

Knop E 1985 Shoot apex development, date of anthesis and grain yield of autumn-sown spring and winter barley (*Hordeum vulgare* L.) after different sowing times. *Zeitschrift für Acker- und Pflanzenbau* **155**: 73–81

Kramer P J 1983 *Water Relations of Plants*. Academic Press

Krampitz M J, Klug K, Fock H P 1984 Rates of photosynthetic $CO_2$ uptake, photorespiratory $CO_2$ evolution and dark respiration in water-stressed sunflower and bean leaves. *Photosynthetica* **18**: 322–8

Kriedemann P E, Loveys B R, Possingham J, Satoh M 1976 Sink effects on stomatal physiology and photosynthesis. In Wardlaw I F, Passioura J B (eds) *Transport and Transfer Processes in Plants*. Academic Press

Krieg D R, Hutmacher R B 1986 Photosynthetic rate control in sorghum: stomatal and nonstomatal factors. *Crop Science* **26**: 112–17

Krijthe K 1962 Observations on the sprouting of seed potatoes. *European Potato Journal* **5**: 316–33

Kruger N J, Bulpin P V, ap Rees T 1983 The extent of starch degradation in the light in pea leaves. *Planta* **157**: 271–3

Ku S-B, Edwards G E 1977 Oxygen inhibition of photosynthesis. II. Kinetic characteristics as affected by temperature. *Plant Physiology* **59**: 991–9

Laing W A, Ogren W L, Hageman R H 1974 Regulation of soybean net photosynthetic $CO_2$ fixation by interaction of $CO_2$, $O_2$ and ribulose 1,5-diphosphate carboxylase. *Plant Physiology* **54**: 678–85

Laloux R, Falisse A, Poelaert J 1980 Nutrition and fertilization of wheat. In Hafliger E (ed) *Wheat*. Ciba-Geigy, Basle, Switzerland, pp 19–24

Lambers H 1980 The physiological significance of cyanide-resistant respiration in higher plants. *Plant, Cell and Environment* **3**: 293–302

Lambers H, Day D A, Azcón-Bieto J 1983 Cyanide-resistant respiration in roots and leaves. Measurements with intact tissues and isolated mitochondria. *Physiologia Plantarum* **58**: 148–54

Lambers H, Szaniawski R K, de Visser R 1983 Respiration for growth, maintenance and ion uptake. An evaluation of concepts, methods, values and their significance. *Physiologia Plantarum* **58**: 556–63

Lambert R J, Johnson R R 1978 Leaf angle, tassel morphology, and the performance of maize hybrids. *Crop Science* **18**: 499–502

Langer R H M 1979 *How Grasses Grow* 2nd edn. Edward Arnold

Langer R H M, Ryle S M, Jewiss O R 1964 The changing plant and tiller populations of timothy and meadow fescue swards. 1. Plant survival and the pattern of tillering. *Journal of Applied Ecology* **1**: 197–208

Large E C 1954 Growth stages in cereals, illustrations of the Feekes scale. *Plant Pathology* **3**: 128–9

Last F T 1957 The effect of date of sowing on the incidence of powdery mildew in spring-sown cereals. *Annals of Applied Biology* **45**: 1–10

Last F T 1962 Analysis of the effects of *Erisiphe graminis* D.C. on the growth of barley. *Annals of Botany* **26**: 279–89

Lawlor D W, Day W, Johnston A E, Legg B J, Parkinson K J 1981 Growth of spring barley under drought: crop development, photosynthesis, dry-matter accumulation and nutrient content. *Journal of Agricultural Science, Cambridge* **96**: 167–86

Lawlor D W, Fock H 1977 Photosynthetic assimilation of $CO_2$ by water-stressed sunflower leaves at two oxygen concentrations and the specific activity of products. *Journal of Experimental Botany* **28**: 320–8

Lazenby A 1981 British grasslands: past, present and future. *Grass and Forage Science* **36**: 243–66

Lazenby A, Down K M 1982 Realizing the potential of British grasslands: some problems and possibilities. *Applied Geography* **2**: 171–88

Leafe E L, Jones M B, Stiles W 1980 The physiological effects of water stress on perennial ryegrass in the field. *Proceedings of the 13th International Grassland Conference* **1**: 253–60

Lehnherr B, Machler F, Nösberger J 1985 Influence of temperature on the ratio of ribulose bisphosphate carboxylase to oxygenase activities and on the ratio of photosynthesis to photorespiration of leaves. *Journal of Experimental Botany* **36**: 1117–25

Levitt J 1980 *Responses of Plants to Environmental Stresses*. Vol II. *Water, Radiation, Salt and Other Stresses* 2nd edn. Academic Press

Lewis D H 1973 Concepts in fungal nutrition and the origin of biotrophy. *Biological Reviews* **48**: 261–78

Lewis D H 1976 Interchange of metabolites in biotrophic symbioses between angiosperms and fungi. In Sunderland N (ed) *Perspectives in Experimental Biology*. Vol 2 *Botany*. Pergamon Press

Lichtner F T, Spanswick F M 1981 Electrogenic sucrose transport in developing soybean cotyledons. *Plant Physiology* **67**: 869–74

Lidgate H J 1984 Nitrogen uptake of winter wheat. In *The Nitrogen Requirements of Cereals*. Reference Book 385, Ministry of Agriculture, Fisheries and Food, HMSO, pp 177–81

Lingle S E, Chevalier P 1985 Development of the vascular tissue of the wheat and barley caryopsis as related to the rate and duration of grain filling. *Crop Science* **25**: 123–8

Littleton E J, Dennett M D, Elston J, Monteith J L 1979 The growth and development of cowpeas (*Vigna unguiculata*) under tropical field conditions. 1. Leaf area. *Journal of Agricultural Science, Cambridge* **93**: 291–307

Livne A, Daly J M 1966 Translocation in healthy and rust-affected beans. *Phytopathology* **56**: 170–5

Long D E, Fung A K, McGee E E M, Cooke R C, Lewis D H 1975 The activity of invertase and its relevance to the accumulation of storage polysaccharides in leaves infected by biotrophic fungi. *New Phytologist* **74**: 173–82

Loomis R S, Gerakis P A 1975 Productivity of agricultural ecosystems. In Cooper J P (ed) *Photosynthesis and Productivity in Different Environments*. Cambridge University Press

Loomis R S, Rabbinge R, Ng E 1979 Explanatory models in crop physiology. *Annual Review of Plant Physiology* **30**: 339–67

Loomis R S, Williams W A 1963 Maximum crop productivity: an estimate. *Crop Science* **3**: 67–72

Loomis R S, Williams W A 1969 Productivity and the morphology of crop stands: patterns with leaves. In Eastin J D, Haskins F A, Sullivan C Y, van Bavel C H M (eds) *Physiological Aspects of Crop Yield*. American Society of Agronomy, pp 27–47

Loomis R S, Williams W A, Duncan W G 1967 Community architecture and the productivity of terrestrial plant communities. In San Pietro A, Greer F A, Army T J (eds) *Harvesting the Sun*. Academic Press

Lovett J V, Kirby E J M 1971 The effect of plant population and CCC on spring wheat varieties with and without a dwarfing gene. *Journal of Agricultural Science, Cambridge* **77**: 499–510

Loveys B R, Kriedemann P E 1974 Internal control of stomatal physiology and photosynthesis. I. Stomatal regulation and associated changes in endogenous levels of abscisic and phaseic acids. *Australian Journal of Plant Physiology* **1**: 407–15

Ludlow M M, Ng T T 1976 Effect of water deficit on carbon dioxide exchange and leaf elongation rate of *Panicum maximum* var. *trichoglume*. *Australian Journal of Plant Physiology* **3**: 401–13

Ludwig L J, Charles-Edwards D A, Withers A C 1975 Tomato leaf photosynthesis and respiration in various light and carbon dioxide environments. In Marcelle R (ed) *Environmental and Biological Control of Photosynthesis*. Dr W Junk, The Hague

Ludwig L J, Saeki T, Evans L T 1965 Photosynthesis in artificial communities of cotton plants in relation to leaf area. I. Experiments with progressive defoliation of mature plants. *Australian Journal of Biological Sciences* **18**: 1103–18

Lynch K W, Stewart R H, White E L 1979 The effect of nitrogen and seed rate on yield and its components in five spring barley cultivars. *Record of Agricultural Research (N.I.)* **27**: 27–32

Macfarlane I, Last F T 1959 Some effects of *Plasmodium brassicae* Woron. on the growth of young cabbage plants. *Annals of Botany* **23**: 547–70

MacGillivray W 1840 *A Manual of Botany: Comprising Vegetable Anatomy and Physiology, or the Structure and Functions of Plants*. Scott, Webster and Geary, London

MacKerron D K L 1985 A simple model of potato growth and yield. Part 2. Validation and external sensitivity. *Agricultural and Forest Meteorology* **34**: 285–300

MacKerron D K L, Heilbronn T D 1985 A method for estimating harvest indices for use in surveys of potato crops. *Potato Research* **28**: 279–82

MacKerron D K L, Waister P D 1985 A simple model of potato growth and yield. Part 1. Model development and sensitivity analysis. *Agricultural and Forest Meteorology* **34**: 241–52

MacKerron D K L, Waister P D, Thompson R 1982 Why we fall short of potential potato yields. *Arable Farming* **April 1982**: 27–9

Mader P, Volfova A 1984 Influence of senescence and nitrogen fertilisation on the ultrastructural characteristics of barley chloroplasts. *Photosynthetica* **18**: 210–18

MAFF 1982a Seed rate for potatoes grown as maincrop. *Leaflet* 653, Ministry of Agriculture, Fisheries and Food. HMSO

MAFF 1982b *Irrigation*. Reference Book 138, Ministry of Agriculture, Fisheries and Food, HMSO

MAFF 1983 *1983–84 Fertiliser Recommendations*. Reference Book 209, Ministry of Agriculture, Fisheries and Food, HMSO

MAFF 1984 *The Nitrogen Requirements of Cereals*. Reference Book 385, Ministry of Agriculture, Fisheries and Food, HMSO

Magyarosy A C, Buchanan B B 1975 Effect of bacterial infiltration on photosynthesis of bean leaves. *Phytopathology* **65**: 777–80

Magyarosy A C, Malkin R 1978 Effect of powdery mildew infection of sugar beet on the content of electron carriers in chloroplasts. *Physiological Plant Pathology* **13**: 183–8

Magyarosy A C, Schurmann P, Buchanan B B 1976 Effect of powdery mildew infection on photosynthesis by leaves and chloroplasts of sugar beets. *Plant Physiology* **57**: 486–9

Makino A, Mae T, Ohira K 1983 Photosynthesis and

ribulose-1,5-bisphosphate carboxylase in rice leaves. Changes in photosynthesis and enzymes involved in carbon assimilation from leaf development through senescence. *Plant Physiology* **73**: 1002–7

Manning K, Maw G A 1975 Distribution of acid invertase in the tomato plant. *Phytochemistry* **14**: 1965–9

Mansfield T A, Davies W J 1985 Mechanisms for leaf control of gas exchange. *Bioscience* **35**: 158–64

Marsh H V, Galmiche J M, Gibbs M 1965 Effects of light on the citric acid cycle in *Scenedesmus*. *Plant Physiology* **40**: 1013–22

Marsh R, Campling R C 1970 Fouling of pastures by dung. *Herbage Abstracts* **40**: 123–30

Marshall B, Biscoe P V 1980 A model for $C_3$ leaves describing the dependence of net photosynthesis on irradiance. I. Derivation. *Journal of Experimental Botany* **31**: 29–39

Mathre D E 1968 Photosynthetic activity of cotton plants infected with *Verticillium albo-atrum*. *Phytopathology* **58**: 137–41

McCree K J 1970 An equation for the rate of respiration of white clover plants grown under controlled conditions. In Setlik I (ed) *Prediction and Measurement of Photosynthetic Productivity*. Pudoc, Wageningen

McCree K J 1974 Equations for the rate of dark respiration of white clover and grain sorghum, as functions of dry weight, photosynthetic rate, and temperature. *Crop Science* **14**: 509–14

McCree K J 1981 Photosynthetically active radiation. In Lange O L, Nobel P, Osmond B, Ziegler H (eds) *Physiological Plant Ecology*. Vol 12A *Encyclopaedia of Plant Physiology*. Springer-Verlag, Berlin

McCree K J 1982 The role of respiration in crop production. *Iowa State Journal of Research* **56**: 291–306

McCree K J, Kresovich S 1978 Growth and maintenance requirements of white clover as a function of daylength. *Crop Science* **18**: 22–5

McCree K J, Troughton J H 1966a Prediction of growth rate at different light levels from measured photosynthesis and respiration rates. *Plant Physiology* **41**: 559–66

McCree K J, Troughton J H 1966b Non-existence of an optimum leaf area index for the production rate of white clover grown under constant conditions. *Plant Physiology* **41**: 1615–22

McCree K J, Van Bavel C H M 1977 Respiration and crop production: a case study with two crops under water stress. In Landsberg J J, Cutting C V (eds) *Environmental Effects on Crop Physiology*. Academic Press

McLaren J S 1981 Field studies on the growth and development of winter wheat. *Journal of Agricultural Science, Cambridge* **97**: 685–97

Menzel C M 1985 The control of storage organ formation in potato and other species. *Field Crop Abstracts* **38**: 527–37, 581–606

Meynard J M 1983 Le raisonnement de la fumure azotée du blé d'hiver par bilan previsionnel – perspectives et limites d'application. *Agro (Belgium)* **2**: 17–33

Michael G, Seiler-Kelbitsch H 1972 Cytokinin content and kernel size of barley grains as affected by environmental and genetic factors. *Crop Science* **12**: 162–5

Midmore D J, Cartwright P M, Fischer R A 1984 Wheat in tropical environments. 2. Crop growth and grain yield *Field Crops Research* **8**: 207–27

Milford G F J, Pocock T O, Riley J 1985 An analysis of leaf growth in sugar beet. 2. Leaf appearance in field crops. *Annals of Applied Biology* **106**: 173–85

Milford G F J, Pocock T O, Riley J, Messem A B 1985 An analysis of leaf growth in sugar beet. 3. Leaf expansion in field crops. *Annals of Applied Biology* **106**: 187–203

Millard P, MacKerron D K L 1986 The effects of nitrogen application on growth and nitrogen distribution within the potato canopy. *Annals of Applied Biology* **109**: 427–37

Milthorpe F L 1963 Some aspects of plant growth. In Ivins J D, Milthorpe F L (eds) *The Growth of the Potato*. Butterworths, pp 3–16

Milthorpe F L, Davidson J L 1966 Physiological aspects of regrowth in grasses. In Milthorpe F L and Ivins J D (eds) *The Growth of Cereals and Grasses*. Butterworths, pp 241–54

Milthorpe F L, Moorby J 1969 Vascular transport and its significance in plant growth. *Annual Review of Plant Physiology* **20**: 117–38

Milthorpe F L, Moorby J 1979 *An Introduction to Crop Physiology* 2nd edn. Cambridge University Press

Minchin P E H, McNaughton G S 1986 Phloem unloading within the seed coat of *Pisum sativum* observed using surgically modified seeds. *Journal of Experimental Botany* **37**: 1151–63

Mitchell D T 1979 Carbon dioxide exchange by infected first leaf tissues susceptible to wheat stem rust. *Transactions of the British Mycological Society* **72**: 63–8

Mitchell R E 1978 Halo blight of beans: toxin production by several *Pseudomonas phaseolicola* isolates. *Physiological Plant Pathology* **13**: 37–49

Monsi M, Saeki T 1953 Über der lichtfaktor in den pflanzengesellschaften und seine bedeutung für die stoffproduktion. *Japanese Journal of Botany* **14**: 22–52

Montalbini P, Buchanan B B 1974 Effect of a rust infection on photophosphorylation by isolated chloroplasts. *Physiological Plant Pathology* **4**: 191–6

Montalbini P, Buchanan B B, Hutcheson S W 1981 Effect of rust infection on rates of photochemical polyphenol oxidation and latent polyphenol oxidase activity of *Vicia faba* chloroplast membranes. *Physiological Plant Pathology* **18**: 51–7

Monteith J L 1965 Light distribution and photosynthesis in field crops. *Annals of Botany* **29**: 17–37

Monteith J L 1969 Light interception and radiative exchange in crop stands. In Eastin J D, Haskins F A, Sullivan C Y, van Bavel C H M (eds) *Physiological Aspects of Crop Yield*. American Society of Agronomy, pp 89–111

Monteith J L 1973 *Environmental Physics*. Edward Arnold

Monteith J L (ed) 1976 *Vegetation and the Atmosphere*, Vol 2. Academic Press

Monteith J L 1977a Climate and the efficiency of crop production in Britain. *Philosophical Transactions of the Royal Society of London* **B281**: 277–94

Monteith J L 1977b. Climate. In Alvim P de T, Kozlowski T T (eds) *Ecophysiology of Tropical Crops*. Academic Press, pp 1–27

Monteith J L 1978 Reassessment of maximum growth rates for $C_3$ and $C_4$ crops. *Experimental Agriculture* **14**: 1–5

Monteith J L 1981 Does light limit crop production? In Johnson C B (ed) *Physiological Processes Limiting Plant Productivity*. Butterworths, pp 23–38

Monteith J L, Elston J 1983 Performance and productivity in the field. In Dale J E, Milthorpe F L (eds) *The Growth and Functioning of Leaves*. Cambridge University Press, pp 499–518

Moorby J 1967 Inter-stem and inter-tuber competition in potatoes. *European Potato Journal* **10**: 189–205

Moorby J 1978 The physiology of growth and tuber yield. In Harris P M (ed) *The Potato Crop*. Chapman and Hall

Moorby J 1985 Wheat growth and modelling: conclusions. In Day W, Atkin R K (eds) *Wheat Growth and Modelling*. Plenum Press, New York

Moorby J, Milthorpe F L 1975 Potato. In Evans L T (ed) *Crop Physiology*. Cambridge University Press, pp 225–57

Moorby J, Munns R, Walcott J 1975 Effect of water deficit on photosynthesis and tuber metabolism in potatoes. *Australian Journal of Plant Physiology* **2**: 323–33

Morgan C L, Austin R B 1983 Respiratory loss of recently assimilated carbon in wheat. *Annals of Botany* **51**: 85–95

Morris D A 1967 Intersprout competition in the potato. II. Competition for nutrients during pre-emergence growth after planting. *European Potato Journal* **10**: 296–311

Morris D A 1982 Hormonal regulation of sink invertase activity: implications for the control of assimilate partitioning. In Wareing P F (ed) *Plant Growth Substances 1982*. Academic Press

Morris D A 1983 Hormonal regulation of assimilate partition: possible mediation by invertase. *News Bulletin of the British Plant Growth Regulator Group* **6**: 23–35

Morrison J 1973 National grassland manuring trials. *Annual Report of the Grassland Research Institute for 1972*, pp 38–9

Morrison J, Jackson M V, Sparrow P E 1980 The response of perennial ryegrass to fertiliser nitrogen in relation to climate and soil. *Grassland Research Institute, Technical Report* **27**

Morton A G, Watson D J 1948 A physiological study of leaf growth. *Annals of Botany* **47**: 281–310

Moser L E, Volenec J J, Nelson C J 1982 Respiration, carbohydrate content, and leaf growth of tall fescue. *Crop Science* **22**: 781–6

Munns R, Pearson C J 1974 Effect of water deficit on translocation of carbohydrate in *Solanum tuberosum*. *Australian Journal of Plant Physiology* **1**: 529–37

Munns R, Weir R 1981 Contributions of sugars to osmotic adjustment in elongating and expanded zones of wheat leaves during moderate water deficits at two light levels. *Australian Journal of Plant Physiology* **8**: 93–105

Murphy M C 1984 *Report on Farming in the Eastern Counties of England 1982/83*. Agricultural Economics Unit, University of Cambridge

Nafziger E D, Koller H R 1976 Influence of leaf starch concentration on $CO_2$ assimilation in soybean. *Plant Physiology* **57**: 560–3

Naidu R A, Krishnan M, Ramanujam P, Gnanam A, Nayudu M V 1984 Studies on peanut green mosaic virus infected peanut (*Arachis hypogaea* L.) leaves. I. Photosynthesis and photochemical reactions. *Physiological Plant Pathology* **25**: 181–90

Natr L 1972 Influence of mineral nutrients on photosynthesis of higher plants. *Photosynthetica* **6**: 80–99

Neales T F, Incoll L D 1968 The control of leaf photosynthesis rate by the level of assimilate concentration in the leaf: a review of the hypothesis. *Botanical Review* **34**: 107–25

Nevins D T, Loomis R S 1970 Nitrogen nutrition and photosynthesis in sugar beet (*Beta vulgaris* L.). *Crop*

*Science* **10**: 21–5

Nicolas M E, Gleadow R M, Dalling M J 1985 Effect of post-anthesis drought on cell division and starch accumulation in developing wheat grains. *Annals of Botany* **55**: 433–44

Niilisk H, Nilson T, Ross J 1970 Radiation in plant canopies and its measurement. In Setlik I (ed) *Prediction and Measurement of Photosynthetic Productivity*. Centre for Agricultural Publishing and Documentation, Wageningen, pp 165–77

Nobel P S 1974 *Introduction to Biophysical Plant Physiology*. W H Freeman and Co.

Norris I B 1982 Soil moisture and growth of contrasting varieties of *Lolium, Dactylis* and *Festuca* species. *Grass and Forage Science* **37**: 273–83

Nösberger J, Joggi D 1981 Canopy structure and photosynthesis of red clover. In Wright C E (ed) *Plant Physiology and Herbage Production. British Grassland Society Occasional Symposium* **13**: 37–40

O'Brien P J, Allen E J 1986 Effects of nitrogen fertilizer applied to seed crops on seed yields and regrowth of progeny tubers in potatoes. *Journal of Agricultural Science, Cambridge* **107**: 103–11

O'Brien P J, Allen E J, Bean J N, Griffith R L, Jones S A, Jones J L 1983 Accumulated day degrees as a measure of physiological age and the relationships with growth and yield in early potatoes. *Journal of Agricultural Science, Cambridge* **101**: 613–31

Offler C E, Patrick J W 1984 Cellular structures, plasma membrane surface areas and plasmodesmatal frequencies of seed coats of *Phaseolus vulgaris* L. in relation to photosynthate transfer. *Australian Journal of Plant Physiology* **11**: 79–99

Okabe K, Lindlar A, Tsuzuki M, Miyachi S 1980 Effects of carbonic anhydrase on ribulose 1,5-bisphosphate carboxylase and oxygenase. *FEBS Letters* **114**: 142–4

O'Leary G J, Connor D J, White D H 1985 A simulation model of the development, growth and yield of the wheat crop. *Agricultural Systems* **17**: 1–26

Oparka K J, Davies H V 1985 Translocation of assimilates within and between potato stems. *Annals of Botany* **56**: 45–54

Öquist G 1983 Effects of low temperature on photosynthesis. *Plant, Cell and Environment* **6**: 281–300

Osmond C B, Winter K, Powles S B 1980 Adaptive significance of carbon dioxide cycling during photosynthesis in water-stressed plants. In Turner N C, Kramer P J (eds) *Adaptation of Plants to Water and High Temperature Stress*. J Wiley and Sons

O'Toole J C, Crookston R K, Treharne K J, Ozbun J L 1976 Mesophyll resistance and carboxylase activity. A comparison under water stress conditions. *Plant Physiology* **57**: 465–8

Outlaw W H, Fisher D B, Christy A L 1975 Compartmentation in *Vicia faba* leaves. II. Kinetics of $^{14}C$-sucrose redistribution among individual tissues following pulse labeling. *Plant Physiology* **55**: 704–11

Owera S A P, Farrar J F, Whitbread R 1981 Growth and photosynthesis in barley infected with brown rust. *Physiological Plant Pathology* **18**: 79–90

Owera S A P, Farrar J F, Whitbread R 1983 Translocation from leaves of barley infected with brown rust. *New Phytologist* **94**: 111–23

Pant M M 1979 Dependence of plant yield on density and planting pattern. *Annals of Botany* **44**: 513–16

Parsons A J 1985 New light on the grass sward and the grazing animal. *Span* **28**: 47–9

Parsons A J, Leafe E L, Collett B, Stiles W 1983a The physiology of grass production under grazing. 1. Characteristics of leaf and canopy photosynthesis of continuously-grazed swards. *Journal of Applied Ecology* **20**: 117–26

Parsons A J, Leafe E L, Collett B, Penning P D, Lewis J 1983b The physiology of grass production under grazing. 2. Photosynthesis, crop growth and animal intake of continuously-grazed swards. *Journal of Applied Ecology* **20**: 127–39

Parsons A J, Robson M J 1980 Seasonal changes in the physiology of S24 perennial ryegrass (*Lolium perenne* L.). 1. Response of leaf extension to temperature during the transition from vegetative to reproductive growth. *Annals of Botany* **46**: 435–44

Parsons A J, Robson M J 1981a Seasonal changes in the physiology of S24 perennial ryegrass (*Lolium perenne* L.). 2. Potential leaf and canopy photosynthesis during the transition from vegetative to reproductive growth. *Annals of Botany* **47**: 249–58

Parsons A J, Robson M J 1981b Seasonal changes in the physiology of S24 perennial ryegrass (*Lolium perenne* L.). 3. Partition of assimilates between root and shoot during the transition from vegetative to reproductive growth. *Annals of Botany* **48**: 733–44

Passioura J B 1979 Accountability, philosophy and plant physiology. *Search* **347**: 347–50

Pate J S, Gunning B E S 1972 Transfer cells. *Annual Review of Plant Physiology* **23**: 173–96

Pate J S, Sharkey P J, Atkins C A 1977 Nutrition of a developing legume fruit. Functional economy in terms of carbon, nitrogen, water. *Plant Physiology* **59**: 506–10

Paterson W G W, Blackett G A, Gill W D 1983 Plant growth regulator trials on spring and winter barley. *Council of the Scottish Agricultural Colleges, Research and Development Note* **16**

Patrick J W 1982 Hormonal control of assimilate transport. In Wareing P F (ed) *Plant Growth Substances 1982*. Academic Press

Patrick J W, McDonald R 1980 Pathway of carbon transport within developing ovules of *Phaseolus vulgaris* L. *Australian Journal of Plant Physiology* **7**: 671–84

Patrick J W, Wareing P F 1980 Hormonal control of assimilate movement and distribution. In Jeffcoat B (ed) *Aspects and Prospects of Plant Growth Regulators*. Monograph 6, British Plant Growth Regulator Group

Peacock J M 1975a Temperature and leaf growth in *Lolium perenne*. 2. The site of temperature perception. *Journal of Applied Ecology* **12**: 115–23

Peacock J M 1975b Temperature and leaf growth in *Lolium perenne*. 3. Factors affecting seasonal differences. *Journal of Applied Ecology* **12**: 685–97

Pearman I, Thomas S M, Thorne G N 1978 Effect of nitrogen fertilizer on growth and yield of semi-dwarf and tall varieties of winter wheat. *Journal of Agricultural Science, Cambridge* **91**: 31–45

Pearman I, Thomas S M, Thorne G N 1979 Effect of nitrogen fertiliser on photosynthesis of several varieties of winter wheat. *Annals of Botany* **43**: 613–21

Pearman I, Thomas S M, Thorne G N 1981 Dark respiration of several varieties of winter wheat given different amounts of nitrogen fertilizer. *Annals of Botany* **47**: 535–46

Pegg G F 1981 Biochemistry and physiology of pathogenesis. In Mace M E, Bell A A, Beckman C H (eds) *Fungal Wilt Diseases of Plants*. Academic Press, New York

Penning de Vries F W T 1972 Respiration and growth. In Rees A R *et al.* (eds) *Crop Processes in Controlled Environments*. Academic Press

Penning de Vries F W T 1975a Use of assimilates in higher plants. In Cooper J P (ed) *Photosynthesis and Productivity in Different Environments*. Cambridge University Press

Penning de Vries F W T 1975b The cost of maintenance processes in plant cells. *Annals of Botany* **39**: 77–92

Penning de Vries F W T, Van Laar H H 1977 Substrate utilization in germinating seeds. In Landsberg J J, Cutting C V (eds) *Environmental Effects on Crop Physiology*. Academic Press

Penny A, Addiscott T M, Widdowson F V 1984 Assessing the need of maincrop potatoes for late nitrogen by using isobutylidene diurea, by injecting nitrification inhibitors with aqueous N fertilizers and by dividing dressings of 'Nitro-Chalk'. *Journal of Agricultural Science, Cambridge* **103**: 577–85

Peoples M B, Beilharz V C, Waters S P, Simpson R J, Dalling M J 1980 Nitrogen redistribution during grain growth in wheat (*Triticum aestivum* L.). II. Chloroplast senescence and the degradation of ribulose 1,5- bisphosphate carboxylase. *Planta* **149**: 241–51

Perrenoud S 1983 Potato – Fertilising for High Yield. *Bulletin* **8**, International Potash Institute, Berne, Switzerland

Pharr D M, Huber S C, Sox H 1985 Leaf carbohydrate status and enzymes of translocate synthesis in fruiting and vegetative plants of *Cucumis sativus* L. *Plant Physiology* **77**: 104–8

Plaut Z 1971 Inhibition of photosynthetic carbon dioxide fixation in isolated spinach chloroplasts exposed to reduced osmotic potentials. *Plant Physiology* **48**: 591–5

Poincelot R P, Day P R 1976 Isolation and bicarbonate transport of chloroplast membranes from species of differing net photosynthetic efficiency. *Plant Physiology* **57**: 334–8

Poovaiah B W, Veluthambi K 1985 Auxin-regulated invertase activity in strawberry fruits. *Journal of the American Society of Horticultural Science* **110**: 258–61

Porter J R 1984 A model of canopy development in winter wheat. *Journal of Agricultural Science, Cambridge* **102**: 383–92

Porter J R 1985 Models and mechanisms in the growth and development of wheat. *Outlook on Agriculture* **14**: 190–6

Porter J R, Klepper B, Belford R K 1986 A model (WHTROOT) which synchronizes root growth and development with shoot development for winter wheat. *Plant and Soil* **92**: 133–45

Porter J R, Kirby E J M, Day W, Adam J S, Appleyard M, Ayling S, Baker C K, Beale P, Belford R K, Biscoe P V, Chapman A, Fuller M P, Hampson J, Hay R K M, Hough M N, Matthews S, Thompson W J, Weir A H, Willington V B A, Wood D W 1987 An analysis of morphological development stages in Avalon winter wheat crops with different sowing dates and at ten sites in England and Scotland. *Journal of Agricultural Science, Cambridge* **109**: 107–21

Porter M A, Grodzinski B 1983 Regulation of chloroplastic carbonic anhydrase. Effect of magnesium. *Plant Physiology* **72**: 604–5

Porter M A, Grodzinski B 1984 Acclimation to high $CO_2$ in bean. Carbonic anhydrase and ribulose bisphosphate carboxylase. *Plant Physiology* **74**: 413–16

Potter J R, Jones J W 1977 Leaf area partitioning as an important factor in growth. *Plant Physiology* **59**: 10–13

Powles S B, Osmond C B 1978 Inhibition of the capacity and efficiency of photosynthesis in bean leaflets illuminated in a $CO_2$-free atmosphere at low oxygen: a possible role for photorespiration. *Australian Journal of Plant Physiology* **5**: 619–29

Preiss J 1982 Regulation of the biosynthesis and degradation of starch. *Annual Review of Plant Physiology* **33**: 431–5

Prew R D, Church B M, Dewar A M, Lacey J, Penny A, Plumb R T, Thorne G N, Todd A D, Williams T D 1983 Effects of eight factors on the growth and nutrient uptake of winter wheat and on the incidence of pests and diseases. *Journal of Agricultural Science, Cambridge* **100**: 363–82

Puckridge D W, Donald C M 1967 Competition amongst wheat plants sown at a wide range of densities. *Australian Journal of Agricultural Research* **18**: 193–211

Purvis O N 1961 The physiological analysis of vernalization. *Encyclopaedia of Plant Physiology* **16**: 76–122

Rademacher W, Graebe J E 1984 Hormonal changes in developing kernels of two spring wheat varieties differing in storage capacity. *Berichte Deutschen Botanischen Gesellschaft* **97**: 167–81

Radley M 1978 Factors affecting grain enlargement in wheat. *Journal of Experimental Botany* **29**: 918–34

Radley R W 1963 The effect of season on growth and development of the potato. In Ivins J D, Milthorpe F L (eds) *The Growth of the Potato*. Butterworths, pp 211–20

Radley R W, Taha M A, Bremner P M 1961 Tuber bulking in the potato crop. *Nature* **191**: 782–3

Raggi V 1978 The $CO_2$ compensation point, photosynthesis and respiration in rust infected bean leaves. *Physiological Plant Pathology* **13**: 135–9

Rahmann M S, Wilson J H 1977 Determination of spikelet number in wheat. 1. Effect of varying photoperiod on ear development. *Australian Journal of Agricultural Research* **28**: 565–74

Raven J A, Glidewell S M 1981 Processes limiting photosynthetic conductance. In Johnson C B (ed) *Physiological Processes Limiting Plant Productivity*. Butterworths

Raymond W F 1969 The nutritive value of forage crops. *Advances in Agronomy* **21**: 1–108

Reed M L, Graham D 1977 Carbon dioxide and the regulation of photosynthesis: activities of photosynthetic enzymes and carbonate dehydratase (carbonic anhydrase) in *Chlorella* after growth or adaptation in different carbon dioxide concentrations. *Australian Journal of Plant Physiology* **4**: 87–98

Reed M L, Graham D 1981 Carbonic anhydrase in plants: distribution, properties and possible physiological roles. *Progress in Phytochemistry* **7**: 47–94

Reid D 1970 The effects of a wide range of nitrogen application rates on the yields from a perennial ryegrass sward with and without white clover. *Journal of Agricultural Science, Cambridge* **74**: 227–40

Reid D 1984 The seasonal distribution of nitrogen fertiliser dressings on pure perennial ryegrass swards. *Journal of Agricultural Science, Cambridge* **103**: 659–69

Rhodes I 1969 The yield, canopy structure and light interception of two ryegrass varieties in mixed culture and monoculture. *Journal of the British Grassland Society* **24**: 123–7

Rhodes I 1970 Competition between herbage grasses. *Herbage Abstracts* **40**: 115–21

Rhodes I 1973 Relationship between canopy structure and productivity in herbage grasses and its implications for plant breeding. *Herbage Abstracts* **43**: 129–33

Richards I R 1977 Influence of soil and sward characteristics on the response to nitrogen. In Gilsenan B (ed) *Proceedings of an International Meeting on Animal Production from Temperate Grassland*. An Forais Taluntais, pp 45–9

Riggs T J 1984 Plant breeding – potential for improvement of yield and grain quality. In *The Nitrogen Requirements of Cereals*. Reference Book 385, Ministry of Agriculture, Fisheries and Food, HMSO, pp 5–18

Riggs T J, Hanson P R, Start N D, Miles D M, Morgan C L, Ford M A 1981 Comparison of spring barley varieties grown in England and Wales between 1880 and 1980. *Journal of Agricultural Science, Cambridge* **97**: 599–610

Rijven A H G C, Gifford R M 1983 Accumulation and conversion of sugars by developing wheat grains. 3. Non-diffusional uptake of sucrose, the substrate preferred by endosperm slices. *Plant, Cell and Environment* **6**: 417–25

Ripley E A, Redman R E 1976 Grassland. In Monteith J L (ed) *Vegetation and the Atmosphere*, Vol. 2. Academic Press, pp 349–98

Robson M J 1973 The growth and development of simulated swards of perennial ryegrass. II. Carbon assimilation and respiration in a seedling sward. *Annals of Botany* **37**: 501–18

Robson M J 1981 Potential production – what is it, and can we increase it? In Wright C E (ed) *Plant Physiology and Herbage Production. Occasional Symposia of the British Grassland Society* **13**: 5–18

Robson M J 1982 The growth and carbon economy of selection lines of *Lolium perenne* cv. S23 with differing rates of dark respiration 1. Grown as simulated swards during a regrowth period. *Annals of Botany* **49**: 321–9

Robson M J, Deacon M J 1978 Nitrogen deficiency in

small closed communities of S24 ryegrass. II Changes in the weight and chemical composition of single leaves during their growth and death. *Annals of Botany* **42**: 1199–213

Robson M J, Parsons A J 1978 Nitrogen deficiency in small closed communities of S24 ryegrass. I. Photosynthesis, respiration and dry matter production and partition. *Annals of Botany* **42**: 1185–97

Robson M J, Stern W R, Davidson I A 1983 Yielding ability in pure swards and mixtures of lines of perennial ryegrass with contrasting rates of 'mature tissue' respiration. In Corral A J (ed) *Efficient Grassland Farming*. British Grassland Society, Hurley

Rodger J B A 1982 Spring barley. East of Scotland Agricultural College, Edinburgh, *Bulletin* **27**

Roebuck J F, Trenerry J 1978 Precision drilling of cereals. *Experimental Husbandry* **34**: 1–11

Ross J 1970 Mathematical models of photosynthesis in a plant stand. In Setlik I (ed) *Prediction and Measurement of Photosynthetic Productivity*. Centre for Agricultural Publishing and Documentation, Wageningen, pp 29–45

Rufty T W, Huber S C 1983 Changes in starch formation and activities of sucrose phosphate synthase and cytoplasmic fructose-1,6-bisphosphatase in response to source-sink alterations. *Plant Physiology* **72**: 474–80

Russell C R, Morris D A 1982 Invertase activity, soluble carbohydrates and inflorescence development in the tomato (*Lycopersicon esculentum* Mill.). *Annals of Botany* **49**: 89–98

Russell E W 1973 *Soil Conditions and Plant Growth* 10th edn. Longman

Ryle G J A, Cobby J M, Powell C E 1976 Synthetic and maintenance respiratory losses of $^{14}CO_2$ in uniculm barley and maize. *Annals of Botany* **40**: 571–86

Ryle G J A, Hesketh J D 1969 Carbon dioxide uptake in nitrogen deficient plants. *Crop Science* **9**: 451–4

Ryrie I J, Scott K J 1968 Metabolic regulation in diseased leaves. II. Changes in nicotinamide coenzymes in barley leaves infected with powdery mildew. *Plant Physiology* **43**: 687–92

SAC 1982 Seed mixtures for Scotland. *Scottish Agricultural Colleges Publication* **86**

SAC 1983 Fertiliser recommendations for grassland. *Scottish Agricultural Colleges Publication* **108**

Saftner R A, Wyse R E 1984 Effect of plant hormones on sucrose uptake by sugar beet root tissue discs. *Plant Physiology* **74**: 951–5

Sale P J M 1974 Productivity of vegetable crops in a region of high solar input. III. Carbon balance of potato crops. *Australian Journal of Plant Physiology* **1**: 283–96

Salter PJ, Goode J E 1967 Crop responses to water at different stages of growth. *Research Review* **2**. Commonwealth Bureau of Horticulture and Plantation Crops

Sambo E Y 1983 Leaf extension rates in temperate pasture grasses in relation to assimilate pool in the extension zone. *Journal of Experimental Botany* **34**: 1281–90

Sanders D C, Nylund R E, Quisumbing E C, Shetty K V P 1972 The influence of mist irrigation on the potato. IV. Tuber quality factors. *American Potato Journal* **49**: 243–54

Sattelmacher B, Marschner M 1979 Tuberization in potato plants as affected by applications of nitrogen to the roots and leaves. *Potato Research* **22**: 49–57

Saurer W, Possingham J V 1970 Studies on the growth of spinach leaves (*Spinacea oleracea*). *Journal of Experimental Botany* **21**: 151–8

Scharen A I, Krupinksy J M 1969 Effect of *Septoria nodorum* infection on $CO_2$ absorption and yield of wheat. *Phytopathology* **59**: 1298–301

Schippers P A 1968 The influence of rates of nitrogen and potassium application on the yield and specific gravity of four potato varities. *European Potato Journal* **11**: 23–33

Schmitt M R, Hitz W D, Lin W, Giaquinta R T 1984 Sugar transport into protoplasts isolated from developing soybean cotyledons. II. Sucrose transport kinetics, selectivity, and modelling studies. *Plant Physiology* **75**: 941–6

Scholes J D, Farrar J F 1986 Photosynthesis and chloroplast functioning within individual pustules of *Uromyces muscari* on bluebell leaves. *Physiological Plant Pathology* **27**: 387–400

Schulze E-D 1986 Carbon dioxide and water vapour exchange in response to drought in the atmosphere and in the soil. *Annual Review of Plant Physiology* **37**: 247–74

Schussler J R, Brenner M L, Brun W A 1984 Abscisic acid and its relationship to seed filling in soybeans. *Plant Physiology* **76**: 301–6

Scott K J 1965 Respiratory enzymic activities in the host and pathogen of barley leaves infected with *Erysiphe graminis*. *Phytopathology* **55**: 438–41

Scott K J 1972 Obligate parasitism by phytopathogenic fungi. *Biological Reviews* **47**: 537–72

Scott R K, Wilcockson S J 1978 Application of physiological and agronomic principles to the development of the potato industry. In Harris P M (ed) *The Potato Crop*. Chapman and Hall, pp 678–704

Scott W R, Appleyard M, Fellowes G, Kirby E J M 1983 Effect of genotype and position in the ear on carpel and grain growth and mature grain weight of spring barley. *Journal of Agricultural Science, Cambridge* **100**: 383–91

Seethambaram Y, Rao A N, Das V S R 1985 The levels

of carbonic anhydrase and of photorespiratory enzymes under zinc deficiency in *Oryza sativa* L. and *Pennisetum americanum* L. Leeke. *Biochemie und Physiologie der Pflanzen* **180**: 107–13

Selman M 1983 Avoiding the great N robbery. *Arable Farming* **May 1983**: 96–7

Servaites J C, Geiger D R 1974 Effects of light intensity and oxygen on photosynthesis and translocation in sugar beet. *Plant Physiology* **54**: 575–8

Setter T L, Brun W A, Brenner M L 1980a Stomatal closure and photosynthetic inhibition in soybean leaves induced by petiole girdling and pod removal. *Plant Physiology* **65**: 884–7

Setter T L, Brun W A, Brenner M L 1980b Effect of obstructed translocation on leaf abscisic acid, and associated stomatal closure and photosynthesis decline. *Plant Physiology* **65**: 1111–15

Shantz H, Piemeisel L N 1927 The water requirements of plants at Akron, Colorado. *Journal of Agricultural Research* **34**: 1093–189

Sharkey T D 1985 Photosynthesis in intact leaves of $C_3$ plants: physics, physiology and rate limitations. *Botanical Review* **51**: 53–105

Shaw M, Samborski O J 1957 The physiology of host-parasite relations. III. The pattern of respiration in rusted and mildewed cereal leaves. *Canadian Journal of Botany* **35**: 389–407

Short J L 1983 Cara responds to ageing treatment. *Arable Farming* **December 1983**: 36–7

Short J L, Shotton F E 1968 Storage conditions affecting the sprouting of seed potatoes and their yield. II. Light. *Experimental Husbandry* **16**: 93–8

Short J L, Shotton F E 1970 Storage conditions affecting the sprouting of seed potatoes and their yield. III. Temperature. *Experimental Husbandry* **19**: 69–87

Sicher R C, Kremer D F, Harris W G 1984 Diurnal carbohydrate metabolism of barley primary leaves. *Plant Physiology* **76**: 165–9

Siddiqui M Q, Manners J G 1971 Some effects of general yellow rust (*Puccinia striiformis*) infection on $^{14}$carbon assimilation and growth in spring wheat. *Journal of Experimental Botany* **22**: 792–9

Siedow J N, Berthold D A 1986 The alternative oxidase: a cyanide resistant respiratory pathway in higher plants. *Physiologia Plantarum* **66**: 569–73

Sigee D C, Epton H A 1976 Ultrastructural changes in resistant and susceptible varieties of *Phaseolus vulgaris*. *Physiological Plant Pathology* **9**: 1–8

Silvey V 1981 The contribution of new wheat, barley and oat varieties to increasing yield in England and Wales 1947–78. *Journal of the National Institute for Agricultural Botany* **15**: 399–412

Sim L C 1982 Some practical and financial aspects of cereal growing. In *Yield of Cereals, Course Papers 1982*. Arable Unit, National Agricultural Centre, Coventry, pp 120–5

Simkin M B, Wheeler B E J 1974 Effects of dual infection of *Puccinia hordei* and *Erisiphe graminis* on barley cv. Zephyr. *Annals of Applied Biology* **78**: 237–50

Simons R G 1982 Tiller and ear production of winter wheat. *Field Crop Abstracts* **35**: 857–70

Simpson K 1962 Effects of soil-moisture tension and fertilisers on the yield, growth and phosphorus uptake of potatoes. *Journal of the Science of Food and Agriculture* **13**: 236–48

Sinclair T R, de Wit C T 1975 Photosynthate and nitrogen requirements for seed production by various crops. *Science* **189**: 565–7

Singh B K, Jenner C F 1982 Association between concentrations of organic nutrients in the grain, endosperm cell number and grain dry weight within the ear of wheat. *Australian Journal of Plant Physiology* **9**: 83–95

Singh B K, Jenner C F 1984 Factors controlling endosperm cell number and grain dry weight in wheat: effects of shading on intact plants and of variation in nutritional supply to detached, cultured ears. *Australian Journal of Plant Physiology* **11**: 151–63

Smedegaard-Petersen V 1980 Enzyme-inhibitor studies of respiratory pathways in barley leaves infected with *Pyrenophora teres* and affected by its isolated toxins. *Royal Veterinary and Agricultural University, Yearbook 1980*: 23–35

Smedegaard-Petersen V 1984 The role of respiration and energy generation in diseased and disease resistant plants. In Wood R K S, Jellis G J (eds) *Plant Diseases: Infection, Damage and Loss*. Blackwell Scientific Publications

Smith A, Allcock P J 1985a Influence of age and year of growth on the botanical composition and productivity of swards. *Journal of Agricultural Science, Cambridge* **105**: 299–325

Smith A, Allcock P J 1985b The influence of species diversity on sward yield and quality. *Journal of Applied Ecology* **22**: 185–98

Smith D, Muscatine L, Lewis D 1969 Carbohydrate movement from autotrophs to heterotrophs in parasitic and mutualistic symbiosis. *Biological Reviews* **44**: 17–90

Smith H 1981 Light quality as an ecological factor. In Grace J, Ford E D, Jarvis P G (eds) *Plants and their Atmospheric Environment*. Blackwell Scientific Publications, pp 93–110

Smith L P 1983 The pattern of British grassland farming – a verification. *Agricultural Meteorology* **30**: 129–34

Smith P R, Neales T F 1977 The growth of young peach trees following infection by the viruses of peach rosette and decline disease. *Australian Journal of Agricultural Research* **28**: 441–4

Snyder F W, Carlson G E 1984 Selecting for partitioning of photosynthetic products in crops. *Advances in Agronomy* **37**: 47–72

Soetono, Donald C M 1980 Emergence, growth and dominance in drilled and square-planted barley crops. *Australian Journal of Agricultural Research* **31** 455–70

Soetono, Puckridge D W 1982 The effect of density and plant arrangement on the performance of individual plants in barley and wheat crops. *Australian Journal of Agricultural Research* **33**: 171–7

Sofield I, Evans L T, Cook M G, Wardlaw I F 1977 Factors influencing the rate and duration of grain filling in wheat. *Australian Journal of Plant Physiology* **4**: 785–97

Somerville C R 1986 Future prospects for genetic manipulation of Rubisco. *Philosophical Transactions of the Royal Society of London* **B313**: 459–69

Spedding C R W 1971 *Grassland Ecology*. Oxford University Press

Spedding C R W, Diekmahns E C 1972 Grasses and legumes in British agriculture. Commonwealth Bureau of Pastures and Field Crops, *Bulletin* **49**

Spiertz J H J 1980 Grain production of wheat in relation to nitrogen, weather and diseases. In Hurd R G, Biscoe P V, Dennis C (eds) *Opportunities for Increasing Crop Yields*. Pitman, pp 97–113

Stern W R, Donald C M 1962a Light relationships in grass–clover swards. *Australian Journal of Agricultural Research* **13**: 599–614

Stern W R, Donald C M 1962b The influence of leaf area and radiation on the growth of clover in swards. *Australian Journal of Agricultural Research* **13**: 615–23

Stern W R, Kirby E J M 1979 Primordium initiation at the shoot apex in four contrasting varieties of spring wheat in response to sowing date. *Journal of Agricultural Science, Cambridge* **93**: 203–15

Stewart T A, Haycock R E 1983 Beef from white clovers/grass or high nitrogen swards – production results from a six-year comparison. *Agriculture in N. Ireland* **58**: 37–43

Stitt M 1984 Degradation of starch in chloroplasts: a buffer to sucrose metabolism. In Lewis D H (ed) *Storage Carbohydrates in Vascular Plants*. Society for Experimental Biology Seminar Series 19. Cambridge University Press

Stitt M 1985 Control of photosynthetic sucrose synthesis by fructose-2,6-bisphosphate: comparative studies in $C_3$ and $C_4$ species. In Jeffcoat B, Hawkins A F, Stead A D (eds) *Regulation of Sources and Sinks in Crop Plants*. Monograph 12, British Plant Growth Regulator Group

Stockman Y M, Fischer R A, Brittain E G 1983 Assimilate supply and floret development within the spike of wheat (*Triticum aestivum* L.). *Australian Journal of Plant Physiology* **10**: 585–94

Stoy V 1980 Grain filling and the properties of the sink. In *Physiological Aspects of Crop Productivity*. International Potash Institute

Sunderland N 1960 Cell division and expansion in the growth of the leaf. *Journal of Experimental Botany* **11**: 68–80

Svensson B 1972 Influence of the place of a stem in the hill on the weight and dry matter content of its tubers. *Potato Research* **15**: 346–53

Swift G, Holmes J C, Cleland A T, Fortune D, Wood J 1983 The grassland of East Scotland – A survey 1976–78. East of Scotland College of Agriculture *Bulletin* **29**

Swift G, Mackie C K, Harkess R D, Franklin M F 1985 Time of fertiliser nitrogen for spring grass. *Scottish Agricultural Colleges Research and Development Note* **24**

Tanner W 1980 On the possible role of ABA on phloem unloading. *Berichte Deutschen Botanischen Gesellschaft* **93**: 349–51

Taylor B R, Blackett G A 1982 Influence of some husbandry practices on the physical properties of spring barley grains. *Journal of the Science of Food and Agriculture* **33**: 133–9

Terry N, Waldron L J, Taylor S E 1983 Environmental influences on leaf expansion. In Dale J E, Milthorpe F L (eds) *The Growth and Functioning of Leaves*. Cambridge University Press, pp 179–205

Thom A S 1971 Momentum absorption by vegetation. *Quarterly Journal of the Royal Meteorological Society* **97**: 414–28

Thomas H, Stoddart J L 1980 Leaf senescence. *Annual Review of Plant Physiology* **31**: 83–111

Thomas S M, Thorne G N 1975 Effect of nitrogen fertilizer on photosynthesis and ribulose 1,5-diphosphate carboxylase activity in spring wheat in the field. *Journal of Experimental Botany* **26**: 43–51

Thomas W D 1982 Plant growth regulators. In *Yield of Cereals, Course Papers 1982*. Arable Unit, National Agricultural Centre, Coventry, pp 78–97

Thompson R, Taylor H 1974 Stem density and maturity studies with the potato cultivars Maris Peer and Pentland Marble. *Potato Research* **17**: 51–63

Thorne G N, Wood D W 1982 Physiological behaviour of the cereal crop. In *Yield of Cereals, Course Papers 1982*. Arable Unit, National Agricultural Centre, Coventry, pp 12–22

Thorne J H 1981 Morphology and ultrastructure of maternal seed tissues of soybean in relation to the import of photosynthate. *Plant Physiology* **67**: 1016–25

Thorne J H 1982 Temperature and oxygen effects on $^{14}C$-photosynthate unloading and accumulation in developing soybean seeds. *Plant Physiology* **69**: 48–53

Thorne J H 1985 Phloem unloading of C and N assimilates in developing seeds. *Annual Review of Plant Physiology* **36**: 317–43

Thorne J H, Koller H R 1974 Influence of assimilate demand on photosynthesis, diffusive resistances, translocation, and carbohydrate levels of soybean leaves. *Plant Physiology* **54**: 201–7

Thorne J H, Rainbird R M 1983 An *in vivo* technique for the study of phloem unloading in seed coats of developing soybean seeds. *Plant Physiology* **72**: 268–71

Thornley J H M 1970 Respiration, growth and maintenance in plants. *Nature* **227**: 304–5

Thornley J H M 1976 *Mathematical Models in Plant Physiology*. Academic Press

Thornley J H M 1980 Research strategy in the plant sciences. *Plant, Cell and Environment* **3**: 233–6

Thrower L B, Thrower S L 1966 The effect of infection with *Uromyces fabae* on translocation in broad bean. *Phytopathologische Zeitschrift* **57**: 269–76

Tinker P B, Widdowson F V 1982 Maximising wheat yields, and some causes of yield variation. *Paper read to the Fertiliser Society of London, 10 December*

Tolbert N E 1980 Photorespiration. In Davies D D (ed) *The Biochemistry of Plants*, Vol 2. Academic Press

Toosey R D 1964 The pre-sprouting of seed potatoes: factors affecting sprout growth and subsequent yield. *Field Crop Abstracts* **17**: 161–8, 239–44

Tottman D R, Makepeace R J, Broad H 1979 An explanation of the decimal code for the growth of cereals, with illustrations. *Annals of Applied Biology* **93**: 221–34

Trenbath B R, Angus J F 1975 Leaf inclination and crop production. *Field Crop Abstracts* **28**: 231–44

Troughton J H 1969 Plant water status and carbon dioxide exchange of cotton leaves. *Australian Journal of Biological Sciences* **22**: 289–302

Tsuzuki M, Miyachi S, Edwards G E 1985 Localization of carbonic anhydrase in mesophyll cells of terrestrial $C_3$ plants in relation to $CO_2$ assimilation. *Plant and Cell Physiology* **26**: 881–91

Turgeon R, Webb J A 1973 Leaf development and phloem transport in *Cucurbita pepo*: transition from import to export. *Planta* **113**: 179–91

Turgeon R, Webb J A 1975 Leaf development and phloem transport in *Cucurbita pepo*: carbon economy. *Planta* **123**: 53–62

Turkington R, Harper J L 1979 The growth, distribution and neighbour relationships of *Trifolium repens* in a permanent pasture. 1. Ordination, pattern and contact. *Journal of Ecology* **67**: 201–18

Turner N C, Kramer P J (eds) 1980 *Adaptation of Plants to Water and High Temperature Stress*. Wiley

Uchijima Z 1970 Carbon dioxide environment and flux within a corn crop canopy. In Setlik I (ed) *Prediction and Measurement of Photosynthetic Productivity*. Pudoc, Wageningen

Uchijima Z 1976 Maize and rice. In Monteith J L (ed) *Vegetation and the Atmosphere*, Vol. 2. Academic Press, pp 33–64

Van Burg P F J 1968 Nitrogen fertilizing of grassland in spring. *Netherlands Nitrogen Technical Bulletin* **6**

Van Burg P F J, t'Hart M L, Thomas H 1980 Nitrogen and grassland – past and present situation in the Netherlands. *Proceedings of an International Symposium of the European Grassland Federation, Wageningen*, pp 15–33

Van der Meer H G 1983 Effective use of nitrogen on grassland farms. *Occasional Symposia of the British Grassland Society* **14**: 61–8

Vine D A 1983 Sward structure changes within a perennial ryegrass sward: leaf appearance and death. *Grass and Forage Science* **38**: 231–42

Volokita M, Kaplan A, Reinhold L 1981 Evidence for mediated $HCO_3$ transport in isolated pea mesophyll protoplasts. *Plant Physiology* **67**: 1119–23

von Caemmerer S, Farquhar G D 1981 Some relationships between the biochemistry of photosynthesis and the gas exchange of leaves. *Planta* **153**: 376–87

von Caemmerer S, Farquhar G D 1984 Effects of partial defoliation, changes of irradiance during growth, short-term water stress and growth at enhanced $p(CO_2)$ on the photosynthetic capacity of leaves of *Phaseolus vulgaris* L. *Planta* **160**: 320–9

Vos J 1979 Effects of temperature and nitrogen on carbon exchange rates and on growth of wheat during kernel filling. In Spiertz J H J, Kramer Th (eds) *Crop Physiology and Cereal Breeding*. Pudoc, Wageningen

Vreugdenhil D 1983 Abscisic acid inhibits phloem loading of sucrose. *Physiologia Plantarum* **57**: 463–7

Waite R 1970 The structural carbohydrates and the *in vitro* digestibility of a ryegrass and a cocksfoot at two levels of nitrogenous fertiliser. *Journal of Agricultural Science, Cambridge* **74**: 457–67

Walker A J, Ho L C 1977 Carbon translocation in the tomato: effects of fruit temperature on carbon metabolism and the rate of translocation. *Annals of Botany* **41**: 825–32

Walker A J, Ho L C, Baker D A 1978 Carbon translocation in the tomato: pathways of carbon metabolism in the fruit. *Annals of Botany* **42**: 901–9

Walker D A 1974 Chloroplast and cell – the movement of key substances across the chloroplast envelope. In Northcote D H (ed) *International Review of Science, Biochemical Series* 1, vol II. Butterworths

Walters D R 1985 Shoot : root interrelationships: the effects of obligately biotrophic fungal pathogens. *Biological Reviews* **60** 47–79

Walters D R, Ayres P G 1984 Ribulose bisphosphate carboxylase and enzymes of $CO_2$ assimilation in a compatible barley/powdery mildew combination. *Phytopathologische Zeitschrift* **109**: 208–18

Walton D C 1980 Biochemistry and physiology of abscisic acid. *Annual Review of Plant Physiology* **31**: 453–89

Wardlaw I F 1976 Assimilate movement in *Lolium* amd *Sorghum* leaves. Irradiance effects on photosynthesis, export and the distribution of assimilates. *Australian Journal of Plant Physiology* **3**: 377–87

Wardlaw I F, Moncur L 1976 Source, sink and hormonal control of translocation in wheat. *Planta* **128**: 93–100

Wareing P F, Khalifa M M, Treharne K J 1968 Rate-limiting processes in photosynthesis at saturating light-intensities. *Nature* **220**: 453–7

Wareing P F, Phillips I D 1981 *Growth and Differentiation in Plants* 3rd edn. Pergamon

Warren Wilson J 1959 Analysis of the distribution of foliage area in grassland. In Ivins J D (ed) *The Measurement of Grassland Productivity*. Butterworths, pp 51–61

Watson D J 1947 Comparative physiological studies on the growth of field crops. 1. Variation in net assimilation rate and leaf area between species and varieties and within and between years. *Annals of Botany* **11**: 41–76

Watson D J 1952 The physiological basis of variation in yield. *Advances in Agronomy* **4**: 101–45

Watson D J 1958 The dependence of net assimilation rate on leaf area index. *Annals of Botany* **22**: 37–54

Watson J D 1968 *The Double Helix*. Weidenfeld and Nicolson

Weatherley P E, Johnson R P C 1968 The form and function of the sieve tube: a problem in reconciliation. *International Review of Cytology* **24**: 149–92

Weir A H, Bragg P L, Porter J R, Rayner J H 1984 A winter wheat crop simulation model without water or nutrient limitations. *Journal of Agricultural Science* **102**: 371–82

Weir A H, Day W, Sastry T G 1985 Using a whole crop model. In Day W, Atkin R K (eds) *Wheat Growth and Modelling*. Plenum Press, New York

Wheeler A W 1972 Changes in growth-substance content during growth of wheat. *Annals of Applied Biology* **72**: 327–34

Whingwiri E E, Kemp D R 1980 Spikelet development and grain yield of the wheat ear in response to applied nitrogen. *Australian Journal of Agricultural Research* **31**: 637–47

Whingwiri E E, Stern W R 1982 Floret survival in wheat: significance of the time of floret initiation relative to terminal spikelet formation. *Journal of Agricultural Science, Cambridge* **98**: 257–68

Whipps J M, Lewis D H 1981 Patterns of translocation, storage and interconversion of carbohydrates. In Ayres P G (ed) *Effects of Disease on the Physiology of the Growing Plant*. Society for Experimental Biology, Seminar Series 11. Cambridge University Press

Whitear J D 1976 Nitrogen: too much at the wrong time? *Journal of the National Institute of Agricultural Botany* **14**: 196–201

Whitehead D C 1970 The role of nitrogen in grassland productivity. *Commonwealth Bureau of Pastures and Field Crops Bulletin* **48**

Whitehead D C, Dawson K P 1984 Nitrogen, including $^{15}N$-labelled nitrogen, in components of a grass sward. *Journal of Applied Ecology* **21**: 983–9

Whitehead T, McIntosh T P, Findlay W M 1953 *The Potato in Health and Disease*, 3rd edn. Oliver and Boyd, Edinburgh

Widdowson F V 1979 The nitrogen requirements of cereals. In *Yield of Cereals, Course Papers 1979*. Cereal Unit, National Agricultural Centre, Coventry, pp 97–104

Wilcox J C 1962 The effect of different dates of grazing and nitrogen top dressing on the subsequent hay crop. *Experimental Husbandry* **8**: 104–12

Wilhelm M W, Nelson C J 1978 Growth analysis of tall fescue genotypes differing in yield and leaf photosynthesis. *Crop Science* **18**: 951–4

Willey R W, Holliday R 1971 Plant population and shading studies on barley. *Journal of Agricultural Science, Cambridge* **77**: 445–52

Williams R F 1975 *The Shoot Apex and Leaf Growth*. Cambridge University Press

Williams R F, Rijven A H G C 1965 The physiology of growth in the wheat plant. 2. The dynamics of leaf growth. *Australian Journal of Biological Sciences* **18**: 721–43

Williams W A, Loomis R S, Duncan W G, Dovrat A, Ninez A 1968 Canopy architecture at various population densities and the growth and grain yield of corn. *Crop Science* **8**: 303–8

Willmer C 1983 *Stomata*. Longman

Wilman D 1965 The effect of nitrogen fertilizer on the

rate of growth of Italian ryegrass. *Journal of the British Grassland Society* **20**: 248–54

Wilman D, Wright P T 1983 Some effects of applied nitrogen on the growth and chemical composition of temperate grasses. *Herbage Abstracts* **53**: 387–93

Wilson D 1972 Overcoming physiological limitations to production from herbage. *Annual Report of the Welsh Plant Breeding Station for 1981*, pp 202–15

Wilson D 1973 Physiology of light utilization by swards. In Butler G W, Bailey R W (eds) *Chemistry and Biochemistry of Herbage*, Vol 2. Academic Press

Wilson D 1975 Variation in leaf respiration in relation to growth and photosynthesis of *Lolium*. *Annals of Applied Biology* **80**: 323–38

Wilson D 1976 Dark respiration. *Annual Report of the Welsh Plant Breeding Station for 1975*: 63–6

Wilson D 1982 Response to selection for dark respiration rate of mature leaves in *Lolium perenne* and its effects on growth of young plants and simulated swards. *Annals of Botany* **49**: 303–12

Wilson D, Jones J G 1982 Effect of selection for dark respiration rate of mature leaves on crop yields of *Lolium perenne* cv. S23. *Annals of Botany* **49**: 313–20

Wilson D, Robson M J 1981 Varietal improvement by selection for reduced dark respiration rate in perennial ryegrass. In Wright C E (ed) *Plant Physiology and Herbage Production. Occasional Symposium* **13**, *British Grassland Society*

Wittenbach V A, Ackerson R C, Giaquinta R T, Hebert R R 1980 Changes in photosynthesis, proteolytic activity, and ultrastructure of soybean leaves during senescence. *Crop Science* **20**: 225–31

Wittenbach V A, Lin W, Hebert R R 1982 Vacuolar localisation of proteinases and degradation of chloroplasts in mesophyll protoplasts from senescing primary wheat leaves. *Plant Physiology* **69**: 98–102

Woledge J 1977 The effects of shading and cutting treatments on the photosynthetic rate of ryegrass leaves. *Annals of Botany* **41**: 1279–86

Woledge J 1986 The effect of age and shade on the photosynthesis of white clover leaves. *Annals of Botany* **57**: 257–62

Woledge J, Leafe E L 1976 Single leaf and canopy photosynthesis in a ryegrass sward. *Annals of Botany* **40**: 773–83

Woledge J, Pearse P J 1985 The effect of nitrogenous fertiliser on the photosynthesis of leaves of a ryegrass sward. *Grass and Forage Science* **40**: 305–9

Wolfe M S, Schwarzbach E 1978 Patterns of race change in powdery mildews. *Annual Review of Phytopathology* **16**: 159–80

Wolswinkel P 1985 Phloem unloading and turgor-sensitive transport: Factors involved in sink control of assimilate partitioning. *Physiologia Plantarum* **65**: 331–9

Wolswinkel P, Ammerlaan A 1983 Phloem unloading in developing seeds of *Vicia faba* L. The role of several inhibitors on the release of sucrose and amino acids by the seed coat. *Planta* **158**: 205–15

Wolswinkel P, Ammerlaan A 1984 Turgor sensitive sucrose and amino acid transport into developing seeds of *Pisum sativum*. Effect of a high sucrose or mannitol concentration in experiments with empty ovules. *Physiologia Plantarum* **61**: 172–82

Wolswinkel P, Ammerlaan A, Kuyvenhoven H 1983 Effect of KCN and p-chloromercuribenzenesulfonic acid on the release of sucrose and 2-amino(1-$^{14}$C)isobutyric acid by the seed coat of *Pisum sativum*. *Physiologia Plantarum* **59**: 375–86

Woodward F I, Sheehy J E 1983 *Principles and Measurements in Environmental Biology*. Butterworths

Woodward R G, Rawson H M 1976 Photosynthesis and translocation in dicotyledenous plants. II. Expanding and senescing leaves of soybean. *Australian Journal of Plant Physiology* **3**: 257–67

Woolley E W 1982 Performance of current growth regulators in cereals. In *Opportunities for Manipulation of Cereal Productivity*. Monograph 7, British Plant Growth Regulator Group, pp 44–50

Wrigley C W, Webster H L 1966 The effect of stem rust infection on the soluble proteins of wheat. *Australian Journal of Biological Sciences* **19**: 895–901

Wurr D C E 1974 Some effects of seed size and spacing on the yield and grading of two maincrop potato varieties. *Journal of Agricultural Science, Cambridge* **82**: 37–45, 47–52

Wurr D C E 1978 Seed tuber production and management. In Harris P M (ed) *The Potato Crop*. Chapman and Hall, pp 327–54

Wurr D C E 1980 Physiological quality of potato seed tubers. *ADAS Quarterly Review* **36**: 27–39

Yabuki K, Miyagawa H 1970 Studies on the effect of wind speed on photosynthesis. 2. The relation between wind speed and photosynthesis. *Journal of Agricultural Meteorology, Tokyo* **26**: 137–41

Zadoks J C 1983 An integrated disease and pest-management scheme, EPIPRE, for wheat. *CIBA Foundation Symposium* **97**: 116–29

Zelitch I 1975a Improving the efficiency of photosynthesis. *Science* **188**: 626–33

Zelitch I 1975b Pathways of carbon fixation in green plants. *Annual Review of Biochemistry* **44**: 123–45

Zelitch I 1980 Basic research in biomass production: scientific opportunities and organizational challenges. In Staples R C, Kuhr R J (eds) *Linking Research to Crop Production*. Plenum Press

# INDEX

LIBRARY
...HORE COLLEGE OF HORT...
PERSHORE
...ORCS WR10 3JP

LIBRARY
...ORE COLLEGE OF H...
PERSHORE
...RCS WR10 3JP

LIBRARY
PROPERTY OF
PERSHORE COLLEGE
AVONBANK
PERSHORE
WORCESTERSHIRE WR10 3JP